Robotics and Automation in Construction

Edited by

Carlos Balaguer and Mohamed Abderrahim

Robotics and Automation in Construction
Edited by Carlos Balaguer and Mohamed Abderrahim

Republished by InTech
Janeza Trdine 9, 51000 Rijeka, Croatia

Technical Editor Teodora Smiljanic

Cover Designer Dino Smrekar

Additional hard copies can be obtained from orders@intechopen.com

Robotics and Automation in Construction, Edited by Carlos Balaguer and Mohamed Abderrahim
p. cm.
ISBN 978-953-7619-13-8

For EU product safety concerns:
IN TECH d.o.o., Prolaz Marije Krucifikse Kozulić 3, 51000 Rijeka, Croatia,
info@intechopen.com or visit our website at intechopen.com.

Meet the editors

Carlos Balaguer received his Ph.D. in Automation from the Polytechnic University of Madrid (UPM), Spain in 1983. From 1983-1994 he was with the Department of Systems Engineering and Automation of the UPM as Associated Professor. Since 1996, he has been a Full Professor of the Robotics Lab at the University Carlos III of Madrid. He has been Director of the Department of Systems Engineering and Automation (2006-2007) and the vice-Rector for Research of the university in the period of 2007-2015. Prof. Balaguer´s research has included, but not limited, humanoids robotics, robots' design and development, robot control, path & task planning, force-torque control, assistive and service robots, rehabilitation and medical robots, climbing robots, robotics and automation in construction, human-robot interaction. He participates in numerous EU projects since 1989, like Eureka projects SAMCA, AMR y GEO; Esprit projects ROCCO and CEROS; Brite project FutureHome; IST project MATS; 6FP IP projects ManuBuild, I3CON, Tunconstruct; Strep project RobotCWE; 7FP project RoboSpect; and H2020 projects STAMS and BADGER (coordinator). He has published more than 200 papers in journals and conference proceedings, and several books in the field of robotics. He is a member of IEEE and IFAC, and former President of IAARC (2001-2004). He participates in the European networks Euron and Clawar. Prof. Balaguer had been also an Associate Editor of IEEE Robotics & Automation magazine (2000-2005) and is member of the Editorial Board of Automation in Construction journal (Elsevier). He was the coordinator of the Spanish Robotic Network (2005-2009) and of the Madrid Community universities' consortium RoboCity2030 on Service Robots (2006-2018), and was the Spanish representative in the European platform EUROP (2006-2008). He is currently member of the euRobotics Board of Directors (2015 -). He is also Chairman of the Council for Science and Technology of the Community of Madrid (2016 -). He received several awards, among them the best book Fundamentos de robotica edited by McGraw-Hill (1988), best paper of the ISARC'2003 in Eindhoven (The Netherlands), IMSERSO´s Award 2004 for assistive robots research, the Industrial Robot journal Innovation Award of the Clawar'2005 in London (UK), Tucker-Hasegawa Award 2006 in Tokyo (Japan) for a major contribution in the field of Robotics & Automation in Construction and FUE´s Award 2014 for AIRBUS-UC3M Joint R&D Center. He was the General Chair of the IEEE-RAS Humanoids'2014 conference and is the General Chair of the IEEE/JRS IROS'2018 to be held in Madrid.

Mohamed Abderrahim graduated in "Optics and Precision Mechanics" from Setif University, Algeria, and received the Ph.D. from the Mechanical Engineering Department of the University of Glasgow, Scotland, in 1996. After spending two year as a Postdoctoral Research Fellow at the Robotics Lab of the University Carlos III of Madrid, he was employed as a Lecturer in the Department of Systems Engineering and Automation. He was involved in the development of the climbing robot ROMA and contributed to the Future Home EU project as a local technical coordinator. Recently, he has been involved in the RISANAR project which deals with the autonomous navigation using computer vision and visual servoing techniques. He is currently the lead reseacher of the ASIROV and MEIGA-3 projects funded by the MICIN, and the HANDLE project funded by the EU within the 7th FP. From April 2003 to February 2008, he was a "Ramón y Cajal" fellow. His general research interests include modelling, simulation and experimental validation of mechatronic systems with application to robotics, Man-Machine Interaction and Interfaces and subjects related to the Automation in the construction sector. On the teaching level, he has been involved in teaching subjects such as Industrial Automation, CIM Laboratory, Programming and Computational Perception (computer vision).

Preface

Construction is a heterogeneous sector that involves several fields of expertise and input from other industries. The construction industry is characterised by almost unique circumstances for each project and a dynamic unstructured environment, with safety hazards, several temporary activities and changing weather conditions. This combination of circumstances holds back the introduction of greater automation and robotization in the construction processes. However, recent years have seen an increase in the use of new technologies such as ICT, sensory systems, automated machines and robots. Examples of these special machines include wall climbing robots for inspection and maintenance, concrete power floating machines, concrete floor surface finishing robots, construction steel frame welding robots, wall panels' bricklaying robots, robotic excavators and automated cranes for the assembly of modular construction elements.

This book addresses several issues related to the introduction of automaton and robotics in the construction industry in a collection of 23 chapters. The chapters are grouped in 3 main sections according to the theme or the type of technology they treat. Section I is dedicated to describe and analyse the main research challenges of Robotics and Automation in Construction (RAC). The second section consists of 12 chapters and is dedicated to the technologies and new developments employed to automate processes in the construction industry. Among these we have examples of ICT technologies used for purposes such as construction visualisation systems, added value management systems, construction materials and elements tracking using multiple IDs devices. This section also deals with Sensorial Systems and software used in the construction to improve the performances of machines such as cranes, and in improving Human-Machine Interfaces (MMI). Authors adopted Mixed and Augmented Reality in the MMI to ease the construction operations. Section III is dedicated to describe case studies of RAC and comprises 8 chapters. Among the eight chapters the section presents a robotic excavator and a semi-automated façade cleaning system. The section also presents work dedicated to enhancing the force of the workers in construction through the use of Robotic-powered exoskeletons and body joint-adapted assistive units, which allow the handling of greater loads.

The new automated machines and processes have the potential to increase productivity by performing tasks efficiently and improving working conditions through applications that limit the exposure of humans to safety hazards.

Editors

Carlos Balaguer and Mohamed Abderrahim
RoboticsLab, Department of Systems Engineering and Automation
University Carlos III of Madrid, Spain
{Carlos.Balaguer, Mohamed.Abderrahim}@uc3m.es

Contents

1

Trends in Robotics and Automation in Construction

Carlos Balaguer and Mohamed Abderrahim
University Carlos III of Madrid
Spain

1. Introduction

Until very recently, the construction industry was one of the most unfamiliar R&D fields for the robotics and automation community, despite the fact that this industry is one of the oldest and represents the largest economic sectors. The construction industry's contribution to the GDP in industrialized countries is about 7-10%. In the US this contribution rises to 12% and in the EU there are about 2.7 M enterprises (most of them Small and Medium Enterprises) involved in the business. This figure is comparable to that of the manufacturing industry. However, the investment in R&D is the double in the case of manufacturing.

The technological level of the construction industry during the old ages was very high for their historical period. The old civilizations have built very long lasting structures like pyramids, acropolis, aqueducts, cathedrals, etc. They used innovative processes and elements for their contemporary normal building procedures. Nevertheless, some of nowadays construction processes have changed little. For example, the building erection process has changed very little over the past eight hundred years. The old ages pulleys are substituted by cranes. These are more sophisticated than centuries ago, but they work with the same principles: manual control, human operator visual feedback, big positioning error, etc. The only elements that have change are: electrical or diesel actuators replaced the human force and steel structures replaced the wooden elements. These two advances allowed increasing the elevation speed, the payload and reachability, but the construction philosophy itself has changed little.

In recent years, the construction industry has become one of the most important research areas in the field of service robotics. The main difficulty of Robotics and Automation in Construction (RAC) is related to the nature of the work environment, which is highly unstructured in general. Working in this environment involves handling heavy objects, elements made with big tolerances, low level of standardization, medium level of industrialization and pre-fabrication, in addition to the intervention of numerous non-coordinated actors (architects, builders, suppliers, etc.). Therefore, a big effort needs to be made to increase the level of automation of this important sector and to coordinate more the involved processes in order to improve its productivity.

During the 90s the R&D activities in the field of RAC were lead by Japanese companies and universities, and were focused on the development of new robotic systems (most of them teleoperated) and in the automation of existing machinery. This era of the RAC research is called *hard robotics* (Balaguer, 2003). These robots tried to automate several construction processes in the house building and the civil construction. These robots were for interior building finishing, brick layer masonry, modular industrialized building's construction, road paver's sensor-based guidance, excavator's control, infrastructure inspection, tunnel and bridge construction among others. The "bubble economy" crisis en Japan among other factors such as the unsatisfied over-expectation of the RAC strongly reduced investment in research activities during the last few years. Only few construction robots had succeeded to make their way to the market. Nevertheless, the situation is changing now and new RAC research trends have been launched. The actual R&D activities are centring more in the *software and IT technologies*. This is not limited to software only but also include hardware, but not in the machinery sense. It includes on-site sensory data acquisition and processing, human operator's field safety and security, chip-based process control and monitoring, automated inventory and shop keeping among many others.

The rest of this chapter presents a comparison between the construction and the car manufacturing industry and discuss some of the issues that limit more technological advances and higher levels of automation in the construction sector. The following sections are dedicated to present and discuss some examples of the state of the art in robotics and automations technologies in construction. The following section discusses some aspects that affect higher implementation of robotics and automation in construction, and the last section presents the conclusions.

3. Comparison between the construction and the automobile industries

As mentioned briefly in the previous sections, the construction industry has already seen the introduction of automated and semi automated means in the production of construction elements. The transition from totally manual process to nowadays semi-automated system permits to increase the productivity. Nevertheless, the advance in construction industry is not comparable to advances in other industries such as manufacturing and especially in the sectors of automobile, electronics, train, aircraft, etc. The car prices, the number of different models and variations, and the concept of mass production make the automobile industry much close to construction than the others.

One of the key factors of any industry's success evaluation is its productivity. Fig. 1 shows the comparison of the construction and automobile industries in the EU. This figure clearly demonstrates that the automobile industry productivity has increased several times more than that of the construction during the last decade. The main reason in this high productivity is the modern manufacturing concept: Computer Integrated Manufacturing (CIM). This concept was developed during the last two decades and has changed not only the manufacturing process itself but the concept of the product (Rembold et al., 1993), (Rehg, 1994). The CIM systems permit to balance the flexibility in the product with the manufacturing productivity. This relationship is one of the key factors of the success of the automobile industry.

While the house-building construction industry continue to be very close to craft work, constructing mostly singular buildings, the automobile industry continuously seek to reduce the cost of product development. This permits also to reduce the cost of the final product. The so called platform concept of the actual automobile industry is one of the newest advances of the CIM system. It is based on the use of a number of elements in various models. The same platform design, engine, electronics, etc. are used not only in different

models of cars of the same company but also in the cars of other companies. This concept reduces a vehicle cost and makes the automobile companies more competitive.
The high level of integration in all the production stages permits to start from the design process taking in mind the manufacturing and market aspects. The platform concept and integration lead to the high level of robotization and automation in automobile industry. In some of the EU plants the level of automation (the number of non-manually made operations respect to the total number of operations) is more than 60%. Mass production brings down the cost not only of the end product (in this case, the cars) but also the cost of manufacturing equipment (robots, machine tools, etc.). This is why during the last decade industrial robot prices in the EU have decreased and their number has increased (Fig. 2).

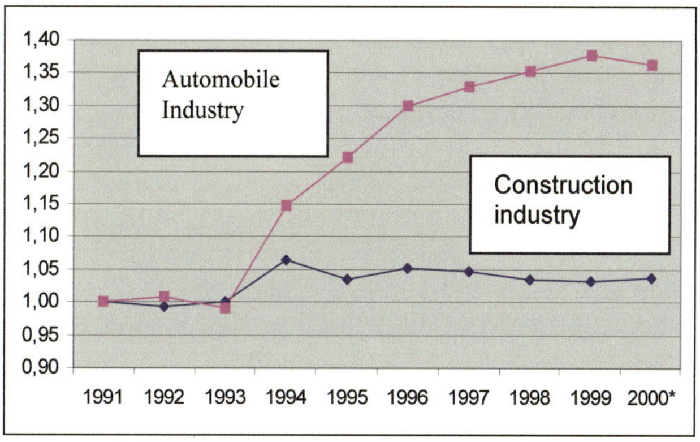

Fig. 1. Productivity of the construction and automobile industries in EU (sources: Euroconstruct, Eurostat, ACEA)

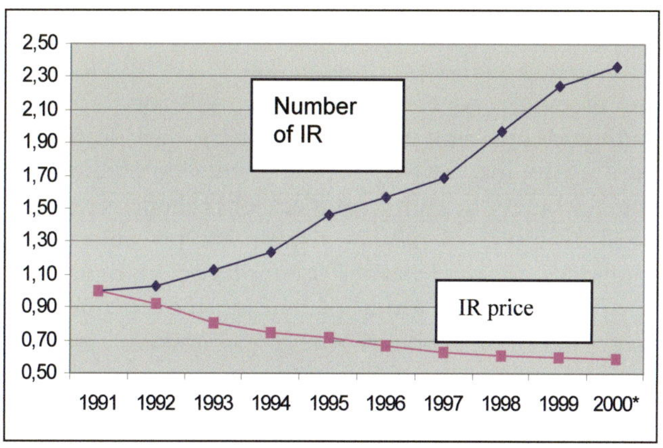

Fig. 2. Number of industrial robots (IR) in EU and its price in US$ (source: IFR)

Robotics in manufacturing industry is an evolution while the robotics in construction industry is the not yet finished revolution. While the number of industrial robots is counted in hundreds of thousands the number of robots in the construction industry is counted in hundreds only. Important efforts have been made to adapt the CIM concept to the construction industry created the Computer Integrated Construction (CIC) (Miyatake & Kangari, 1993) (Balaguer et al. 2002). Unfortunately, this effort has better results only in the

IT related stages of the construction process (planning, suppliers' relationship, etc.) but not as good results in the production stages (pre-fabrication technology, building erection, masonry, on-site automation, etc.). Despite the recent development in RAC, the gap between the technological levels of both industries is still very high

The CIM concept permits to reduce not only the cost of manufacturing but also changes the corporate culture (Kangarii, 1996). It is easier to introduce the new technologies in automobile industry than in the construction. In general, the construction industry continues to be very conservative. In many cases when the new automatic products are not complementary to the old ones, they are hardly implemented and their use is kept to minimum. Moreover, if these products introduce inconveniences to the whole construction cycle, they are openly refused. To the contrary, in the manufacturing industry the people and the environment respond very positively to technological innovation. Researchers and end users speak the same "language" and share the same objective, which allows introducing these new technologies very quickly.

According to ACEA, in 1999 the EU automobile industry investments in R&D were over 5% of the turnover while the construction industry investments in house-building technology were less than 3% (Euroconstruct, 1998). In the construction industry the big companies tend to limit their capacity to invest in "tomorrow's construction robots" from which return on investment is uncertain and too far in the future. This is also the case of the big construction machines companies, which tend to invest more in civil engineering equipments than in development of equipments dedicated for house-building.

2. State of the art in construction robotics

The main research activities of the RAC in the past decade were divided accordingly to applications into two large groups: civil infrastructure and house building. Typical civil infrastructure robot applications are the automation of road, tunnel and bridge construction, earthwork, etc. In the group of house construction, main applications include building skeleton erection and assembly, concrete compaction, interior finishing, etc. Classification according to applications is consistent with other possible classifications, which divide RAC R&D activities according to the development of new equipment and processes or the adaptation of existing machinery to transform them into robotic system.

In this section several examples of *robotics systems* are presented. In the group of civil infrastructure the examples are road pavers' sensor-based guidance, earthmoving control and infrastructure inspection. In the group of housing the examples are interior building finishing, brick layer masonry, column welding, modular industrialized building's construction.

2.1 Civil infrastructures

In the field of road construction, several projects had been developed over the last decade. They were mainly focused in the development of the new generation of semi-autonomous road pavers and asphalt compactors. The EU projects CIRC (Peyret et al., 2003) and latter OSYRIS (www.osyris.org) had as the main objectives, based in the GPS and laser data, the semi-autonomous guidance of the machines and the quality control of pavers and roller processes by controlling the speed, temperature, layer thickness, travelled distance, etc. (Fig. 3). The coordination of several machines in order to improve productivity is also the objective of the project.

Fig. 3. OSYRIS project sensor-based compactor

In the field of earthwork the research is centred in the introduction of new control techniques to existing machinery like excavators, bulldozers, draglines, etc. One of the major exponents of this research area is the control by CSIRO of the 100-m tall walking crane used in surface coal mining (Corke et al., 2006). The swing cycle of the dragline accounts for about 80 percent of time taken. The automatic swing cycle improves the efficiency of the machine, taking in mind that the bucket which weighs around 40 tonnes when empty and up to 120 tonnes when full, acts as a large pendulum and requires operator skill to control well (Fig. 4). The torque-force control during the excavation is also improving the productivity of the processes. The University of Sydney project (Ha et al., 2000) developed an automated excavator that accounts for interaction forces in analysing the required bucket motion therefore seems promising. As the bucket comes in contact with its environment, the contact force must be regulated such that it remains within a specific range by using specific control strategy (Fig. 5).

Fig. 4. CSIRO's dragline project

Fig. 5. the Automated excavator of the University of Sidney

The periodic inspection and maintenance of the civil infrastructures was another important research activity. The inspection of building skeletons, complex roofs, off-shore platforms, bridges, etc. represents an extensive and valuable field of work. It is estimated that in the EU there are over 42.000 steel bridges with a replacement cost of 350 M€. The ROMA family climbing robots (Balaguer et al., 2000) able to travel in a complex 3D environment carry out several inspection sensors (laser telemeters, colour cameras) in order to transmit the field data to the "ground" system (Fig. 6). The key issue of these robots is the grasping method (grippers, electromagnets, suction cups, etc.).

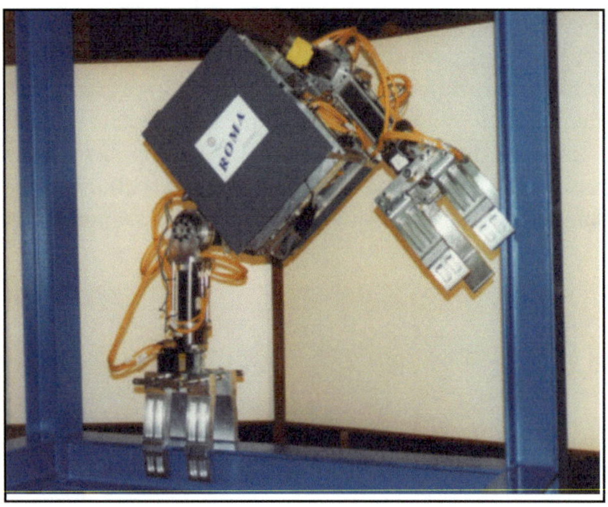

Fig. 6. ROMA climbing inspection robot

2.2 House building

Interior-finishing operations in the building are very time consuming and requires high degree of accuracy. There are several mobile manipulators able to perform variety of operations like extend, compact and control the thickness of the floor concrete, painting and steel column fire protection spraying, assembly of interior walls and ceilings, etc. Most of these robots are teloperated and perform only simple operations. The most representatives' robots of this type are Japanese ones. Three examples are presented: the "Mighty Hand" robot from Kajima (www.kajima.co.jp), which lifts heavy elements in construction as concrete walls, etc. (Fig. 7), and the SurfRobo from Takenaka (www.takenaka.co.jp), which automatically compact the concrete floor by using two sets of rotary floats (Fig. 8). The right hand side of the figure shows Kajima´s concrete floor surface finishing robots. These robots are already used in several building construction sites where they succeeded in releasing workers from thousands of operations (Hasegawa, 2006).

Fig. 7. Kajima's interior wall assembly robot

The last decade has witnessed the development of several robots for automatic assembly of buildings. An effort had been done in the brick laying masonry and the development of robotic prefabrication of façade and wall elements. The EU project ROCCO developed a large-range (10 m reach) and high payload (up to 500 kg) hydraulic 6 DOF robot for brick assembly (Gambao et al., 1997). The robot is equipped with auto-tracking laser telemeter in the tip in order to perform prices (up to 5 cm) brick assembly. In this way the control system avoid important arm flexion. The robot performs the assembly sequence obtained by the planning software and needs an initialization process in order to know the bricks pallet position (Fig. 9).

Fig. 8. Takenaka's concrete compactor robot and Kajima´s concrete finishing robot

Fig. 9. ROCCO project brick assembly robot

During the last few years a tendency to develop wearable robots for different applications immerged. First this type of robot were thought of from a military point of view, and that is to provide soldiers with powered exoskeletons to allow them handle heavy loads and resist longer periods without being exhausted. The main limitation of these robots is their power supply, but in the construction site this should not be a serious problem, since the robot can be iambically connected to a power source while being wearied by an operator. A wearable/exoskeletons robot is able to endow the operator with more strength beyond his natural limits and allow him/her to handle heavy objects during their construction activities such as carpentry or fitting ceiling boards as they require large muscular power. The prototype developed in (Naito et al., 2007) is an example of such application (fig. 10).

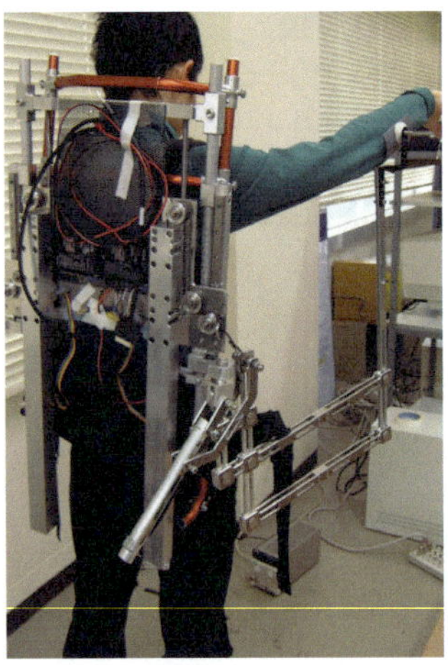

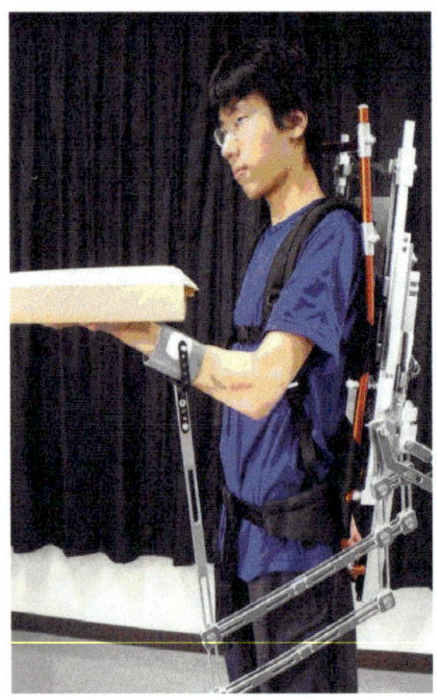

Fig. 10. An example of a carpentry aid wearable robot under development

The assembly of steel-based buildings is performing by welding, such as column-to-column and column-to-beam joints. The Japanese WR mobile robot performs a variety of column-to-column welding (Fig. 11). The steel columns of up to 100 mm thickness can be round-, square-, or H-shaped, as well as box-sectional members. For column-to-beam welding, there is a combination of welder/transport type which can run on decks and a type which can weld lower flanges from below.

Fig. 11. WR mobile robot for column welding

Automation and robotization of the complete building erection is the most exciting experience. Applying to the high-rise building there were several Japanese projects. The most significant is the SMAT system developed by Shimizu (Miyatake, 1993). It was used for construction of more than 30 stories office building. It consists of all-wheatear, full-robotic factory on the top of the building. The lift-up mechanism automatically raises the construction plant and at the same time raises the on-site factory, called *field factory* (Fig. 12). More recently the Dutch companies develop the new whole building erection technology but in opposite way of the previous system. The building is totally constructed in like factory environment and then transported to the final location. The 10 floor building called Bolder was transported by water in a three day operation (Fig. 13).

Fig. 12. SMART factory in the top of the building

Fig. 13. Bolder building transportation by water

3. Software and IT technology in RAC

As discussed in earlier sections, software and IT technologies (also called soft robotics in earlier publications by the authors) is not only limited to software itself but also includes other related technology as sensory data acquisition and processing, human operator's field safety and security, chip-based process control, etc. This section describes the main applications and some examples of the actual software and IT trends in the automation of the construction industry.

3.1 Software integration

Software integration in the field of RAC is crucial for implementing the concept of the Computer Integrated Construction (CIC). The idea is to integrate in a common exchange format all the stages of the construction, i.e. from architect's desk and planning tools to site robots. The EU FutureHome projects develop the AUTOMOD3 system (Fig. 14) that integrates in a common CAD environment several tools like design, planning and automatic robot and machine programming (Balaguer et al., 2002). Due the high level of conservatism of building designers, the main idea is to use the common 2D architectural design (drawings) and automatically transform it into 3D drawings. In this way it is possible to perform also automatically the modularization of the traditionally designed buildings. This process permits to industrialize the house-building by modular pre-fabricated construction.

Schedule management software packages are used more and more in construction. Nevertheless, its dynamic integration whit all the actors participate in the construction is not yet done. In a construction project, although the completion day is clearly decided, construction schedule is often changed by the weather or the actual progress situation of the project. When a difference arise between present state and the master schedule, it is necessary to adjust the construction schedule and to execute it immediately. The communication with part's produced factory, transport agents, stores and other suppliers is performing in real-time and in automatic way (Lipman & Reed, 2000).

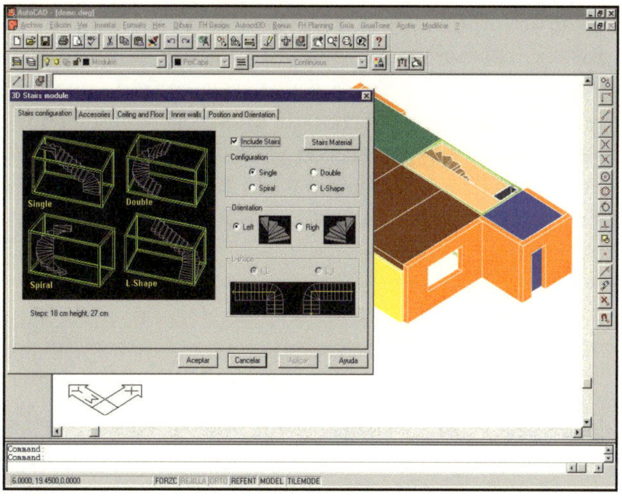

Fig. 14. FutureHome integrated design and planning tools

Mobile computing systems for data transfer between constructor managers and different web-sites have been implementing. The progress monitoring wireless mobile system permits to check the progress of the work. At the same time field note system is used to note unacceptable parts of works (Fig. 15). Inspection system is also used for inspect the result of construction. The document management system not only can communicate with the designers DBs in order to download the CAD drawings, but also permits the on-site modifications of these drawings. This soft technology is very useful and has a low cost which make it candidate for massive introduction in the site environment. The day when construction managers and operators carry only some paper drawings will be finished soon.

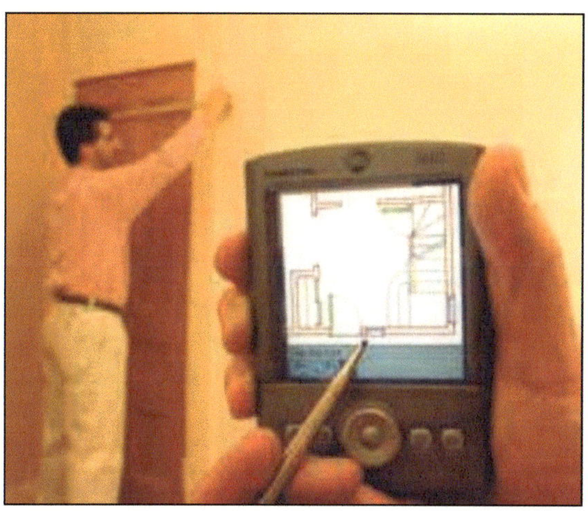

Fig. 15. PDA-based document management system in the site environment

3.2 Virtual Reality systems

The Virtual Reality (VR) software together with an immersive projection display (IPD) allows construction managers to enter and interact with the contents of a full-scale building, before start of the construction or during the execution of the project. The virtual mock-up offers first person presence, or the feeling that you're actually in the room when you're just standing in a space bounded by five large screens that surround you with a projected image. The virtual mock-up experience is real enough to enable welders, for example, to crawl under virtual structures and hit their heads on virtual pipes to determine if there's enough room to work. Several immersive VR systems were developed during the last years, like at the Penn State University (http://www.arl.psu.edu), at the NIST (http://cic.nist.gov/vrml/equip.html), etc.

a)

b)

Fig. 16. VR environment for excavators training system: a) from inside the cabin, and b) outside the cabin

In the world of construction operations analysis, the ability to see a 3D dynamic animation of an operation that has been simulated allows the experts, field personnel, and decision makers can discover differences between the way they understand the operation and the way the model developer understands it. The dynamic VR is more close to animation than geometrical visualization. The actual research is focusing on designing automated, discrete-event process simulation-driven methods to visualize construction operations and the resulting evolving products in dynamic, smooth, continuous, 3D virtual worlds. The discrete-event simulation systems, allows a computer to create a world that is accurate in time and space; and which shows people, machines, and materials interacting as they build constructed facilities.

Using VR system for simulation and training is another software and IT technology (also designated soft robotics by some authors) area. For complex machines like excavators, the VR system needs not only to simulate the geometry and kinematics of the machine but also the terrain and the interaction between the machine and terrain (Lipman & Reed, 2000). The simulation of digging and driving over the terrain is the crucial test. The terrain model is generated with an elevation grid technique which specifies a height field over a uniform grid. If the size of an individual grid in the simulation is smaller than the footprint of the excavator the system will work correctly and the operator's sensation will be good. The system permits the simulation of the view from inside or outside the excavator cabin. To visually represent the digging process, the location of the bucket relative to the terrain and relative to the excavator needs to be known. A complete Caterpillar 3D backhoe simulator, can be consulted at the site page http://www.howstuffworks.com/backhoe-loader.htm.

3.3 Sensory data acquisition and processing

One of the most promising areas of research and development in software and IT in RAC is the sensory data acquisition and processing. The use of sensors for modelling the environment and then use this data for processing is much valuable for the control of automatic construction machinery or robots. The LADAR (Laser Radar) on-site data acquisition was one of traditional research area at NIST (Cheok at al., 2000). Nevertheless, only recently this technology has produced important applications in the automatic excavation, truck guidance, topography and inspection. LADAR technology is based on the high precision pan-and-tilt mounted laser rangefinder with the frame rate at least 10 Hz (commonly 25 Hz). The range of the laser scanner is up to 150 m for objects with reflection coefficient grater than 80% and 50 m for objects with reflection coefficient grater than 10%. Once the data are registered, they are used to generate 3D models or surfaces (Fig. 17).

Evaluation of surface generation algorithms involves a three parts process. In the first part, the characteristics (accuracy, noise, and related uncertainties) of the sensor would be determined. This set up or calibration would be performed in an indoor facility, which allows for a controlled environment. In the second part, mathematical procedures are used to determine the statistical uncertainties of particular calculations (e.g., volume) based on the results of the instrument calibration. In the third part, the characteristics of the algorithms used to generate the 3-D model would be determined. These characteristics determine how well the algorithms handle missing points, outliers, discontinuities, vertical surfaces, etc.

The using of GPS for data collection has become very common. Some applications are very well known such as automatic truck guidance, topography, etc. But nowadays low-cost facilities of using PDA-based GPS and web data transmission and collection make new applications possible. One of these applications is the GPS-aided earthquake monitoring. The data is collected via a GPS station with a circular antenna firmly fixed in a 4.5 tonne slab of 300 million year old sandstone from Yorkshire, which is in turn embedded almost three metres into the earth. This natural landmark is monitored every 15 second via web.

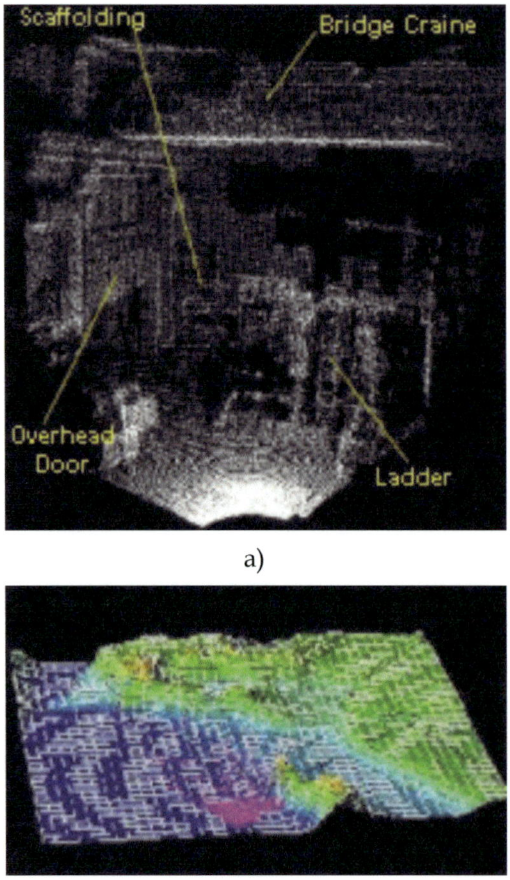

a)

b)

Fig. 17. NIST's LADAR data: a) environment map, and b) 3D reconstruction

3.4 Safety of operators and machines

Thousands of construction workers are injured or killed in construction accidents each year. Researches and development efforts have been made in the last few years to look into new ways of improving the security and developing methods and reliable systems to detect possible failures and to avoid any harm to the workers, machines and installations. Studies showed that the main risk sources in on-site environment are collision with the machine transporting heavy and big objects, fallings, machine running over, and therefore these have to be taken into consideration when designing preventive security systems (Abderrahim et al., 2003)

In this research and development effort based on *IT and mechatronics systems*, the compulsory safety helmet required for all workers in the site can be used as the base to hold miniature positioning and communication instruments (Fig. 18) (Abderrahim at al., 2003). In the work described by the authors, bidirectional voice channel, portable GPS and micro-camera with video link have been integrated in the security the helmet. The position and ID of each worker is communicated periodically via radio link to a monitoring station. This information is compared with a dynamics Data Base containing the tasks and processes to be perform in the site. If a given worker is at what the system considers a hazard source it acts according to the nature of source.

There are two basic security levels: machine and human ones. Machine level refers to the failures in the machinery, possible erroneous operation, bad condition of the components, etc. As far as the human level is concerned, the objective is to prevent the operatives from suffering the accidents. The strategy to adopt consist in the definition of different safe and prohibited zones around the workers and the sources of danger, so that in the moment in which these areas comes into contact a danger situation is triggered and warning is generated. There are several actions to be done in this situation such as advising the worker thought the voice instructions, halting a machine movement via central computer among others. The proposed prototype systems records all the detected risk situations for later examination and is able to be used for monitoring of some activities of the site as it records the position of workers and automated machines continuously.

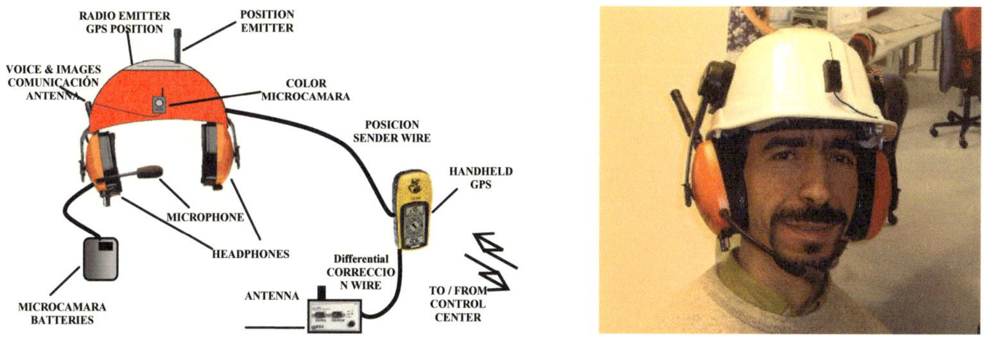

Fig. 18. UC3M's active security system's elements and a picture of the equipped helmet

3.5 Identification tags and part-oriented construction

One of the most innovative *software and IT Technologies* applications in the construction industry is the so called *part-oriented construction* (Yagi, 2003). The idea is very compatible with the above mentioned security system, where The main idea is to link the fixed and temporary facilities of construction site (ground, cranes, field factories, etc.) with the parts-peoples dynamically changing world via information network. It means that the status, position and timing of parts and human operators in the site are known in every moment. Moreover, it is possible to dynamically plan, command, tracking and monitoring all the construction recourses.

To perform this monitoring each part, machine and operator has assigned an identification chip which wireless connects with external devices. This chip is a wireless semiconductor integrated circuit that stores an ID number in its memory (Radio Frequency Identification Device -RFID). The μ-chip developed by Hitachi is a micro-device with square of 0.4 mm that uses the frequency of 2.45GHz. It has a 128-bit ROM for storing unique ID (Fig. 19).

The system which controls whole the construction process is called *glue logic*. This system binds multiple application software modules, referred as "agents", developed and compiled separately, and coordinates those agents. As "glue logic" supports even notification and conditioning monitoring features based on active data scheme, users can easy build real-time event-driven application agents. The *glue logic* consists of two major parts:

communication subsystem exchanging data with concurrently running agents and the data management subsystem.
Every part has attached μ-chip (which includes antenna). When a chip-implanted part passes through the gate, the gate reads the product URL. It determines what it is, when and in what state it is. The corresponding data point in the *glue logic* is then altered, which generates an event and a chain of succeeding actions. For communicates with human operators some of them carry a wireless PDA which is capable to connected to the main web server where *glue logic* is running and, at the same time, is capable to read μ-chip attached to the part. Moreover, the system is applied to automated handling of devices, by communication with automatic cranes and *glue logic* DB. The correct assembly of parts is also be monitoring.

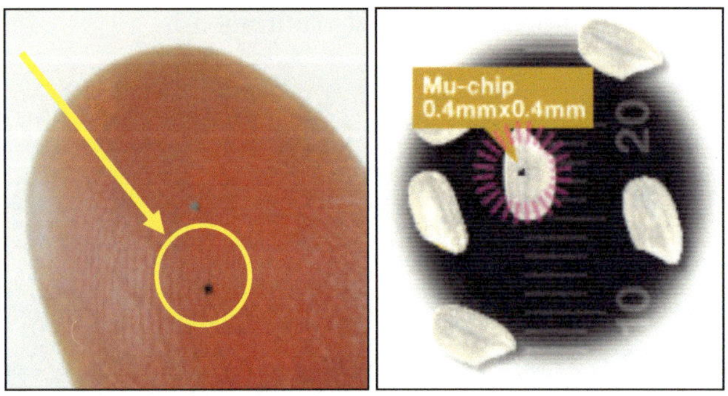

a)

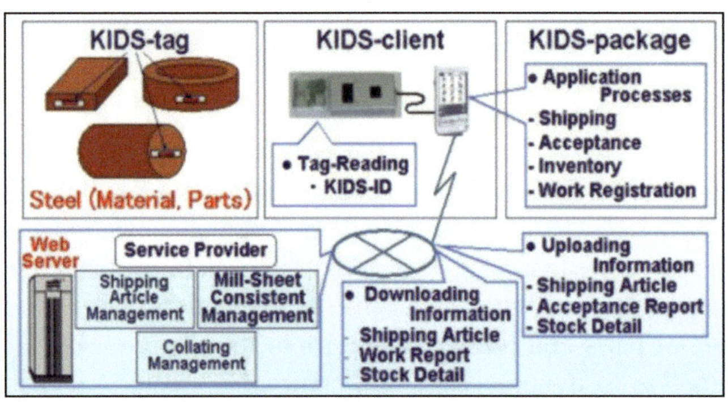

b)

Fig. 19. Hitachi's RFID μ-chip a) in comparison with the human finger and rice grain, and b) their application for steel-based parts' tracking

4. The future of robotics and automation in construction

The automation in construction has moved through several historical periods, according the ISARC trends (Ueno, 1998): a) cradle (1984-85), b) growing (1986-89), and c) developing (1990-98). However, even at the time of preparing this document one can assert that the consolidated has not been achieved yet. This may happen in the near future, but it is difficult to imagine that the houses will be built in the future like today's cars. However, as illustrated in the examples above the automation in construction is increasing and many of the developed prototypes will see their way to real application. Some factors are very important and will affect the way to real implementation in the near future. These factor are summarised as follow:

- Change of attitude in the construction companies, the machinery industry, the research centres and the government R&D officials, in order to develop new high tech commercial products and pass the phase of prototypes.
- Implementation of new IT and telecommunications technologies is already changing the work process in all the social segments, including the construction people. Today's form of work is unimaginable only a few years ago.
- Globalization of the market and consequently adaptation of the commercial structure in today's construction sector introduces a very high level of competitiveness, which urges companies to adopt more automated and efficient means.

To achieve the consolidation period in the construction automation big efforts need to be made in different fronts:

1. **Integration**. This is one of the key issues which are necessary to be consolidated during the next years, being the main lemma "from architect's desk to site robots". For this purpose three main actions should be taken:
   1.1 Feedback design of houses, taking into account the prefabrication, erection, assembly, transportation and other stages of the construction process.
   1.2 Diversity of the design using the highest number of the similar standard pre-fabricated elements (i.e. building different houses with the same parts).
   1.3 Software standardization which permit the easy and fast data exchange between architects, civil engineers, electrical engineers and computer science experts.
2. **Pre-fabrication**. Expand this technology to cover other materials other than the concrete (including composites), which shall immediately boost the productivity. Three main actions are:
   2.1 Mass production using pre-fabrication in order to select the parts from a catalogue. This means that CIM concept must be introduced, including JIT production.
   2.2 Standardization of the maximum number of parts through the use of grid dimensions, common joints, connections, etc.
   2.3 2.3. New materials for pre-fabricated parts which make them lighter, maintaining the same mechanical features.
3. **Robots and automated machines**. The robots and highly automated machines are the key issue. Using them ensures a high level of productivity. Some of the main actions are:
   3.1 "Easy" to use robots. Develop robust robots which are easy to control and program through friendly human machine interfaces.
   3.2 Cheap robots. Develop cheap robots which cover single type of application, being not general. This will permit to increase the sales of units.
   3.3 Increasing the level of automation of existing machinery. Modify the conventional construction machines (cranes, compactors, etc.) in order to convert them into robotic system.
4. **Investment in R&D**. More research and developed investment in RAC both in basic and applied research through national and international targeted programs, such as the EU research frameworks. One of the main objectives has to be targeted also at changing

the culture of the operators directly involved in the construction process, through education and training. Otherwise the operators would resist the introduction of innovation.

5. Conclusion

This chapter presented the summarised -state-of-the-art in the area of *robotics and automation in construction* focusing on the new robots development and machine automation. This area of robot development was very strong during 90s. However actual research and development in the RAC is more focused on new emerging technologies and mainly based on *software and IT technologies.* This is based on the software integration, simulation and Virtual Reality environments, sensor-based monitoring and tracking, part-oriented construction, etc. These examples are the most representatives but are not exclusive of others. It is important to mention that *software and It technologies* in RAC include other important applications such as artificial life modelling of the construction process, life cycle engineering, RFID chip-robot interaction, etc.

The research in RAC' focus on *software and IT technologies* does not mean that *construction robotics* development from the hardware point of view has seen a cessation, but their development is actually slow. Integration and coordination of both *hard* and *soft* areas is the objective of the long-term research in the field of RAC. It is important to note that this research focus strategically appeared in several national and regional research programs, like the EU 6th Frame Program and was also supported in the inter-regional global program IMS. In order to rapidly advance in RAC and reach the consolidation period the ideas discussed in the previous section need to be adopted and especially new national and international research programs have to be established.

6. Acknowledgments

The authors would like to acknowledge the support of the Robotics Lab team of the University Carlos III of Madrid who assisted in the RAC activities, the funding organizations as EU Brite, Esprit, and IST programs, and Spanish scientific agency CICYT.

7. References

Abderrahim, M.; Garcia, E.; Diez R. & Balaguer, C. (2003). A mechatronic security system for construction site, *Journal of Automation in Construction,* Vol. 14, No. 4, pp. 460-466, ISSN 0926-5805

Balaguer, C.; Giménez, A.; Padron, V. & Abderrahim, M. (2000). A climbing autonomous robot for inspection applications in 3D complex environment, *Robotica,* Vol. 18, No. 3, pp. 287-297, ISSN 0263-5747

Balaguer, C.; Abderrahim, M. et al., (2002). FutureHome: An integrated construction automation approach, *IEEE Robotics and Automation magazine,* Vol. 9, No 1, pp. 55-65, ISSN 1070-9932

Balaguer, C. (2003). EU FutureHome project results (key note paper), *20th International Symposium on Robotics and Automation in Construction (ISARC'03)*, The Netherlands, September 2003, Eindhoven

Cheok, G. S.; Lipman, R. R.; Witzgall, C.; Bernal, J. & Stone, W. C. (2000). Field Demonstration of Laser Scanning for Excavation Measurement, *17th International Symposium on Automation and Robotics in Construction (ISARC'2000)*, pp. 683-688, September 2000, Taipei, (Taiwan)

Corke, P.; Winstanley, G.; Dunlin, M. & Robert, J. (2006). Dragline Automation: Experimental Evaluation through Productivity Trial, In: *Advances in Tele-Robotics*, Siciliano,B.; Khatib, O. & Groen, F.(Ed.), pp. 459-468, Springer, ISBN: 78-3-540-32801-8, Berlin

Gambao, E.; Balaguer, C.; Barrientos, A.; Saltaren, R. & Puente, E. (1997). Robot assembly system for the construction process automation, *IEEE international Conference on Robotics and Automation (ICRA'97)*, pp. 46-51, ISBN 0-7803-3612-7, Albuquerque (USA), April 1997, IEEE (USA)

Ha., Q. P., Nguyen Q. H.; Rye D. C. & Durrant-Whyte, F. (2000). Impedance control of a hydraulically-actuated robotic excavator, *Journal of Automation in Construction*, Vol. 9, No 5, 421-435, ISSN 0926-5805

Hasegawa, Y. (2006). Construction Automation and Robotics in the 21th century, *23rd International Symposium on Robotics and Automation in Construction (ISARC'06)*, Japan, October 2006, Tokyo

Kangari, R. (1996). Re-engineering Construction Work-process for Buiding Automation, *13rd International Symposium on Robotics and Automation in Construction (ISARC'96)*, (Japan), month of the year, Tokyo

Lipman, R. and Reed, K. (2000). "Using VRML in construction industry applications", *Virtual Reality Modelling Language Symposium (VRML'2000)*, Monterey (USA)

Miyatake, Y. & Kangari, R. (1993). Experiencing Computer Integrated Construction, *ASCE Journal of Construction Engineering and Management*, Vol. 119, No 2, pp. 307-322, ISSN 0733-9364

Miyatake, Y. (1993). SMART system: A full-scale implementation of construction integrated construction, *10th International Symposium on Robotics and Automation in Construction (ISARC'93)*, Houston (USA)

Naito, J.; Obinta, G.; Nakayama, A. & Hase, K. (2007). Development of a Wearable Robot for Assisting Carpentry Workers, *International Journal of Advanced Robotic Systems*, Vol. 4, No. 4, pp. 431-436, ISSN-1729-8806

Peyret, F.; Jurasz, J.; Carrel, A.; Zekri, E. & Gorham, B. (2000). The Computer Integrated Road Construction Project, *Journal of Automation in Construction*, Vol. 9, No 5-6, pp. 447-462, ISSN 0926-5805

Rehg, J.A. (1994). *Computer-Integrated Manufacturing*, Prentice Hall, ISBN 0134638867, Englewood Cliffs

Rembold U.; Nnaji, B. O. & Storr, A. (1993). *Computer Integrated Manufacturing and Engineering*, Addison-Wesley, ISBN 0 201 56541 2

Ueno, T. (1998). Trend Analysis of ISARC, *15th International Symposium on Robotics and Automation in Construction (ISARC'98)*, March 1998, Munich

Yagi, J. (2003). Robotics Construction in the 21th century in Japan IF7II, *20th International Symposium on Robotics and Automation in Construction (ISARC'03)*, The Netherlands, September 2003, Eindhoven

2

Construction Automation and Robotics

Thomas Bock
TU München
Germany

1. Introduction

A competitive, market oriented and rationalized construction tomorrow requires developing of automated and robotized construction system today. This includes industrialized process originating in a mining, construction material production, prefabrication of construction components, on site construction, facility management, rehabilitation and recycling.
Today's construction projects are characterizing by short design and build period, increased demands of quality and low cost. These problems can be approached by a flexible automation using robots based on computer assisted planning, engineering and construction management. Especially in high labor cost countries, automated and robotized construction technologies can compensate increasing demand on construction projects. I consider construction robotics technology as a key to rationalization. By automation, increased productivity could reduce high labor cost share of 40 or more percent. Automated and robotized construction process lead to a continuous working time through the year. Introduction of robotic technology would result in better working and health conditions, and advanced mechatronics know how and skills. The reduction of construction time would improve cost benefit analysis of construction project due to faster availability and return on investment of real estate.

2. Automation and robotics in Masonry prefabrication

German brickwork construction is characterised by the high percentage of private builders and an extreme orientation to the craft trade. Machines and auxiliary devices are applied at a larger extent on widespread level, and hardly contribute to increasing productivity, but relieve the worker of physical load. Pre-fabricated brickwork elements are often not applied on smaller construction sites due to lacking crane capacity.
Brickwork construction in Germany reflects two development trends:
Brickwork without dislocating aids will no longer be acceptable in the near future and therefore reduced to exceptions in modernisation, renovation or reconstruction. The monolithic production of brickwork at the construction site with increasingly large-format bricks and further developed dislocating aids or robots will therefore be a continuing development.
On the other hand the pre-fabricated brickwork elements will gain increasing significance, as they represent the economic solution due to the improved working conditions and independence of weather conditions and because they are the more economic solution for many construction companies due to more reliable calculations. For stationary production of pre-fabricated brickwork parts various machines are known.

With stationary brickwork machines drastically raised production capacities can be achieved. Moreover, they lead to considerable relief in labour and manpower savings.

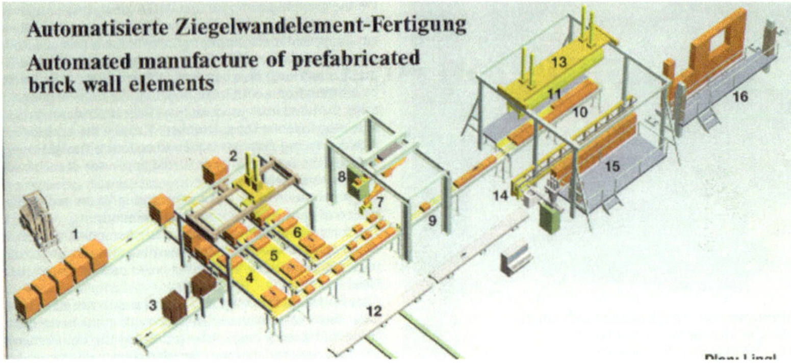

Fig. 1. Prebricated Masonry

The deliberations to manufacture pre-fabricated brickwork elements have taken a whole series of ideas in the mechanical-technical development into account so that individually planned brickwork elements can now be manufactured in a wide variety of production plants with semi-automatic production systems or fully automatic brickwork robots under industrial conditions.

The high capital involvement is one difficulty which production outlets of brickwork elements have to face in view of economic fluctuations and which is caused by establishing the expensive production systems.

There is also a certain dependency on the supplied exactness of contractor products which have to observe certain tolerances not only in a geometric, but also in a physical respect to meet the exactness in robot dislocating.

An additional difficulty is the necessary standardisation of software and hardware used by the architects, engineers, building material manufacturers and constructing companies involved in the building process. Prerequisite for automated construction technologies is in other words the exact definition of all software and hardware standards as well as the respective interfaces.

In conventional on site masonry construction competitivity is achieved by using large masonry elements of weight up to 40kg. Since further productivity can hardly be achieved by masonry wall elements produced in a similar way as a pre cast concrete elements which has been described above. All kinds of existing masonry such as bricks lime stone etc. can be horizontally positioned on a pallet. Reinforcement bars placed between the bricks in order to allow safe transportation and assembly. Concrete is placed between the bricks. All necessary wall openings such as windows, doors, water closet unit, plumbing, service pipes and wiring are also placed between the bricks on the pallet. Interior and exterior plaster is sprayed on the prefab masonry walls. Roof panels already contain roof windows, insulation and roof tiles. In addition there is vertical masonry production robot which place the bricks layer by layer in between distribute mortar and put vertical horizontal reinforcement bars.

Similar as their precast concrete panels, houses made of prefab masonry elements can be assembled in one day. It takes one to two weeks to finish a 120m^2 house.

Besides these automated and robotized facility are half mechanized masonry wall prefabrication units which are controlled by one or two workers.

In the case of these mechanized units the automation just includes distribution of mortar. The productivity of these facility ranges from 3.5 to 50 m² per hour.

Fig. 2. Factory based mobile semi automated masonry wall production unit including automated mortar distribution

3. Automation and robotics in the construction sector and precast concrete industry

In mining, tunneling, earthworks, road construction etc. we have reached a high degree of mechanization with partial automation. In the production of construction materials such as cement, steel, aluminum, glass and wood etc., the degree of automation is very high - almost up to 100%. Automation and robotics in constructions and building components prefabrication is high in pre cast concrete element production where we moved from a mass production to mass customization. This development was enabled by flexible production system using robotic cells which could execute various tasks such as setting molds, placing reinforcement bars or mats distributing concrete for various products such as floor, roof, wall, beam and column elements.

A high degree of automation has been prevailing for several years in stationary systems to mix building materials (concrete and tar) and in serial production of standardised concrete products. The automated pre-fabrication of large-scale pre-fabricated reinforced concrete parts has experienced a fluctuating development. After a gradual increase of the automation degree to the early seventies the beginning recession forced several pre-fabrication plants with higher degree of automation to discontinue manufacturing their products.

A new development surge has been under way for some years which in individual cases has already led to Computer-Aided Manufacturing (CAM) and in some approaches to Computer-Integrated Manufacturing (CIM) of pre-fabricated concrete parts for ceilings, walls and roofs.

In the industry for pre-assembled units the term „Construction System" is frequently viewed as a competitive instrument. Every manufacturer states that he has his own special systems to grant the contractor maximum advantages. This is often misleading and leads to reservations among planners who are not familiar with the construction of pre-fabricated parts. The reality is, however, different: On the market there is a large number of systems for more or less complete solutions; these systems, however, almost all belong to a limited number of pre-fabricated systems with a more or less identical basis. In addition, these systems are as a rule limited to the carcass and the preliminary electrical installation. That means that the production systems are able to manufacture raw construction parts with the exactness of millimetres which are combined on conventional building sites with traditional building systems. Construction systems with a high, trans-trade pre-fabrication degree are generally non-customary.

The further developments from the generally offered raw construction product to the finished wall or ceiling/roof product pave the way to diverse possibilities with the product and product technology to manufacture pre-fabricated construction parts at low prices according to individual requirements for the housing construction industry.

For example, portal robots as they are already used as formwork robots could be further developed and transformed into installation robots for electrical cabling operations. In connection with that aspect the surfaces and assembly engineering should also be further developped. Finished roughcast and insulated wall surfaces could be manufactured in partially automated processes with systems already available on the market. With the increasing production depth, e.g. by installing windows, blinds, cabling etc. the added valuation at the production plant is enhanced. Suitable transport and assembly systems which supply and assemble just in time with optimised logistics are required for such products.

The use of robotic technology in pre cast concrete element production also resulted in constant quality of products and less waste in factories, because due to computer assisted planning and programming only the necessary amount concrete is being provided from the batcher plant. The computer assisted planning and engineering provides the necessary data for the production of all elements such as reinforcement bars or mats originating from the architectural design of floor plans, elevation sections, HVAC plans and structural calculation.

Compared to conventional prefabrication there are less mistakes in transferring data because of defined interfaces between planning, engineering and production.

Fig. 3. Multifunctional robot placing magneto moulds on 3 to 12 m dimensioned steel pallet for PC panel production

The various elements are produced on steel pallets which have dimensions of about 3 to 12 meters. On these pallets the production management system optimizes layout and arrange went of the concrete elements to be produced depending on the priority of the factory manager. For expel if there urgent customer order, all panels will be produced at one time. If there is normal production run, then panels of different construction projects can be produced on one pallet in order to use pallet surface efficiently. The pallets run from the station to station where various robots do the collection of previously used mold, cleaning of pallets, plotting of a panel production layout; gantry type robots place the mold, reinforcement and distribute concrete. The curing station works like big automated warehouse.
You can find the highest degree of prefabrication in the production of concrete box units with a prefabrication ratio of 85% and 6 hours on site assembly time for a 120 m^2house.

Fig. 4. Teleoperated concrete distribution

4. Automation and robotics in timber construction

The prefabrication degree in wood construction can be characterised as favourable according to the current state-of-the-art prevailing in technology in Germany in comparison with other European countries. All conventional wood construction systems are applied. In the recent past focus has particularly been on the „novel block construction“ system (glued laminated wood, bulk wood and log wood construction). Perhaps also because this - in the form of massive constructions normally made of bonded two-dimensional wood - associates wood construction more intensively with massive construction („knock test“, wood/massive compound constructions).
The processing technology in wood construction is developping continuously from manual processing with small machines to full-scope processing on CNC machines.
The requirements with regard to flexibility in processing are noticeably rising.

The division between raw construction and interior design no longer exists. Wood constructions are transformed into pieces of furniture. The standards required with regard to precision in production exceed the general level of a carpenter by far.

In production there is an enormous difference whether raw wood constructions, construction parts for prefabricated houses, staircases or winter gardens or even all together have to be processed on one machine. In serial production the aim is to manufacture the largest possible quantity of identical or similar parts within the shortest possible space of time. For the wood construction worker the most important aspect is traditionally bonding construction wood. For these operations optimally functioning and reliable bonding systems have been on the market for many years. They are characterised by high performance and relatively low programming requirements.

The processing liberty is nevertheless limited: Only construction wood for roof construction, layers of beams or timber framework can be processed. Additional manual processing is in many cases essential; the dimension and form of the parts to be manufactured is also restricted to straight timbers in the majority of cases.

The technical evolution in the production sector indicates a development which will make the application of CNC systems with up to five axes the state-of-the-art in technology in a few years. Above all in the sector of CAD/CAM solutions there still seems to be a great deal of concealed development potential. In the field of prefabricated house manufacturing almost fully automatic plants in production belts are available in individual cases which leave only very few supplementary operations and finishing the surfaces to be performed by hand.

The intensified use of machines with several degrees of freedom has paved the way to new fields of operation for the wood processing companies, also beyond the construction wood sector, whereby new sales options and a higher diversity for the customer can also arise.

The further development of the software required will be a key field of tasks to exhaust the capacity of the machines and the diversity of the product. Direct machine monitoring on the basis of architecture plans without converting efforts by an additional engineer will be a cost factor of rising significance in the future. The advantage in comparison with competitors in this sector may result from the fact that due to the almost complete automation it is possible to manufacture in line with specific customer requirements and individual needs. In particular in the sector of prefabricated wooden house construction the aim of mass individualisation now seems to have come within reach.

Processing technologies gradually shift from handheld tools to precut CNC machines. Increasing flexibility and accuracy in timber processing is achieved by robotic and automated technologies. Functions of primary, secondary and tertiary building system merge by integrating structural components with fitting out functions of interior finishing and building service functions such as plumbing, wiring and HVAC. Carpenters who previously build just timber roofs are now offering complete buildings. They were enabled by multi functional CNC precut machines which could automatically produce any wooden joint based on architectural floor plans, elevations, sections, structural plans and HVAC CAD data. Modular home makers take advantage of these precut CMC machines by combining them with automated and robotic 5 axis assembly and transfer productions lines allowing an output of more than 1000 units and capital investment of about 10 million Euros or more depending on the value added within the prefabrication plant.

Fig. 5. Multifunctional robotic wooden wall production unit

Automatic timber positioning systems and laser assisted marking devices allow flexibility within automated CADCAM timber element production. The highest degree of prefabrication is achieved by the mobile home prefabrication with a prefabrication ratio of 95% and by box unit prefabrication where a prefabrication ratio of up to 85% can be achieved.

5. Automation and robotics in steel component production

From a technical point of view, in any case, it is hardly possible to explain the difference of the development between Germany and Japan in prefabrication automation in steel housing market. The current situation in steel construction and assembly can be characterised as follows: The building market mainly demands solutions from the steel construction companies which fulfil the clients' individual needs and therefore only conditionally allow rational standardisation with regard to production and assembly. This applies to all fields of steel construction, e.g. bridge construction, multi-storey building and hall construction, container construction, compound construction and steel machine and plant construction. As the percentage of steel in the housing sector is low, the steel frames applied in prefabrication for room cells are to a large extent welded or screwed manually.
If we wonder as to how the acceptance of steel in the housing industry can be enhanced in Germany, then Japan could be given as a good example. We see a possibility to learn from the experience made by Japan in the way the building material steel has been supported consistently and with perseverance by direct marketing with united forces.
Today the material steel offers a variety of new possibilities in comparison with the first steel enterprises. Material and production technology have gone through enormous

developments, whereby technological developments were in the majority of cases initiated by other branches (automobile industry). It can, however, be imagined that as a result of a new intensified use of steel in the housing industry innovation potential for the material will arise. By research, experiments and applications, steel can be improved in its capacities and characteristics so that any possible objections raised against steel in the housing sector will lose their validity.

CAD/CAM solutions are the state-of-the-art in steel construction companies to ensure the flexibility required from projecting via CNC production to delivery (logistics) to the construction site and, if applicable, to assembly organisation.

The aim is to produce constructions tuned to manufacturing and assembly requirements to a large extent without reworking at the construction site (e.g. adapting resp. cutting operations) enabling short assembly or construction operations. The construction parts are cut by laser, gas burner cutting, sawing, drilling, before undergoing straightening including metallic cleaning, interim and end coating and complete corrosive protection which are normally processes applied in pre-fabrication. These operations are performed with consistent high quality.

Here you can find a level of automation and robotics similar to the car industry. Factories churning out 5-10 thousands houses a year whether it is a panel based or box unit based system offer not only highest and constant production and product quality but also custom made houses where the client can choose from up to 2 million parts.

Fig. 6. Automated and robotic steel panel production facility

Production cycle time for box unit is down to 2,5 minutes and 120 m^2 houses can be assembled in 4 hours. Customers enjoy 10 or 20 years garanty. Suppliers provide the modular house factory in a 4 day cycle. One day for order output, two days for production at the supplier and one day for delivery to final assembly factory. A house is produced within a week after order intake. If you want to exchange your old by a new house this can be done in two weeks. Within the first week the furniture is moved from the old house, stored and then the old house is disassembled and recycled. During the second week the new house is built and furnished. These customer friendly services made possible by extensive automation and robotics in production are very beneficial for the client since the purchased product is available within one or two weeks resulting in reduced financial burden.

6. International comparative developments: automation and robotics in Germany

To compare the status of automation in housing construction in Japan with the situation in Germany and to derive further findings for possible development in automation, the Federal Ministry for Regional Planning, Construction and Town Planning have commissioned a survey within the scope of a research project[1].

The focus of the current development covers primarily all fields of mechanical engineering and process engineering, e.g. manufacturing building materials, concrete products and prefabricated concrete products, brickwork machines and brickwork robots, controlling and monitoring mobile construction machines, as well as tunnel and microtunnel construction.

Automation and robotics have long found their way into the building industry in actual fact due to a variety of elements which can only be automatically manufactured and without which nowadays construction would not exist at all.

Building materials, construction boards, construction parts, installations, windows, fittings etc. would always have remained high-priced luxury articles, if it had not been possible to manufacture them in fully automated processes.

6.1 Development in Germany

The majority of German building machine manufacturers and construction companies accompany these activities with an only moderate degree of interest.

As this is a part of the future building industry which is highly research- and development-intensive, there is the danger that this market with its long-term and probably existential technical and economic possibility will probably to a major extent be lost to foreign competitors without any resistance.

As a result of the violent technical development in the electronic age more and more focus is being devoted to the need to redefine the opinions regarding the building standards.

The increasing discrepancy between the performance of tools, machines and robots and small tools in general creates an increasingly unstable situation between the craft trade and industrial branches. This development is now intensified by the increasing application of low wages and the pending EU extension to the east, as a result of which it will become more and more difficult to survive in view of European competition. Due to subcontracting low wage workers companies are heading for the innovation and qualification trap. Instead of new technologies being developped and construction workers being further trained and educated they subcontract to low cost / low wage companies.

When examining the construction methods applied in Germany, the building methods and building systems used in concrete construction, brickwork construction, wood construction and in steel construction were investigated.

7. From factory to site automation and robotics

Since the 80s this procedure has led to the fact that the prefabricated house in comparison with the past enjoys a far better image than conventionally constructed buildings.
The annually recurring international symposiums on automation and robotics in the building sector underline the fact that considerable efforts have been undertaken world-wide in Japan and in the USA to utilise automation in all fields of the building sector.
In Japan automation and robotics have been operated consistently for many years on a widespread basis in cooperation with building enterprises, manufacturers, research institutes and national authorities.
In Japan robots of the third and fourth generation have also been presented. As argumentation for these activities the same conventional reasons are stated world-wide, e.g. lack of qualified workers, facilitation in working, quality enhancement, labour protection, environmental protection and productivity improvements.
A highly important reason for the Japanese enterprises is, however, the enhanced image in the building branch, which as low-tech industry enjoys hardly any prestige.
The developments of the last ten to fifteen years show that the Japanese building industry has achieved remarkable success with this strategy.
Impressive examples are 20 partly automatic superstructure systems with which the key building companies Obayashi, Shimizu, Taisei and Takenaka have been constructing buildings in Japan since 1992.
Lack of skilled workers is a coercive reason for forcing such measures which exists in no other industrial state other than Japan where restrictive immigration regulations largely prevent the employment of guest workers.
The reason why the Japanese have not yet offensively offered their construction robots on the world's major building markets is no proof for their assumed unsuitability. The fully automatic superstructure systems cannot be dismissed with the statement that their economic application presumes serial production either.
The automation of building processes has been the object of research and development by key Japanese building corporations since the end of the seventies. Japan started off with the development of individual robots and remote controlled manipulators for certain processes at the building site. These include robots for concreting, concrete treatment, applying fire protection measures to steel constructions, handling and positioning large-scale parts and facade robots for applying plaster and paint.
To date over 200 different prototypes of robotic solutions have been developped in the construction industry and tested on building sites. One common factor is that they have all been determined for specifically defined tasks under construction site conditions and moreover designed to prevent the building site workers' activities from being disturbed.
Experience has shown that under these premises only a few robots can be applied economically. The restrictions for workers, the necessary safety regulations paired with the unforeseeable and unplanned influences at the building site impose restrictions on the application of individual robots in parallel to normal construction site operation. Only a few

are currently in economic operation or are offered on the market for sale. These comprise, for example, the concrete smoothing robots manufactured by Kajima or Shimizu.
The outcome of this development is the finding that it is not possible to transfer production situations similar to those prevailing in the production hall to the construction site either without having to face difficulties or economic drawbacks. This may seem to be a trivial and foreseeable result, but it is necessary to realise that these developments were seen at the beginning of work only as a way into the automation of construction processes and that their economic use was not the foremost goal to be achieved. Two other results which play a key role in the future of Japan's building industry were moreover decisive. On the one hand these were the findings and capacities acquired in the field of automation and robotics resp. sensitisation of the employees for innovation in the building sector. On the other hand, preparation of the actual goal, this being the fully automatic production of a terrain on the building site under application of the regularities known from serial production.
About 200 different robotic devices had been developed, tested on site and improved. The highest degree of automation had been reached in tunneling from the prefabrication of tunnel sections, its transportation and assembly.

Fig. 7. Modular mobile light weight concrete finishing robot

Since sites and its conditions greatly vary at each project, the processes have to be well defined in order to be robotized. Furthermore the planning and design has to facilitate robotic construction by robot oriented design methodology.

The full potential of robotics will unfold as soon as robots do not just copy human work but rather be enhanced by robot oriented planning, engineering, management, labor training and qualification. Robots will probably not being used if total hourly labor cost is below 35, -euro. Another positive side effect of these high labor cost and high labor productivity would be wealth generation for construction workers as consumers.

Financing of expensive robotic equipment must be supported by financial institutions. Investors should appreciate the immediate availability of their real estate by forwarding their higher and earlier return of investment in the form of higher construction project costs.

About 20 integrated automated and robotic building construction systems were running between early nineties till year 2008. Some companies developed systems that pushed the building up to ten floors, others had climbing systems with one to three gantry cranes or about 22 trolleys simultaneously transporting and assembling columns, beams, floor, interior wall and exterior wall panels and sanitary or installation units. The machine reutilization ratio was about 95%.

It took about a week for one floor and the finishing ratio reached 85% by using prefabricated and highly integrated components.

Working conditions on site became similar to factories and there were no accidents or quality problems.

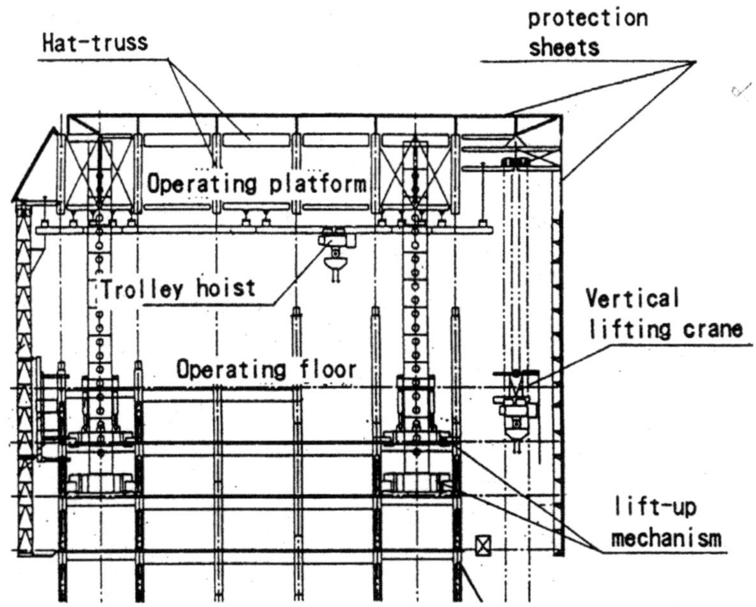

Fig. 8. Section of an integrated automated building construction unit with about 20 robotic trolley hoists for logistics and positioning

Similar as the JIT just in time of the Toyota production system the factories supplied building components in a ten minute cycle to the site. Since there were no storage areas for construction materials on site, construction materials were directly grasped by the robotic trolley hoists from the truck. Some systems could also adjust to non rectangular floor plan lay out proving that flexibility in design can be achieved by constantly improving robotic technologies.

Fig. 9. Transportable welding robot in an integrated automated building construction site

The work force reduction initially was 30%, then 50 % and can reach up to 70%. It takes between 3 to 6 weeks to construct and disassemble an integrated automated and robotized building construction system. Investment cost for the on site integrated automated building construction units are about 5 to 10 mio Euro or more depending on its functionalities and performance. These additional costs have to be recovered from faster return of investment by earlier availability of rental space.

Service robot systems and humanoid construction robots

Building automation systems are state of art.

Energy costs had been successfully reduced due to efficient HVAC systems and energy saving facades. This leaves cleaning, building servicing and rehabilitation costs with a considerable impact on life cycle costs of real estate.

Since the real estate servicing costs during the building life cycle is several times -up to 6-30 times depending on the building type- of the initial investment or construction cost, it is obvious to rationalize this significant cost factor by automation and robotics.

There are many examples of façade cleaning robots, interior cleaning robots, security robots, transport robots, service robots for hospitals, elderly and physically disabled.

Since aproximately six years humanoid construction robots were developed to drive fork lifts, excavators or carry building parts jointly with a construction worker. These humanoid robot technology transfer to construction is based on nearly two decades of humanoid robot subsystem technology development.

Humanoid construction robots vary from teleoperated devices through autonomous ones that can walk on 5 degrees inclined slopes, compensate 2cm high obstacles and are able to get up by itself once they fell down. Positioning is achieved by vision systems, force sensors in the feet recognize inclined slopes, balancing sensors detect the body's inclination versus the surface slope and an autopilot controls its attitude.

8. Guidelines for construction robot development

Focus in robot development in Germany is to be mainly determined from the viewpoint of the workers. It is necessary to inquire in which sectors high or unacceptable burdens are registered and it is exactly there that analyses should take place to find out which technical aids are required. An analysis of requirements based on the types of load is therefore urgently necessary.

Robots are primarily developed for the sectors in which poor labour conditions prevail and in which a reduction of the load is possible. The comparatively high frequency of accidents as well as the high statistics of labour-related sickness and premature retirement in the building industry are an indication for the special requirements. Robot systems should take over the task of handling heavy loads, of performing dirty or dangerous work or of working at hardly accessible locations and in unfavourable physical positions.

Above all robots should function as tools of the human being. They are to be developped as intelligent tools and must not force the human being to the limits of working activities. It must be possible to integrate the robot systems into labour procedures. These must not disturb the existing communications structures and cooperation, for example, within the scope of a gang. Robot development should therefore be implemented together with those persons who will operate these systems at the building site at a later point of time. Changes in the labour environment and labour organisation by the application of robot systems must be primarily oriented to the working people in the first step and then in the second to technology.

One important aspect is high system flexibility to adapt the robots to the prevailing structures. Fully automatic systems are therefore only suitable in exceptional cases, for examples in areas with high safety risk. Semi-automated machines, in contrast, can be flexibly monitored and applied. The focus of development must therefore lie on semi-automated systems. Other industrial sectors have in the mean time also withdrawn from the aim to achieve inappropriate full automation. Semi-automated systems are by far cheaper and more flexible than fully automatic systems. They can be applied by smaller-sized and medium-sized building companies to improve their competitiveness.

9. Development of integrated construction automation and robotics building processes

The building processes and systems to be automated and furnished with robotic controls have to be redeveloped. The existing management methods require revising before qualifying the staff involved according to the application of new technologies. A successful implementation of robot technology is enabled with a robot-oriented construction industry which reflects certain characteristic features: flexible industrial pre-fabrication, flexible production of different building parts on the site and project management enabling the application of construction robots.

Automated building comprises industrial and flexible prefabrication of complicated standardised building parts and their automatic construction and maintenance using construction robots. Automated building production enterprises are able to achieve a high level of variations with a wide range of construction parts. With the help of freely programmable robots a flexible production of a wide variety of building parts is enabled and administered using the suitable software. The industry manufacturing prefabricated parts should benefit in particular from the advantage that it is possible to flexibly manufacture with a large degree of automation by aligning production technology in order to meet the requirements of mass individualisation in the housing construction sector.

As far as automation of the construction company is concerned, the development of an integrated system to plan and produce buildings should be envisaged. This system can be

used not only in drafting buildings, but in operational planning for robots and in logistics for building sites.
Due to the high wage costs in executing construction work the largest rationalisation effects are achieved by an intensified rationalisation of the construction work with the help of automation components. On-site construction work has to be aligned to subsequent robot operation in the planning and construction phases. That means that all construction planning phases have to be integrated into the computer systems before being processed. The conventional building processes have to be tuned to automation requirements. These new building processes will differ fundamentally from the known building processes. The normal sequential procedures of building production will also be replaced by parallel procedures. Partial systems from prefabrication will also be integrated into building operation and will therefore drastically reduce the construction period.
By contracting a project for an automated building construction, the whole activity has to be furnished with robotic controls, planning, construction and manufacturing of construction parts. These parts will have been largely prepared and completed so that after signing the contract the construction project only represents a geometric configuration problem, timely organisation problem and a physical implementation problem.
The corporate structure is transformed from the current assembly company to a future service company.
Contemporary buildings consist in comparison with pre-industrial buildings of many partial systems. Planning, production and the product are increasingly mechanised and will be additionally mechatronised. This fundamental development in the building sector requires an integrated and interdisciplinary problem-solving approach. In implementing building management this means the specification of conditions for operating robots on the site with a geometric, physical and timely definition of the elements for every constructional subsystem.
That requires an interlinking of the data and information flow from the draft to design, manufacture, assembly and facility management. The interlinking in the prefabrication of partial systems and their integration into the building processes plays a decisive role hereby.

10. Strategies for an automation and robot oriented construction process

Systems able for automation in construction should satisfy the needs of all parties participating in the construction process. Suppliers have to increase the quality of the construction materials and products under the geometrical, physical and design aspects, to fulfill the conditions for automation oriented design methods. The present production sequences in the construction process have to be adapted to an automation concept. A construction system able for automation should contain, additionally to the conventional properties like stability or economic efficiency, also a flexible strategy, which includes all participating parties and allows a future reusability.
The goal of an automated construction will be achieved, if following parameters will be met simultaneously without excluding each other:

- Freedom in esthetic and design
- Determination of the production costs before the execution of work
- Determination of the production time before the execution of work
- Guaranty and transparency of the price
- Continuos production

- Definition of quality
- Control and transparency of the quality

10.1 Information integration during the construction process.

The task is not only the automation of the componentÆs pre-fabrication and assembly, but also the coordination and connection of all processes by means of computer integration and interface management. Prerequisite for an automated construction technology is the exact definition of all soft- and hardware standards, all interfaces and communication protocols used for the project.

One essential base for the computer integrated construction process is the design of the building based on a 3-dimensional geometry model representation in the necessarily used CAD system. The second column of a successful construction is the exact description of the properties of the different sub-systems. According to the definitions, the suppliers are chosen. This information can be provided separately to the geometric information only referring to the single positions in the design. Also the third column can be generated separately: a time scheduling of the different construction processes on the level of the sub-systems provided by the respective suppliers. Under another point of view, the scheduling can be regarded as a rated assembly precedence graph for the different assembly procedures on-site, which can be referred to the different geometrical descriptions in the design database.

Through an intelligent interface management each supplier is able to get the necessary information concerning the geometries, the material properties and the time scheduling. It is not necessary to hold all information in one system or on one database. It is even not necessary to have all information at all on the computer. The only purpose is to create an internal production plan, whereas many informations can be used as easy and as fast as possible. The geometrical and the material information can be used for the automated production of the components of the demanded subsystem. The time and sequence information can be used to optimize the production sequence concerning delays and delivery time security. Additional, all three informations can be used for refining: more detailed plans of the sub-system, more detailed material information of the parts of the sub-system and a more detailed assembly sequence of the parts can be generated. This can be delivered back to the coordinator, who can use the information for the logistics on-site and the assembly of the parts as far as the supplier does not assemble the parts by himself.

After integrating all the information through the coordinator of the construction site, the information is ready to be used for the assembly of the sub-systems or of the parts of the sub-systems. The goal must be to use the information out of the three areas refined by the suppliers to generate the control sequences for the automated construction tools available on the construction site respective to generate the assembly instructions for the manual part of work. For that it is necessary to have all information in the access of the coordinator, where the different necessary interfaces and information flows and directions should be well and flexible defined. It is also necessary to have the possibility to integrate information in the databases at each level manually or via a defined interface.

The integration of all participants into an information network guarantees an individual and simultaneous efficient construction of buildings, which is able to cope with short term changes without increasing the costs or decreasing the quality and design freedom.

The gained data should be used also for the optimized operation and the recycling of the building, since the costs of a building concern not only the erection but also in a considerable amount the operation and recycling.

Fig. 10. NCC Komplett wall assembly in on site assembly hall

In the NCC Komplett factory 60 operators work on job rotation time schedule. The yearly capacity is 1000 apartments and each worker is producing 17 apartments yearly. Automation and mechanization are ergonomically designed to reduce labour fatigue. Every 15 Minutes a truck leaves the factory. The apartments are 90% prefabricated. The investment was about 30 million euro. The on site assembly factory is all weather proofed enabling ergonomic working conditions all year around.

Fig. 11. On site factory of NCC Komplett

10.2 Transfer of construction processes to the pre-fabrication.

The automation and industrialization of the construction process will have the first break through in areas, where already a high mechanization rate is already existing. This is the stationary pre-fabrication of construction parts in contrary to the processes on-site, particularly in the building construction.
Through an intelligent shift of process parts from the site to the pre-fabrication, it is possible to achieve a higher automation rate together with a higher integration rate of the construction parts in an earlier and more defined stage of the construction process. This increases the quality and decreases the transport costs.

10.3 Increase of efficiency and quality.

Through the higher integration of the parts, the more sophisticated design and the information integration, different effects occur during the construction process: In the pre-fabrication it is possible to apply industrial production methods for a higher integration of the parts. This results in lower production and transport costs with a simultaneous increase quality. Also the design flexibility of the produced parts is increased through automated production technologies and the integration in an information network. Additionally the integration of the complete construction process enables the suppliers to participate in the complete design and construction process. With this approach also "Just-In-Time" and "Simultaneous Engineering" concepts can be realized.
The advantages of the individual design of a single-piece production and of the industrial mass production can be combined. So the construction time can be reduced through higher parallelization, the flexibility increased without higher costs for short term changes and the quality enhanced through better quality control and new construction processes.
A hybrid high-rise construction site is understood as the semi-automated storage, transport and assembly equipment and/or robots used to erect a building almost completely automatically. It is the attempt to improve the sequencing of construction processes and construction site management by using real-time computerized control systems. This includes an unbroken flow of information from planning and designing the building through programming the robots with this data to using computers to control and monitor building operations on site.
After the foundations have been laid, the production equipment, on which the steel construction has been installed with assembly and transport robots, is covered completely with a roof of plastic film. Depending on the system, this takes from three to six weeks. Then the robots go into action. Two steel and ten concrete plants supply parts in ten-minute cycles on a just-in-time basis. This approach to supplying is not necessarily part of the system, but is due more to the lack of space around building sites in large Japanese cities. The prefabricated parts are checked and then placed in specific depots at the foot of the building or in the building itself to be available to the robots.
Once a story has been finished, the whole support structure which rests on four columns is pushed upwards by 12 hydraulic presses to the next story. Three 132 ton presses in each pillar are required to achieve this in 1.5 hours. Fully extended, the support structure is 25 meters high; retracted it measures 4.5 meters. Once everything has been moved up, work starts on the next story. By fitting out the topmost story of the high-rise as the roof at the beginning of the building process, the site is closed off in all directions, considerably reducing the effect of the weather and any damage it might cause.

Fig. 12. Hybrid Construction System

This system reduces labor requirements by around 30%. Future projects are expected to achieve a labor saving of around 50%. The building consists of a remarkably high proportion of prefabricated parts. Once the foundations have been laid, the remaining construction procedure can be described as a matter of configuring transport and geometry. All the elements are prefabricated; only some of the fitting, joint insulation and other minor works need to be carried out by hand. Problems with the construction arise less from the timing of deliveries of materials or from the choice of processes and/or machines but more from the need for accurate planning, from programming the robots or from the just-in-time supply of parts.

10.4 Construction components for an robot oriented design & construction

10.4.1 Clear product structure.

The clearer the structure of the products, the simpler a realization of automated construction becomes. It is necessary to insert the product system elements into clear hierarchical levels. The product has to be divided into "sub"-products according to the construction process.
In the development phase it is necessary to check all product features according to their ability to be used for an automated application. With the following procedure, existing

products can be redesigned according to the rules of automated construction. After the first analysis of the product function, sub-functions are defined within each effecting area and each effecting direction. During further analysis all sub-functions are eliminated which refer not directly to the sub-functions of assembling and connecting. Elements with similar effects can be summarized to defined classes of functional solutions. In the next step the effects of these elements concerning the assembly process are analyzed. After the analysis a simplification of the functions through further sub-dividing and the search for solutions should follow, which are then combined again to a global solution.

10.4.2 Design focused on simplicity, handling and assembly.

Construction elements should be designed clear and simple. If an element has a specific assembly alignment, the design should support the easy alignment of the element. So problems concerning the handling, orientation and identification are reduced. If the assembly direction stays the same during the whole construction process, the costs are less than if several alignments have to be made. If a part is asymmetric, the asymmetry has to be emphasized to find the assembly direction easier.

10.4.3 Families of components.

The sequence of construction elements should be rationalized through grouping technology. Grouping creates families of parts that can be assembled or processed with the same or similar tools. Unnecessary distinction of construction components should be avoided. Necessary distinctions should be reduced to the system specific principle.

10.4.5 Component interfaces.

To get defined construction systems, the components have to be standardized to a certain extent. To realize all possible combinations it is necessary to confine oneself to few basic parts. The connections should be compatible to enable the mutual assembly. They have to be defined geometrically and physically. Similar to the "open systems" standards, the construction components of different suppliers should be compatible.

Connection surfaces and points should be compatible. The construction parts need an exact defined connection zone. If the structure becomes static effective, the importance of the connection zones increases. The assembly of complex structures may need specific parts for the connection zones.

10.4.6 Integral product structures.

The number of element connections and in parallel the connection work should be reduced. To achieve a lower number of components and sub-assemblies the product structure should be summarized. This leads also to lower assembly times.

The pre-assembly of complex structures is possible thanks new flexible production technologies. The cost intensive handling of small parts during positioning, adjusting and fixing can be avoided, if pre-assembled units are brought to the construction site.

10.4.7 The accuracy problem.

During the design process, one has to pay attention to the tolerance system, too:

- To avoid geometrical faults between to elements, which have to be connected.
- To compensate accuracy deviations in the control system.

- To handle the production tolerances of the construction elements.

Everybody dealing with construction is aware of the accuracy problem. The most inaccuracies can be recognized and corrected. But for automated equipment the increased necessity of sensors increases the complexity, susceptibility and price and reduces considerably the working speed.

To position components precisely, the structure must be produced and assembled with minimum tolerances. The statistical average of the accuracy, on which the construction is usually based, cannot be applied to automated construction processes since not all cases of minimum and maximum accuracy can be considered equivalently. Therefore the accuracy and its problems have to be described in categories, which are easy to integrate into the construction process.

The tolerances of the different trades vary very much: The low accuracy of foundation work is insufficient for other trades and the high accuracy of installation work is not necessary for the foundation. The different categories of accuracy for different trades have to be defined exactly and considered during the development of sub-systems.

11. Summary

Due to the high complexity of the construction process and the stagnating technological development a long-term preparation is necessary to adapt it to advanced construction methods. Architects, engineers and all other participants of the construction process have to be integrated in this adaptation process.

The short- and long-term development of automation will take place step-by-step and will be oriented to the respective application and requirements. In the initial phase existing building machines will be automated step-by-step. In the medium term a mixed concept consisting on the one hand of manual operation with programmable partial processes and on the other hand of automatic operation with manual monitoring options including all controlling concepts lying in between will gain ground.

In the end phase the CIC concept (Computer Integrated Construction) will be implemented. The use of robots will be more effective, the more appropriately it is integrated into a CIC production chain.

Not only in stationary industry, but also on-site the computer-supported building production of the future could be monitored by the human being in a control room, whereby a qualified building worker can simultaneously control several building machines. All that is needed is an effective communications system between the control officer and autonomous building machines. Application planning and monitoring will be automatically controlled, whereby every individual building machine will constantly communicate with the central control room. In the event of irregularities which are not stated in the program, automatic operation can be manually monitored by the control officer.

The performance of robotic technology is increasing rapidly and we can support its advancement by designing, engineering, managing the construction processes and products in a robot oriented way. On the engineer level we need robotic and mechatronic construction engineers, managers and architects education. The workers need mechatronic and robotic training and qualifications. For real estate to be build by robots and integrated automated construction systems we need additional investment in order to cover the higher construction costs caused by greater capital investment in construction equipment. One way could be borrowing financing methods from the leasing sector, aircraft or car industry,

which often offer 0% interest loans to attract new customers. Towards the investor we have to communicate the advantages of constant construction quality and faster availability of rental space resulting in higher return on investment.
The realization of automation and integration of advanced technologies in the construction field can be supported, if the guidelines for automation oriented construction systems are followed and took into the thinking process. Together with a slightly modified design, the effective pre-fabrication and automated assembly on-site are processes, which can be linked together through a sophisticated computer integration and interface management.

12. References

Pictures 1-9,12, Thomas Bock, TU Munich, Germany
Pictures 10,11, NCC Sweden
Guest Editorial in: Autonomous Robots: Special Issue on Construction Robots, Nov. 2007, Springer ed., T. Bock, guest editor

3

Mechanising, Robotising and Automating Construction Processes

Frans van Gassel and Ger Maas
Eindhoven University of Technology
The Netherlands

1. Introduction

Building objects are produced by people who perform the necessary tasks using equipment. On the basis of preconditions, the process designer can have a particular task performed by a specific combination of a worker and equipment. The worker performs a number of tasks and the equipment does the rest.

Nowadays, newer, more suitable technologies are becoming available. In order to use these technologies successfully, it is essential to have a good understanding of the work processes of an object that is to be built.

The terms mechanising, robotising and automating are defined in order to be able to describe the physical, cognitive and organising tasks in relation to the possible use of human-machine technologies. It sometimes makes more sense to redesign the building products to achieve a more effective and efficient building process using workers and additional tools or machines.

Mechanising, robotising and automating construction processes is necessary in order to reduce production times and costs, improve working conditions, avoid dangerous work, allow work to be performed that people cannot do and increase performance. For the construction industry, more and more human-machine technologies are becoming available, but their use does not automatically lead to more effective and efficient construction processes.

Building expertise is the domain of the professional builder and not of the process engineers who look to apply the technologies in the construction industry. The implicit know-how of the builders and construction process designers regarding the execution of construction processes has to be made explicit. The builder's implicit know-how comprises knowing how to choose the sequence of the building elements, how to join the elements, where the elements fit in the construction as a whole and how they have to be positioned.

This chapter contains a systematic definition of the terms mechanising, robotising and automating and explains an analysis method with which a worker-equipment system that produces better performance can be designed.

2. An automated construction system

In Japan, construction process designers have upscaled the worker-equipment system into a cohesive building production system to find solutions to problems such as the aging of workers, a higher training level for employees and the low numbers of young people looking for jobs in construction (Obayashi, 1999). A building production system can be defined as a technical installation that assembles construction elements into a building. In this context, an installation can be seen as a collection of people, tools and machines, computers and telecommunications equipment that may all be working together. If we couple this definition to the various tasks required for the performance of a building activity - physical, cognitive and organising tasks - we see production systems subdivided into traditional, mechanised, robotised and automated building production systems. Table 1 shows the relationships between the different parameters using human-machine technologies.

Construction system type	Physical tasks	Cognitive tasks	Organising tasks
Traditional	Workers Equipment	Workers	Workers
Mechanised	Equipment	Workers	Workers
Robotised	Equipment	Computers and software. Means of communication.	Workers
Automated	Equipment	Computers and software. Means of communication.	Computers and software. Means of communication.

Table 1. Types of construction systems in relation to various tasks (Van Gassel, 2003).

An automated construction system consists of an assembly area where building work can be carried out regardless of the weather, an automatic hoisting system for the assembly area, an automatic vertical and horizontal conveyor system and a centralised information system to execute and manage organisation tasks (see Fig. 1).

Fig. 1. An automated construction system in Japan.

3. Developments in construction processes

Developments in construction processes are the result of a set of changing circumstances and conditions, such as enormous migration into the cities. The forecast is that in 2015, 55% of the world's population will live in urban areas. These metropolises impose their own

requirements on construction management and production systems. These changes encourage the development of technologies to ensure the creation of a process that leads to improved performance for the client. These developments are based on an analysis of the Status Report issued in 2001 by the CIB Task Group TG27 'Human-Machine Technologies for Construction Sites' (Maas & Van Gassel, 2001) and of the proceedings of the ISARC2003 Symposium: The Future Site (Maas & Van Gassel, 2003).
When all building production is ultimately designed to lead to improved performance and a satisfied client, it is always difficult to keep sight of the overall picture and this final goal. The overview in Fig. 2 shows the relationship between the various aspects of automation in construction: construction management, construction engineering and performance management help the process designers to meet the needs of the client and society.

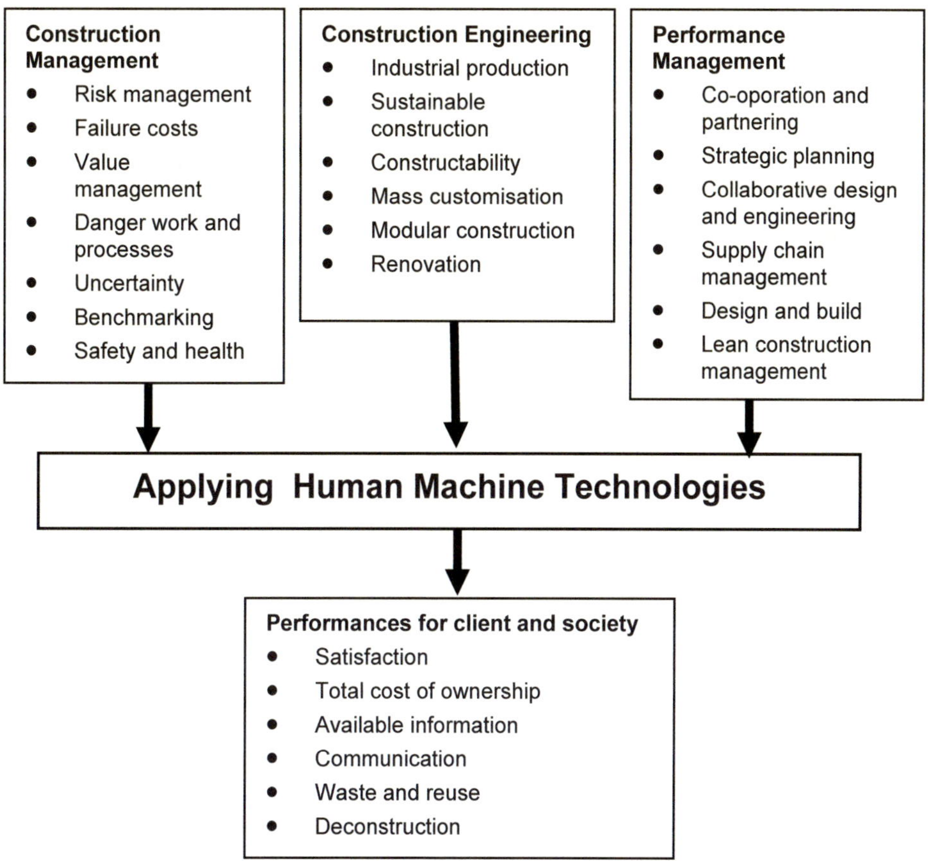

Fig. 2. Relationship between management, engineering and performance (based on Maas & Van Gassel, 2005).

The building assignment will focus on metropolises, which sets specific requirements for performance management, construction management and construction engineering.
Clients need individual treatment and a specific approach designed to solve their problem and meet their demands. They are less concerned with the size of the investment, but are becoming more and more interested in the total cost of ownership and life cycle costs. Nowadays, clients are less concerned with the structure itself. They pay more attention to its functional use, primarily encouraged by the use of information and communication technology in the projects.

Construction engineering has been changed by the application of more industrial production, sustainable construction, mass customisation, and modular construction to improve constructability.
Construction management has to deal with health and safety, uncertainty and danger. Developments are taking place in risk management and value management, supported by partnering, collaborative design and supply chain management.
These developments demonstrate that there is plenty of room for improvement in all process elements of construction projects in metropolises (Maas & Van Gassel, 2005).

4. Worker-equipment system

To produce a building object, three types of task have to be performed: (i) provide strength and energy (physical tasks), (ii) receive and issue information (cognitive tasks) and (iii) make decisions (organising tasks). The human body has a number of suitable parts and society has developed equipment designed to perform the tasks more effectively (see Table 2). As human beings, our speed and power are limited to what equipment can do, but people are far more sensitive to input and have a large, versatile memory.

Tasks	Human body	Equipment
Provide strength and energy	Movement system: muscles lungs	Power tools: energy sources transmissions
Receive and issue Information	Senses: eyes ears voice hands	Telecommunications tools: scanner microphone monitor keyboard
Make decisions	Thought system: brain memory	Computer equipment: computer software artificial intelligence

Table 2. Human body parts and equipment to fulfil tasks.

To perform specific tasks, the process designer chooses the right combination of worker and equipment. Describing such a combination is possible using the basic diagrams of the worker-equipment system (see Fig. 3 and 4).

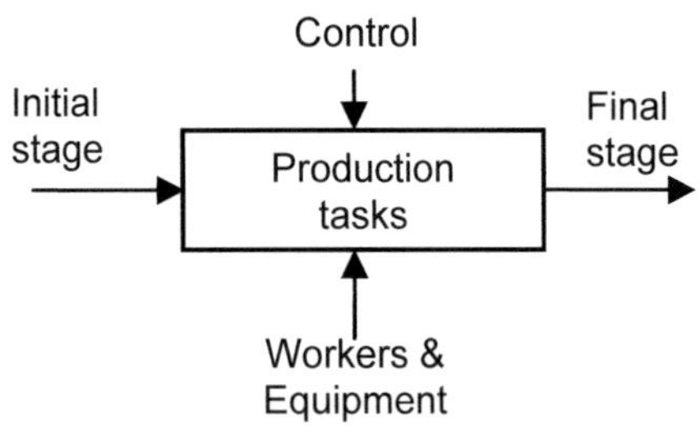

Fig. 3. Basic diagram of the process.

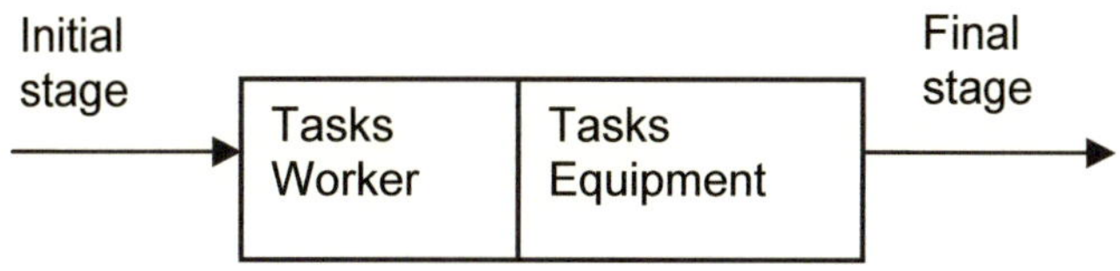

Fig. 4. Basic diagram of the worker-equipment system.

The diagrams used here are based on system analysis. Materials are transformed by the worker-equipment system from an initial to a final situation (Maas, 1991). That part of the tasks to be performed by the equipment and that by the worker are represented by the size of the surface of the rectangle. The rectangles can be divided into activities that can take place in sequence and/or at the same time, so a building activity can be divided into subactivities.

5. Mechanisation and robotisation concepts

The mechanisation concept is defined on the basis of the diagram in Fig. 4: 'Mechanisation is the shift of tasks from worker to equipment'. The concept is shown in a diagram in Fig. 5.

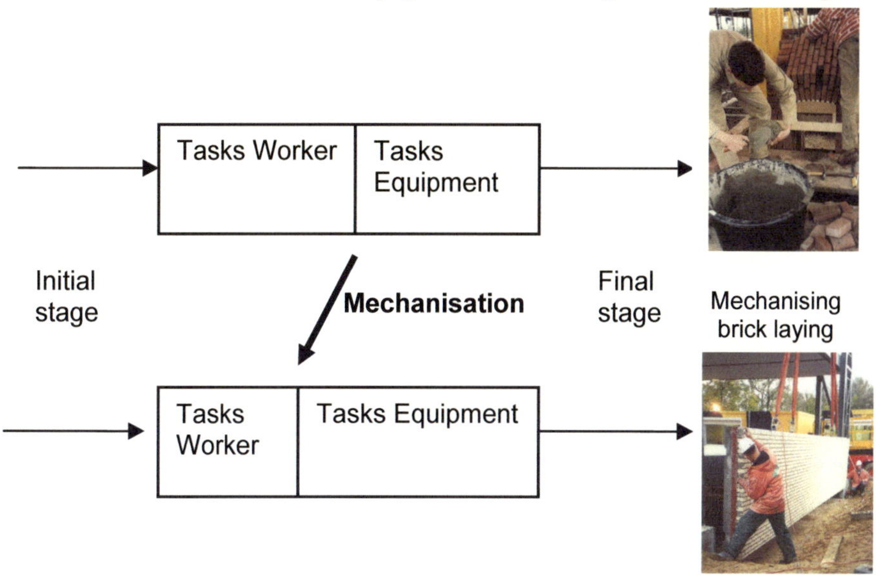

Fig. 5. The mechanisation concept.

Robotisation is a special type of mechanisation in which all tasks are shifted from the worker to the equipment (see Fig. 6). Control and support activities are not included in these tasks, because they are not directly related to the specific production activity.

6. The mechanisation graph

The tasks that workers and equipment carry out can be divided into energy tasks and control tasks.

Three situations are considered for the performance of energy tasks:

1. the equipment does not supply energy
2. the equipment supplies a certain amount of driving energy
3. the equipment supplies all the driving energy and the worker only has to operate the controls.

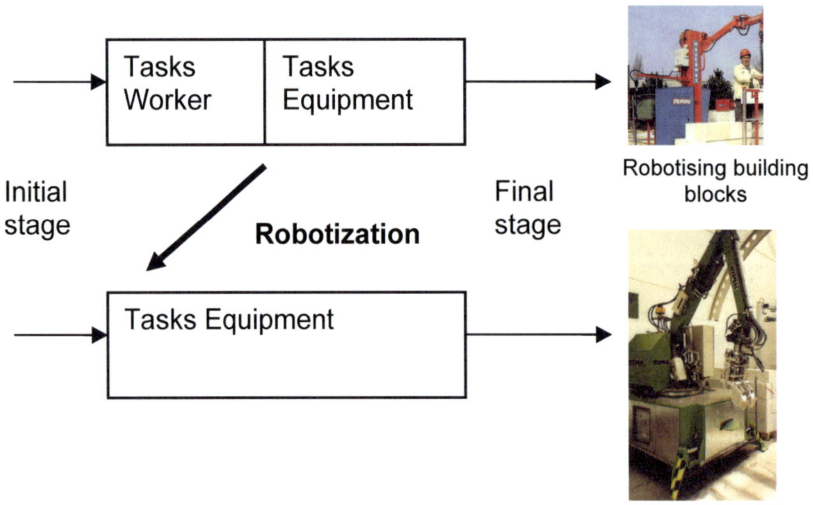

Fig. 6. The robotisation concept.

Driving energy means, for example, the rotating and linear motion of a drill and the supply of energy necessary to hold the equipment in a steady position. The latter can be achieved by placing the equipment on a stand.

The classification of control tasks is based on Guo and Tucker's machine line arrangement (Guo & Tucker 1993). This classification comprises the following levels: hand tools, manually controlled devices, telecontrolled devices, pre-programmed devices and cognitive robots. At the beginning of the line, all tasks are carried out by people, and at the end of the line by machines.

On a graph, energy tasks are placed on the vertical axis and control tasks on the horizontal axis, resulting in a mechanisation graph (see Fig. 7).

A worker-equipment system or elements of such a system can be placed in the squares of the graph.

Examples:

A. The laying of bricks with the aid of a mason's trowel.
B. The drilling of a hole in a wall with an electric drill.
C. The drilling of a hole in a wall with an electric drill that has been placed on a guide and that is started electrically.
D. The riding of an operator on an excavator.
E. The moving of loads with a crane that is remotely controlled.
F. The laying of bricks by a robot that places bricks and mortar according to instructions.
G. The digging of a trench in the ground by an intelligent excavator that can make decisions on the basis of observations.

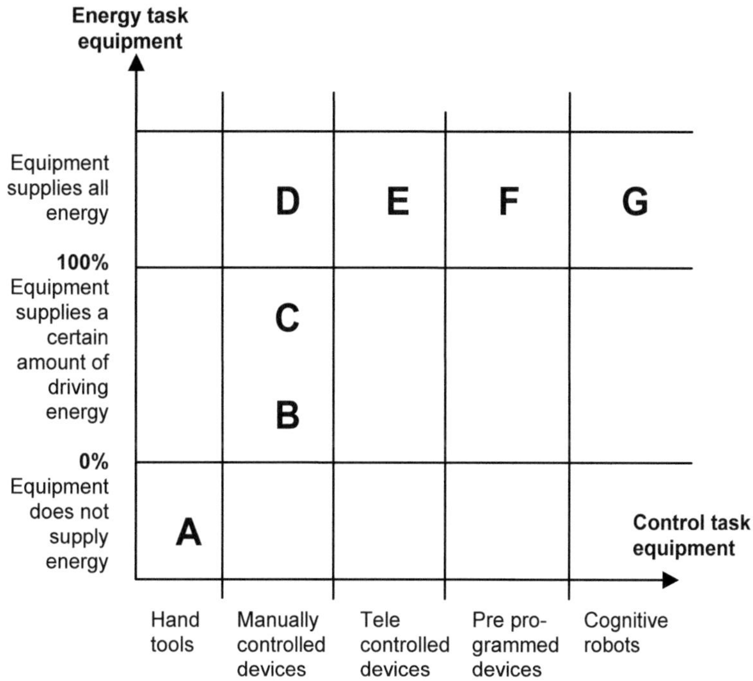

Fig. 7. Mechanisation graph.

7. Mechanisation phases

It is now possible to indicate the mechanisation phases on the mechanisation graph.
The following phases can be distinguished:

1. optimisation of tools
2. use of drives
3. use of guides
4. use of control equipment
5. use of remote control
6. use of computers
7. use of artificial intelligence

The mechanisation phases are represented in the mechanisation graph in Figure 8. The phases above are based on the mechanisation of existing situations. Entirely new production processes, however, can be designed as well. Here, it is only possible to represent the end of the mechanisation phase on the graph.
An increasing degree of technology is necessary in order to complete the mechanisation phases. This requires knowledge of:

- Materials and construction products
- What workers can do (ergonomics)
- Drive technology, guides and manipulators
- Machine controls, remote controls and programming technologies
- Sensors and artificial intelligence

A chain of mechanisation phases can also be represented on the graph. The smoothing of a poured concrete floor is used as an example. Traditionally, these concrete floors were smoothed with a hand trowel, which involved manually moving the power trowel over the floor. The next mechanisation phase enabled the floor-layer to sit on the machine, so that only operation of the controls was necessary. Next, the machine was equipped with remote control. The machine was then provided with computer controls that can finish the floor according to programmed instructions. The subsequent step would be to provide computer controls that make observations and a number of decisions on the basis of these observations.

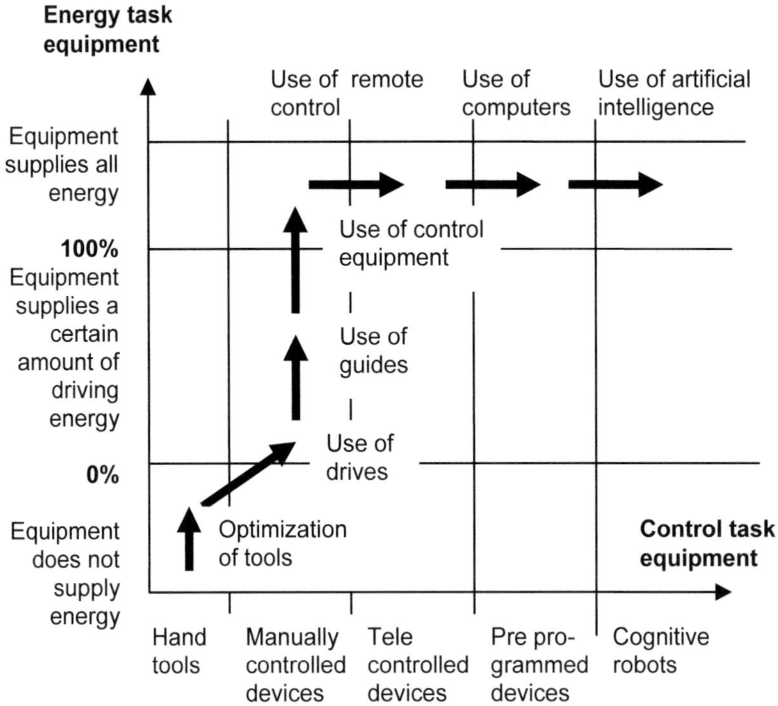

Fig. 8. Mechanisation phases.

A distinction can be made between the mechanisation of production processes of existing products and those of modified or new products. Mechanisation of existing products is the mechanisation of traditional production processes carried out by craftsmen, such as masonry, plastering and carpentry. Generally it is quite possible to mechanise the energy task, while the mechanisation of control tasks is highly complex.

Just analyse the laying of bricks by a bricklayer and try to put these activities in the operating program of a machine. The mechanisation of modified or new products has more of a chance of being put into practice.

8. Example

As an example, the basic diagram is applied to a production process in which large, heavy wall elements are taken from a lorry and mounted on a façade. This can be divided into a number of constituent processes: attachment of the element to the hoisting hook, transportation to the location where it is to be mounted and turning the element to the desired position (see Fig. 9). The tasks performed by equipment and those performed by people can be described (see Fig. 10).

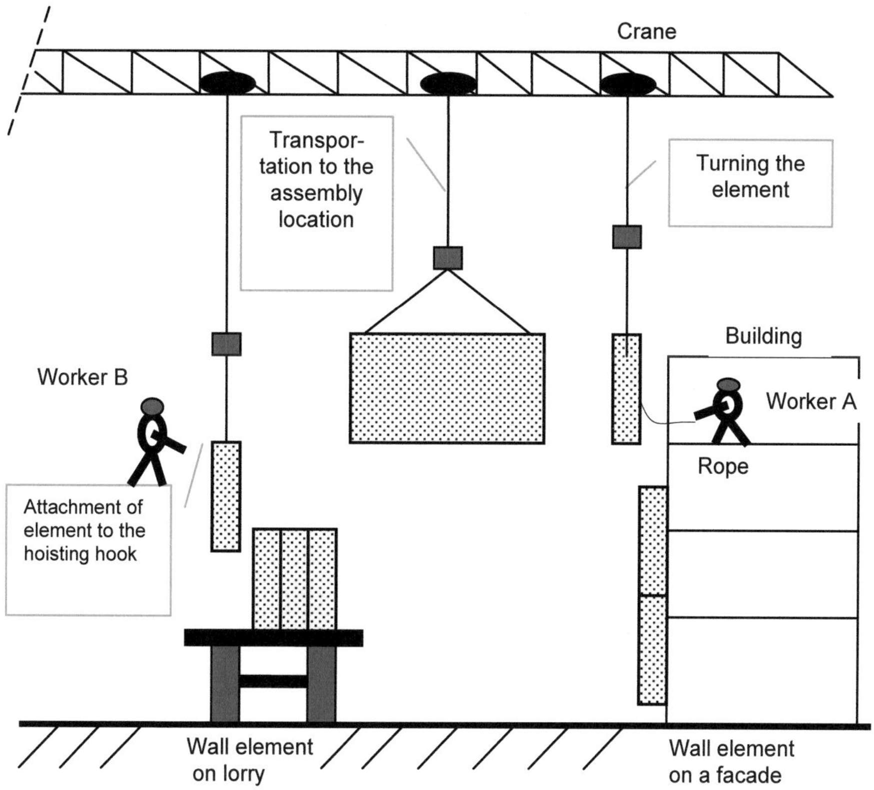

Fig. 9. The mounting of large wall elements.

With the help of technologies, some tasks of worker A and B can be done by equipment (see Fig. 10). For instance, the wall element can be guided automatically by equipment between the hoist hook and the wall element (see Fig. 11 and 12).

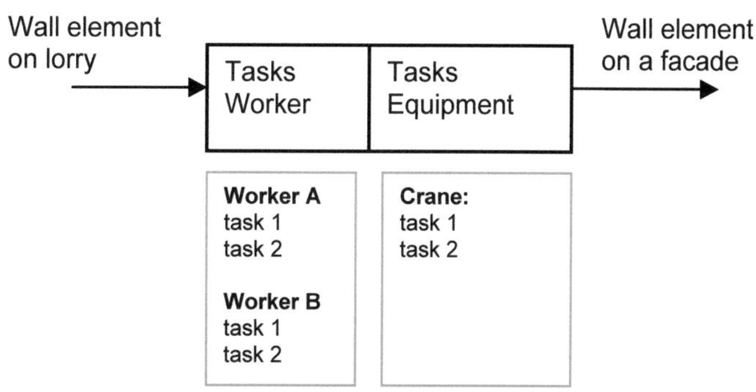

Fig. 10. Analysing the tasks before mechanising.

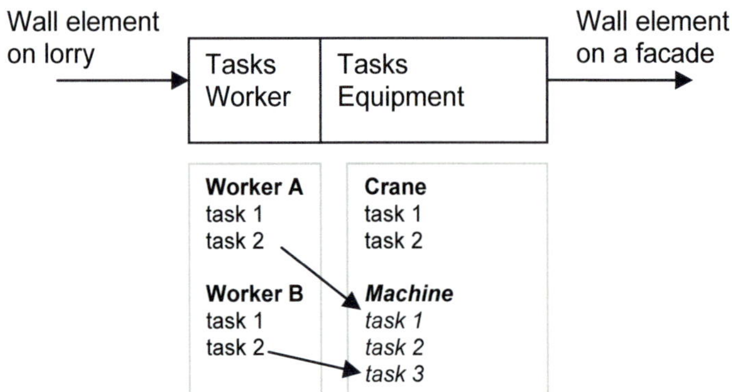

Fig. 11. Desired situation after mechanising.

Fig. 12. Mechanizing the position of a wall element by special equipment.

9. References

Van Gassel, F. and G. Maas, (2001) *International Status Report on Aspects of Future Sites*, CIB TG27 'Human-Machine Technologies for Construction Sites.' CIB 265. ISBN 90-6814-119-8. CD-ROM.

Van Gassel, Frans, 2003, *Bouwproductiesystemen in Japan* in Bouwen in Japan (*Building Production Systems in Japan)* in Building in Japan. ARKO, The Netherlands.

Guo, S. and Tucker, R. (1993) *Automation needs determination using the AHP approach*, in the proceedings of the 10th International Symposium on Automation and Robotics in Construction (ISARC) , Elsevier Amsterdam.

Maas, G.J. *Modernisation by considering the total building process*. De Ingenieur, (Engineer) no. 3 (March 1991).

Maas, G. and F. van Gassel (editors), 2003, *Proceedings The Future Site ISARC2003*, Bouwstenen 75, 675 p. Eindhoven University of Technology, The Netherlands.

Maas, G. and F. van Gassel (2005) *The influence of automation and robotics on the performance construction*. Automation in Construction 14 435 - 441.

Obayashi, Shigeyuki, 1999, *Construction Robot System Catalogue*, Council for Construction Robot Research, Tokyo.

4

Powerline Communication in Home-Building Automation Systems

Elena Mainardi and Marcello Bonfè
Engineering Department, University of Ferrara
Italy

1. Introduction

Domotics, Smart Home Systems, Ambient Intelligence are all terms that describe the intelligent cooperation of several different equipments to manage the home environment in an intelligent, safe and comfortable way. The same idea is also applicable to bigger constructions, and in that case it takes the name of Building Automation.

Whatever term one wants to use, it refers to a multidisciplinary field that includes informatics, electronics, automation and telecommunication, and also touches fields like building constructions and architecture. In fact, during the process of designing a building, people have to consider appropriate spaces for the electric plant, and if the presence of a domotic system is planned, it is better to take it into account during the design phase, just to optimize spaces, the amount of used wires, the position of the modules and so on.

There are really many home system producers in the world (Smart Home Systems, EIB-Konnex, Lonworks, Bticino, Vimar, Duemmegi, EasyDom Corporation, Futurware, Digital Cybermasters, Hills Home Systems, Intellihome etc, just to mention a few), and their products differ from each others in many characteristics, such as functionality, dimension, weight, typology of installation, materials, net topology, power consumes, aesthetic appearance, communication protocol and communication mean.

Regarding this last point, the majority of the domotic systems, especially in Europe, tend to use a dedicated bus cable to exchange data among modules, to make the communication link more robust and reliable. Lately, using radio communication is in fashion, but radio modules, respect to their equivalent standard ones, are more expensive, and in the bargain many people don't want to use them due to the fear of radio signals (even if it were proved that they are completely harmless).

Another communication mean, that is often not taken into account, is the powerline.

In point of fact, using the installed poweline wires to send information is a very smart idea: there is no additional cost to install other dozens of meters of wires, there is not the necessity to break the walls and to do building works at home, there are no interferences with other devices (like in the radio communication case) or reflection problems (like in the infrared case), there is the possibility to put the modules in every place (it is sufficient to have an electrical socket in the nearness, or to use an extension cable), there is no need to have an extra power source (usually, in a bus cable domotic system, there is a direct voltage generated by a power supply and distribuited on the whole domotic net).

Moreover, powerline communication (PLC, also called BPL in the USA, where the acronym stays for Broadband over Power Line, or NPL, Narroband over Power Line) is not only used in a home environment to create a virtual net among domotic modules, but is also used on the power distribution net to perform actions like reading the electricity meter, monitoring the power consumes and the state of a building, finding faults along the net, detecting illegal electricity usages and to solve the so-called *last mile problem* , that is the problem related to the final leg of delivering connectivity from a communications provider to a customer. In fact a cheap possibility to cover this final leg is using powerline communication.

Of course there are not only positive factors, but also some troublesome aspects that have to be considered. For example, if a noisy home appliance is connected to the powerline, it can disturb it and create problems over powerline communication (this can be easily solved using an appropriate powerline filter). Again, if a house is near an electric closet, the powerline may be unstable, with the same problems of the previous case.

The intent of this work is therefore to illustrate, going into more details, advantages and disadvantages of the powerline communication systems (PLCS), to show the differences between PLCS for power distribution net and PLCS for home and building environments, to indagate the methods to send data over the powerline, to explain which are the automations that is possible to connect and to control in a powerline domotic system and to show some case studies tackled by the authors.

2. Advantages and disadvantages of power line communication

Most private dwellings do not have dedicated neither low nor high-speed network cabling installed, and the labor costs required to install such wiring is often quite high. Power line communication is an emerging home networking technology that allows consumers to use their already existing electrical wiring systems to connect home appliances to each other and to the Internet (Dhir, A.& Mousavi, S.,2001).

Home networks power-line technology can control anything that plugs into an outlet, including lights, televisions, thermostats, alarms, home automation modules and so on. If there is the availability of multiple power outlets in every room, the home power line infrastructure represents an excellent network to share data among intelligent devices, also with high data transfer rate, up to a few hundreds of Mbps (mega bit per second).

Unfortunately, there are some problems that have to be overcome and some aspects that have to be taken into account to realize a successfull data communication. They are listed in the following:

- *minimum-security levels*: powerlines do not necessarily provide a secure media
- *data attenuation*: due to the presence of numerous elements on a powerline network, data attenuation is likely to be an issue
- *high costs of residential appliances*: the cost of a powerline network modem is not always competitive with the cost of a standard modem used to connect to a phoneline network
- *lack of global standards*: there are several different standards for powerline communication, and the development of a global standard for distributing data over existing in-home powerline systems does not seem to be the trend of the international market
- *noise*: the greater amount of electrical noise on the line limits practical transmission speed (vacuum cleaners, light dimmers, kitchen appliances and drills are examples of noise sources that affect the performance of a powerline-based home network).

3. The noise problem

Since the powerline distribution grid is present in almost every dwelling and building (on the contrary of other cabling, like xDSL - digital subscriber line - and similar), it represents a mean that offers a huge potential of communication services. Thus the electric power supply system is evolving from a pure energy distribution network to a multi-purpose medium, delivering energy, voice and other data communication services (Cuncic & Bazant, 2003).

Unfortunately the power line channel has not been thought and designed to carry information other than electrical energy, so it has not good characteristics in terms of noise robustness and attenuation of high frequency signals. Due to this, power line represents a particularly difficult communication environment (Ferreira et al, 1996).

According to the measurements of Vines et al (1984), that are still valid nowadays, the sources of noise over the power line can be summarized as depicted in figure 1.

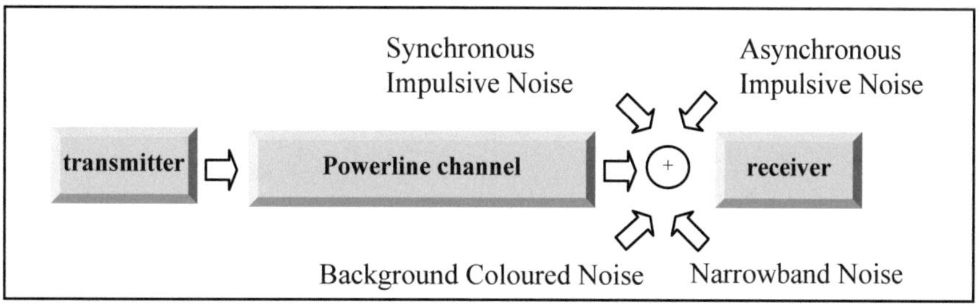

Fig. 1. Noise contributions over a power line network

- *Background Coloured Noise* shows a spectral power density that is low at medium frequency (around 20-30 MHz) but increases toward lower frequencies. It can be thought as the sum of many contributions, every one of each is a white noise acting in a non-overlapped frequency band. Its power spectral density (PSD) varies over time in terms of minutes or even hours (Zimmermann, 2002). This kind of noise can be caused by appliances like dimmers, hair dryers, computers and so on.
- *Narrowband Noise* is represented by portions of the spectrum with a very high level of PSD, but relegated in very thin and limited frequency ranges. In the frequencies beyond 150 kHz it is mainly caused by amplitude modulated sinusoidal signals of broadcast radio stations in the medium and short wave bands (1-22 MHz typical), while up to approximately 150 kHz it can be generated by switching power supplies, frequency converters, fluorescent lamps and television sets. The level is generally varying with daytime (Majumder & Caffery, 2004).
- *Synchronous Impulsive Noise* is characterized by short voltage peaks, which are rare single events caused mainly by on and off switching actions, and which appear at frequencies that are multipliers of the main frequency (50 Hz for Europe, 60 for the USA). The pulses are of short duration (a few microseconds) and have a PSD decreasing with frequency. They are caused by rectifiers within DC power supplies and appliances with thiristor- or triac-based light dimmers.
- *Asynchronous Impulsive Noise* is said to be the worst kind of noise that afflicts powerline communications. In fact its time duration varies from few microseconds to milliseconds, and besides the uncertainty over this parameter, it is characterized by other three

random variables: amplitude, pulse width and interarrival time (the time between two consequent noise pulses). The PSD of this noise can be grater than 50 dB above the background noise spectrum, and it can have contributions also up to 20MHz. Due to its characteristics, this kind of noise can corrupt long blocks of data transmitted over the powerline. It is mainly caused by switching transients generated by switching power supplies.

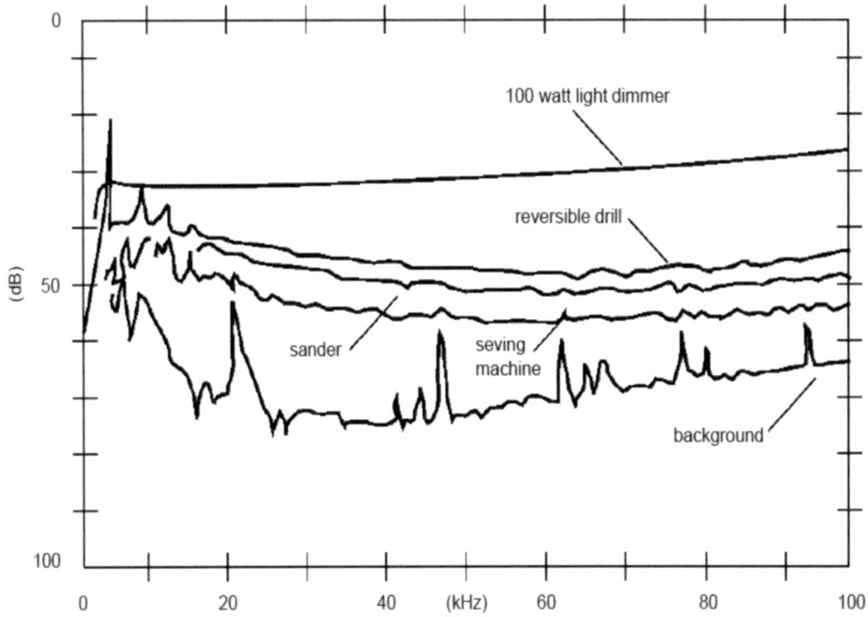

Fig. 2. Voltage spectra for three universal motors compared to light dimmers operating into the 60 Hz power circuit (Vines et al, 1984)

From what mentioned above, it can be seen that the powerline is a communication mean difficult to be treated. Common analog modulation schemes like AM (amplitude modulation) or PM (phase modulation) are not so robust to be used in powerline communication systems, so other more complicated methods are exploited. They will be briefly illustrated in section 4.

Besides being disturbed from several noise sources, the powerline itself creates some distrubances over other communication means. This is the case of the HF (high frequency) radio band, that can be upset by powerline systems in the range 1- 30 MHz. (Rhee et al 2008, Hansen 2002). Due to this phenomena and considering that powerline communications must not disturb the powerline mean, that is already upset by its self, some regulations became necessary to manage the powerline communication field. This subject will be treated in section 5.

4. Modulation schemes

Power line cables represent a very hostile environment, due to the fact that several different equipments connected to them can generate a wide variety of disturbances, causing the powerline to exhibit unpredictable and variable attenuation with time-variant and frequency selective behaviour. So transmitting data over them is not an easy task. Depending on the bit rate and on the frequency range used to injected data, the modulation methods are different.

Modulation is the teqnique that allows to send a signal (called *modulating*) by means of another signal, called *carrier*. The result of the interation between the two creates a new signal, called *modulated*, with different characteristics, more suitable to the behaviour of the transmitting channel (in this case the powerline). In general, the reason why the modulation is necessary is the low pass or band pass characteristic of the channel.

There are several modulation techniques. Besides the very simple and already mentioned analog modulation schemes, like AM, FM (frequency modulation) and PM, where the information resides respectively in the value of amplitude, frequency or phase of the modulated signal, for digital information, like in the case of PLC, other cathegories of modulation techniques are more appropriate: among them, the most simple are ASK (amplitude shift keying), where the information resides in the presence or absence of the modulated signal, FSK (frequency shift keying), where logic 0 and 1 are codified with two signals with different frequency, and PSK (Phase shift keying), where the information stays in signals with different phases.

All the mentioned modulation methods use a single and fixed carrier signal, and can be suitable for low trasmitting bit rates (up to a few hundreds of kbits/s) in domotic environments and also in rural telephony, to transmit voice with ranges of several hundred kilometers without repeaters.

Increasing the bit rate, noise is not the only problem, but also other disturbances have to be taken into account, for instance the intersymbol interference (ISI). This phenomena is due to the low pass behaviour of the powerline channel, which smooths the very ripid fronts of the digital information, like shown in figure 2.

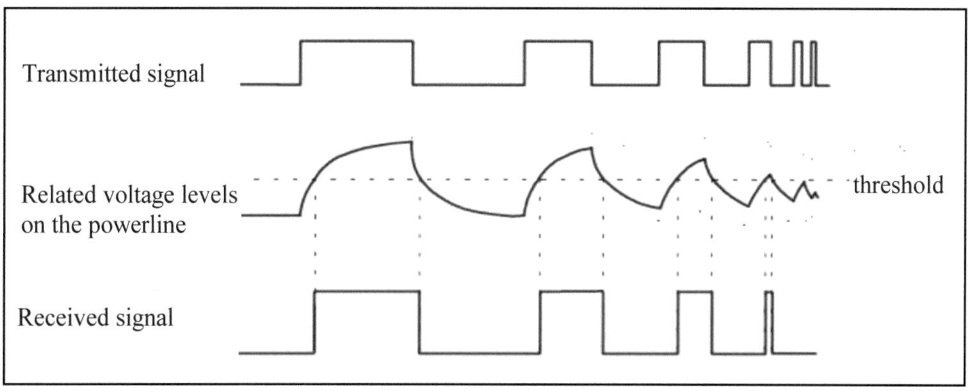

Fig. 3. The ISI problem

This is true for all kind of digital signals at every bit rate, but the difference between low and high bit rates is that if the delay caused by the low pass characteristic is greater than the symbol time duration and two consecutive bits are transmitted too close to each other, the receiver can be confused and can recognize a bit sequence different from the real one.

From what just mentioned, it is clear that ISI limits the maximum bit rate.

Several field tests have demonstrated that distributing information to separated carriers within the available bandwith is a successful way to lower the vulnerability against interference, attenuation and ISI problems. The most used modulation methods in this case are SST (spread spectrum techniques) and OFDM (orthogonal frequency division multiplexing).

Coming to an end, it can be stated that for low bit rates (up to some hundreds kbits/s) narroband modulation methods like PSK, FSK or also MSK (Minimum shift keying) and GMSK (Gaussian Minimum shift keying) can be used. For data rates around 1Mbits/s the CDMA (Code-division multiple access) technique may provide an effective solution, while for high-speed services, with data rates of dozens or hundred of Mbits/s, only techniques like SST, resistant to all kind of narrowband interference, or OFDM, resistant to frequency selective noise, are suitable.

In any case, to increase the robustness of the communication link, besides the adopted modulation technique there are also other digital messages manipolations methods, like ARQ (Automatic Repeat reQuest), FEC (Forward Error Correction) and CRC (Cyclic Redundancy Check).

Interesting dissertations about modulation schemes and error detection methods are illustrated in Karl & Dostert (1996), Waldeck & Dostert (1996), Han Vinck at al (2000), Lin et al (2002), Cuncic & Bazant (2003), Hakki Cavdar (2004).

5. Regulations

Concerning regulations, since PLC devices are electric equipments that transmit over the power line (being both in the electric and in the telecommunication field), powerline networks, products and services have to work under regulations regarding (Napolitano, 2004):

- electric safety
- electromagnetic compatibility
- nets and communication services.

Different regions of the world have different kind of directives and tolerated levels.

In the European Community, regarding the first point, everything connected to the powerline cables must respect the low voltage directives, that in Europe are Regulations *73/23/EEC, 93/68/EEC* and *EN 60950* (Safety of information technology equipment – ITE- also regarding devices destinated to use the power line network as a transmission mean).

Regarding the third point, the European Regulations are:

- *Directive 2002/19/EC* on access to, and interconnection of, electronic communications networks and associated facilities (**Access Directive**)
- *Directive 2002/20/EC* on the authorization of electronic communications networks and services (**Authorization Directive**)
- *Directive 2002/21/EC* on a common regulatory framework for electronic communications networks and services (**Framework Directive**)
- *Directive 2002/22/EC* on universal service and users' rights relating to electronic communication networks and services (**Universal Service Directive**).

Regarding the second point, we have to distinguish between PLC used at low frequencies to send small amount of data (like in the case of home and building automation) and PLC used to realize an internet connection, used at high frequency to share big amount of data.

In the first case (Lauder& Sun, 1999) regulation is *EN 50065-1* (in the band 3–148.5 kHz), in the second one is *EN 55022*, that specifies limits for conducted disturbance at the mains ports in the frequency range 150kHz - 30 MHz.

Outside the Region 1 of the International Telecommunications Union (ITU), which covers Europe, there is no Long Wave broadcast band and the applicable standard for mains-borne communications systems is *IEC 61000-3-8* (regarding Electromagnetic compatibility). This standard permits significant levels of signal injection up to 525 kHz, just below the start of the Medium Wave broadcast band.
Regarding the immunity, and not the radiated emissions, *CENELEC EN 50412-1:2004* (Immunity requirements for power line communication apparatus and systems used in low-voltage installations in the frequency range 1,6 MHz to 30 MHz -- Part 1: Residential, commercial and industrial environment) and *EN 55024* (Information technology equipment - Immunity characteristics - Limits and methods of measurement) are reference regulations.

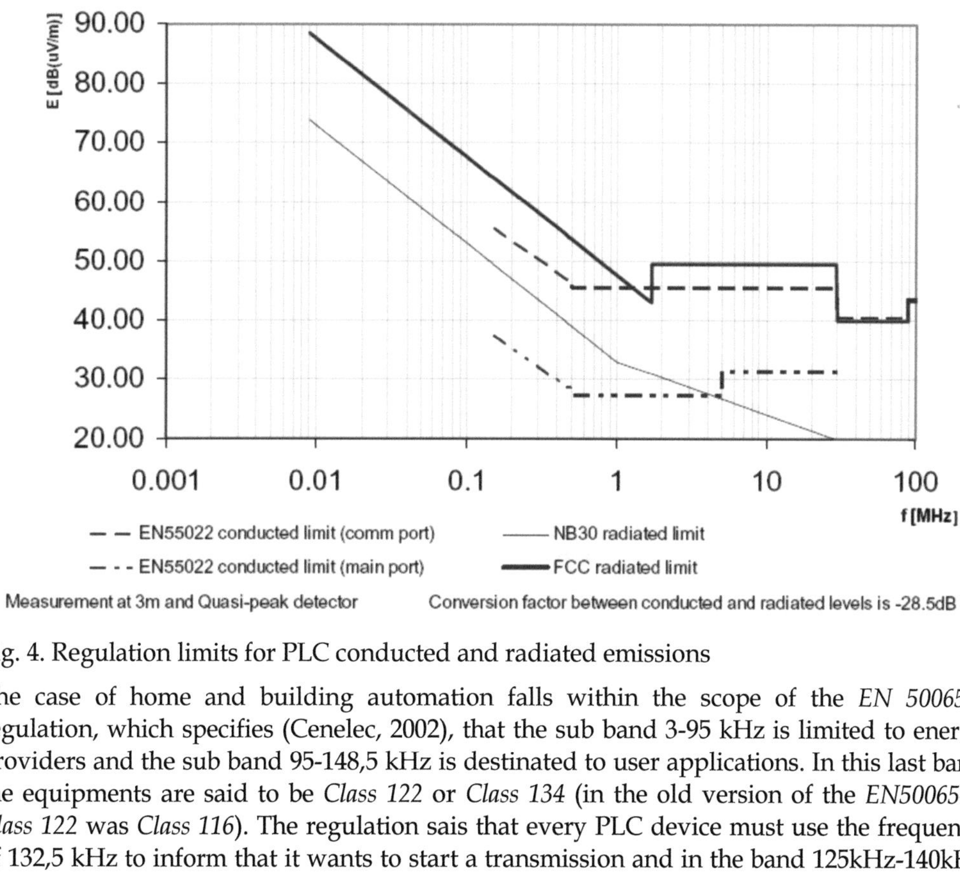

Fig. 4. Regulation limits for PLC conducted and radiated emissions

The case of home and building automation falls within the scope of the *EN 50065-1* regulation, which specifies (Cenelec, 2002), that the sub band 3-95 kHz is limited to energy providers and the sub band 95-148,5 kHz is destinated to user applications. In this last band the equipments are said to be *Class 122* or *Class 134* (in the old version of the *EN50065-1*, *Class 122* was *Class 116*). The regulation sais that every PLC device must use the frequency of 132,5 kHz to inform that it wants to start a transmission and in the band 125kHz-140kHz a CSMA (Carrier Sense Multiple Access) access protocol is mandatory.
In the USA regulations for radiated limits are enacted by *FCC* (Federal Communications Commission). They are more permissive and allow plc modem to attain higher data rate than the European Standards (Schneider et al, 2004).
The German Regulatory Authority for Telecommunications and Posts (called Reg TP) established the *NB30* limits which are based on radiated measurements. Contrary to the previously discussed standards, the limits are given in peak levels. It is difficult to make a direct comparison with other standards since the conversion factor depends on the measurement setup.

United Kingdom in the past used the *MPT1570* standard, that specified limits up to 30 MHz until January 2003. The current version specifies radiated limits below 1.6 MHz only. PLC modems do not need to be compliant with this standard if they operate in another frequency band.

Recently also the IEEE Society (Institute of Electrical and Electronic Engineers) has shown some interest in standardization of the PLC field, so a new commettee (TC-PLC: Technical Committee on Power Line Communications) is born, sponsorising two standards:

- *IEEE P1901*: Standard for Broadband over Power Line Networks: Medium Access Control and Physical Layer Specifications
- *IEEE P1775*: Powerline Communication Equipment - Electromagnetic Compatibility (EMC) Requirements - Testing and Measurement Methods.

The areas of interest of the Technical Committee are the analytical, theoretical, simulation, experimental, and practical aspects of digital communications over power lines.

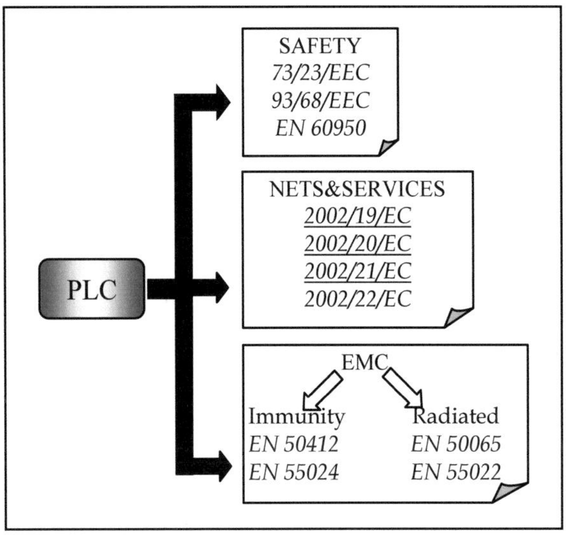

Fig. 5. PLC European regulations

6. Powerline applications

Powerline communication systems can be classified into four distinct cathegories, concerning the type of services that they can provide: communication over high voltage grid, access to an internet provider, in-home networking with high data rate and in-home simple control application with low bit rate.

Regarding the first cathegory, the available services for electricity provider companies, as mentioned in the first section of this paper, concern reading the electricity meter, monitoring the power consumes and the state of a building, finding faults along the net, detecting illegal electricity usages (Hakki Cavdar *b*, 2004), trasmitting telephone signals in rural regions.

Access powerline technologies send data over the low-voltage electric network that connects the consumer's home to the electric utility provider. The powerline access technologies enable a *last mile* local loop solution that provides individual homes with broadband connectivity to the Internet.

In-home powerline technology communicates data exclusively *within* the consumer's dwelling and extends the connectivity to all the electrical outlets within the home. The same electrical outlets that provide power will also serve as access points for the network devices. Whereas the access technologies focus on delivering a long-distance solution, competing with xDSL and broadband cable technologies, the in-home powerline technology focuses on delivering a short-distance solution.

Services like powerline intranet solutions and powerline music distribution (that have to follow the *EN 55022* directive) belong to the in-home networking cathegory. Their typical bit rate can achieve values of 200 Mbps (declared). These kind of services compete against other in-home interconnection technologies such as wireless, even if up to now the bit rates are definitely lower (Lin, Y et al, 2003).

Regarding the last cathegory, it can provide services like sending small amount of data with low bit rate (in fact to open an automatic door or to control the switching on and off of a light not big amount of data are required). Belong to this cathegory the services of the home and building automation. They have to follow the *EN 50065* directive.

7. Home-Building automation and powerline

Many, but not all, Home and Building Automation system producers sell modules that can communicate over the powerline.

A few examples are mentioned in the following.

Konnex, the European consortium that was born from the join of Batibus, EHS (European Home System) and EIB (European Installation Bus) , has defined a standard that up to now covers three physical mediums. They are Twisted Pair (called TP0 and TP1 variants with data rates of 1200 and 2400 baud), Radio frequency (RF with 868 MHz carrier and data rate 16.384 baud) and Power line (called PL110 specification with data rate of 1200 baud and PL132 variant with center frequency of 132.5 kHz, FSK modulation with CSMA/CA, FEC , 16 bit CRC and data rate equal to 2400 baud).

X10 is an open standard very widespread in the USA and in Europe. It uses an ASK modulation, synchronized with the zero crossing of the power line signal (110V, 60Hz in the USA and 220V, 50Hz in Europe). It has an effective bit rate of 60 bps and it does not implement FEC or CRC techniques. A binary 1 is represented by a 1 millisecond burst of 120 kHz at the zero crossing point, and a binary 0 by the absence of the 120 kHz signal.

CEBus is an open standard proposed by the Electronic Industries Association (EIA). It defines a set of rules for consumer products to communicate with each other (Dhir, A.& Mousavi, S., 2001). The CEBus base products consist of two fundamental components, a transceiver and a microcontroller. Data packets are transmitted by the transceiver at about 10 Kbps. The CEBus standard includes commands such as volume up, fast forward, rewind, pause, skip, and temperature up or down one degree. These commands are based on a language called Common Application Language (CAL). CEBus uses spread spectrum technology to overcome communication impediments found within the home's electrical powerline . The CEBus powerline carrier spreads its signal over a range from 100 Hz to 400 Hz during each bit in the packet.

To avoid data collisions, CEBus uses a Carrier Sense Multiple Access/Collision Detection and Resolution (CSMA/CDCR) protocol.

A CEBus-based home network is comprised of a control channel and potentially multiple data channels on each of the CEBus media. CEBus control channel is used exclusively to control devices and resources of the network, including data channel allocations. Data channels typically provide selectable bandwidths that can support high data rates and are used to send data such as audio, video, or computer files over the network. The characteristics of a data channel can vary greatly depending upon the medium and connected device requirements. All data channel assignments and functions are managed by CEBus control messages sent via the control channel.

LONWorks technology is a solution for control networks developed by Echelon Corporation. It has been estimated that more than six million LONWork based appliances are installed worldwide. A LONWorks system includes all the necessary hardware and software components for implementing complete end-to-end control systems. In a LONWorks network, intelligent control devices, called *nodes*, communicate with one another using a common protocol. Each node in the network contains embedded intelligence that implements the protocol and performs control functions. In addition, each node includes a physical interface (transceiver) that couples the node *microcontroller* with the communications medium. A typical node in a LONWorks control network performs a simple task. Devices such as proximity sensors, switches, motion detectors and sprinkler systems may all be nodes on a home network. The core of a node is the Neuron chip, which is basically a microcontroller specifically designed to offer a good cost-effective solution available for home control devices. Appliances on a LONWork enabled home network use a protocol to communicate with each other. This protocol is known as *LonTalk* and has been approved as an open industry standard by the American National Standards Institute (ANSI)—EIA 709.1.

The LONWorks Network Services (LNS) architecture provides a range of network services to appliances that are connected to a control system. A couple of examples of Lonworks powerline modem are PL3120 and PL3150. They work under *EN 50065* regulation, using a BPSK (binary phase shift keying) modulation, a primary frequency of 132kHz in band C (125-140 kHz) of the *EN 50065* and of 86kHz in band A (40-90 kHz). If these primary frequencies do not work properly, they can switch over a secondary frequency (115kHz in banc C and 75kHz in band A).

Another important player in the PLC world is the Home Plug Alliance, formed by very important industry leaders such as Intel, LG, Cisco Systems, Motorola, Samsung, Sharp, Texas Instruments and others. The Alliance created three *Promoters' Groups* focused on specific standards initiatives within the Alliance. Initially, the Promoters' Groups will address three key technology areas:

- HomePlug 1.0 + AV (in-home connectivity, including digital home and consumer electronics applications)
- HomePlug BPL (to-the-home, Broadband-over-Powerline applications)
- HomePlug Home Automation (command-and-control applications)

The alliance's effort to develop a specification for an advanced command and control technology for home automation, known as HPCC (HomePlug Command & Control) is well underway. Command and Control is a low-speed, very low-cost technology intended to complement the alliance's higher-speed powerline communications technologies, which is specified in the Homeplug 1.0 standard and is not intendend for home automation purposes but for home networking, allowing multiple home desktop and notebook computers to be networked to share an Internet connection, printers, files, to play games and so on. OFDM is the basic transmission technique used by the HomePlug 1.0. Cyclic prefix and differential modulation techniques (DBPSK, DQPSK) are used to completely eliminate the need for any equalization. Impulsive noise events are overcome by means of forward error correction (FEC) and data interleaving. Privacy is established by creating logical networks through the use of encryption . HomePlug's privacy and security scheme is based on the 56-bit Data Encryption Standard (56-bit DES). The maximum baud rates depends on the level they are considered. The physical layer can achieve a rate of 13.78 Mbps, and the MAC (Medium Access Control) layer 8.2 Mbps.

8. Experiments and case studies

The interest of our research group is focused on the field of low cost home automation systems, mainly devoted to disabled people.
First of all, at the beginning of our interest in this field, we tried to explore applications regarding the implementation of powerline node by our selves.
This has been done both designing the interface electronics with the powerline and using consumer off the shelf powerline modems with a TTL (Transistor Transistor Logic) interface for our embedded hardware.
Regarding the first case, at the beginning we developed a simple two points communication link with an ASK modulation with a frequency of 75KHz, and after that a more robust communication link with a BFSK modulation, using the ST7540 power line transceiver produced by ST Microelectronics.

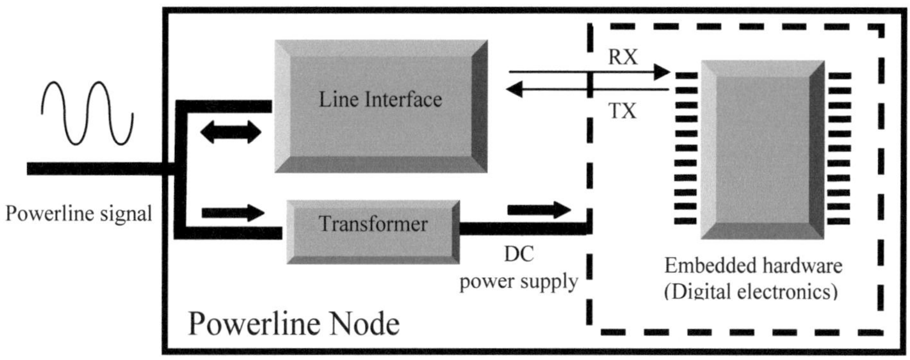

Fig. 6. General structure of a powerline node

A few tests had been made on these simple electronics, revealing that the performances of the ASK link in terms of noise robustness were considerably under the ones of a commercial power line modem, instead the link built using the ST 7540 was robust and affordable.
In any case, whatever may be the powerline node hardware, the general structure is depicted in figure 6. The node needs some electronic components (line interface) to be coupled to the powerline. The line interface is typically formed by an isolation transformer and some capacitors and inductances to implement selective filters. The DC power supply for the digital section of the node can also be obtained from the powerline voltage, typically by means of a transformer and a rectifier. The embedded hardware is generally a digital microcontroller that manages the transmission protocol and executes or superintends to its own particular domotic function (controlling a light, opening a door or a window and so on).
After this practical trials our efforts have been focused on a higher level.

Regarding the second case, we implemented a communication link between two FSK commercial powerline modems (Mainardi et al, 2005) with the usage of the SNAP (Scaleable Node Address Protocol) protocol (SNAP, 2002). The final demonstrative application was the opening and closing of a window in an environment designed for disabled people (the Centro H, a centre for information about disability in the city of Ferrara).

The reason why we decided to use this protocol reside in the fact that it is a scalable protocol, depending on the user's requests, it is very light regarding the necessary CPU resources, it can addresses 16,7 milion nodes, it can support 8 different check error methods and also broadcast messages, it is mean independent (it does not matter if one uses a powerline or an RF link) and can be implemented both for synchronous and asynchronous communication channels.

The main achieved goal of this first step was to demonstrate that is possible to implement a very cheap powerline application using off the shelf hardware and a free protocol, without any license cost.

The aim of this work was to demonstrate that is not necessary to spend a lot of money to build a home automation system, and that is possible to leave behind the troubles on choosing among a lot of possible commercial solutions using very few, simple, cheep and off the shelf components.

The used powerline modem was a PLM-01, produced by High Tech Horizon, with a baud rate of 2400 bps, which utilized a ST7535 integrated circuit (IC), with a frequency of 131.85 kHz to transmit a logic 1 and 133.05 kHz to transmit a logic 0.

After these two simple experiments, we decided to test a commercial powerline system, to verify its potentialities.

Our choice fell on the Marmitek X10 powerline system.

X10 is a standard and open protocol for powerline communications, that allows to address up to 256 devices (enough for a standard apartment). Every address code is formed by a number between 1 and 16 and a letter between A and P (for instance A3 or P14 are valide codes). The used modulation technique is an ASK, and the frequency of the signal is 120kHz. With this kind of system we equipped a small flat of about 80 mq (Mainardi & Banzi 2007), owned by the already mentioned Centro H, with automations for doors, windows, rolling shutters and lights. The advantages of a X10 system concern the fact that it uses an open protocol, it does not require connecting other cables, respect to those already present for the standard power line, is very cheap, it can be used with a low cost just to implement basic domotic functions (often disabled people do not need the entire house to be automatized, but is sufficient for them to have a few rooms, typically the bed room, the bath room and the dining room, with a minimum level of technology and an adequate cost), it can be easily extended to more sophisticated functions, is provided by a lot of electronic firms, with dozens of different modules, is very easy to use, also for inexperienced (unskilled) people, it can be easily used to implement experimental devices in the educational field.

After having installed this system in the apartment and having verified its excellent and stable functioning for more than two months, we decided to develop a user friendly interface for disabled people (Mainardi, 2008). Another key point of domotics concerns in fact the control interface (Thornett & Brown, 1988, Vaughan et al, 2006). Domotic systems are usually controlled by means of switches, in-wall touchscreens, remote controllers or PDAs. These interfaces are useful for people with no health problems, no handicaps, no restrictions in mobility, but creates many difficulties for people that for example have low vision, or hand trembling, people with severe spinal cord injuries and so on. It is true that there are a few dedicated devices for disabled people to control domotic systems with alternative inputs like the voice (Sicare Pilote and Easy by Voice System are two examples), but they are still expensive and represent a very small part of the market, with a consequent little possibility of choice, or they are still research projects, not available on the market. Besides this, there are many disabled that cannot use the voice, but for instance they can move a little part of their body.

In other words, domotic systems and products are thought mainly for *able-bodied* people, with attentions to many expensive details of aesthetics and ergonomics, and only at a second time for disabled people, that maybe can take more advantages of them in terms of autonomy and indipendence. So the project was focused on the design and the realization of a portable touchscreen interface for disabled people, used to control a low cost powerline domotic system. The interface has been implemented for Marmitek modules, but of course, knowing the infrared protocol of other domotic systems, it will be possible to make the device working also with other products, simply adding a software routine.

This interface has a touchscreen input, with icons big and separated enough for people with low vision or hand trembling (in any case the icons, their shape, size and subject, were approved by personnel and disabled users of the Centro H). The icons are organized in a three level menu, like shown in figure 7: the first general menu has an icon for every of the six rooms (the kitchen, the dining room, the bathroom, the corridor and two bedrooms). Touching on a room image the second menu will open, with an icon for every automation of that room (light, window, door or rollingshutter). Touching again the third menu will appear, with the options for every automation (turn on/off the light, rise/fall the rolling shutter and so on).

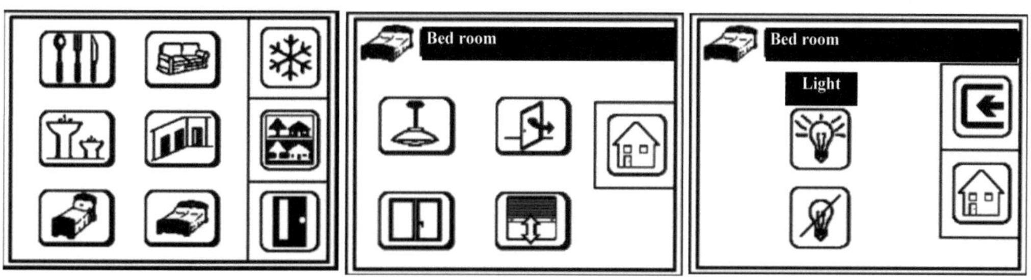

Fig. 7. Three levels of the portable touchscreen interface menu

The general structure of the interface is shown in figure 8.

The interface communicates with the Marmitek powerline system by means of an IR (infrared) link, sending IR messages to the Marmitek IR7243 device, that translates them into X10 commands over the powerline. The modules used to control the automations were the Marmitek UM7206, AWM2 and LW12, while to control scenarios (sets of X10 commands) a CM11 module was utilized, as shown in figure 9. More details of this implementation can be found in Mainardi, 2008.

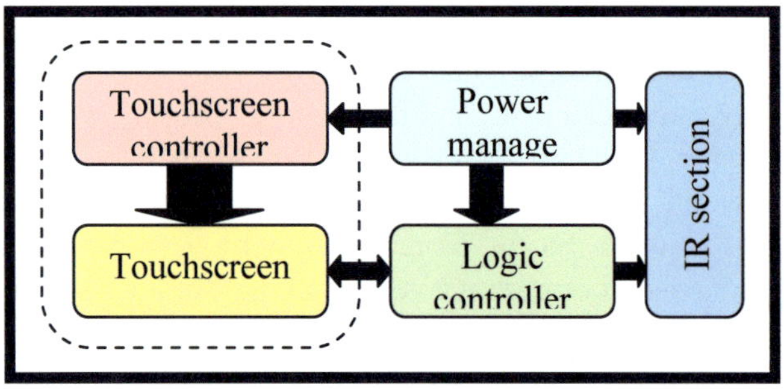

Fig. 8. Portable touchscreen interface structure

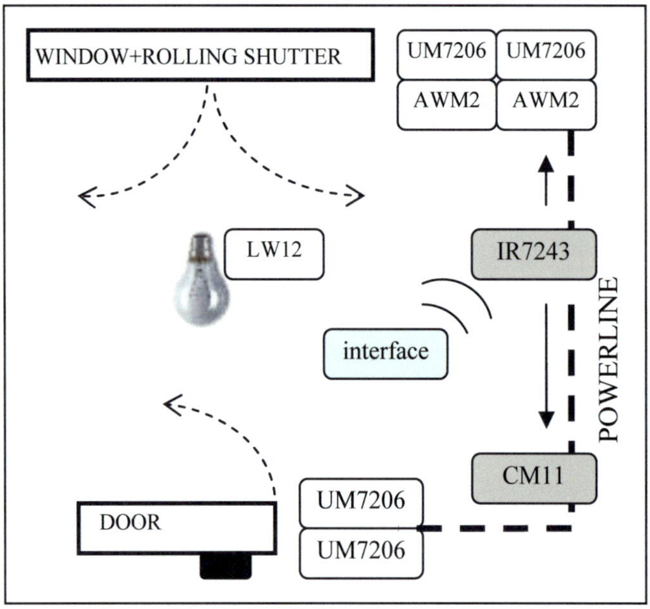

Fig. 9. Example of a room controlled with the Marmitek X10 domotic system

Another interesting Marmitek module is the XM10E, that is an interface between the X10 codes for powerline and the TTL electronic world. Using this device is possible to develop several applications, if they are not available in the wide X10 world. Using the XM10E module, everyone can create a simple hardware with a microcontroller that reads and sends every X10 code over the powerline, being able to control every possible automation connected to a Marmitek powerline module.

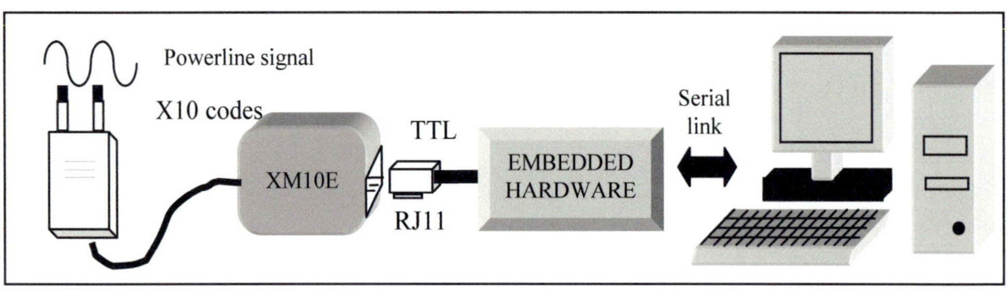

Fig. 10. Creating X10 applications with the XM10E module

After having design an embedded TTL electronics able to communicate with the XM10E, the further step has been creating a serial link with a computer, so that by means of a graphical interface every user can manage the whole domotic X10 system. In fact the user can write a code and a command using the keyboard of the PC, (for instance: turn on the ligth managed by the X10 modules with A1 address), this command is sent by means of the serial link to the embedded hardware, that converts it following a certain protocol and communicates a X10 code to the XM10E, that translates it over the powerline, as shown in figure 10.
Anyway, independently from the described application, the XM10E allows the implementation of every desired function or device, and for this reason is a tool that provides excellent flexibility to the Marmitek system.

9. How to set up an environment for a domotic powerline system

To arrange a home environment for a conventional home automation system, cabled with a standard bus cable, is necessary to create separated electrical tubings for the bus or to think about larger conventional electrical tubings, able to carry both the powerline cables and the bus cable and totally interconnected from each other. Is necessary also to take the bus everywhere in the home environment (also near the windows and the doors, considering the possibility to motorize them) and to arrange a power supply (or more than one, if the bus line is longer than a fixed length), typically with a so called SELV voltage (Safety Extra Low Voltage) of about 27-29V. If the same home automation system is realized with the powerline technique, the electric plant has not to be modified. The electrical tubings can have the standard diameter and no other electrical traces have to be installed. The only essential things are taking the powerline cables near doors and windows, arranging several outlets in every room, at least three or four, and being sure that all the sockets are served by the same powerline chain, without thermal magnetic or differential switches among them, because these devices, filtering the line, block every signal potentially overlapped to the powerline.

10. Powerline as a successful educational tool

Powerline world is definitely an interdisciplinary field, that involves electronics, telecommuncation, informatics and automation. Thus, from an academical point of view, is an excellent tool in the educational effort for engineering students.
Regarding our experience, several students of the Automation Group of the University of Ferrara took their degree with a graduation thesis based on projects concerning powerline and domotics applications. A few students designed the interface electronics with the powerline mean, a few studied the powerline communications, others designed X10 systems, others again designed customer electronics to develop new tools for the domotic X10 system. These activities are very useful in the sense that the students can touch with their own hands the real working of what they study in theory, also cooperating with the final users of this domotic systems, that in our case are disabled people that live and work inside the apartments, kindly placed at our disposal from the Centro H.

11. Conclusion

Power line communication is a valid technique that allows the exchange of data by means of the powerline cables that are present in every dwelling and in every building. Even if there are some problems of noise and possible interferences, modern modulation methods make PLC systems in general affordable and reliable. Information transmitted through the powerline can be used to share data with high bit rates (generally used for home intranets, for the last mile problem or for in-home music distribution), or also to control home and building automation systems, that typically need lower bit rates, up to a few kbps (kilo bit per second).

Several home system producers, like Lonworks and Konnex, besides the standard devices that communicate by means of a bus cable, sell powerline modules, but the most cheapest solution for a powerline system is represented by the X10 standard, that allows to control small domestic environments. Setting some basic functions in a X10 system is very easy. There are 256 available addresses, that one can set directly on the module, simply with a screwdriver. Two minutes are enough to control a light with an IR or a RF remote controller. We can't say the same for the other systems, where an expert is necessary to install and program (via software or directly on the back of the equipments) the modules and their functions.

Equipping a home environment with a smart powerline domotic system, beside its possibilities of energy saving and comfort increasing, is also one of the way to try to give back a certain level of autonomy to people with special needs. A smart home system can improve their independence in the every day's activities, in a comfortable environment which is very personal and peculiar for everyone, in any case different from a hospital-like setting.

12. Future research

After having tried with several practical X10 experiments, further steps will be explore other powerline systems (like Lonwork or Konnex) to test their potentialities, also conducting a comparative study on the power consumes of home and building automation systems.

Another key point remains in any case the world of domotic control interfaces for disabled people, so our effforts will be again concentrated on this topic.

13. References

Cenelec (Comitè Européen de Normalisation Electrotechnique), (2002). EN 50065-1 Signalling on low-voltage electrical installations in the frequency range 3KHz to 148,5 kHz - Part 1: General requirements, frequency bands and electromagnetic disturbances, July 2007

Cuncic, P. & Bazant, A. (2003). Analysis of Modulation Methods for Data Communications over the LOW-voltage Grid, *Proceedings of 7th International Conference on Telecommunications,* pp.643-648, ISBN 953-184-052-0, Croatia, June 2003, Zagreb

Dhir, A.& Mousavi, S. (2001). Home Networking Using "No New Wires" Phoneline and Powerline Interconnection Technologies, available at http://www.xilinx.com/support/ documentation/white_papers/wp133.pdf

Ferreira, H.C.; Grovè, H. M.; Hooijen, O. & Vinck, A.J. (1996). Powerline Communications: an Overview, *Proceedings of 4th AFRICON Conference,* Volume 2, pp. 558 - 563, 24-27 Sept. 1996

Gelman, A. (2006). Communications Society and Standards, *Proceedings of* ICC 2006 (International Conference on Communications), 11-15 June 2006, Istanbul, TURKEY

Gotz, M.; Rapp, M. & Dostert. K. (2004). Power Line Channel Characteristics and Their Effect on Communication System Design, *IEEE Communications Magazine*, April 2004, pp. 78-86

Hakki Cavdar. (2004), Performance analysis of FSK power line communication systems over the time-varying channels: measurements and modeling, *IEEE Transaction on Power Delivery*, VOL. 19, No.1, January 2004, pp. 111-117

Hakki Cavdar, I. (2004), A Solution To Remote Detection Of Illegal Electricity Usage Via Power Line Communications, *IEEE Transactions On Power Delivery*, Vol. 19, No. 4, October 2004

Hansen, D. (2002). Update on Power Line Telecommunication Activities in Europe, *IEEE International Symposiumon Electromagnetic Compatibility*, 19-23 Aug.2002 , Vol.1, pp:17-22, Minneapolis, USA, ISBN: 0-7803-7264-6

Han Vinck, A. J.; Haering, J. & Wadayama, T. (2000). Coded M-FSK for Power Line Communications, *Proceedings of ISIT 2000 (IEEE International Symposium on Information Theory)*, pp. 137, June 2000, Sorrento, Italy

Karl, M. & Dostert, K. (1996). Selection of optimal modulation scheme for digital communications over low voltage power lines, *Proceedings of IEEE 4th International Symposium on Spread Spectrum Techniques and Applications*, pp. 1087-1091, Vol. 3, September 1996

Lauder, D. & Sun, I. (1999). Modelling and Measurement of Radiated Emission Characteristics of Power Line Communications Systems for Standards Development, *Proceedings of IEEE International Symposium on Power Line Communications and Its Applications*, Lancaster UK, 30 March - 1 April 1999

Lin, Y.J.; Latchman, H. A. & Lee, M. (2002). A power line communication network infrastructure for the smart home, *IEEE Wireless Communications*, December 2002, Vol.9, pp:104-111, ISSN: 1536-1284

Lin, Y.; Latchman, A. & Newman, R. E. (2003). A comparative performance study of wireless and power line networks, *IEEE Communications Magazine*, April 2003, Volume 41, Issue 4, pp: 54 - 63

Mainardi, E.; Banzi, S.; Bonfè, M. & Beghelli, S. (2005) A low-cost Home Automation System based on Power-Line Communication Links, *Proceeding of ISARC 2005 (International Symposium on Automation and Robotics in Construction)*, September 2005, Ferrara (Italy)

Mainardi, E. & Banzi, S. (2007). The Centro-H smart homes, a point of convergence between technology and social sciences, *Proceedings of AAATE 2007*, pp:96-100, 3-5 October 2007, San Sebastian, Spain

Mainardi, E. (2008). Design of a portable touchscreen interface for powerline domotic systems, *accepted for CASE 2008 (Conference on Automation, Science and Engineering)*, August 2008, Washington (USA)

Majumder, A. & Caffery, J. (2004). Powerline Communications: an overview, *IEEE Potentials*, Vol. 23, Issue 4, Oct.-Nov. 2004, pp: 4 - 8

Napolitano R. (2004). PowerLine Telecommunications: National and International Regulamentations, *available at* http://www.aei.it/ ita/federAEIT/ PLC_01_12_2004 / 11.Napolitano.pdf, Milan 01.12.2004 (in Italian)

Rhee, J.; Rhee, E. & Park, J. (2008); Electromagnetic interferences caused by power line communications in the HF bands, *Proceedings of IEEE International Symposium on*

Power Line Communications and Its Applications (ISPLC), 4 April 2008, pp. 249 - 252, Korea

Schneider, M.; Favre, P. & Rubinstein, M. (2004). Emc Analysis Of Powerline Systems, *Final Report of IEEE International Symposium on Power Line Communications and Its Applications,* July 2004

S.N.A.P. Scaleable Node Address Protocol, © 1998-2002 HTH, Document revision 1.03, Document revision 1.03, © 1998-2002 HTH, Document revision 1.03

Sun, M.C. & Lun, D.P.K. (2002). Power-line communications using DWMT modulation, *Proceedings of the IEEE International Symposium on Circuits and Systems, ISCAS 2002,* Vol.4, pp:493-496, ISBN: 0-7803-7448-7

Thornett, C. E. & Brown,. A. W. (1988). Interfacing modern technology to disabled patients, *IEEE Colloquium on Engineering Design for the Disabled,* 10 Nov 1988, pp:9/1 - 9/2

Vaughan, T.; McFarland, D. J.; Schalk, G.; Sarnacki, W.A.; Krusienski, D.J.; Sellers,E.W. & Wolpaw, J.R. (2006). The wadsworth BCI research and development program: at home with BCI, *IEEE Transactions on Neural Systems and Rehabilitation Engineering,* Volume 14, Issue 2, June 2006 pp:229 - 233

Vines, R.M.; Trussell, H.J.; Gale, L.J. & Ben O'neal, J. (1984). Noise on Residential Power Distribution Circuits, *IEEE Transactions on Electromagnetic Compatibility,* Volume EMC-26, Issue4, Nov.1984, pp:161-168

Waldeck, T. & Dostert, K. (1996). Comparison of modulation schemes with frequency agility for application in power line communication systems, *Proceedings of IEEE 4th International Symposium on Spread Spectrum Techniques and Applications,* pp. 821-825,Vol . 2, Issue , 22-25 September 1996

Zimmermann, M. & Dostert, K. (2002). Analysis and modeling of impulsive noise in broad-band powerline communications, *IEEE Transactions on Electromagnetic Compatibility,* Vol . 44, Issue1, Feb.2002, pp:249-258

5

Towards n-D Construction Visualization: Cost Integration into 4D Models

Katherine A. Liapi and Theodora Paschoudi
University of Patras
Greece

1. Introduction

Computer Aided Design, in addition to its use in the planning process, has often been used during construction in order to mitigate various problems. Taking into account that the major general problem the construction industry faces is constructability, or, otherwise, the ability to convert a design document into a real project, for a constructor, a proper design document is one that reflects an inherent knowledge of how the project will be built; that is, one that reflects the details that support the construction process and that includes sufficient information to avoid ambiguity and thus reduce claims and disputes. However, despite the fact that many contractors use 3D modeling to better visualize complex geometry, very often the capabilities of state-of-the-art 3D software are only marginally exploited during construction. Data that could be extracted from the model, or data that is already included in the software, is not used for solving constructability issues, developing or updating cost estimates etc; consequently many times data is re-entered in new and different formats wasting time and money. In other cases data is lost or corrupted resulting to loss of vital information.

In advanced project management processes, contractors can choose one construction plan over another based on construction process visualization (4D CAD). Yet, though 4D CAD models can provide invaluable information for the construction process, do not display information regarding the effect of alternative construction schedules on cash flows. Reversing the problem, although contractors have access to cost estimates, their usefulness in the constructability review process is really questionable since they do not reflect or highlight specific costs related to constructability issues.

Responding to the general need for visualization of building construction related information, the research discussed in this chapter focuses on the emerging need for displaying cash flow information in parallel with the construction sequence visualization.

2. From 3D to n-D

The increasing utilization of Information Technology applications in the construction industry is explained by the rising need for competitive and cost-efficient practices at a global level. (http://www.cmit.csiro.au/innovation/1998-08/pdf/innovation_ITconstruct.pdf). Information Technology in construction contributes primarily to improved planning, design, construction and maintenance methods. Computerized approaches to automate the planning process were proposed since the early nineties (Bennett & Ditlinger 1994; Varghese & O'Connor 1995). Accordingly Computer Aided Design (CAD) project models, specifications of the task, and other supportive non-graphical information had to be tightly integrated to fully describe planning situations in an interactive planning environment (Ling & Haas, 1996).

In advanced management processes, contractors make use of 4D CAD models. Construction Sequence Visualization or Schedule Simulations display the progression of the construction over time by integrating 3D engineering data(CAD models) with a construction schedule. Practically a Schedule Simulation relates graphic model objects and schedule activity objects, and such models provide an effective decision making tool that can facilitate the construction of large scale complex projects. There are many state-of-the-art 4D software available for the construction industry such as the Bentley Navigator and PM Vision. With the use of 4D CAD, project managers can create and update realistic schedules, rapidly re-sequence activities when necessary while reducing the risk of running out of space. Changes in the schedule are automatically updated by updating the design and schedule links.

Feedback from construction to the design team, resulting from 4D model reviews, leads to a more readily constructible, operable and maintainable project. 4D simulations and virtual reality techniques as part of a Detailed Planning Module can be used in order to allow site supervisors to visualize, rehearse, and develop construction plans (McKinney & Fischer, 1998; Retik & Shapira, 1999). 4D simulations have been shown to considerably improve the quality of construction plans and schedules (Lee et al. 2005). Examples of projects that benefited from the use of 4D CAD for Construction include the San Mateo County Health Facility, the Experience Music Project at Seattle, the Roof for Haas School of Business at UC Berkeley and others (Fischer & Kam, 2003).

The capability and functionality of these applications are often limited when required to fully assist the generation of detailed execution plans; i.e. though 4D models can provide invaluable information for the construction process, information regarding the effect of alternative construction schedules on cash flows is scarcely supported. Specifically a 4D CAD model is not designed to support the time sequenced display of cost related information.

Since most 3D software allow for information exchange with innovative software used within the AEC industry, in recent state of the art applications the CAD model is not only used as a base for the 4D CAD model but also as an effective platform for the integration of more project parameters. Current methods of Construction Integrated Design do allow linking cost data to a 3D software, but most of the times such applications are data oriented, and do not support information visualization. Applications that take the concept of 3D computer modeling on an almost infinite number of dimensions, in order to address the entire life cycle of a construction project (Lee et al, 2002.) have also been developed. Such advanced applications do support cost integration and visualization and have been successfully applied to large scale projects (Fischer, 2005). However the costly and time consuming development of 4D, n-D models, which are mostly in house developed by the research team, is often a serious drawback in the utilization of n-D automation processes. Finding personnel with the skills to develop the interface that connects the 4D model to the databases, as well as superintendents and foremen to operate all databases consist additional difficulties.

Based on the above, a main objective of this research was to develop a simple application which could be used with most of the commercial software that support the development of 4D CAD models. Accordingly emphasis was placed on the development of an application that integrates material take-off with labor cost and takes advantage of the precision in quantity calculation resulting from detailed 3D models.

3. Interoperability and IFC compliant software

The International Alliance for Interoperability holds the leading edge in "collaborative-design frameworks" that allow data transfer between building model software and analytic applications, including energy simulation, cost estimating, and construction scheduling. Indeed the main objective and task of the International Alliance for Interoperability (IAI), formed in 1994, was to define specifications for Industry Foundation Classes (IFC) to enable information sharing throughout the project lifecycle and across all disciplines and technical applications in the building industry (Staub-French & Fischer, 2000). The Industry Foundation Classes, developed by IAI and defined as international standards for the building model data exchanging can be described as "a class library for Architecture, Engineering and Construction (AEC) objects" (IAI website, 2003). The first version of the IFC was released in 1995 and in subsequent years, many design software companies have added functionality to export IFC-based product models. The implementation of IFC specifications, available since late 1996, was very limited until the Building Lifecycle Interoperable Software (BLIS) project was conceived as a way to initiate the next phase in the widespread adoption of an object data model standard for the AEC/FM industry (BLIS, 2002).

The BLIS group that supported the implementation of IFC specifications in software products demonstrated software-interoperability among CAD design software, estimating, design spell-checking, code-compliance checking, production-schedule simulation (4D), and model-viewer applications (Khemlani, 2003). As IFC is non-proprietary and available globally to AEC/FM software developers, it enables interoperability among software applications, which means the end users in AEC/FM area can share the model data by IFC. This helped to make interoperability a reality for AEC firms, building owners and authorities that use IFC-compliant software (IAI, 2003). More recent efforts of the IAI are focusing on data share and seamless transfer of data between platforms and among applications so that each discipline contributes and withdraws information from the core model. As a result errors and omissions are reduced, and money and time are saved (IAI North America, 2003).

All major AEC CAD vendors such as Bentley's TriForma; Graphisoft's ArchiCad; Autodesk Architectural Desktop; Nemetschek AG's Allplan, Allplot, Allfem, and Allfa have been IFC-compliant already before 2000 (Staub-French & Fischer, 2000). Working with AEC design software of multiple disciplines requires interoperability frameworks. One part of these interoperability frameworks relates to cross-file interchange of model geometry. These developments do not include though cost and schedule data in their interoperability framework, and are only part of larger interoperability concerns. A feature of IFC compliant software that is of primary importance for this research is that they support the generation of cost estimates and assist estimators with the maintenance of cost estimates throughout the project's life cycle changing the expectations of engineers from the existing cost estimating software.

The evolution of the Internet and the ability to access the World Wide Web with fast connections tend to change the way that Construction Cost Estimating Software Systems (CCESS) are used. Current estimating software help estimators take off quantities automatically by taking advantage of the IFC concept and establishing an electronic link between a component in a 3D-product model and a cost item in a cost estimating database to create a cost estimate. On average, each item in a cost estimate is created seven times throughout the life of a project (Laitinen, 1998). Consequently, it is critical that IFC support the generation of cost estimates and also assist estimators by maintaining the cost estimates throughout the project life cycle (Staub-French et al, 2000).

A widescpread IFC compliant software available in the market is Timberline Estimating. It includes a CAD Integrator which incorporates support for the IFC standards thus becoming a useful tool to support cost estimators. Within the software environment a designer who uses IFC-compliant software, such as a CAD program, can save designs in IFC format. No additional measuring, digitizing, or manual takeoff is needed for the IFC objects. Should the design change, a new estimate can be generated in minutes. However, when the users need assistance in maintaining cost estimates throughout the project, the existing estimating software lack the necessary functionality. This problem is even more acute in the case of design changes. The reason is that construction costs are not only affected by material quantities but also by component properties (e.g. size and shape), and product features (e.g. repetition and openings) (Staub-French & Fischer, 2000).

Except from the direct material cost deriving from a quantity take off, a cost estimate generally includes all costs related to site preparation, project financing costs, temporary facilities, project definition costs, insurance, acquisition/rent of land and property, taxes, project management, engineering costs, direct construction labor costs, quality assurance costs, start-up costs, escalation, contingency, operations and maintenance. Only direct material and labor cost are considered in the scope of this research.

4. Cost integration into 4D models

The research described in this chapter focuses on the development of an application that integrates material take-off with labor costs and takes advantage of the precision in quantity calculation resulting from the detailed 3D models (Paschoudi, 2003). The effort extends to the integration of the generated cash-flows within a 4D CAD model which animates the construction sequence by keeping the same timeline of the 4D CAD model.

The approach towards the fulfillment of this main objective is better described by the following sub-objectives that also reflect the process followed in this research:

- Identify methods to calculate the quantities and the cost of materials from the 3D model
- Identify methods to calculate the cost of resources (labor and engineering) from the resource loading on the schedule.
- Develop a method to generate a cash flow for materials and resources
- Develop a method to display the changes in the cash flow by keeping the same timeline of the 4D CAD model.

In order to generate the 4D model, information that originated from a 3D design needs to be combined with information developed from the scheduling. Since the schedulers in traditional project management do not develop the scheduling activities to accommodate the names and the ID's of the 3D model, the information provided for the development of the 4D model cannot be used without being first analyzed and appropriately modified.

Essentially, during the development of the 3D model, data is entered on materials, simultaneously or later, while the scheduler creates his files based on 2D plans, or the work breakdown structure (WBS) provided by the contractor. Therefore, when the information for the generation of the 4D model is finally delivered, it often has to be re-entered or adjusted. Apparently, when a construction schedule simulation is produced, the information provided by the scheduler cannot be fully exploited because the resource loading he has already developed is not being used.

The developed method (Figure 1) reroutes the flow of information and benefits from this procedure by elaborating more data. Specifically the scheduler, instead of working independently for the development of the scheduling model, will first filter the information included in the 3D model, and then provide all necessary data for the development of the cash flow for materials and resources. The next step will be to integrate the cash flows into the 4D model and to visualize the changes in the cash flow while running the 4D model.

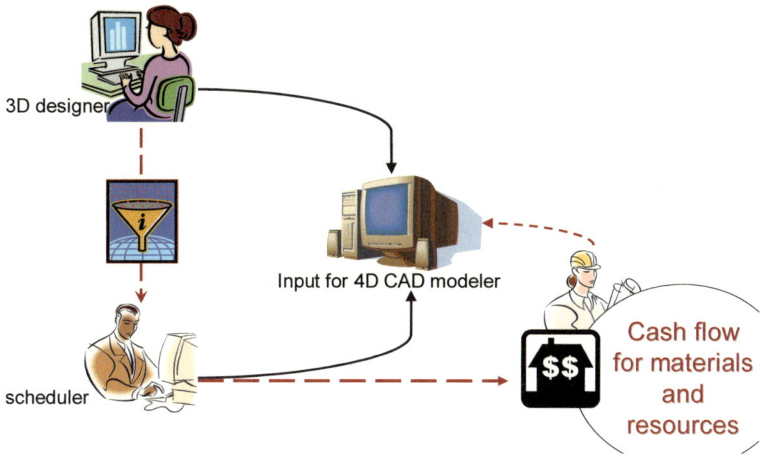

Figure 1. Standard vs. Proposed process for 4D modeling

The process consists of five basic steps:
STEP 1: Material take-off
STEP 2: Resource Loading
STEP 3: Generation of 4D model
STEP 4: Generation of cash flow for materials, labor and engineering costs
STEP 5: Visualization/Decision-making
Each step is further investigated and analyzed

4.1 STEP 1: Material take-off

This step leads to the generation of a bill of materials. In order to perform the quantity take-off, the MicroStation TriForma has been used since: (a) allows for easy quantity take-off and (b) is compatible with commercially available 4D construction visualization software.

MicroStation TriForma, a software application for 3D solid modeling, is used to generate: a) "intelligent presentation drawings" (plans, building sections, elevations, details, perspectives) and b) quantification reports (material quantities, schedules, specifications, and cost estimates) from an integrated 3D project model. TriForma maintains the link between the model and all related documents and consequently once a change is made on the 3D model all the deriving documents update automatically. Practically, in TriForma all the information that needs to be tracked about a project can be inbuilt in the intelligent 3D model and extracted as presentation documents and/or as quantification reports .

The 3D model consists of objects that understand how to behave (e.g. a door can automatically cut the proper opening in a wall, and a form can be made to reshape its top to match the profile of another form). The dataset consists of intelligent pieces that reside in xml format and can be modified as regular text files. These text files are located in multiple directories. The directories are hierarchically structured within one main directory and the use of directories allows for use of the information from different users, in different computers and for more that one projects. Two of the most critical pieces of the MicroStation TriForma dataset are the parts and the components and are further described.

A part is a name given to all model elements that are intelligent when created and is linked to additional model and project information. The part definition serves in determining the symbology of the element at the time it was placed and in the extracted documents. In order to attach to a part supplementary information that concerns model materials, associated unit costs, and specifications it is necessary to attach a component. Components are linked to parts through a formula. Parts and components are both leveraged in the extraction of presentation materials and reports (TriForma User Guide, 2003).

The following sequence should be followed in order to generate a report:

- Build necessary families-parts-components
- Assign every element to be calculated to a part
- Link at least one component to every part that is assigned to an element
- Enter a valid Formula to every Component that is linked to a Part

As shown on the example in Figure 2, once a wall is designed, a part family is attached to it, and then a component family is specified (e.g. exterior, interior). By specifying the component, the unit of measurement, and the unit price that is used to apply material take-offs are determined.

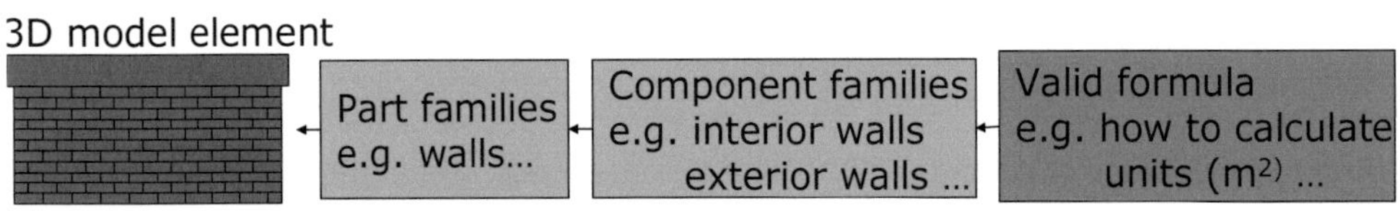

Figure 2. Generating the material take-off

The most important outcome from this step is the detailed report (in xls format) that contains all the quantities generated by each element in the 3D model (Table 1). The generated report in the Microsoft Excel file is dynamically linked to the model. "This link enables Element Selection and the creation of selection sets within the report. The model can then be opened within Excel to view the selection set of elements. The detailed report extracted from TriForma into Microsoft Excel file (Table 1) lists the quantities of the materials sorted first according to the component and then according to their IDs (e.g. the amount of cubic yards for every column). It can also list the unit price of the materials and finally the total cost per component.

Component number	ID	Quantity	Unit price
Component 1	1	q1	p1
	2	q2	p1
		Total for component 1	Total for component 1
Component 2	3	q3	p2
	4	q4	p2
		Total for component 2	Total for component 2

Table 1. Report generated from MicroStation TriForma

4.2 STEP 2: resource loading

At the completion of this step the bill of labor costs is generated.

Primavera Project Planner, one of the most popular, powerful and widely used scheduling and cost estimating software by the AEC industry is used.

Scheduling software is broadly used in construction industry to facilitate effective management of projects. Practically it comprises part of the project controls defined as the process of planning, directing, and controlling the project from the start of construction through its completion. One of the essential concerns is that the information developed for the estimate should be presented in a way that can be easily used during the actual construction project (Feigenbaum, 2002). Material deliveries, equipment and people can all be managed through the schedule. Moreover advanced scheduling software are used for cost estimating as they allow for development of estimate information that can be easily used during the actual construction project. To generate cost estimates for labor costs the estimator/scheduler needs to have a completed quantity take-off, a list of crews and their activities, a list of crafts included in the crews, the wage rates for all of the crafts and the productivity rate for each crew (Feigenbaum, 2002). The efforts of these software focus basically on improving the resource management, cutting costs and making employee scheduling, room scheduling and shared equipment scheduling easy for everyone for companies of all sizes. Some of the features available include: simplicity of use, minimal server requirements, no need of additional software, web and browser based, remote access, day / week / month views for schedules, customizable time display, resource utilization and costing reports.

Most Construction Cost Estimating software companies have incorporated the spreadsheet functionality in their software. Spreadsheet interface, allowance for modification of the cost database, Open Database Connectivity (ODBC) with spreadsheet software, or even Construction Cost Estimating add-ins for spreadsheet software are very common in today's Construction Cost Estimating Software Systems (CCESS).

The following process clarifies the tasks the scheduler should perform while executing the resource loading, in order that the information provided by him can be used in the next steps:

- Retrieve the IDs of the elements from TriForma
- Add scheduling activities for all desired objects of the 3D model
- Add resources and their unit costs for each scheduling activity

Ultimately the detailed report (csv) contains all labor and engineering costs for each activity/element of the 3D mode, and appears in the format shown in Table 2: for every resource, all the correspondent IDs are listed and then cost per ID is calculated for every resource.

Resources	ID	Quantity	Unit price
Labor	1	q1	p1
	2	q2	p1
Field Engineer	1	q3	p2

Table 2. Report generated from Primavera Project Planner

4.3 STEP 3: generation of the 4D model

For the generation of the 4D model, Bentley's Navigator was chosen as a state-of-the-art software that encompasses the concept of classes and is compatible with both Bentley's TriForma and Primavera Project Planner.

The schedule objects, as defined within Primavera Project Planner, are linked to the graphic objects that were previously created with Bentley's TriForma. The linking occurs through an interface that is easy and user friendly that enables creating, breaking, editing, and reviewing multiple relationship links between objects.

When a Schedule Simulation is being replayed, the Navigator obtains graphic or geometric data from the graphic object in the model. Then, in order to display the graphic object at the right time during the schedule simulation, the Navigator finds the link between the graphic model object and the corresponding schedule activity.

Another very important concept the Navigator encompasses, essential for creating the links between the graphical and the scheduling objects is the JSpace modeling, specifically, defined as the "general purpose object-oriented system for creating data files and interacting with the large and complex sets of data encountered in the architecture, engineering, and construction industry (AEC) (JSpace Class Editor manual, 2002).The class is the template used for the JSpace model and from its definition, the shape, structure, data storage, behaviors, and knowledge for an object of that class is provided. The class definitions are stored in files called class libraries, and can be organized into one or more libraries depending on the project.

Moreover, within the JSpace environment that allows for intelligent interfaces with CAD systems, such as MicroStation and AutoCAD, and with database systems, such as Microsoft Access, Oracle, and SQL Server, the project participants can work together and customize the environment and the applications. Besides, as shown in this step of the progress, JSpace is used for real-time visualization as it can integrate real-time visualization, animation, and simulation within the modeling environment. Different views of a 3D model are available with the use of multiple windows which additionally allow display and view in a wire-frame or rendered display.

The OLE links enable to create JSpace schedule models by using data that is imported from external schedules. Whether new objects are added, or simply the properties of the existing objects are updated, an appropriate JSpace class must be used. For this purpose, two classes are provided in the standard class library, one for the Primavera Project Planner and one for the Microsoft Project: P3_activity class is designed for importing Primavera Project Planner schedule activities.

Before linking the 3D model to the scheduling activities, in Figure 3, the following tasks must be completed:

- Extract the JSPACE model from Triforma and open it in Navigator
- Insert the activities from P3
- Create the new JSPACE model from P3
- Link the objects of the 3D model to the activities of the P3

Once the graphic objects are linked to the scheduling activities, the 4D model is automatically generated, and in the next step the cash flows are generated in order to integrate them in the 4D model.

4.4 STEP 4: Generation of cash flow

Once the material take-off and the resource loading are complete, the next goal is to generate the cash flow that includes the material and labor costs. The combination of this information since it is deriving from different software is not feasible with the currently available software.

Hence there is a need for the development of a new tool that will allow the merging of the two spreadsheet files and will generate the cash flows. Therefore a commercial software based application that allows for linking the labor costs with the material costs on a predefined timeline is developed. The conceptual approach and use of this application as part of the discussed process are explained in the following section.

A simple and user friendly application was developed to facilitate the needs of this step of the process as the files generated from MicroStation TriForma and Primavera Project Planner needed to be linked. The application is called TriVera (since it is combining files generated from TriForma and Primavera) and is written in Visual Basic. Basically it is an add-in for Microsoft Excel and can be easily installed (Figure 3).

The conceptual approach, shown in Figure 4, depicts the procedure repeated every time the application is used. First, the unique ID generated within TriForma for every element chosen to display is retrieved from the detailed report. This ID is used by the scheduler to schedule the respective activity. The two files are combined within TriVera to create a new Microsoft Excel file and then generate the cash flow

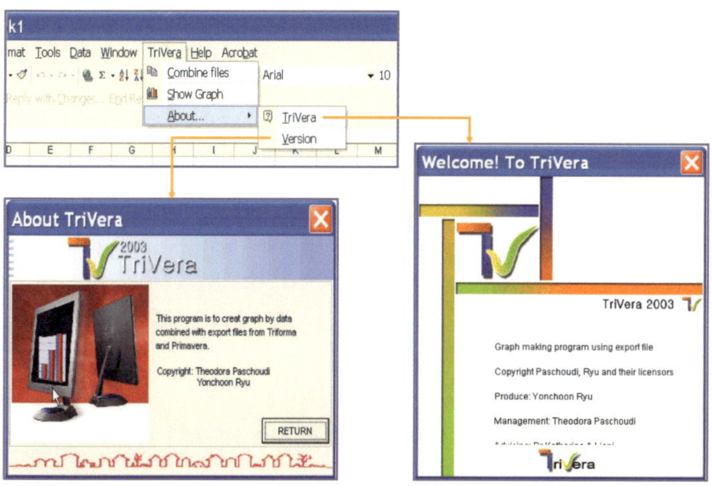

Figure 3. TriVera's menu interface

The procedure performed by TriVera in order to create this new Microsoft Excel file is the following:

- The application defines the ID of the first scheduled activity listed in the first pre-defined time period (e.g. a week) in the file created from Primavera.
- This ID from the Primavera Excel file is matched with the ID in the TriForma Excel file and the cost of labor and materials is identified, respectively.
- Once the same procedure is repeated and completed for all the IDs listed under the first pre-defined period, the two types of costs are plotted on the cash flow output.
- The same procedure is repeated for every time period reflected on the cash flow.

It is important to note that the cost of material is credited at the beginning of the activity, while the cost of resources appears on the cash flow during the entire time required to complete the construction activity.
Once the cash flow is generated the next step focuses on its integration in the 4D model.

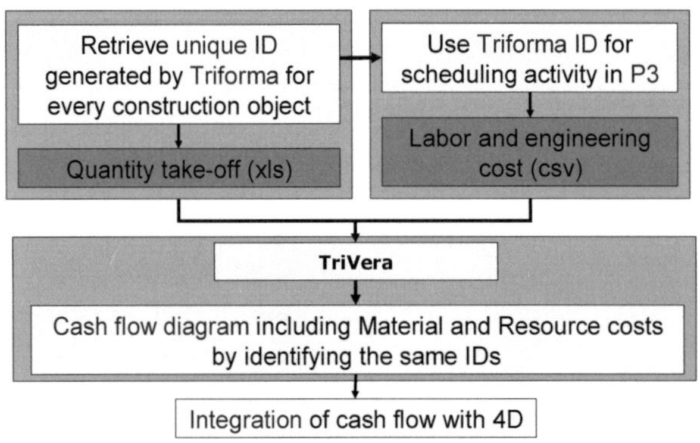

Figure 4. TriVera - Conceptual approach

4.5 STEP 5: visualization/decision-making

In this step the animation of the cash flows is generated establishing the integration of cost into the 4D model. A user friendly animation creation software, VEGAS, that creates animations from a series of picture files (gif, jpeg, bitmap etc) has been chosen.
Before using the animation creation software the user has to generate from TriVera all the necessary cash flows he desires to display simultaneously to the display of the 4D model. Once all the cash flows are generated for regular time frames throughout the progress of the project (Figure 5) then Vegas can be used and the next actions to be taken are:

- Define the timeline of the animation of the 4D CAD
- Use the same timeline for the animation of the cash flows

At this point of the research it s not possible to have the animation in the same screen as the 4D CAD display; the two animations are viewed on the same time but from different monitors.

5. Case study: Dallas High Five

In support of this research an important infrastructure project was chosen for a small scale application.
In 2002 the Texas Department of Transportation started the construction of a 5-level Interchange (intersection of I-635 and U.S. 75) to accommodate approximately 500,000

vehicles per day. The cost of the project was estimated to $261 Million. It started on January 2, 2002 and was scheduled to be completed in 2006. The so called "Dallas High Five" was a very challenging project as it had schedule and budget constraints and severe business impacts.

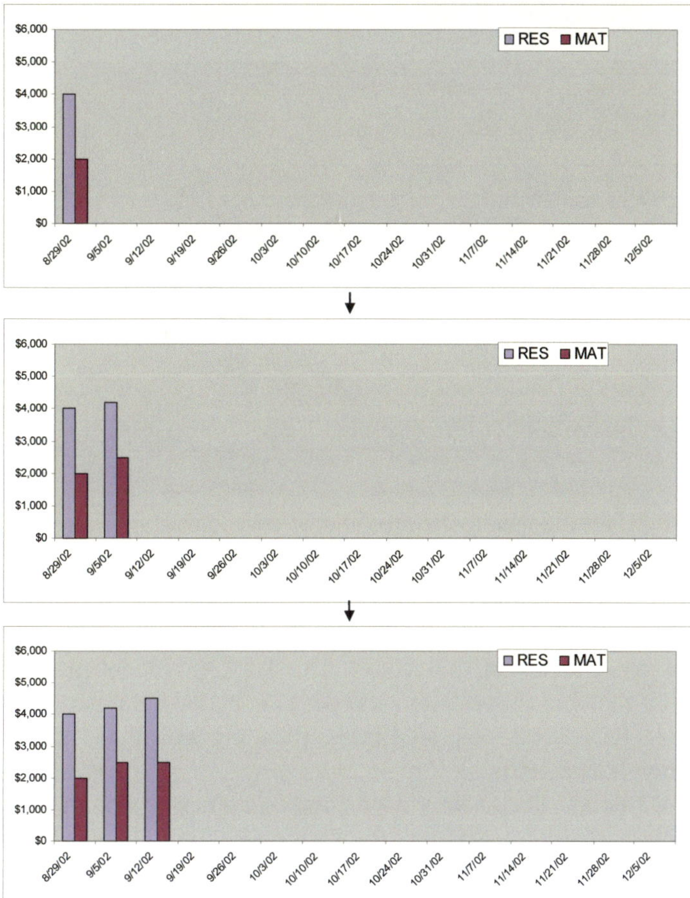

Figure 5. Cash flow animation

The continuously changing geometric configurations during its construction affected traffic planning decisions and complicate the overall project management. To facilitate decision making regarding both construction sequence and traffic planning, a visualization system that effectively uses 3D and 4D CAD technologies to support the planning, construction, and execution of the Dallas High Five project was proposed and developed by a UT research team (Liapi et al, 2003). A section of the project that was particularly complex in which the 4D helped a lot was the design for the Coit Road Bridge (Figure 6). The reconstruction of the Coit Road Bridge that had two lanes and two left-turn lanes in both directions was completed in August 2003. An important feature of the developed system was a comprehensive 3D database with three distinct levels of geometric configurations: a) current as-built, b) intermediate planned/as-built (including temporary structures), and c) final as-designed.

Figure 6. Dallas High Five interchange (source http://www.dallashighfive.org. November 2003)

The development and usage of the 4D model to support construction improved communication between project teams (Department of Transportation Engineers, contractors or sub-contractors, etc.) by visually displaying alternative schedule and traffic plans during the construction of the project. The 4D CAD model provided the project's participants with a valuable visualization and decision making aid. An important feature of the 4D CAD model was that it integrated traffic planning which was of critical importance during construction (Liapi, 2003).

The developed 4D model, albeit a very useful and cost efficient tool, did not reflect the cost implications of alternative construction plans. Though for the specific case the cost was not a determinant factor, the project has been selected for demonstrating how costs could be taken and displayed, because of the overall interest it presents.

For the purpose of this research a simplified demonstration for some construction elements of the Coit Road Bridge is performed. So for the sake of this study, a part, a family and a component are attached to a set of cylindrical columns in the project. A component is attached to the columns including a formula that calculates volume in cubic meters for the quantity take-off. The IDs for every graphic object are displayed as well.

These IDs are then used for the creation of the Primavera Project Planner file. As shown in Figure 7 for every ID generated from TriForma for the Coit-side cylinders, an activity with the same ID and a short description is added in Primavera. In this case, since all the IDs shown are referring to columns, they all have the same description: cylinder column-coit.

Then, the necessary labor resources for every activity are uploaded in the Primavera Project Planner and the report is generated. Once the two spreadsheet files are generated, the TriVera application is used to combine them and create the cash flow diagram displayed in Figure 8.

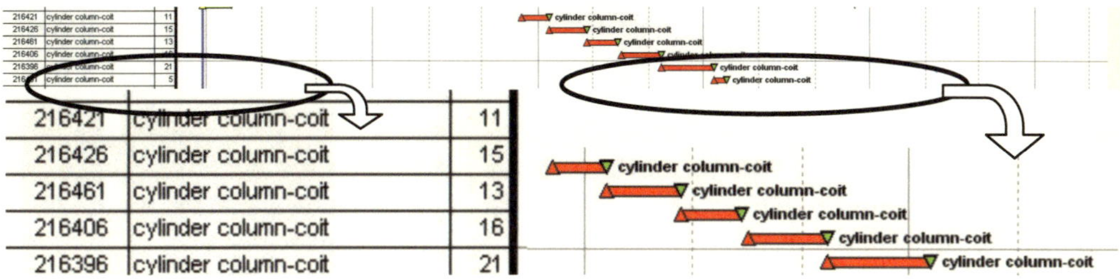

Figure 7. Activity scheduling in Primavera Project Planner

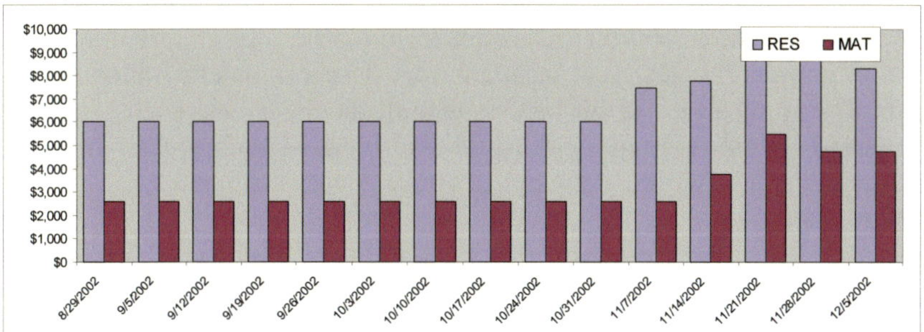

Figure 8. Cash flow generation within TriVera

In Figure 9, the final result is displayed. On the left side the 4D model is showing the animation of the two constructability scenarios and on the right side, and in reality on a different screen, the animation of the two corresponding cash flows is displayed. As a result the decision making process is enhanced with new tools that visualize the construction sequence and simultaneously allow for evaluation of the costs incurred.

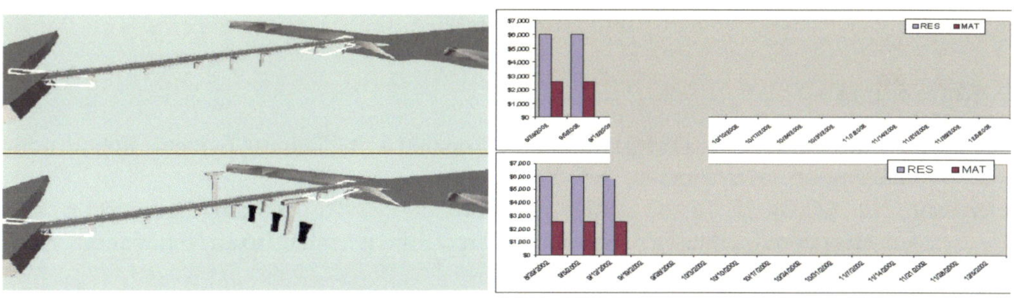

Figure 9. Two snapshots of the 4D model animation that display construction progress at the same time according to two different schedules and the respective cash flows.

6. Conclusions and recommendations

An efficient method for integrating cost to 4D CAD models for construction has been developed. A key feature of the method is a software application, TriVera, that facilitates the integration process and allows for the display of cost at the same time sequence as the construction schedule animation. The proposed method consists of a 5-step process that allows for easy development of cost estimates that can be used both for cost estimating and cost control.

The process provides the project participants, and primarily the project managers, with the ability to analyze and visualize multiple design alternatives in order to develop the most cost-effective and constructable solutions. It consequently allows for better control and decision making over different constructability issues and schedule scenarios, providing in this manner a linkage between constructability, 4D, and cost estimating.

The time and the cost required for the development initially of the 3D model and sequentially the 4D model restrict the use of the process to projects with constructability and/or visualization issues. In these cases it is considered necessary for the understanding of the construction sequence and budget issues from all the project participants. Especially in large scale projects it can also facilitate the decision-making, allowing for faster authorization. The applications of this process include all civil works such as buildings, civil infrastructure and industrial projects. It could also be used both within an owner and a contractor organization while developing their cost estimates and/or reviewing constructability plans.

The anticipated benefits and the long term contributions of this research are expected to be numerous. The proposed process allows for improving the information exchange within the AEC industry by providing a better communication of building related information between the design and construction phases in a project. Since the scheduler uses the data generated by the designer, cost estimates become more accurate and margins for errors and omissions in schedule are reduced. Avoiding reentering data and filling the communication gaps, money and time are saved, as the information is directly received from the 3D model.

The proposed method addresses interoperability and brings the AEC industry one step closer to n-D CAD. Ultimately the proposed process for integrating cost into 4D models will contribute to the development of infrastructure methodologies and technologies that allow for the integration of construction parameters, such as buildability, accessibility, sustainability, and maintainability into the 3D model.

7. References

Bennet, C., and Ditlinger, S. (1994). Bechtel Automated Lift Planning System. Robotics for challenging environments. New York, N.Y.: ASCE, 1994.

Bjornsson, H. (2003) IT Revolutionises Construction. 1998. Building Innovation and Construction Technology. Available: http://www.cmit.csiro.au/innovation/1998-08/itrevolution.htm. October 2003.

Brown & Root Braun. C.A.R.: Computer Aided Rigging Manual. Houston, TX.

Building Lifecycle Interoperable Software (BLIS) (2003). Blis Home Page. Available: http://www.blis-project.org/index2.html. November 2003

Dallas High Five (2003). Available: http://www.dallashighfive.org. November 2003.

Elzarke, H. (2001). Computer Integrated Construction for Small and Medium Contractors. *ASC 37th Annual Conference*. Denver, Colorado, 255-62.

Feigenbaum, L. (2002). Construction Scheduling with Primavera Project Planner. Second ed. Upper Saddle River, N.J.: Pearson Education.

Fischer, M., and Kam, C. (2003) 4D Modeling Application Case Studies. SIENE Workshop on 4D modeling. Available: http://www.scpm.salford.ac.uk/ siene/ 4D%20Application%20Case%20Studies-3.pdf.

Gao, J. and Fischer, M. (2006). 4D CAD on Building Construction Projects: Benefits for Project Success and Controllable Implementation Factors, *Construction Informatics Digital Library* http://itc.scix.net

Goldberg, H. E. (2003). Estimating Software Taps into Cad Building Model Data. *CADALYST.* Available: http://www.cadalyst.com/solutions/aectools/0902aectools/ 0902 aectools.htm. November 2003.

Gould, F. E., and Nancy E. J. (2003). Construction Project Management. Second ed. Upper Saddle River, N.J.: Pearson Education, 2003.

Holness, G. (2003). V.R. Smart IM for Project Man. January 2003. *ASHRAE.* Available: http:// www.leanconstruction.org/pdf/holness_1.pdf. November 2003.

IAI North America (2003). New Member Brief. 2003. Available: http://www.iai-na.org/ membership/membership_brief.php. November 2003.

International Alliance for Interoperability (2003). I.F.C. Resources. 2003. Available: http:// www.corenet.gov.sg/it_standards/iai/5_IFC_Resources.htm. November 2003.

Jspace Class Editor User Guide. Bentley, 2002.

Khemlani, L.(2003). Interoperability and the Solibri Model Checker. November 2002. CADENCE AEC Tech News. Available: http://www.cadenceweb.com/ newsletter/aec/1102_2.html. November 2003.

Lee, A., et al. (2002). Developing a Vision for an nD Modeling Tool, *CIB w78 Conference.* Aarthus School of Architecture, Denmark: International Council for Research and Innovation in Building and Construction, 2002.

Lee, A., Wu, S., Marshall-Ponting, A., Aouad, G., Tah, J., Cooper, R., and Fu, C. (2005) n- D modelling - a driver or enabler for construction improvement, *RICS Research paper series* , University of Salford,United Kingdom.

Liapi, K., Kwaja, N., O' Connor, J. (2003). Highway Interchanges: Construction Schedule and Traffic Planning Visualization, 2003 Transportation Research Board (TRB) Annual Meeting, Proceedings, Liapi January 18, 2003, Washington DC. CD ROM.

Liapi, K. (2003). 4D Visualization of Highway Construction Projects , *IEEE, Seventh International Conference on Information Visualization, Proceedings,* July 14-17, 2003, London, GB, 639-644.

Ling, K-L., and Haas, C.T. (1996). An Interactive Planning Environment for Critical Operations, *Journal of Construction Engineering and Management* , 212-22.

McKinney, K., and Fischer, M. (1998). Generating, Evaluating and Visualizing Construction Schedules with Cad Tools., *Automation in Construction* 7.6 , 433-47.

Microstation Triforma User Guide. Bentley, 2003.

Navigator User Guide. Bentley, 2003.

Paschoudi, Th.(2003). Cost Integration into 4D models, *Thesis,*University of Texas, at Austin.

Retik, A., and Shapira A. (1999). VR-Based Planning of Construction Site Activities. *Automation in Construction* 8.6 (1999): 671-80.

Saad, I.M., and Batie, D. (2002). The Science and Technology Building 4D Construction Model. 2002. *Southeast Section Conference, ASEE.* Available: http://www.ecu.edu /ciitr/ ASEE2002.pdf. December 2002.

Sanvido, V.E., and Madeiros, D.J. (1990). Applying Computer-Integrated Manufacturing Concepts to Construction, *Journal of Construction Engineering and Management* 116.2 (1990): 365-79.

Sriprasert, E., and Dawood, N.(2001). Potential of Integrated Digital Technologies (IDT) for Construction Workforce Instruction, *Conference at Chalmers*, Gothenburg Sweden: AVR II, 2001. 136-45.

Staub-French, S., and Fischer, M. (2003). Practical and Research Issues Using Industry Foundation Classes, Center for Integrated Facility Engineering. Available: http://www.civil.ubc.ca/faculty/Staub-French/index.html.November 2003.

Staub-French, S., and Fischer, M. (2000). Formalisms and Mechanisms Needed to Maintain Cost Estimates Based on an IFC Product Model. *ICCCBE-VIII*, Stanford University.

TIMBERLINE. Cad Integrator (2003). Available: http://www.timberline.com/include/pdfs/cad_integrator.pdf. October 2003.

Tollefsen, T. & Haugen, T. (2007). 3D AND 4D Modeling for Design and Construction coordination: Issues and lessons learned, EDITOR: B-C Björkhttp://itcon.org/2007/26/

TxDOT (2003). TxDOT's High Five Interchange Project Ahead of Schedule. August 2003. TxDOTExpressway. Available : http://www.dot.state.tx.us/dal/newsrel/058%2D2003.htm. October 2003.

Vargese, K., and O'Connor, J.T.(1995). Routing Large Vehicles on Industrial Construction Site. *Journal of Construction Engineering and Management* 121.1 (1995): 1-12.

6

Developing Construction CAD-Based Experience Management System

Yu-Cheng Lin
National Taipei University of Technology/ Civil Engineering
Taiwan

1. Introduction

Experience is valuable, stored specific knowledge obtained by a problem-solving agent in a problem-solving situation (Bergmann, 2002). Construction experience is knowledge that is based on construction methods, field operations and results of prior projects. Construction experience transfer is the use of knowledge gained in previous projects to maximize achievement of current project objectives (Reuss &Tatum, 1993). Although knowledge management is already well established in the construction industry, experience management (EM) is a new concept in information systems. Knowledge management (KM) is the collection of processes governing the creation, storage, reuse, maintenance, dissemination and utilization of knowledge. Experience is the life blood of individuals and organizations, and EM, a sub-discipline of KM, refers to the collection of processes controlling the creation, storage, reuse, evaluation and usage of experience in a particular situation or problem solving context. To transfer experience between similar projects, construction professionals have traditionally used techniques ranging from formal annual meetings to face-to-face interviews (Reuss and Tatum, 1993). To realize potential benefits, construction experience should influence all phases of a project (Tatum, 1993). Furthermore, knowledge gained from experience often requires action and may add cost-effective scope to other functional actions to avoid repeating past problems (Tatum, 1993). EM focuses on the acquisition and management of important issues and experience from participating engineers. Useful experience can be recorded in different forms and media, such as in the minds of experts, in operating procedures or in documents, databases and intranets. EM in the construction field aims to effectively and systematically transfer and share experience among engineers.

This study views experience as the knowledge gained by executing construction projects. To enhance the quality of EM gained by engineers involved in construction projects, this study proposes a Computer-aided Design (CAD)-based Maps (CBM) approach to achieving EM solutions in the construction industry. Combined with web-based technology and CBM, this study proposes a Construction Web-based Dynamic CAD-based Maps Experience Management (CBMEM) system enabling engineers to reuse domain knowledge and experience by dynamically exchanging and managing experience during the construction phase of a project. In the proposed CBMEM system, the map-based experience exchange environment enables engineers to manage and dynamically share their experience with other engineers in current projects. Engineers are, thus, invited to exchange and share their experience, and construct valued content through their own experience. In this study of a Taiwan construction building project, the survey (questionnaire) results indicated that the CBMEM system, integrated with a CBM approach is effective for construction experience exchange and management.

2. Problem satement

Unlike manufacturing, each construction project is designed and executed to serve specific needs of the owner. The nature of the work and the constitution of the work force in a construction project change with time (Manavazhi, 1995). Experience provides strength in a competitive business environment. Thus, effectively leveraging experience is essential to business success. The complicated nature of the construction industry makes it an important field for experience management (EM), particularly regarding experience gained from completed projects. Sharing experiences between engineers can improve construction management during the construction phases of projects, thus helping avoid mistakes that past projects have already encountered. Transferring construction experience between projects can significantly contribute to achieving project objectives such as cost, schedule, quality and safety (Reuss & Tatum, 1993). Learning from experience, also, avoids problem-solving from scratch, i.e., problems that have already been solved need not be solved repeatedly. However, no effective platforms are available to assist engineers or experts in exchanging and sharing their know-how and experiences when contractors execute construction projects. The inability to share the experience of engineers and experts represents a major loss for contractors in the construction industry. When completing projects, these engineers and experts typically accumulate domain knowledge and valued experience, but share little or no experience with others. In view of EM, these significant issues and experiences of construction engineers and experts are particularly valuable due to associated factors such as manpower, significant cost and time.

The primary problems derived from the questionnaire survey of twenty junior and senior engineers from five participating construction building projects, in the sharing and exchanging of experience, specifically during the construction phase of projects, are as follows: (1) difficulty in determining which engineers and experts have helpful and relevant experience; (2) limited efficiency and quality when using only document-based media for experience management; (3) difficulty in finding engineers with relevant experience in similar projects; (4) inadequate documentation of unofficial discussion and communication regarding problem solving for future reuse; (5) tendency for engineers to communicate orally in person or by telephone; and (6) unease with illustrating experience in current commercial information management systems. Documenting and applying experience may avoid problem-solving from the outset, i.e., problems already solved need not be solved repeatedly. However, few suitable design platforms have been developed to assist engineers in illustrating and sharing their experiences when needed. Although enterprises in the A/E/C industry have begun to collect and store explicit information in enterprise databases, they have not always been successful at retrieving and sharing tacit knowledge (Woo et al., 2004). Sharing and using previous tacit experiences in construction projects is, therefore, the primary and significant challenge of this study.

3. Research objectives

This study proposes a novel and practical methodology for capturing and representing the experience and project knowledge of engineers by utilizing a Computer-aided Design (CAD)-based Maps (CBM) approach. Furthermore, this study develops a Construction Dynamic CAD-based Maps Experience Management (CBMEM) system for engineers. The CBMEM provides a dynamical experience exchange and management service in the construction phase of a project for the reuse of domain knowledge and experience (see Fig. 1). Contractors often execute similar projects; accordingly, the problems encountered in like projects can provide a reference for comparable projects in the future. The capture, transfer, reuse and maintenance of construction project experience are, thus, critical (Kamara et al., 2002). To be competitive, a contractor needs to make innovative use of knowledge, accumulate experience through previous projects and apply it in relevant projects. Senior engineers that participate in projects act as knowledge workers; they facilitate the collection and management of experience between current and past projects.

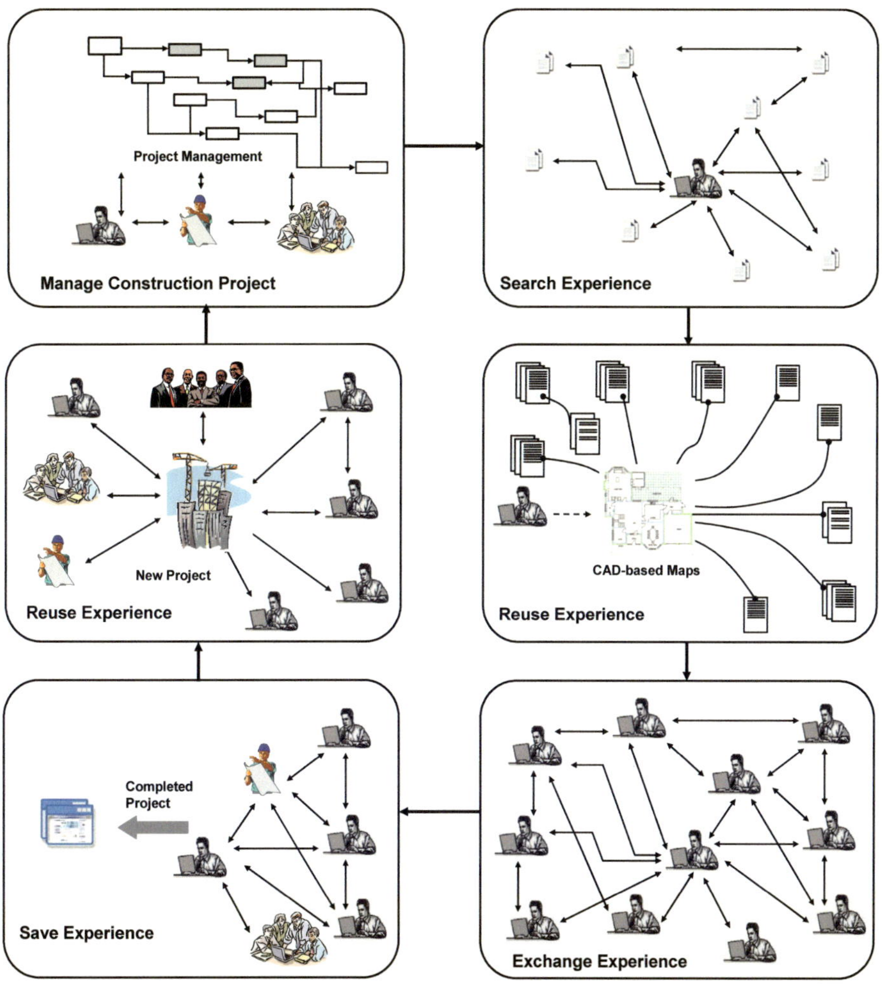

Fig. 1. The application of experience management in construction projects.

This study concentrates on new approaches for managing and reusing past specific experience for a construction project framework. With the newly proposed CBM approach and integration of web-based technology using EM techniques, service engineers and practitioners can exchange original ideas, experience, knowledge and commands. By integrating CBM and web-based technology, engineers can obtain problem solutions and experience directly from senior engineers, decreasing the time and reducing the cost of on-the-job training. By exchanging and sharing previous experiences among engineers, similar and related experiences used to execute similar projects can clarify domain knowledge and enable the exchange of knowledge through web-based EM. The CBMEM system provides a service to users who can request assistance from selected or all engineers in the enterprise who have relevant experience. The user can also submit a problem description through CBM. Moreover, senior and junior engineers can effectively and easily exchange concepts and experience regarding a specific aspect of their current construction project.

To apply EM to new or other construction projects, the process and content of project experience must be collected, recorded and stored effectively in the CBMEM system. To assist the participating engineers in illustrating and managing their own project experience, CAD-based mapping is presented to help them explore their acquired experience. The main objectives of this study are as follows: (1) enhance the illustration capabilities using the CBM approach of captured experience of engineers and experts related to construction projects; (2) optimize the communication of tacit experience among participating engineers in the exchanging environment; and (3) design an efficient web-based platform and maps for users to effectively locate parallel experience from relative engineers. The CBMEM system is then applied in selected case studies of a Taiwan construction building project to verify the proposed approach and demonstrate the value of sharing experience in the construction phase.

4. Background research

4.1 Previous research in experience management In the construction industry

Experience management (EM) deals with collecting, modeling, storing, reusing, evaluating and maintaining experiences (Bergmann, 2002). In the construction industry, EM is a discipline that promotes an integrated approach to the creation, capture, sharing and reuse of the domain knowledge of a profession obtained from projects that have been previously undertaken. Most project-related problems, solutions, experience and know-how are in the minds of individual engineers and experts during the construction phase of a project. Implicit experience is generally undocumented or stored in a system database. To preserve implicit experience as corporate property, capturing the implicit experience and making it in the form of explicit experience is a vital aspect of EM. Two broad categories of experience are tacit experience and explicit experience. Tacit experience is personal, context-specific experience that is difficult to formalize, record or articulate; it is stored in the minds of people (Malhotra, 2000). Tacit experience is personal knowledge acquired through individual experience, which is shared and exchanged through direct, face-to-face contact (Malhotra, 2001). Explicit experience can be codified and transmitted in a systematic and formal language, and can be obtained from documents, including reports, articles, manuals, patents, pictures, images and video (Malhotra, 2000; Tiwana, 2000).

Numerous research efforts have focused on applications of knowledge management in construction. A Hong Kong study examined the main barriers to effective knowledge sharing, as well as critical factors and benefits in the construction companies in Hong Kong and the United Kingdom (Fong & Chu, 2006). Intelligent representation structures store and access construction domain knowledge and couple it with advanced planning tools to facilitate rapid formulation and assessment of initial construction project plans (Udaipurwala & Russell, 2002). Fong et al. (2007) pointed out that the knowledge-creating capability of value management teams not only enhances the reputation of value management, but also, helps to dispel the perception of value management as an outdated problem-solving tool.

4.2 Previous research on knowledge maps in construction

A knowledge map includes the sources, flows, and points of knowledge within an organization (Liebowitz, 2005). All captured knowledge can be summarized and abstracted through the knowledge map. The knowledge map, also, provides a blueprint for implementing a knowledge management system. Well-developed knowledge maps help users identify intellectual capital, socialize new members and enhance organizational learning (Wexler, 2001). A knowledge map is a consciously designed medium for communication between makers and users of knowledge by a graphical presentation of text, model numbers or symbols (Wexler, 2001). Knowledge mapping helps users understand the relationship between stored knowledge and dynamics. Knowledge maps have been applied in various applications, including development of knowledge maps for knowledge management software tools (Noll et al., 2002).

Numerous research efforts have focused on the use of knowledge maps to support various knowledge management tasks (McAleese, 1998). Davenport & Prusak (1998) observed that developing a knowledge map involves locating significant knowledge in an organization and publishing a list or image that indicates a roadmap to locate it. Mind maps (Buzan & Buzan, 1993) illustrate the structure of ideas in an associative manner which attempts to represent how ideas are stored in the brain. A concept map provides a structure for conceptualization by groups developing a concept framework that can be evaluated by others (Trochim, 1989). Dynamic knowledge mapping can assist in the reuse of experts' tacit knowledge (Woo et al., 2004).

5. Methodology- CAD-based maps

Although maps of knowledge representation have been developed for knowledge-based applications, no knowledge map has been developed for experience management (EM) in construction. To assist engineers in extracting the knowledge gained from their own experience in projects with which they have been involved, this study proposes a novel dynamical Computer-aided Design (CAD)-based Maps (CBM) approach for the application of EM in construction. Dynamical CBM help to efficiently illustrate the experiences in the minds of engineers to generate and organize experience within a construction project framework. Dynamical CBM are based on associations flowing outward from a central image in a free-flowing, yet organized, and coherent way. The above content also functions as the experience acquisition tool in the Construction Dynamic CAD-based Maps Experience Management (CBMEM) system. Furthermore, engineers may access and edit many resources, as attachments, in the system. Hence, the CBMEM system can provide engineers with an experience exchanging environment, as well as a web-based platform for acquiring experience from more seasoned engineers.

5.1 Concept of CAD-based maps

The proposed CBM are specific approaches to EM in the construction field. Although knowledge and concept maps are easily recognized in knowledge management, the proposed dynamical CBM approach is a novel concept and is specific to construction EM. CBM can be defined as a diagrammatic and graphic representation of experience linking relationships between experience and attributes of CAD. The CBM mainly provide assistance for easily and effectively obtaining the necessary experience of users. The primary advantages of CBM are as follows: (1) CBM are simply, clearly and dynamically represented in the CBMEM (the EM system); (2) users can easily navigate the CBMEM in order to: a) understand and determine which engineers and experts own special experience related to a problem as it arises, and b) edit their experience based on what the situation may require; (3) CBM enable users' ability to expand flexible experience illustration and linkage; and (4) CBM enhance the available visual experience illustration in the CAD maps.

CBM are designed to be easily integrated with CAD and their construction experience. The key reason for using CBM is the ease with which the combined experience can be understood and reapplied. Figure 2 illustrates an overview and conceptual framework of CBM utilized in construction EM. Like construction project management, EM is based on the concept of undertaking project planning and control activities. Experience gained from activities in previous projects can be collected, managed and applied in future projects. Acquired experience from participating engineers can be accessed and saved as map units in categories for efficient collection, management and finally, retrieval for use in the current projects.

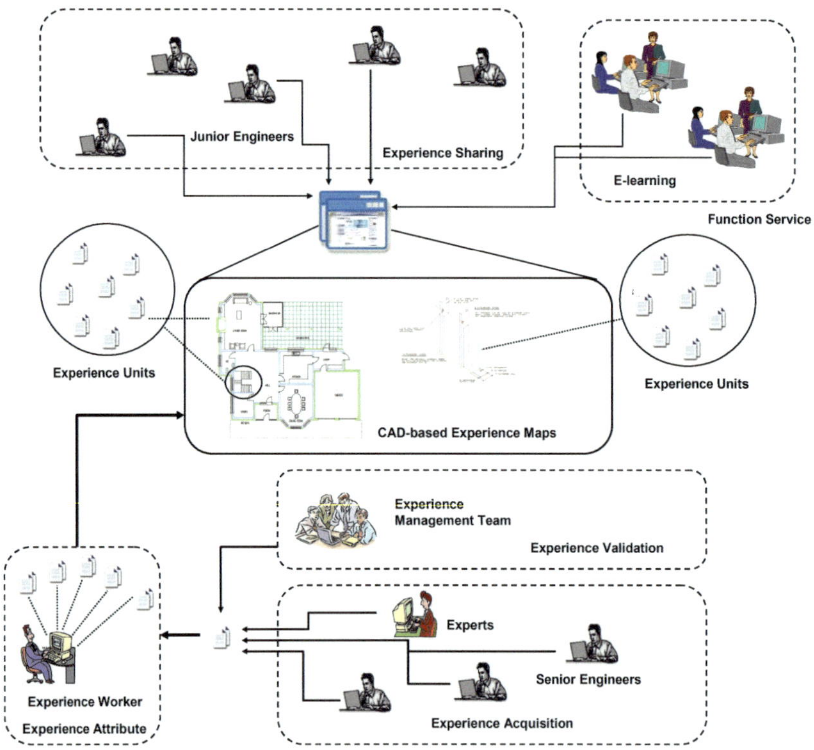

Fig. 2. The application of CAD-based Maps in experience management

5.2 Framework of CAD-based maps

CAD-based Maps (CBM) are defined in multiple levels, and constructed from variables which can be broken down by decomposing the experience units into smaller map units into which the acquired experience is stored. CBM may be comprised of several layers. The project unit is modelled in the first layer. The second-level layers model CAD units (drawing illustration). The lower-level layers model experience units. Similarly, any map unit in this lower layer can be broken down further to incorporate other components in lower layers. The map contents can be viewed as either a single point or as ranges. The structure of CBM enables users to access stored experience through layers based on the attributes and types of acquired experience. Experience stored in map units of a project map includes both tacit and explicit experience. Explicit experience may be comprised of an experience topic, experience description and experience attachments (documents, reports, drawing and other explicit sources). Tacit experience may include problems-faced descriptions, problems-solved explanations, solution suggestions, and know-how explanations. Additionally, CBM give users an overview of available and unavailable experience in core project areas, enabling effective management of tacit and explicit experience. The tacit and explicit experience of map-based experience management (EM) is likened to the duration and relationship of stages-based project management. Identifying the relationship between current and past map units is significant for users to link related experience together. The system is naturally designed to automatically or manually link activities which are highly similar. For example, the experience of a current project can be utilized, and the same or similar map units contributed by past projects can be accessed while the experience of current users is being recorded. There are some cases in which the overall project experience may be captured; however, it may not be clearly classified in project units.

CBM have components and procedures based on construction project management and, thus, differ from existing knowledge maps. The proposed CBM consist of eight components. These ten components are number of experiences, experience topics, experience relationships, experience owners, experience diagrams, experience packages, experience attributes, and similar experiences. Procedures are presented for constructing CBM based on an experience management framework. The procedure consists of the following six primary phases: experience determination; experience extraction; experience attribute; experience linking; experience validation; and, experience sharing.

6. System implementation

This section describes the details of the Construction Dynamic CAD-based Maps Experience Management (CBMEM) system. The CBMEM system is based on the Microsoft Windows 2003 operating system with Internet Information Server (IIS) as the web server. The prototype is developed using Java Server Pages (JSP), which are easily incorporated with HTML and JavaScript technologies to transform an Internet browser into a user-friendly interface.

Three search functions are supported in the system. The server of the CBMEM system supports four distinct layers: interface, access, application and database layers; each has its own responsibilities. The interface layer defines administrative and end-user interfaces. Users can access information through web browsers such as Microsoft Internet Explorer or FireFox. Administrators can control and manage information via the web browser or by using a separate server interface. The access layer provides system security and restricted access, firewall services and system administration functions. The application layer defines various applications for collecting and managing information. These applications offer indexing, experience map edition, digital photo/video management functions, full text search, collaborative work and document management functions. The database layer consists of a primary SQL Server 2003 database and a backup database (also based on SQL Server 2003).

All experience information in the CBMEM system is centralized in a system database. Project participants may access some or all of these documents through the Internet according to their levels of access authorization. Any information/experience about the project can be obtained from and deposited into the system database only through a secure interface. The web and database servers are distributed on different computers, between which a firewall and virus scans can be built to protect the system database against intrusion.

The CBMEM system provides project category search, keyword search and expert category search. The project category and keyword search functions enable users to find the knowledge they need directly from the activities of selected projects. The system, also, provides another function in the expert category for users to find related knowledge according to domain experts. The information held by each domain expert is provided to the users seeking the domain knowledge-related experts. One of the main features of the CBMEM system is enabling users to request assistance in experience support and exchange from specific selected engineers or all engineers in the enterprise through the CAD-based Maps (CBM).

7. Case study

The following case study involves a contractor with seven years of specific experience in Taiwan construction building project. The contractor hoped to take full advantage of experience management (EM) to obtain the valued experience from participating engineers and effectively manage it for exchange and reuse in other comparable projects. The contractor, therefore, announced that all engineers would be encouraged to use the CBMEM system to apply EM to effectively manage acquired experience from participating engineers. The CBMEM system was utilized in the Taiwan construction building project to verify the proposed methodology and demonstrate the effectiveness of sharing previous experience in the construction phase. The case study was undertaken in a 8-month construction project with a schedule including approximately 2,108 activities. Moreover, all engineers were encouraged to explore and edit their own experience in the CBMEM system.

In the experience acquisition phase, senior engineers and knowledge workers undertook most work experience acquisition, since tacit experience must be acquired directly from the minds of engineers. Further, the tacit experience may be transferred into explicit experience by senior engineers and knowledge workers themselves. Most tacit experience extracted for reuse and storage may be available from the memories of experts and engineers. In a broader view, experience extraction may also include capturing knowledge from other sources such as from problem-solution descriptions, suggestions, innovation and collaboration.

In the case study, the senior engineer attempted to edit domain knowledge and experience in the "Interface management among subcontractors" learning lesson. The learning lesson experience in interface problem-facing description among subcontractors, detailed situation description and problem-solution explanations. The knowledge workers and senior engineer initially sketched the main experience map based on the original project network-based schedule plan. After the main map was identified, the five experienced senior engineers were invited to edit their experience in the map regarding interface problem-facing. Related information/documentation was then collected and converted into a digital format. The attached files included digital documents, video and photo files. After the related attached files were digitized, the senior engineer packaged them as an experience set for submission. The knowledge workers, also, assisted the senior engineers in completing the above digitization work and conferred with them weekly to accelerate the problem solving process. The project activities continued for ten months. All engineers were required to provide their own experience regarding the tasks for which they were responsible. Each engineer created an experience map and summarized his experience and domain knowledge in the map to enable the reuse of the solution process for future projects. The experience map included: the experience topic, experience descriptions, experience diagram, experience attribute, experience packages and linkage, the solution to the problem, including related documents, photographs and videos of processes, and expert suggestions, including notes, discussions and meeting records. Experience was extracted based on every process defined as it related to the map units of a project. Domain knowledge and experience were organized according to the attributes of the map units concerned. When the submitted experience set was approved, the system illustrated the process automatically, and an assistant in the EM team attributed the knowledge and classified the experience by placing it in an appropriate position (map units in the experience map) in the system. Restated, users can locate and directly access related experience simply by clicking on these map units located on the multilevel experience maps. In the experience storage phase, all experience was centralized and stored in the central database to avoid duplicating data. All experience can be stored in the system by ensuring that data are all electronic and in a standard format for each file type such as a specific document or drawing format. All experience maps must be validated to perform well before the experience maps are published. All validation is performed in enterprise EM terms by domain experts, knowledge workers and experience map makers. Finally, the experience set is automatically backed up from the experience database to another database. The system automatically sends a message confirming the update to the appropriate users after approving and storing the experience.

A new project is started after completion of the construction project ten months earlier. A senior engineer encounters three different problems in a new project whose information is unavailable in the CBMEM System. The engineer requests suggestions and assistance from other senior engineers involved in the international project to handle the problems directly using communication services in the CBMEM system. After referring suggestions and assistance from senior engineers, the senior engineer solves the problem and shares the new solution with senior engineers. Finally, the senior engineer creates a new map unit and experience package, and submits the obtained suggestion and experience to the map unit of the experience map, linked with the related experience topics. Moreover, the experience is later updated when further feedback and another solution to the same problem are added. The updated experience set is republished in the map units of the experience map after completing the approval process, and a notice is transmitted to the authorized members.

8. Field tests and results

During the field trials, verification and validation tests were performed to evaluate the system. The verification process was proposed to determine whether the system operated as intended while validation was performed to evaluate the system's usefulness. The verification test was conducted by checking whether the CBMEM system could perform tasks specified in the system analysis and design. The validation test involved asking selected case participants to use the system, who then provided feedback via questionnaire. The seventeen respondents included four project managers with 5 years of experience; three senior engineers with 20 years of experience; three engineers with 10 years of experience; four junior engineers with 1 year of test experience; two knowledge workers with 5 years of experience; and one Chief Knowledge Officer (CKO) with 3 years of experience. The CBMEM System was demonstrated to the respondents, who were then requested to express their opinions of the system via the questionnaire.

To evaluate system function and satisfaction with system capabilities, questionnaires were distributed, and the system users were asked to separately rate the conditions of system, system function and system capability, in comparison with the previous system using a five-point Likert scale. A 1, 3 and 5 on the Likert scale corresponded with "not useful", "moderately useful" and "very useful," respectively. Table 1 shows system evaluation result. Some comments for future improvements in the CBMEM system were also obtained from the project participants.

The functionality of system	**Mean Score**
Ease of acquiring experience	4.7
Reliability	4.3
Applicable to Construction Industry	4.8

The use of system	**Mean Score**
Ease of Use	4.8
User Interface	4.5
Over System Usefulness	4.4

The capability of system	**Mean Score**
Reduce Unnecessary time	4.6
Reduce Unnecessary Costs	4.4
Reduce Happening Mistake Percentage	4.6
Ease of finding related experience	4.7
Enhance Experience Updating Problems	4.3
Improve Experience Sharing Problems	4.4

Note: the mean score is calculated from respondents' feedback on fivescale questionnaire: 1(Strongly Disagree), 2, 3, 4 and 5 (Strongly Agree)

Table 1. System Evaluation Result

The use of web technology and CAD-based Maps (CBM) to share and illustrate available experience significantly enhanced the efficiency of experience management (EM) processes. Based on the user satisfaction survey, most users agreed that the CBMEM system enables engineers to exchange and share previous experience using CBM to express their ideas and thoughts. Furthermore, the CBM provided clear and dynamic representations of experience and effectively identified CAD units with experience and knowledge related to the project. The survey revealed a user satisfaction rate of 91 %, indicating that the CBMEM system is useful for assisting engineers in editing their previous experience through the CBM approach to enhance experience acquisition and management. The experimental results showed that the CBMEM system significantly enhanced progress in the construction experience exchange progress and management. Overall, the use of CBMEM system minimized ineffective experience communication and exchange among engineers.

The significant findings of the case study are summarized as follows: (1) the total number of experience units in the system was 1,437 experience units with 129 experience packages during execution of the project; (2) most senior engineers and experts considered recording and editing their experience to be too time consuming; (3) assisting more senior engineers in transferring tacit experience can be problematic, because most senior engineers cannot type their experience by themselves, and (4) most engineers agreed that the CBM approach and CBMEM system are helpful to enabling experience sharing and management in construction projects.

9. Conclusions

This study proposed a novel and practical methodology for capturing and representing the experience and project knowledge of engineers by utilizing a CAD-based Maps (CBM) approach. Furthermore, this study developed a Construction Dynamic CAD-based Map Experience Management (CBMEM) system for engineers which provides a dynamical experience exchange and management service for the reuse of domain knowledge and experience. CBM divide experience into map units, thus forming an effective experience management tool in construction projects. Effective integration of web technology in CBMEM system has been demonstrated in the case study in the Taiwan construction building project. The CBMEM system enables engineers to exchange and share previous experience using CBM to express their ideas and experience. Furthermore, the CBMEM system enables users to request experience support and to exchange experience with selected engineers or all enterprise engineers by submitting problem descriptions through CBM. Novice engineers directly accessing the system can effectively share and exchange experience. The integration of experience management (EM) and the CBM approach appears to be a promising means of enhancing construction EM during the construction phase of a project. In summary, the CBMEM system can assist engineers in illustrating their ideas clearly and sharing their experience. Furthermore, CBMEM system and CBM approach enable users to survey and access effectively the tacit and explicit experience of previous engineers and experts in similar projects.

Although further effort is needed to update the explicit/tacit experience related to various projects, the proposed system benefits construction experience management by (1) providing an effective and efficient web-based environment for exchanging experience specifically regarding construction projects; and (2) providing users options by requesting assistance from selected engineers or all engineers in the enterprise who have relevant experience by submitting a problem description.

The use of the CBM approach in the system mainly provides assistance to help engineers illustrate their own knowledge easily and effectively. The questionnaire results indicate that the primary advantages of CBM in the system are as follows: (1) the CBM provide clear and dynamic representations, thus identifying the experience and knowledge of engineers relevant to the project, (2) the CBM clearly identify the available engineers or experience to request for experience exchange regarding the special experience and knowledge in the current project and (3) users can locate needed experience easily and effectively based on CBM illustration.

10. References

Bergmann, Ralph (2002), Experience Management: Foundations, Development Methodology, and Internet-Based Applications, Springer, Germany.

Buzan T. & Buzan B., (1993), The mind map book: How to use radiant thinking to maximize your brain's untapped potential, New York; Plume.

Davenport, T.H. and Prusak, L. (1998), Working Knowledge, Harvard Business School Press

Fong, P. S. W. and Chu, L. (2006), Exploratory study of knowledge sharing in contracting companies: a sociotechnical perspective. Journal of Construction Engineering and Management, 132(9), 928-939.

Fong, P. S. W., Hayles, C. S., and Hills, M. J. (2007), Dynamic knowledge creation through value management teams. Journal of Management in Engineering, 23(1), 40-49.

Kamara, J.M., Augenbroe, G., Anumba, C.J., and Carrillo, P.M. (2002), Knowledge management in the architecture, engineering and construction industry. Construction Innovation, 2, 53-67.

Liebowitz, J. (2005), Linking social network analysis with the analytic hierarchy process for knowledge mapping in organizations, Journal of Knowledge Management, Vol. 9 No.1, 76-86.

Malhotra, Y. (2000), Knowledge management and virtual organizations. Idea Group Publishing, Hershey, PA.

Malhotra, Y. (2001), Knowledge management and business model innovation. Idea Group Publishing, Hershey, PA.

Manavazhi, M. R. (1995). "Case-based Reasoning and Hypermedia: Enabling Technologies for Construction Experience Transfer." Technical Report # 102, Center for Integrated Facility Engineering, Stanford University, CA, August 1995.

McAleese, R. (1998), The knowledge arena as an extension the concept map: reflection in action, Interactive Learning Environments, Vol. 6 No. 1, 1-22.

Noll, M., Frohlich, D. and Schiebel, E. (2002), Knowledge maps of knowledge management tools: information visualization with BibTechMon, in Karagiannis, D. and Reimer, U. (Eds), Practical Applications of Knowledge Management 2002 Conference Proceedings, Springer-Verlag, New York, NY.

Reuss, Mark C. and Tatum, C. B. (1993). "Requirements and Tools for Transferring Construction Experience between Projects." Technical Report # 78, Center for Integrated Facility Engineering, Stanford University, CA, Feb 1993.

Tatum, C. B. (1993). "Structure and Characteristics of Knowledge from Construction Experience." Technical Report # 81, Center for Integrated Facility Engineering, Stanford University, CA, Feb 1993.

Tiwana, A. (2000), The knowledge Management Toolkit - practical techniques for building a knowledge management system. Prentice-Hall, New Jersey.

Trochim, W.M. (1989), An Introduction concept mapping for planning and evaluation, Evaluation and Program Planning, 12(1), 1-16.

Udaipurwala, A. and Russell, A.D. (2002), Computer-assisted construction methods knowledge management and selection. Canadian Journal of Civil Engineering, 29(3), 499-516.

Wexler, M. (2001), The who, what, and why of knowledge mapping. Journal of knowledge management, 5(3), 249-263.

Woo, Jeong-Han, Clayton, Mark J, Johnson, Robert E., Flores, Benito E., and Ellis, Christopher (2004), Dynamic Knowledge Map: reusing experts' tacit knowledge in the AEC industry. Journal of Automation in Construction, 13(2), 203-207.

7

Applications of Computer Aided Design to Evaluate the Zoning of Hazard Prevention in Community Neighbours

Kuo-Chung Wen
Institute of Architecture and Urban Planning, Chinese Culture University
Taiwan, R.O.C.

1. Introduction

The city government will provide the enough emergence routes, parks, and so on to reduce the hurtful accidents during the escape by making the urban plan. The proportions of the Zoning of Hazard prevention will be influenced by some main policy such as the develop directions, population and some effects, and sometimes get a poor proportions. So in this study we want to use some methods such as spatial and Network Analysis to set up the Zoning of Hazard prevention and estimate the safety of these area (Li, 1997).

So in this study we use Geographic Information System (GIS) to combine with the spatial information, systematize, and escape behaviour theory to simulate the escape situations. Spatial information talks about characteristic in community layout. Systematize talks about the relationship of the open space. Escape behaviour theory talks about the actions of evacuation people and simulate the escape path in the community escape path. We aimed at the community neighbours for study area. At first, we assess the escape paths and establish the relationship of the street space. Second, we set up the cell to interpret the spatial environment. Third, we suppose some types of people to simulate the escape path evacuation.

The theoretical basis of the research included Genetic Algorithms (GA). The biological evolution aroused GA, which is a kind of optimization search model within natural choice process. It operates by the way of the encoding gathered by parameter and gets rid of restrictions of seeking space analysis. For this reason, we can get the Global Optimum faster, and prevent it become the Local Optimum (Blanco, et al. 2000). Therefore, the study uses the dynamic process of the genetic calculation, and goes on the choice of the evacuation path. Receiving the batter population, we combine the function of the GIS Spatial Analysis (SA), under the disaster prevention theories, it can simulate and present a more safe model of Dynamic Path Choose Model (DPCM) that near to the behaviour of the really evacuation in mankind.

The part of DPCM could be divided into four parts. (i), is to set the population of GA operation. (ii), is to choose crossover and mutation. (iii), is to calculate the fitness function of each generation and to select the better gene arrangement. (iv), is to reproduce, after evolution, we can establish evacuation path that more reflect really human action and choice when hazard takes place.

Finally, we compare the NA with GA calculation, and Establish real model of DPCM to choosing evacuation path. The results are three parts: 1, measure the safety of the community neighbours. 2, simulate of escape path evacuation. 3, estimate the safety area of community.

2. Theory

The theoretical basis of the research included GA, GIS, Network Analysis (NA) and disaster prevention theories combine with the spatial information, systematize, and escape behaviour theory to simulate the escape situations. Spatial information talks about characteristic in urban area. Systematize explain the relationship of the street networks. Escape behaviour theory talks about the actions of evacuation people and simulate the escape path in the community escape path (Breaden, 1973). The biological evolution aroused GA, which is a kind of optimization search model within natural choice processes. It operates by the way of the encoding gathered by parameter and gets rid of restrictions of seeking space analysis. For this reason, we can get the Global Optimum faster, and prevent it become the Local Optimum. Therefore, the study uses the GA and NA to goes on the choice of the dynamic flooding evacuation path. By the way, we can display the more real human behaviour and find the least cost evacuation path by the dynamic program of the data base in time. Receiving the batter population, we combine the function of the GIS Spatial Analysis, under the disaster prevention theories, it can present a more safe model that near to the behaviour of the really evacuation in mankind. The structure of combined GA with GIS is like Fig. 1.

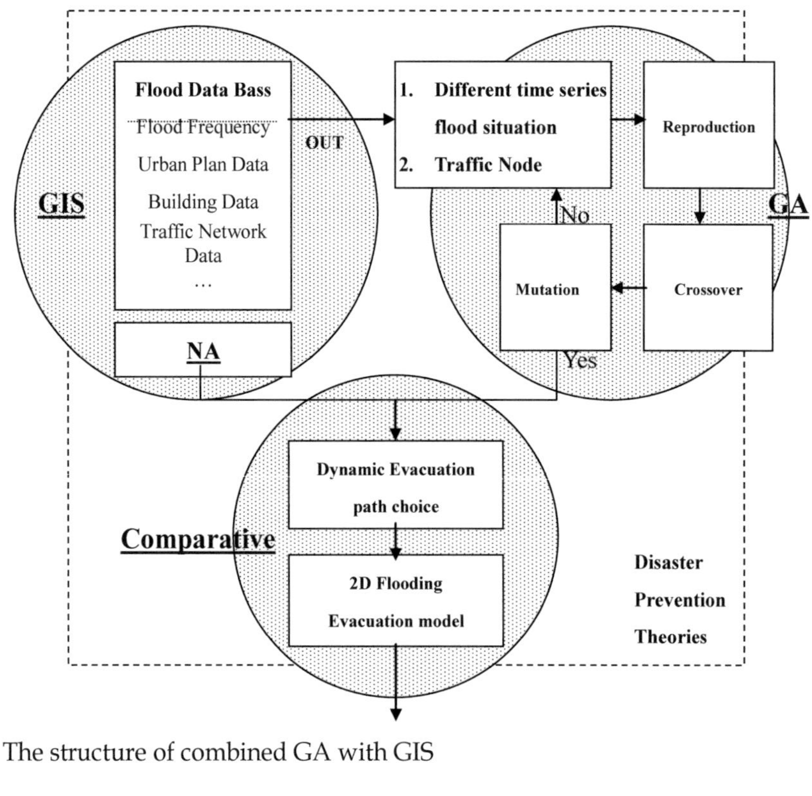

Fig. 1. The structure of combined GA with GIS

2.1 Disaster prevention systems

According to the level of disaster and place, the emergence shelter should proved the functions like command post, information, disaster prevention, medical treatment, goods saving and so on. So the setting of emergence shelter should choice the local government, lodge to be the command post and the park, elementary school, community centre to be the emergence shelter.

The planning of the emergence routes are depended on the different situation. The most important function is to provide people to escape to the safety place. And according to the Taipei Disaster-Prevention Planning, The roads were classified into emergency path system, rescue transport path system, fire control path system, and assist path system (8m) (Tseng, 2000).

Zoning of hazard prevention is an independence area which is not be influence from next area and when disaster occurred the people in the area dose not escape to other ones. And the zoning area can accept enough people.

The walk speed will be closed to normal speed if there is enough space. On the contrary, if there is not enough space, walk speed will be slow down even closed to stop depending on the increasing density. Dr. Tanaboriboon and Dr. Guyano think about that walk speed and body characteristics of western is differ with oriental. At the centre street of Bangkok city in Thailand it was studied to survey location the ambulation of people (Tanaboriboon & Guyano, 1989). They divide service level of ambulation into 6 rankings (A, B, C, D, E and F). And they convert walk speed base on the relationship of density and discharge like Table 1.

Services level	Density (Person / Square metre)	Velocity (Metre /Second)	Flow (Person / Metre * Second)	Condition
A	<=0.42	>=1.12	<=0.47	* Don't generate conflict each other
B	0.43~0.63	1.06~1.11	0.48~0.67	* The velocity and flows become slightly slow
C	0.64~1.02	1.00~1.05	0.68~1.02	* The pedestrian needs to adjust the velocity and directions
D	1.031~1.54	0.88~0.99	1.03~1.35	* Difficult to change the direction and cross
E	1.55~2.70	0.62~0.87	1.36~1.68	* Extremely difficult to change the direction and cross
F	>2.71	<0.61	>1.69	* Can't reverse direction and cross

(Tanaboriboon & Guyano, 1989)

Table 1. The walking services level that Tanaboriboon and Dr. Guyano established

2.2 Dynamic programming

We can use dynamic programming to find the suitable evacuation path. With the dynamic programming, we can establish the decide node in the time and the node of the traffic street network. And we can combine the attribute of the traffic network and some important information with GA to find the optimum Escape Path.

2.3 Network analyze

Network Analyze is a way can get the optimum solution by some designate standards of the traffic network database. Networks are making up with some information which are expressing with points and lines. It is suitable to modelling roads, pipes, facilities and finds the optimum answers. Network analysis is one kind of analysis which in depended on the model of GIS system. The analysis way is to connect the spatial feature such as polygon, line and point and calculate the type and characteristic with the connecting networks (Djokie & Maidment, 1996).

$$\min z(X) = \sum_{(i,j)\in A} \mathrm{C}_{\mathrm{ij}} X_{ij} \tag{1}$$

A: The set of arc in street network
Xij: The flow of arc (i,j)
Cij: The cast of arc (i,j)

2.4 Genetic algorithms

John Holland proposed genetic algorithms (GA) in 1795. This is an optimization of problem solving and technologic of machine learning. It is enlightenment from creature evolution process. The answer of every problem expresses a chromosome that present an individual creature. A group of creature were evolution by Darwin's evolutionism compete and select. The fitting creature exists that present the good solution survival the bad eliminates through competition. The new solution of new generation also to model creature propagates by survival's individual copulation and mutation (Bullock, 1995).

There are four different points between GA and traditional way of optimization and search (Woodbury 1993).

1. GA deals with whole set of solution, not only solution itself.
2. The search of GA starts from a group of population fitting well and scattering beginning, not from a point.
3. GA is objective function, not differentiation or others assist knowledge.
4. GA leads the direction of search only by hands around rule of probability.

It is a series process of self adjusts in search control of design reasoning of GA (Jo & Gero, 1995). The combination of design reasoning rules could be a chromosome of one of evacuation path solution. Every set of chromosome is whole result of inference path generated by probability. These evolution from parents and generate next generation were selected by environment conditions. Those are constantly adjusted through heuristic rules and search strategy, stop until solution fit need. The whole process of evolution is the process of finding out answer. The final result of inference paths, evaluative rules and solution is important knowledge of evacuation path. There are three process follow:

1. Reproduction

The probability of copy from parents is derived from fitness degree of the chromosome. The common method is Roulette Wheel Selection by the percentage of its fitness degree of the chromosome over summarization of all fitness degree. That is the more high fitness degree the more opportunity to duplicate from parents.

2. Crossover

After reproduction the crossover provide for exchange chromosomes between mother generations in order to get the befitted chromosome from parents (Chan & Tansri, 1994).

3. Mutation
It may change some genes form some chromosomes to avoid lost the befit information by reproduction and crossover in the genetic process. So we can extend the searching space to escape from the local optimum to the global optimum.

3. Modelling

3.1 Establish the platform

This research carries on the construction of the development platform of the database with the geographical information system. Collect and set up relevant databases at first, then derive the model assessed in simulation and set up the data to export into interfaces. In collection of the database, will mainly collect the urban planning map and relevant attribute data, in order to build the basic geographical information system of database which constructs the calamity area, to offer basic spatial analysis and application, but the digit picture that this stage needs to finish, change of the scope of activities via presenting the time array of this area after network analysis. So must turn attribute data and spatial data into information forms of GIS, for systematic operation, it is mainly attribute data to move the urban street network, the attribute data are with the basic graph: To spread out with point ,line and polygon, each of spatial data all has specific codes and corresponding attribute value, taking network layout of traffic way as an example, its attribute data include: Serial number, length, driving speed, etc. the attribute data are stored in the data form of attribute, elected fetching the data record When, figure when it is corresponding choosing. Attribute data form and basic map is construct for escape simulation of flood disaster to take special database, digital elevation model urban planning street map, floods water possibilities map, traffic network data, etc.

We use GIS to establish the system which combining the data base of the flood information. At first, we search and collect the flooding data base. And than we infer the estimating model, and set up the in put and out put of the system. About the base data, we collect the urban planning map and some correlating data to be the digital data. It can provide some applications and display the variations of the activity of the time series in the area. We divide the format into three parts, the data base are display in the shape of point, line and polygon. There are its own coding and data in each spatial object. For example, the traffic network has its own data just like coding, length, speed and so on. These data are written in the table of the data base. When you select the record, the corresponding shape will be selected. In this study, we establish these spatial data which are the topographic chart, the data of traffic network, the block of urban planning, and the flood frequency, and so on to model the evacuation path.

We study with the community neighbors. At first, we assess the Emergence route and establish the relationship of the street space. Second, we set up the cell and street networks to interpret the spatial environment. Third, we suppose some methods to simulate the Zoning of Hazard prevention evacuation and to divide the different of the Zoning of Hazard prevention (Li, 1999).

3.2 Establish the rescue refuge rings

3.2.1 Emergence shelter

After the basic data are collected, to city taking escape stronghold is it select to go on, as take escape to choose stronghold mainly with school, park open space mainly here of the area.

3.2.2 Emergence route

Planer will take escape in the route and divide into four grades mainly while urban planning of the road value and to assess proceed of present situation, and classify the traffic network, in order to use as the follow-up network analysis.

3.2.3 Zoning of hazard prevention

The escape area is divide four part to mainly to set up, the major is divide into spatial geometry and network analysis two ways.

1. Spatial Geometry Analysis
With analysis and geometry of ring land that is regarded as the calculation that is rowed and had in space geometry is analyses, analysis in accordance with making and rowed and set up the space equidistantly, but several geometry analyses that separate and take escape the area with the central line in the ring land.
2. Network Analysis
Network analysis is make mathematical calculations to go on with time and distance two impedance, it is analysis to go on with unit length by each impedance value of route to hinder to resist, regard place of making it as with school export when the it can be regarded, in area can reached is it appear to go on so as to a data.

3.3 Establish the dynamic evacuation path model
3.3.1 Dynamic evacuation path

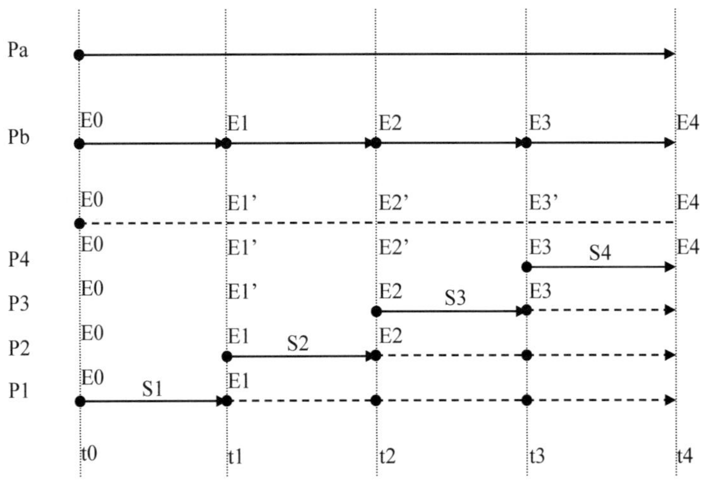

Fig. 2. The dynamic evacuation path model

We establish the evacuation path by the data of different time series. We suppose that the depth of the flood get an even rising. So we divide the time into some parts of time series. Upon the data of the time series, we can get the flood frequency in the different time series and help us to make some decision. In this study, we used different decision node in the traffic network and different time series to select the evacuation path like Fig. 2. In Fig. 2, DP is combined with S1, S2, S3 and S4, and according to the different data bases in each time series these evacuation paths. E is the decision nodes of path.
The dynamic evacuation path is defined by

$$DP = \left(P_{s1d1}\right)_{t1} + \left(P_{s2d2}\right)_{t2} + \ldots + \left(P_{sndm}\right)_{to}$$

$$DP = \sum_{\substack{t=1 \\ d=1 \\ s=1}}^{\substack{o \\ m \\ n}} \left(P_{sndm}\right)_{to} \tag{2}$$

d: the depth of the flood ; t: the time series;
p: the moving path; s: segment of path
DP: Path Distance.

We use Best Route (BR) to calculate the optimum in this study. BR is one kind of the network analysis. It uses the minimum cumulative impedance to find the optimum with two or more traffic nodes in the traffic network. These path nodes can be sequence. And the response unite can be selected in the traffic network data items. For example, we can use the distance and time as the response unites to simulating the more real situation. So we use distance and deliver time to calculate the optimum in this study.

3.3.2 Genetic algorithms

The knowledge representation is the key of whole system of Evacuation Path Model (EPM). There are chromosome, environmental parameters and fitness function. These derived from path table, node table, choose table, dynamic function and GA table in GIS.

Method \ Point	P1	P2	P3	P4	P5	P6	P7	P8	P9	P10	P11	P12	....
0	0	0	0	0	0	0	0	0	0	0	0	0	...
1	P4	P1	P2	P1	P4	P5	P6	P10	P8	P8	P16	P4	...
2	P2	P3	P8	P5	P13	P2	P14	P9		P7		P13	...
3		P6		P12	P6	P7	P10	P3		P16			...

Table 2. Choose Table

1. Code of Chromosome

The concept is developed by initial evacuation path's idea. The result of chosen path could be transformed a serious genes to combine chromosome. Assuming one area has many nodes (for example… P1, P2, P3, …, Pn.), each node has a lot of path to be chosen. The Fig. 3 shows the node P1 has two paths can be chosen, there are two chosen method. As the same way, the P2 has three chosen method. Thus we can establish the attributes of choose table (like table 2) from node table. The table 2 presents the spatial relationship and chosen method of each node.

The dynamic function is to transform the gene code into the evacuation path (like Fig. 4). The gene code follows the id to be a chromosome in the GA. The id just the sequence number there is no any means in this table. If we decide the start node is (P1) and the end node is (P14) of evacuation path. The first id just is in the name of start node (P1), the gene code is (1). Then we can choose the next node by index the gene code from choose table. For the start node (P1, 1) we can index P1 choose the (1) method to find out the P4. So the next node is P4, we transform the gene code from (1, 1) to (P1, 1) to (P1, P4). Then repeat the steps above until the next node is just equal the end node. Final we can spatial join table to GIS to draw out the evacuation path (like the red line in Fig. 4).

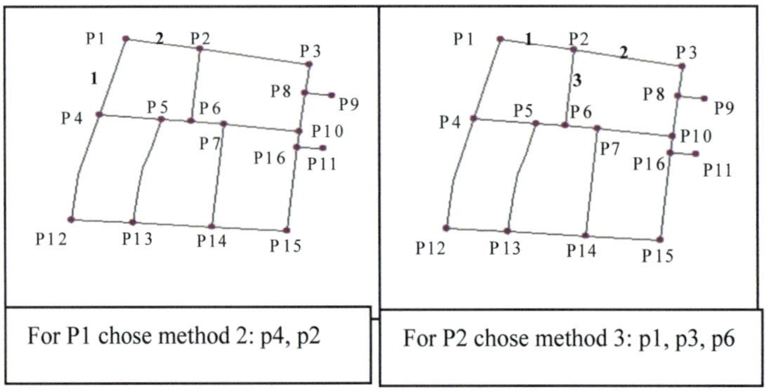

Fig. 3. The method of node choose

ID	1	2	3	4	5	6	7	8	9
Gene	1	2	3	3	2	0	0	0	0

ID	1	2	3	4	5	6	7	8	9
Gene	2	2	2	1	3	3	1	0	0

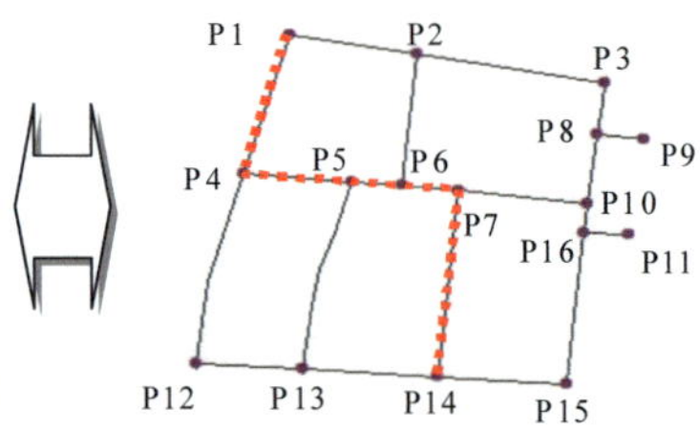

Fig. 4. The dynamic function to transform the gene code into the evacuation path

The rule of crossover between two chromosomes is before we cut any segment by random, after we crossover them.

Before crossover PathA: 1,1,2,……,2,1 PathB: 1,1,4,……,2,1 (3)

After crossover PathA: 2,1,4,……,3,1 PathB: 2,1,2,……,3,1

2. Environmental Parameter

- DF: Degree of Fitness. The value was calculated by the fitness function. Then it transfers each case's subsistence probability. The function follows:

$$ALIVE_i = DF_i / \sum_{j=1}^{n} DF_j \quad (4)$$

DF: Degree of Fitness.
i: the i'th case
j=1 to n, n is the total cases

- PN: Population Number. The numbers of total individual, the max living numbers of controlled environment

- RR: Reproduction Rate is the copy rate of mother generation. Whether the individual child will be reproduction, it depended on its subsistence probability. If the subsistence probability is higher, it will be more chance copied. It will have more opportunity to evolve.
- CR: Crossover Rate is the exchange percentage between any two chromosomes of parents.
- MR: Mutation Rate is self-change probability of any chromosome.

3. Fitness Function

The fitness function is the rules to estimate cases and give weight score. It is the tool to judge the better or worse one. It can decide to eliminate the unsuitable case. It including evaluation the rank of evacuation path, the successive nodes of evacuation path, the number of nodes, the length of evacuation path and the number of repeat path in evacuation path choose.

4. Result

4.1 Computing rescue refuge rings in shiji area

This chapter will take the Shiji City in the Keelung River Basin for case in this study. The boundary is like Fig. 5. We apply river digital topographic map, Digital Elevation Model, Traffic Network Data, Urban Planning Map, etc. According the functions required. We can analyze the demand of data, and build database.

Fig. 5. the urban plan of Shiji City

4.1.1 Emergence shelter setting

To define the boundary of Urban Disaster Prevention & Rescue Refuge Rings, we take the service radius (600m) of high school elementary schools, and the range of refuge rings is about 300m~500m. The walk time of refuge rings is about 5~10 minutes and to consider other resource of Disaster Prevention & Rescue. The construction of the disaster prevention network model is like Fig. 6.

Fig. 6. the Emergence shelter in Shiji City

4.1.2 Emergence route setting

We adopt Taipei Disaster-Prevention Planning for the setting of road class. The roads were classified into four classes: emergency path system (20m) rescue transport path system (15m) fire control path system (8m) and assist path system (8m), like Fig. 7.

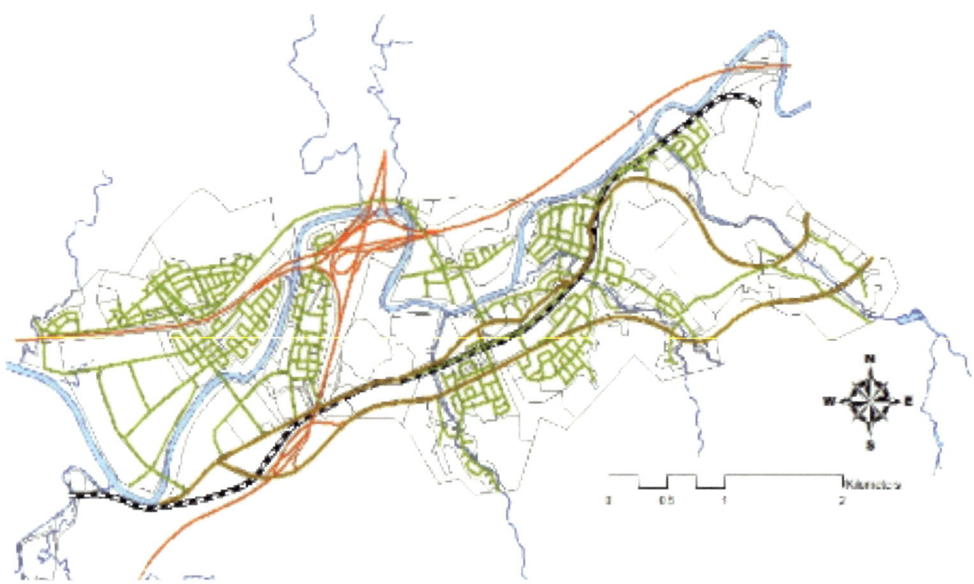

Fig. 7. the Emergence route in Shiji City

4.1.3 Zoning of hazard prevention setting

The zoning of hazard prevention are calculate by spatial and network analysis. We use schools to be the Emergence shelter and the emergence routes to calculate the area of network.

1. Spatial Geometry Analysis

We use the center of schools to be the point and calculate the perpendicular bisector in the area you can use the shortest distance to the Emergence shelter. Like Fig. 8.

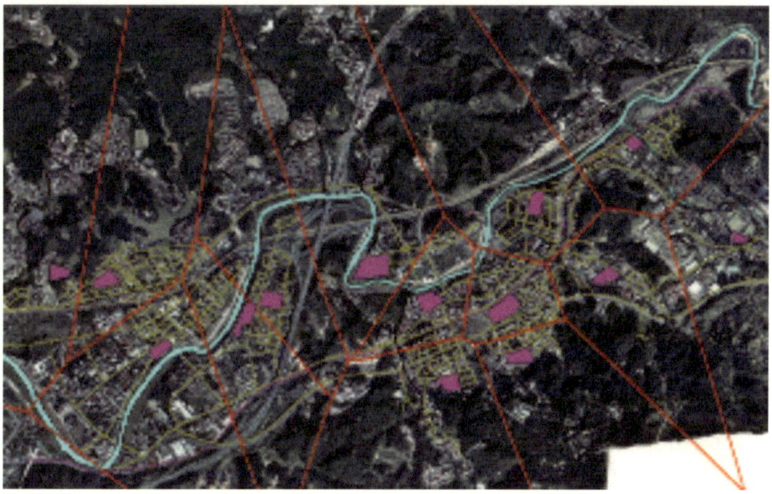

Fig. 8. Calculate by Spatial Geometry Analysis

We also use buffer to calculate the same distance area to be the refuge rings. In Fig. 9, we use the distance of 500m to display the area of near ones. Some area cover the same place and some place are not covered.

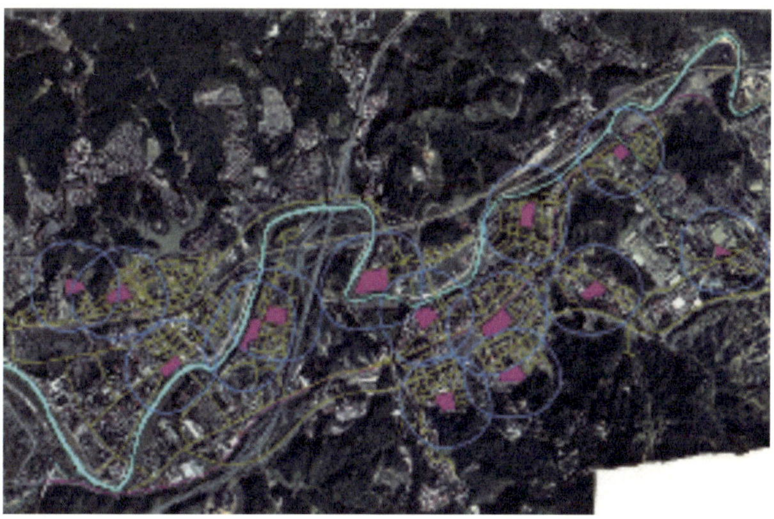

Fig. 9. Calculate by buffer analysis

2. Network Analysis

In this study, we use distance and deliver time to be the cast of the weighting. By distance we calculate the network with 500m, like Fig. 10.

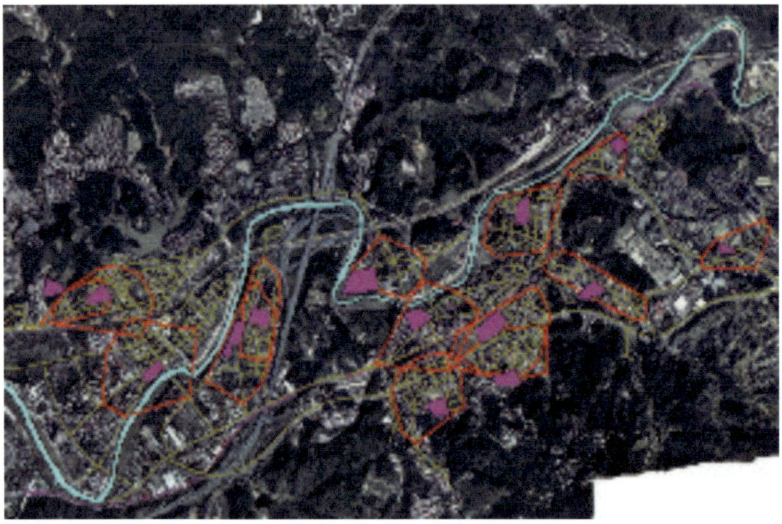

Fig. 10. Calculate by network Analysis (distance)

About the deliver time, we use 10 minutes to display the service area, like Fig. 11. some area have different level emergence routes, so in this service area the distance of road will be change by difference level routes.

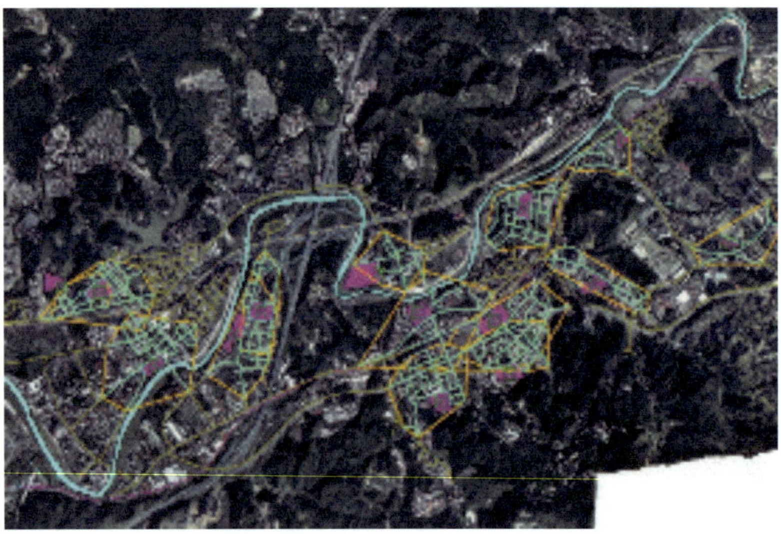

Fig. 11. Calculate by network Analysis (deliver time)

3. Zoning of Hazard Prevention

When look into the large scale, we can find the refuge rings with different shapes. In Fig. 12-15, we can find the relationship between the Emergence shelter and Zoning of Hazard prevention.

Fig. 12. Calculate by spatial geom.etory analysis

Fig. 13. Calculate by buffer analysis (300, 500m)

Fig. 14. Calculate by network analysis (10min)

Fig. 15. Calculate by network analysis (500m)

In Fig. 16. we can compare with this area of Zoning of Hazard prevention, we can find the different area and facility.

Fig. 16. Compare the plan of Zoning of Hazard prevention

With the data of geometry and facility, we can compare the value of plan property like table 3.

Type	AREA	Perimeter	Road	Building	Zoning
Geometry	496759	2835	7811.9	141089	576189
Buffer-300	556097	2681	8370.1	176605	730407
NA-D500	455940	2828	9086.2	135375	604954
NA-T10	379754	2467	8234.9	119762	531208

Table 3. the data of spatial information

We can use the data to calculate the efficiency of the plans, in table 4, we use some formula to estimate the emergence system. We use the building coverage ratio and zoning area to calculate the service efficiency. By using the floor space index floor area ratio and Emergence shelter area to presume the population. The road and area and rate can talk about the efficiency and the area type also can be assessed to the potency.

TYPA	P*P/A	R/A	B/A	Z/A
geometry	16.17	0.0157	0.28	1.15
Buffer-300	12.92	0.0150	0.31	1.31
Buffer-600	12.78	0.0123	0.24	1.15
NA-D500	17.54	0.0199	0.29	1.32
NA-T10	16.02	0.0216	0.31	1.39

Table 4. the estimate of emergence system

We establish the master place to calculate the variation of the Emergence shelter like Fig. 17-20.

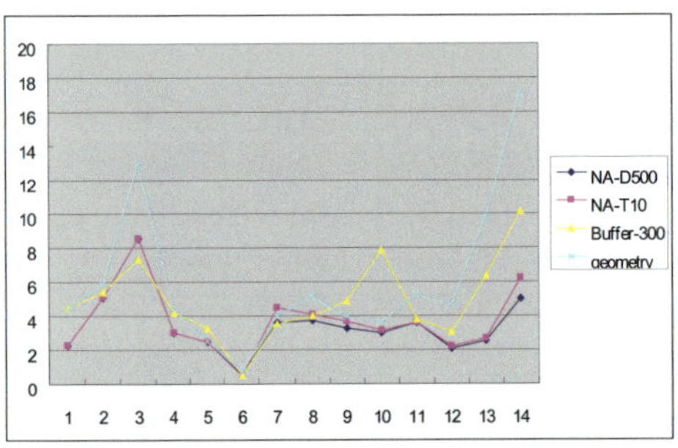

Fig. 17. the variation of plan efficiency

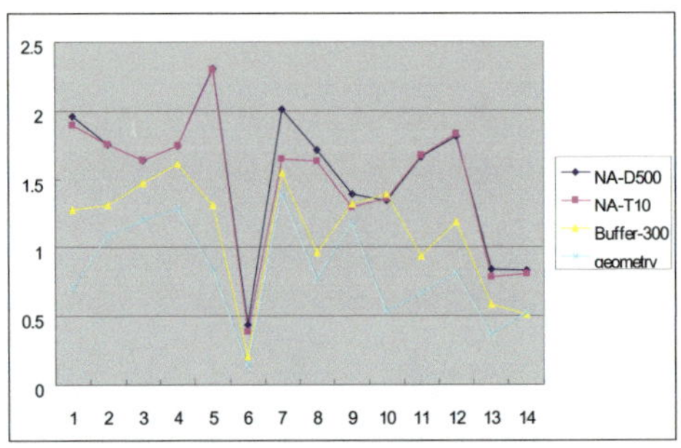

Fig. 18. the variation of population

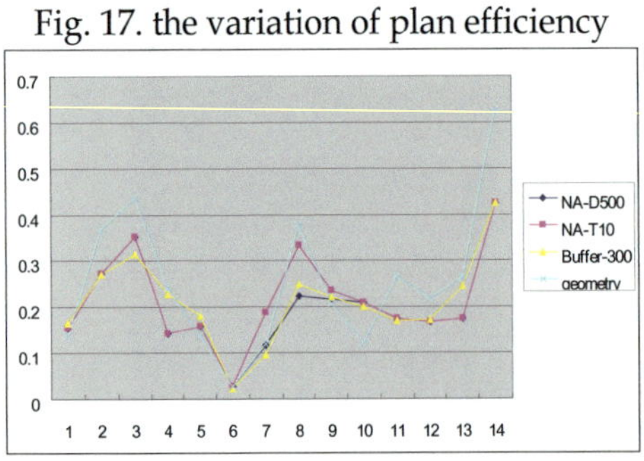

Fig. 19. the variation of acceptability

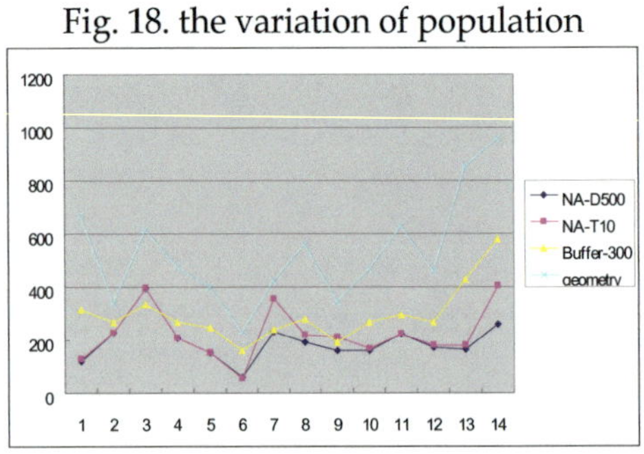

Fig. 20. the variation of spatial service

4.2 Computing evacuation path in shiji area

This chapter will take the Shiji City in the Keelung River Basin for case in this study. The boundary is like Fig. 21. We apply river digital topographic map, Digital Elevation Model, Traffic Network Data, Urban Planning Map, etc. According the functions required. We can analyze the demand of data, and build database.

Fig. 21. The Region of Shiji

4.2.1 Establish the traffic street network

We adopt Taipei Disaster-Prevention Planning for the setting of road class. The roads were classified into four classes: emergency path system (20m) rescue transport path system (15m) fire control path system (8m) and assist path system (8m). To define the boundary of Urban Disaster Prevention & Rescue Refuge Rings, we take the service radius (600m) of high school elementary schools, and the range of refuge rings is about 300m~500m. The walk time of refuge rings is about 5~10 minutes and to consider other resource of Disaster Prevention & Rescue. The construction of the disaster prevention network model is like Fig. 22.

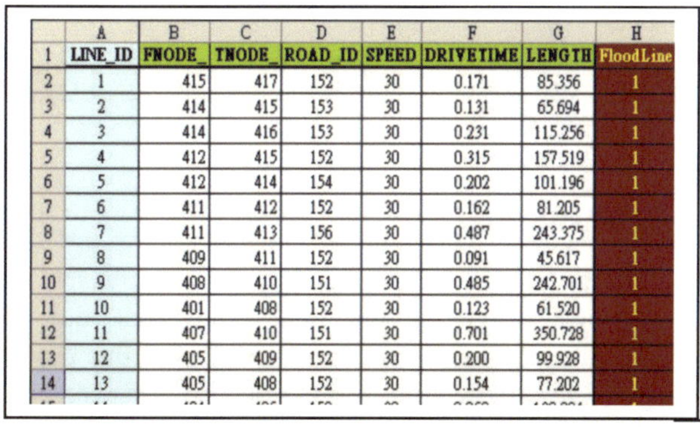

	A	B	C	D	E	F	G	H
1	LINE_ID	FNODE_	TNODE_	ROAD_ID	SPEED	DRIVETIME	LENGTH	FloodLine
2	1	415	417	152	30	0.171	85.356	1
3	2	414	415	153	30	0.131	65.694	1
4	3	414	416	153	30	0.231	115.256	1
5	4	412	415	152	30	0.315	157.519	1
6	5	412	414	154	30	0.202	101.196	1
7	6	411	412	152	30	0.162	81.205	1
8	7	411	413	156	30	0.487	243.375	1
9	8	409	411	152	30	0.091	45.617	1
10	9	408	410	151	30	0.485	242.701	1
11	10	401	408	152	30	0.123	61.520	1
12	11	407	410	151	30	0.701	350.728	1
13	12	405	409	152	30	0.200	99.928	1
14	13	405	408	152	30	0.154	77.202	1

Fig. 22. Disaster prevention network data establishment

4.2.2 Network analysis of the evacuation path

This part we use Best Route to calculate the optimum. We use distance and deliver time to search for the least accumulative cost like Fig. 23 and Fig. 24. In Fig. 23, Best Route calculates the least accumulative cost by the distance. But in Fig. 24 is depends on deliver time. There are some different evacuation paths in the tow figure. The reason of the different is the class of the path. The high level paths get the short deliver time, but these cost more distance. So we get the different optimum with Best Route.

Fig. 23. & Fig. 24. Network Analysis of the Evacuation Path

4.2.3 Working data and parameter setting

Before process the GA calculation, we must to precede the pattern of Gene Coding. Let the variables indicate the suitable sequence in the computer operating. And we decoding it and return the result (like Fig. 25).

Final we set the parameter like Initial population, crossover rate and mutation rate. After we coding the refuge node, we can create initial population and choose the start node. This study on GA's parameter set up 1500 initial populations, and it has 0.5% crossover rate and 0.1% mutation rate.

	A	B	C	D	E	F	G	H	I	J	K
1	Node_ID	PathNum	TNODE1	TNODE2	TNODE3	TNODE4	FloodNode	LINE_ID1	LINE_ID2	LINE_ID3	LINE_ID4
2	1	1	2				1	584	0	0	0
3	2	3	1	10	11		1	584	583	582	0
4	3	1	6				1	581	0	0	0
5	4	1	10				1	580	0	0	0
6	5	1	6				1	579	0	0	0
7	6	3	3	5	8		1	581	579	578	0
8	7	1	8				1	576	0	0	0
9	8	3	6	7	9		1	578	576	574	0
10	9	3	8	12	12		1	574	573	575	0
11	10	4	55	2	4	19	1	571	583	580	572
12	11	3	13	36	2		1	570	577	582	0
13	12	3	9	9	15		1	573	575	569	0
14	13	4	125	21	31	11	1	407	567	568	570
15	14	1	15				1	566	0	0	0
16	15	3	12	14	17		1	569	566	565	0
17	16	1	22				1	564	0	0	0
18	17	3	18	15	20		1	562	565	563	0
19	18	1	17				1	562	0	0	0
20	19	3	25	10	70		1	560	572	561	0
21	20	3	24	17	22		1	558	563	559	0

Fig. 25. The refuge node coding

To search evacuation path, we use GA technique to get an answer belong to the problem form of the limited type model. The region of answer could be very small. The result could be segment to several areas. It would have low rate to get optimization answer with this model, and the rate of best answer also obvious level down. Generally speak the best answer often appearance on cape area that on the boundary region of the feasible solution. If we only adopt the information of the feasible solution, it would increase search time and difficulty.

Gen (1997) use GA to solve the limited type of problem model, it will often appear the result that not falls into feasible solution region. Gen solve these problems by four kinds of strategy, we use two kinds of methods in the following (Gen, M. and Miller, 1997).

1. Reject Strategy

Once the answer of GA output in not feasible solution region, we throw down that chromosome right away. Make sure the chromosome that making duplicate always in the feasible solution region.

2. Penalty Strategy

At original target function, increase a penalty item. The penalties items will check by the level of individual act against restrict. The degree of act against is more. The penalty function is bigger. Whereas is smaller. These study give different degrees of penalty function with have inundation or not. So we can make the limit question into in limit.

4.2.4 Operation interface and process

On the process of searching the best evacuation path, we adopt two different methods to find the solution. First, it is on the condition of evacuation path continuous each other and processes the optimization of path. Second, it is on the unlimited condition, so all influential factor proceed in different indicators weights. The first method has better searching speed, the second method has longer time to calculate, but it is flexible. In this study, we take the first method to simulating. The operation interface is like Fig. 26. The Population Results and Progress Graph like Fig. 27.

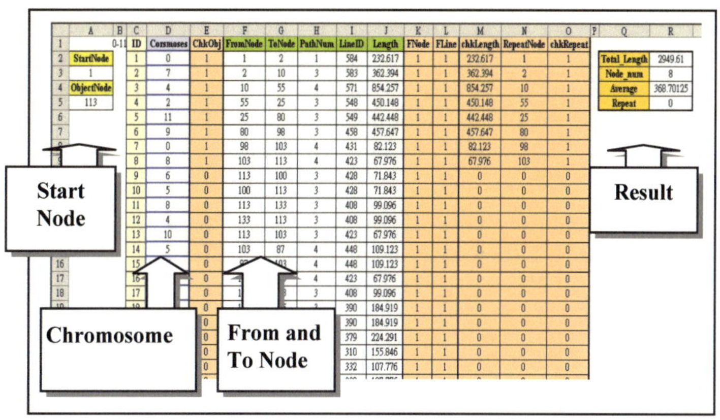

Fig. 26. Operation interface

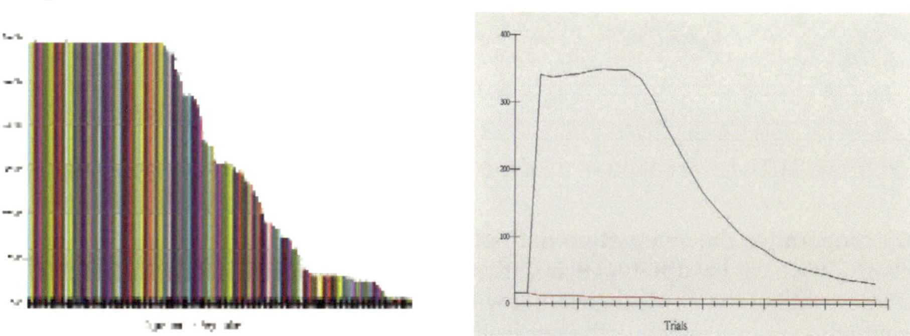

Fig. 27. Population Results and Progress Graph

4.2.5 The Simulation of the dynamic evacuation path by GA

We calculate the different evacuation path with the data base in the first and sixteen time series by GA. According to the depth of the flood frequency of the time series, GA search for the optimum are distinct like Fig. 28 and Fig. 29. In the first time series GA get the smooth evacuation path. In the Fig. 28 GA calculate the evacuation path with the first time series data, and some data base of the traffic network are unhindered. But in the sixteen time series the data of traffic network get more resistance. So the optimums of evacuation path get a more distance like Fig. 29.

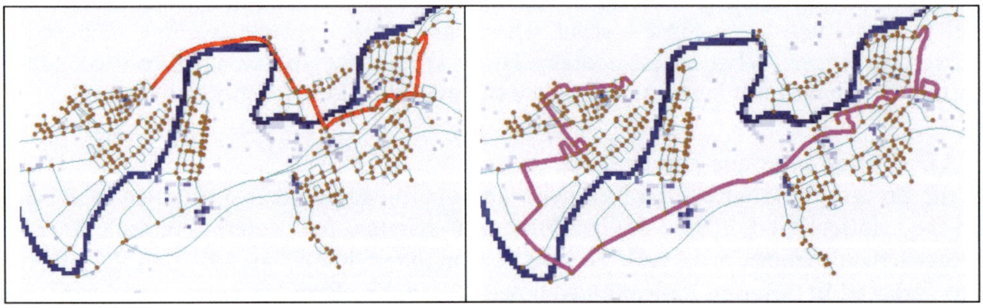

Fig. 28 & Fig. 29. Population Results and Progress Graph

We use the dynamic program to calculate the sequence evacuation paths in different time series. If we set up the more decision nodes, we will get the more real Dynamic Evacuation Path. With the different data base of time series, we divided the time series into three parts. At first, we set up the same destination. We use the data base of the first hour. And it gets the first part of evacuation path like Fig. 30. Second, we try to set up the traffic node to be the first decision node in the first part of evacuation path. Third, we use the fifth hour data to be the second time series. And calculates the evacuation path from the first decision node and get the second part of evacuation path. Forth like Fig. 31, we set up the second decision node from the second part of evacuation path, and use the data base of ninth hour to be the third time series. We use GA to calculate the evacuation path from the second decision node and get the second part of evacuation path like Fig. 32. Finally, we combined with the three parts of evacuation path to be the Devacuation path like Fig. 33.

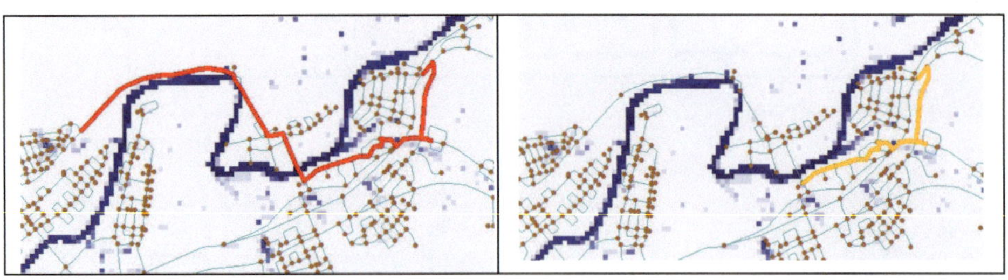

Fig. 30 & Fig. 31.The evacuation path of the First Time Series and the Second Time Series

4.2.6 Comparative the evacuation path of NA and GA

In this study we get the different evacuation path by using the NA and GA calculations. The evacuation path of the NA is depending with the least accumulative cost by deliver time. So the simulation of evacuation path choices the fast moving path which is not depends on the least distance. The evacuation path of NA is green color in the Fig. 34.

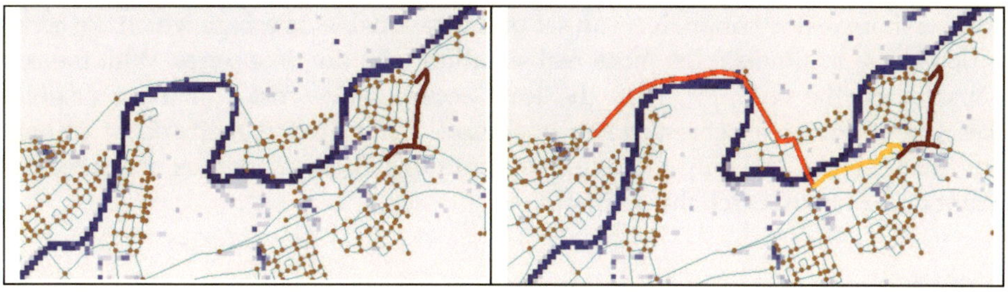

Fig. 32 & Fig. 33 .The evacuation path of the Third Time Series and the dynamic evacuation path of GA

The GA searches the optimum by coding, the weight of the data base of the traffic network and penalty function. So the simulation of the GA's evacuation path in some path avoids the depth of flood. The evacuation paths of GA are brown, yellow and red color in the Fig. 34.

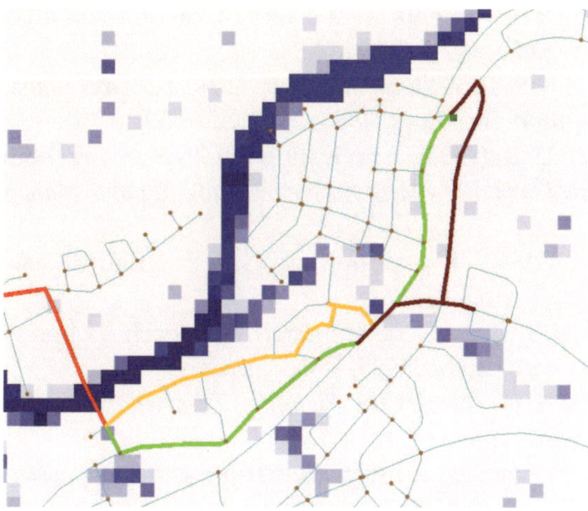

Fig. 34. The Comparative the evacuation path of GA and NA

5. Conclusion

In this study, we use the spatial information, systematize, and escape behaviour theory to establish the zoning of Hazard prevention. And compare the spatial information and some data of facility. By using this number we can understand the plan of the place. We just treat the shape of the zoning Hazard prevention, some area should regulate in some spatial objects to conform the more real situation. Also, we establish disaster databases to proceed with case study and bring up the preliminary analysis result, Combining GA and GIS to deal with the dynamic time space data, we point on the different selections of the path with the GA and NA, and the simulation can offer the better hermeneutic capability to process dynamic flooding evacuation path modal. We constructing the database of dynamic time and spatial and the pattern of analyzing evacuation path, and to propose the method of combination further, and analyze the process of the combination of spatial and time information. Using dynamic program to simulate the evacuation path by calculating with the different time series with these decision nodes which are in the traffic network can provide the more real situation. NA can set up more suitable data base which are according to the flood data to simulate the more real situation with the time series. With the suitable data NA search the optimum with the least accumulative cost will more flexible. GA searches the optimum by chromosome operation. The different methods of coding and penalty function may make up the different optimums. So taking a look at the methods is important operation to search the optimum.

6. References

Blanco, A. ; Delgado, M. & Pegalajar, M. C., (2000). A Genetic Algorithm to Obtain the Optimal Recurrent Neural Network, *International Journal of Approximate Reasoning*, pp.67-83.

Breaden, J. P. (1973). The Generation of Flood Damage Time Sequences, *University of Kentucky Water Resources Institute Paper*, NO.32.

Bullock, G. N. (1995). Developments in the use of the genetic algorithm in engineering design, *Design Studies*, 16: 507-524.

Chan, K. C. & Tansri, H. (1994). *A Study of Genetic Crossover Operations on the Facilities Layout Problem*, Computers Ind. Engr. 1994, 26(3): 537-550.

Djokie, D. & Maidment, D. R. (1996). Application of GIS Network Rountines for Water Flow and Transport, *Journal of Water Resources Planning and Management*, ASCE, 119(2): 229-241.

Gen, M. & Miller, L. (1997). Foundation of Genetic Algorithms, *Genetic Algorithms & Engineering Design*, pp.1-41.

Jo, J. H. & Gero, J. S. (1995). A Genetic Search Approach to Space Layout Planning, in *Architectural Science Review*, 1995, Vol.38, pp.37-46.

Li, W. (1997). *The layout of Taipei City Planning Disaster Prevention System*, R.O.C. city planning academic association.

Li, W. (1999). *Study on the functions of urban disaster-prevention of physical- environment in a city though comparing with the urban disaster prevention system* (Ⅱ), Architecture & Building Research Institute Ministry of interior, Research Project report, Taipei.

Tanaboriboon, Y. & Guyano, J. (1989). Level of Services Standards for Pedestrian Facilities in Bangkok: A Case Study, *ITE Journal*, pp. 39-41.

Tseng, M. & Chen, S. (2000). *A study on the evaluation methods of the emergency routes in the urban area* (Ⅱ), Architecture & Building Research Institute Ministry of interior, Research Project report, Taipei.

Woodbury, R. F. (1993). A Genetic Approach to Creative Design, in *Modeling Creativity and Knowledge-Based Creative Design*, edits Gero, J. S. and Maher, M. L., pp.211-232.

8

Adding Value in Construction Design Management by using Simulation Approach

Hemanta Doloi
The University of Melbourne
Australia

1. Introduction

This chapter focuses on a technique for integrating upstream and downstream project information from the conceptualisation, planning and implementation to the operational phases of projects. A new perspective for adding value in design management practices has been presented by emphasising a whole of project lifecycle approach. An appropriate mechanism for supporting design management practices at an early stage of project is crucial in terms of adding value over scope, time and total investment decisions. Simulation technology acts as a vehicle for analysing the strategic change management of the projects' scope and helps fine-tuning the dynamic business environments.

Increasing complexity and sophistications in construction create new challenges in design management practices. The clients are not only interested in value for money in relation to the investment in project development but costs associated in operation and maintenance over project life cycle as well. While the client's interests may be profit driven in the competitive market, the design professionals have to understand the commercial aspects in terms of design innovations, sophistications and cost effectiveness of the project. Coping with these challenges requires a full understanding of the wide variety of design parameters and technical expertise of each party to deliver the project as per original project objectives.

Most project fails due to an inadequate definition of project objective at the early stage of the project. Due to involvement of various stakeholders in the decision making process, the public sector projects are even more vulnerable compared to the private sector projects. Increasing complexity and requirements for continuous improvement of capital projects exert further constraints for adding values in both construction and project management disciplines in the competitive global environment.

Within the construction industry, there is a definite trend towards outsourcing specialise work to subcontractors, and thereby pushing the liability from one party to another. As such, with each construction project, the need for good design management and appropriate design communication between the designers, the main contractors and subcontractors is becoming increasingly important. Various methods of design management have been emerging with technology, to increase efficiency and reduce the costs and incrased values. Computers/IT has become a huge influence in this regard. The outsourcing of the design has also become a cheaper and more efficient approach to construction industry. This increases the need for efficient design development, effective design quality, information sharing and dealing with constructibity issues in deliverying the projects. The increased trend of procuring public projects with Public-Private partnerships (PPP) procurement methods, such as schools, roads, social infrastructure etc. requires furhter attention on value for money outcomes in projects. Under the PPP contract, the contractor's resposibility extendes over substantial period of project life cycle and the impacts of design and the performane of overall project filter down to the subcontractor, engineers, architects, consultants and project end users. This greatly influences on the upstream design management process for meeting or exceeding expected benefits of project downstream. Based on research undertaken by the author over last eight years, it has been evident that the simulation is one of the best options in adding value in design management practice and to sustain in the emerging complexity in competitive project environment.

2. Objectives

Poor design management practice often leads to confusions and conflicts in complex engineering projects. Innovations in engineering design, construction and operational processes along with increasing regulations have significant contributions in resulting complexity of projects (Nicholson & Naamani, 1992). This chapter portrays how an appropriate analysis of design at an early stage and proactive management practices increase chances for adding values in projects from the operation and end users perspectives. An integrated design management framework has been presented to holistic evaluation of project selection and investment decisions based on functionality and operability of the end facility over operational phase of projects. In the evaluation process, selection of design configuration must enable meeting the target associated with business and strategic objectives of the organisation. A thorough analysis of these objectives is an important requirement to determine the optimum project selection from the available competing alternatives. Simulation based project evaluation and decision analysis adds significant value in evaluating such alternatives by reducing uncertainties in design, implementation and operations with a greater confidence (Jaafari & Doloi, 2002; Doloi, 2007).

Use of process simulation technique assists in analysing feasible design solutions based on technical, functional and operational aspects of projects. Simulation techniques allow design of mathematical-logical models of a real world system and experimentation with different alternatives digitally. It provides a basis for real time scenario analysis by analysing process level decisions at a lower level in the project hierarchy followed by the evaluation of conflicting criteria for making holistic decisions at the project level. A new design management framework, dubbed as Lifecycle Design Management (LCDM) has been discussed with examples where a set of lifecycle objective functions (LCOFs) are employed as the basis for decision making to determine the optimised solution throughout the project's life.

3. Life cycle management

Generally, life cycle management refers to management of systems, products, or projects throughout their useful economical lives. Projects pass through a succession of phases throughout their lives, each with their own characteristics and requiring different types of management. There is no complete agreement on the identification of these phases but they usually entail the following, as described by Morris (1983):

1. *Conceptual phase* - where projects are first identified and feasibility is established (financial, non-financial, and technical). This phase is subject to high-risk levels and should be examined before detailed planning. Consequently this stage includes the analysis of alternatives, development of budgets, setting up of a preliminary organisation, definition of size and location (facility site), and arrangement of preliminary financial and marketing contacts;
2. *Planning/design phase* - when all work from the conceptual phase is detailed and produced further. All major contracts are defined, and prototypes may be built;
3. *Execution/implementation phase* - when plans developed in the previous phases are turned into reality. At this stage, the number of people and organisations involved would have increased, requiring a redefinition of the project organisational structure. Estimation is replaced by performance monitoring. All construction works and major installation activities are completed; and
4. *Handover and start-up phase* - when installation is completed, final testing is done, and resources are released for the start of business operations.

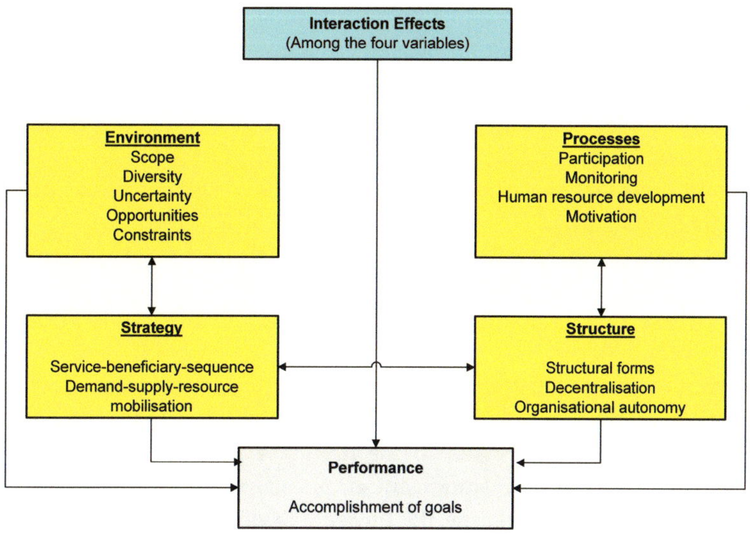

Fig. 1. Key Variables and Performances

In practice, normally these phases overlap. At the end of each phase, the project can progress forward or backward (i.e. a recursive process) depending on the amount of information gathered, produced and utilised (PMBOK, 2004). In LCDM approach as discussed in next section, the project life cycle has been extended to cover the operation and maintenance and disposal phases as well. All these phases are influenced by external and internal variables over the project life cycle (Paul, 1982). Paul (1982) identified four key variables influencing a project in his project management view. As shown in Fig.1, the four key variables are *environment*, *strategy*, *structure* and *process* (Paul, 1982). The interaction among these variables affects the project performances over the entire life cycle. The adequate interventions to these four variables of the project, and according to the specific type of project and environment, project performance can be positively influenced. It is clear that a design management approach requires well-defined strategic objectives, as highlighted in the following sections.

4. Lifecycle design management (LCDM)

Design professionals and project managers are involved in each phase of the project life cycle that entails distinct activities and skills. Failure to properly address the design issues and their underlying impacts over successive phases of the project life cycle can jeopardise the ultimate success of the project. In typical project delivery approach, there is a heavy concentration on the analysis of design and setting objectives for success in terms of three main parameters: time, cost and quality. Time with respect to project start and finish dates, cost with respect to cash flow and the project budget, and quality with respect to pre-defined standards and specifications laid down by the client or the relevant classification society.

LCDM installs a set of business and strategic objectives for decision making throughout the project life cycle in place of the traditional project development protocols. It employs an integrated and concurrent design management approach to substitute the process-based and activity-driven traditional management approach (illustrated in the current practice) for innovative strategy-based and outcome-driven project outcomes. LCDM components comprise:

- A culture of collaboration based on strategic partnership and unity of purpose;
- A life cycle philosophy and framework and an integrated single phase approach;
- An integrated project organisation structure and real time communication system among the design professionals;
- An integrated design management system linked with project information and development systems ; and
- A set of project strategic objectives, known as Life Cycle Objective Functions (LCOFs) for assessing and evaluating holistic project outcomes based in downstream operational conditions. These LCOFs are usually derived based on the Triple Bottom Line (TBL) principles (Doloi, 2007).

Fig.2 represents the perspective that Lifecycle Design Management (LCDM) takes, as opposed to the perspective adopted by the traditional design management practices. As seen, the LCDM framework embraces all the life cycle phases from conceptualisation to demolition (re-cycle) phase with a significant emphasis on the operation and maintenance phase. Such holistic view encapsulating the lifecycle in design management is a major shift in the new LCDM approach.

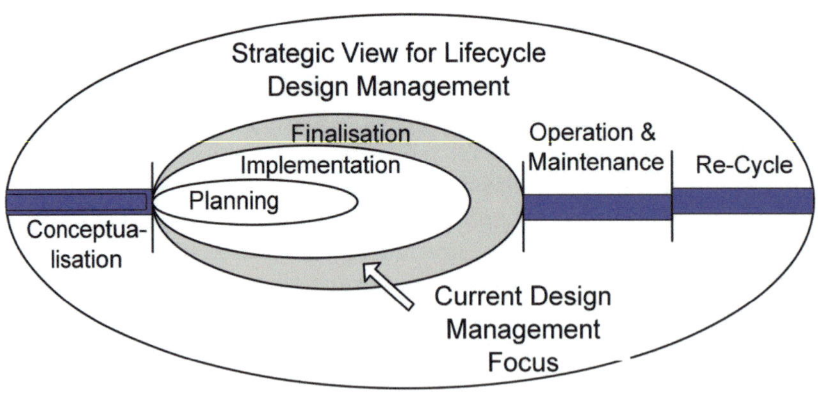

Fig. 2. Lifecycle view of design management

5. Importance of design management

Design management is a leadership activity, focused on managing the creation of an entity. An entity may be an object (motor car, building, etc.), an event (wedding, conference, etc.), a concept (such as the theory of relativity), or a relationship (such as that between employer and employee). Based on this definition of "entity", literally anything can be the focus of design management. The design manager's role includes establishing and clarifying a shared vision of the entity, defining, acquiring and allocating the resources needed to create this entity, managing the effective use of those resources, and monitoring the design team's performance (Chaaya & Jaafari, 2000).

"Design management" and "design managers" are popular expressions in most industries except the construction industry where they have been realised relatively late (mid 1980's). There is nothing innovative in the notion of design management. However, the separation of powers between designers and design managers is clearly a new synthesis in design management practice (Berk, 1994). In the construction industry, the architect used to be at the same time architect, project manager, cost manger, design manager, principal consultant and the undisputed leader of the building procurement team. Specialisation and evolution of professions led the way to a variety of consultants doing much of what architects used to do, including now the design management services.

"We are witnessing a fast migration of the value of architectural services from strictly Information Creation to the incorporation of Information, Management and Distribution. Over 25 years ago, architects gave up certain risks, rights and responsibilities of construction supervision and a new profession emerged to fill those needs of the client, the Construction Manager. Construction management has blossomed into a profession that most projects use today. We are seeing history repeat itself as most architects and other design professionals are fast losing control of their main asset, their information" (Cyberplaces, 1998).

Separating design from management is not a straightforward task since design is a process of decision-making and decision-making is a key process in management. Decision-making often involves defining a list of objectives, analysing the information, considering the alternatives, assessing the consequences of the options, judging the risks, costs, penalties and bonuses, and selling the decision. These steps are naturally reflected in management. Hence, a good designer is envisaged as a good manager and it is often concluded that bad designers are bad managers.

If it is acknowledged that design management is neither a process of managing a design consultancy or practice, nor the education of designers about the importance of the management world, then the importance of defining design management becomes apparent. Throughout this chapter, design management is defined as the effective deployment by the project management team of the design resources available to them in the pursuance of the overall project and business objectives defined at the outset of project.

The growth in new knowledge and increased customer focus has increased the design complexity in projects. Customers no longer simply settle for generic product but want customised product design and services that cater for their ever increasing needs. In today's digital age with an ever growing of consumers' appetite for more sophisticated products and services, increasing product complexity significantly impacts on design management practice. The need to integrate diverge technologies, and thus project management, has emerged as an important discipline for achieving these objectives. The functionality of new production systems to service the changing markets is crucial in responding to shorter product life cycles and market dynamics. The definition of a product directs the added knowledge in scope management, and provides challenges for operative tools that are designed for putting the component parts and processes of the project together (Jaafari, 2000).

The need for better design management in the architectural, engineering and construction (AEC) industry has never been so high. This is due to emerging factors that reflect both changing market conditions, advent of new materials and new procurement processes (Nicholson & Naamani, 1992). To maintain profit margins, the industry needs to focus on the improvement of the design process, especially to cope with tougher competition and tighter fee scales.

Capital projects have necessitated design input from an increasing range of specialists. The increased emphasis for keeping the construction projects on time and within budget has required effective management of project scope associated with multifaceted stakeholder groups in the project (Cleland, 2004). Thus, definition of project's scope in the concept phase vastly influences the project development and its overall business outcomes. Understanding the complexity of design in both functional and operational contexts at the early stage is important in defining appropriate facility of the project.

The primary objective of this chapter is to discuss how to enhance the project's operational performance and increase project's business outcomes from an effective design management perspective. Inherent in this issue are the several sub questions such as: 1) how does the design management impact on setting a benchmark on appropriate project management practices? 2) how the process simulation approach can be used for integrating operational processes and managing design complexity upfront? 3) what will be the consequences of applying project simulation in decision making and overall business outcomes?

Focusing on the above questions, author's research resulted in a new model of project design management that can deliver a view and an understanding of the strategic objectives of projects in a proactive and explicit manner. Process simulation is employed for evaluating operational performances and managing the process complexity at the early phase of the project. Simulation based project evaluation and decision analysis allows evaluating project alternatives by reducing uncertainties with a greater confidence (Artto et al., 2001; Puthamont & Charoenngam, 2007). The approach provides a platform for real time project definition based on technical, functional and operational aspects of projects.

6. Proactive design selection and project performance

Many organisations have found design to be the key to project success in meeting growing and changing conditions. Growing pressure on design innovation and timely delivery is a fact of life for project managers and architects (Heath et al., 1994). The design phase of a project offers the greatest scope for reduction in overall project costs and adds maximum values in the project. The size and complexity of modern design with increased uncertainty requires front-end planning throughout the life of a project. Design management is an incremental continuous iterative process and as the project moves on, it provides feedback points for new information and the flexibility to assimilate and act on it. Thus initial design and planning must concentrate on building viable project bases for each principal subsystem in the context of life cycle planning of projects (Cleland, 2004). In the case of strategic planning, one takes a set of fixed interests, juxtaposes them within a fixed environment (or world, or set of conditions), and then invents a strategy for attaining one's interests given the constraints imposed by the environment (Doloi & Jaafari, 2002).

Current project management philosophy tends to concentrate on the delivery processes and associated functions of contractual scope, time and cost management (Jamieson & Morris, 2004). Traditional design selection and investment decisions in projects are based on static and simplified assumptions regarding the functionality and operability of the production processes. Economic analysis, reflecting the final customer's or investor's life cycle costs is important during decision making, particularly at the early phase of projects (Jaafari, 2000). This is because solutions devised and commitments made at the early phases constitute a major part of the downstream project costs. Modelling of technical and operational functionalities of the end deliverable supports strategic decision making in the early phase of the project. Thus, appropriate design and optimal scope definition considering the entire life cycle are the key for overall project success.

7. Design complexity and process simulation

In recent years, the concept of a modelling has become increasingly important in engineering design management practices. It is no longer sufficient to pay detailed attention to the design of the various elements of a project individually, rather, all elements must be considered in relation to others in order to make the overall system effective. However, good project design is not restricted to detailed design coupled with attention to interrelationships between physical parts and elements. Design must be analysed and evaluated at a deeper level and in relation to the project's operational environments (Cleland, 2004; Doloi, 2007). Design configuration and scope of projects must reassess and readjust to ensure that the objectives are met at the end. As a result, the overall process to reach these goals becomes iterative, involving in the design of each of the parts and products, which constitute the overall project. Simulation approach allows building a model of the proposed system capturing the salient features of the overall system.

Digital computer models facilitate analysis of complex processes associated in projects. A simulation model is a means for collecting information about the likely performance of a system, based upon user-defined conditions (Marmon, 1991). Simulation models can improve the planner's understanding of the real life situation during conceptualisation and final design or actual construction (Luk, 1990). By using the simulation model, the effect of changes in process design can be justified and fine-tuned and investment decisions are optimised over the project life cycle. The life cycle project management (LCPM) model is indeed capable of responding to the global challenges and achieving the true value on investment in the integrated project development.

8. Project development in design management context

A typical project life cycle includes phases such as feasibility, planning and design, execution, commissioning and handover (PMBOK, 2004). As revealed by Artto et al. (2001), the investment project phases are preparation, execution and operation, whereas the phases associated with the post project implementation are sales and marketing, execution and after-sales services. In front end planning, the investment project phases must be integrated with the post project implementation phase (Shi & Abourizk, 1998). Fig.3 depicts the links of three board criteria over project life cycle phases. As seen, the three broad criteria associated with project investment are Risk and Uncertainty, Financial Objectives and Facility Performance. It is important to understand that the impact of the technical and operational functionality of the final deliverable on the end users is an important parameter that contributes to the benefits obtained from the investment (Artto et al., 2001; Dikmen et al., 2005). All these three criteria should be analysed upfront before making the final decisions on project investment and development.

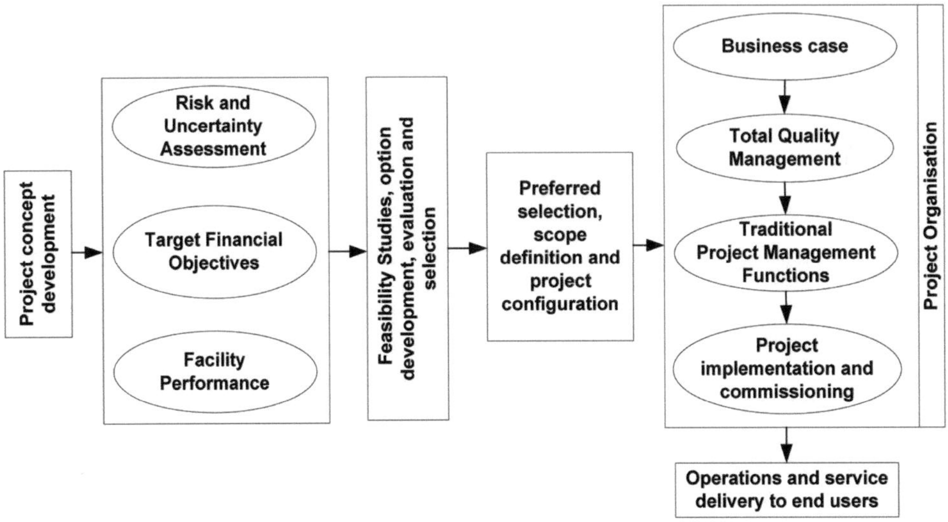

Fig. 3. Broad phases in project development

The three criteria are highly interrelated from the project's end product performance point of view. The role of total quality management (TQM) along with the traditional project management functions intrinsically governs the project development process in delivering the end product. Thus, the scope of the project is the sum of products and services produced in the project. The term 'project product' is used as a synonym to scope of the project. The purpose and benefit of project is realised only when an appropriate scope configuration is achieved. The process includes aspects of 1) quality of the project product; and 2) performance, functionality and technical characteristics of the project product (Jamieson & Morris, 2004). The implications of the scope definition are that the project scope management should focus on fulfilling individual needs of the end users of the project.

Decisions and information generated over feasibility (or conceptual design) and planning phases of projects have a great impact on the downstream activities and consequently on the overall cost (Artto et al., 1991). Understanding the project and its underlying processes, supported by relevant information and tools leads to better decisions on projects. Integration of implications of investment on product life cycle with project development cost is an important consideration in front-end planning of project (Laufer, 1999). Thus the validity of the hypothesis that the contemporary project management approach embodying process simulation technique helps proactive decision making on optimal design, scope definition and overall operating processes to achieve optimality across all phases is a significant advancement in the LCDM concept.

9. Process simulation and decision making in project lifecycle

The simulation is a numerical technique for conducting experiments on digital computers involving certain types of mathematical and logical models to describe the behaviour of a system over extended periods of the real time (Pidd, 1984). During the last decade, discrete event simulation has gained a significant role in engineering planning and design (Doloi & Jaafari, 2002). Numerous examples reported in the literature, provide evidence how organisations can save millions of dollars and avoid major risks using process simulation (Irani et al., 2000). For instance, in early 1993, the IBM PC Company in Europe faced a number of challenges that were eroding its market share, such as frequent price cuts, rapid customer order response times, and a steady arrival of new products by aggressive competitors. The IBM management reacted to record corporate losses by emphasising the necessity of reducing operational costs and inventory throughout the company. The process simulation technique was used to evaluate different manufacturing execution strategies and to identify the lower-cost distribution policies. A strategic distribution policy was adopted based on the analysis of alternative scenarios which resulted an estimated $40 million per year savings in the distribution costs of the company (Artto et al., 2001; Kirkham, 2005).

The research on how the discrete event simulation works is not embryonic. Development of computer-aided process simulation techniques have accelerated in recent years. However, its use for project definition, management practices and life cycle investment decisions is not widespread (Doloi, 2007). The application and influence on setting the benchmarks for management practices within the complex project management framework has proven to be a significant contribution in this research. Table 1 shows how the simulation can be applied as a tool for appropriate front-end management of respective objectives over the project life cycle. As seen, most of the project objectives and the decision making subjects have a natural link to the process simulation outputs.

Definition and effective management of project scope, as well as management of the investment life cycle incorporating the dynamic considerations of the market and customers needs is a challenge within project management practice. Furthermore, simulating an individual process within a project does not add significant value for the evaluation of project level decisions in real life situations. Thus an integrated model embodying simulation capability within the hierarchical project structure simplifies the task of project managers for making strategic decisions on complex projects (O'Kane, 2003). The framework facilitates strategic decision making by defining facility characteristics and improved process design on fluctuating operational environments over the entire life of projects.

10. Project decision framework

Given the increasing use of computers as management and evaluation tools, it is natural to consider their potential applications to design information management. Much valuation work has already been done on the application of computers to understand and modelling design processes and mechanising design tasks. The attempt to reduce design complexity, increase functionality, clarity and constructability at an early stage has now been the focus among researchers in the field. Selection of an appropriate design and configuration of operational processes of project facility is an important consideration in competitive project development environment. Project level decisions are greatly influenced by the feasible alternative designs and their consequences (Goldschmidt, 1992).

Life Cycle Design Management (LCDM), as subset of the Life Cycle Project Management (LCMP) is an approach for integrating business and strategic objectives of projects throughout the project life cycle phases (Doloi & Jaafari, 2002; Jaafari, 2000). The LCPM approach employs an integrated and concurrent project management principle to substitute the process-based and activity-driven approach in the project management paradigm. Much work has already been published LCPM methodology in project evaluation and management contexts (Doloi & Jaafari, 2002; Jaafari, 2000).

Project objectives	Subjects for decision making	Usability of simulation tools in front end management
Project concept development	- Market need analysis - Project option analysis	Supply-demand planning, optimum utilisation of resources
Project facility planning	- Decision process for project development - Product design - Project Management functions	Capacity planning and scope definition
Project implementation	- Scope control and management - Time management - Cost management	Constructability analysis, change control and alternative planning
Project operation and maintenance	- Market economics and changes - Facility operation and flexible production	Project functionality and operability of the end project product
Sales and Marketing	- Market consumption - Customers satisfaction and acceptance	Supply-demand analysis, evaluation of logistics
Research and Development	- Product design and redesign - Product innovation and process reengineering	Simulation model for 'what-if' analysis, process reengineering.
IT/IS support	- Process automation and optimum facility utilization - Waste reduction, cost minimization	Simulation model for evaluating facility utilisation, activity-based costing
Project Organisation	- Resources and skills requirements and utilisation - Self managing teams and cross cultural integration - Key performance measures and controlling - Risk resilient and uncertainty management - Change management	Simulation model for resource planning, resources levelling and optimisation

Table 1. Project objectives and front-end management tools

Fig.4 depicts an overall design management framework embodying the phases over project lifecycle. As seen, selection of design alternatives and investment decision has direct influence on the strategic project objectives and overall performance of projects (Irani et al., 2000). Thus the project's design and their underlying capability should be defined integrating optimum project's configuration and inherent business intents.

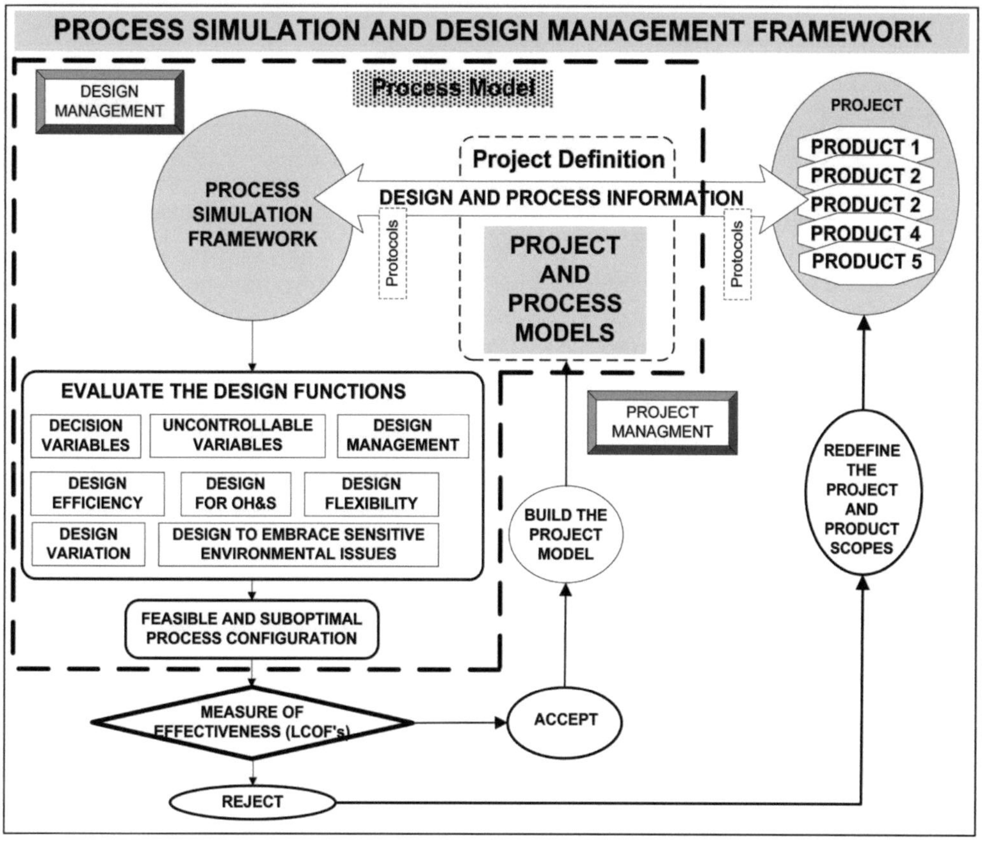

Fig. 4. Process simulation and measure of the project's effectiveness

Once the initial decision on a feasible design is made and project products are established, the underlying processes are identified for analysing feasible alternatives, selection and allocation of appropriate resources and establishment of the best project option for development. The processes of analysing alternative product configuration and selecting best project option are facilitated by the simulation technology. The projects are broken down in smaller products and process models are constructed incorporating operational scenarios for simulation analysis (Doloi & Jaafari, 2002). The outcome of simulation forms the basis for evaluation of the suboptimal configuration against the target LCOFs of the project. After the project is developed and commissioned, operation is monitored based on the performance on LCOFs, organizational strategy and competitive advantages. The dynamic scanning and assessment processes are then continued in the project operating environment.

11. Framework for simulation analysis

The simulation assists management on analysing the functionality and operability of project deliverables by focusing on the business objectives in the early phase of the project. The platform allows a real time project definition based on technical, functional and operational aspects of the project (Doloi & Jaafari, 2002).

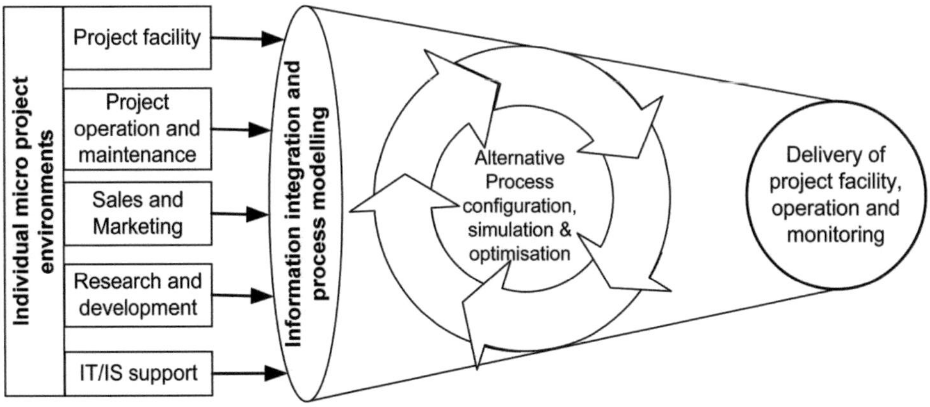

Fig. 5. Integration of functional disciplines within project operation

Fig. 5 shows how the micro project environment and their functional disciplines are scanned and relevant process information is integrated over project life cycle. Hierarchical process models are built and simulated by linking the processes and allocating available resources across all disciplines. Alternative processes are identified and tested for optimal project configuration. The project level decisions on operability, functionality, quality or performance issues are then optimised using a set of criteria known as life cycle objective functions (LCOF) (Jaafari et al., 2004).

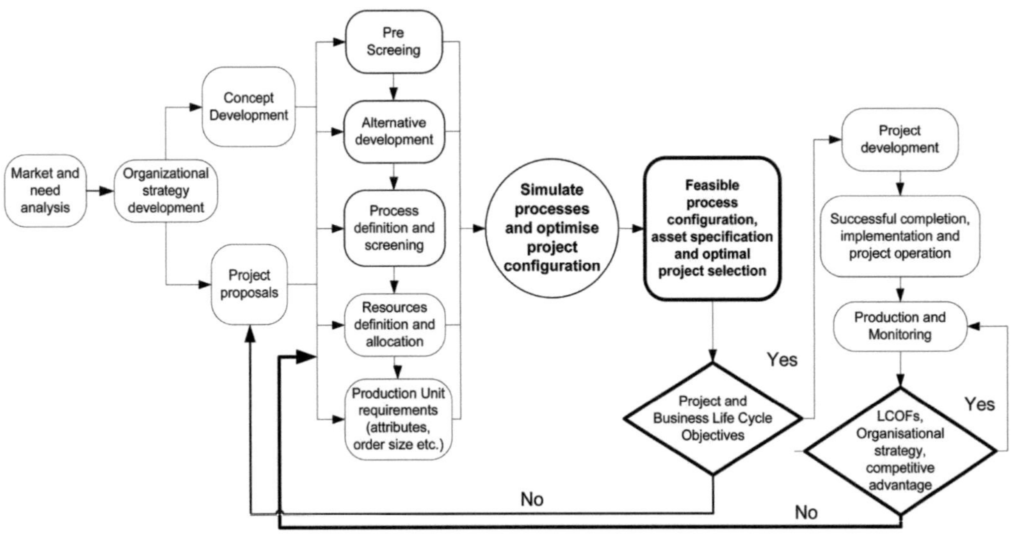

Fig. 6. Framework for life cycle decision analysis

Fig. 6 depicts the overall decision process over the life of the project. Project investment decision and organisational business intents have direct influence on the strategic planning and development of the project (Yeo, 1995). The project concept and alternatives are then identified and resources and product specifications are defined for feasible project solution. The outcomes from simulation modelling on project configuration, operational requirements and resource utilisation are used as input for analysing required management capabilities and transformation for a project specific environment. Continuous assessment on the functionality and operability of the project product and feedback mechanism allow dynamic interaction and evaluation of the project's performance over the life cycle of projects (O'Kane, 2003).

12. Simulation enabled design management – a practical example

In order to demonstrate the use and benefits of the process simulation, a case study is presented in this section. The simulation model representation provided a key decision making platform that quantified the effectiveness of varying level of design and planning to support an optimum operational plan. A significant implementation challenge during a planning level study can be understood from the analysis flow chart shown in Fig. 7.

The selected project was a commercial Ductile Iron manufacturing plant (named hypothetically as XYZ manufacturing plant) located in a regional area of Sydney in Australia. The manufacturing plant was due for a major overhaul for which a front-end decision analysis was quite appropriate to support the strategic management decisions. The ability to quantify the impacts on alternative process design is a huge benefit of using a simulation model. Once the design is altered to suit the required service requirements, the project's life cycle objectives are assessed and validated. The framework presented provides the functionality of make such changes and adjust related variables at project levels impacted by the changes.

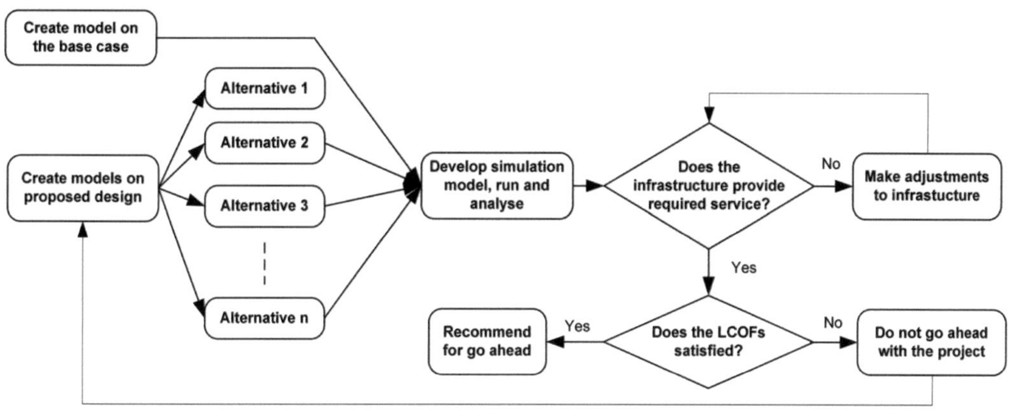

Fig. 7. Typical planning level analysis

12.1 Client project brief

The XYZ manufacturing plant commenced production in 1962 making grey cast iron pipes and was later converted to ductile iron pipe manufacturing in 1976 to take advantage of superior mechanical properties of ductile iron. Ductile Iron Pipeline Systems represent significant improvements in terms of waste recycling. Pipes and fittings are manufactured from 100% scrap steel. Raw materials used in production are selected scrap steel, ductile iron returns, ferro silicon, coke, limestone and fluorspar. Thus steel scraps are converted into valuable assets using less energy and thereby minimizing greenhouse gas emissions during the manufacturing process.

The ductile joint pipes in XYZ manufacturing plant are produced by the centrifugal casting process to a standard length of 5.5 meters in diameters of 100mm to 800mm. The overall project can be described in terms of major processes from the crane operating in the scrap storage area feeding the raw materials onto a conveyor to the final production of pipes after undergoing hydrostatic pressure testing before going through weighing and inspection processes.

12.2 Target production, budgets, and LCOFs

The main stakeholders for the XYZ Manufacturing plant is Tyco Water and the targeted customers, who are both local (40%) and overseas markets (60%). The use of the ductile iron pipes is mainly for transportation of potable water and sewage. Ductile iron pipe standards for the domestic market are the Australian/New Zealand Standard AS/NZS 2280 and for the international market is the British/European Nations Standard BS/EN 545. The total investment on assets in present worth terms is about 100 million dollars. Yearly turnover was not disclosed due to the competitive market. However, it was known that the 60% of the overseas market share was not producing any profit to the company but meeting the running cost of the plant. Current project facility and management capability have long been under increasing scrutiny for its strategic existence in the global business environment. Fig. 8 shows the current trend of utilization of the project facility and resources throughout a calendar year. According to the production manager, the plant is currently running at about 80% of its capacity on average due to falling market demand. However, there is an increasing threat for plant breakdown and higher maintenance cost due to aging facilities in the plant. The simulation study was conducted to see how the overall facility and the existing project business could respond to variable demands and how to make best use of the exiting facility optimally. Table 2 depicts the target LCOFs derived from the available financial data used for decision making at the project level (refer to Fig. 6). The target equity internal rate of return of 30% is the focus of all the decision making on this project.

12.3 Simplified case data and analysis

The case study processes have been designed in order to understand the operational context and utilization of existing facilities. Various products and major processes have been identified from information provided by the production manager and onsite data collection. It is worthwhile to mention that among many functional disciplines within the micro project environment, only the plant operation has been considered for simulation here. The plant produces a number of different size pipes on demand. Production rates vary with internal pipe diameter: smaller diameters have faster production rates than larger diameter pipes. For example, 100 mm diameter pipes can be produced at 50 pipes per hour and 800 mm pipes can be produced at 17 pipes per hour.

12.4 Scenario 1: process network

Figs. 9, 10 and 11 depict process network diagrams built on the existing capability, an alternative and the optimised alternative of the plant respectively. Fig. 9 shows part of the model for a few key processes involved in manufacturing the pipes. Overall, there are four lines of centrifugal casting machines with two annealing furnaces. After annealing, testing and finishing processes take place in three parallel lines. The workflow sequencing and connectivity between processes are shown in the figures.

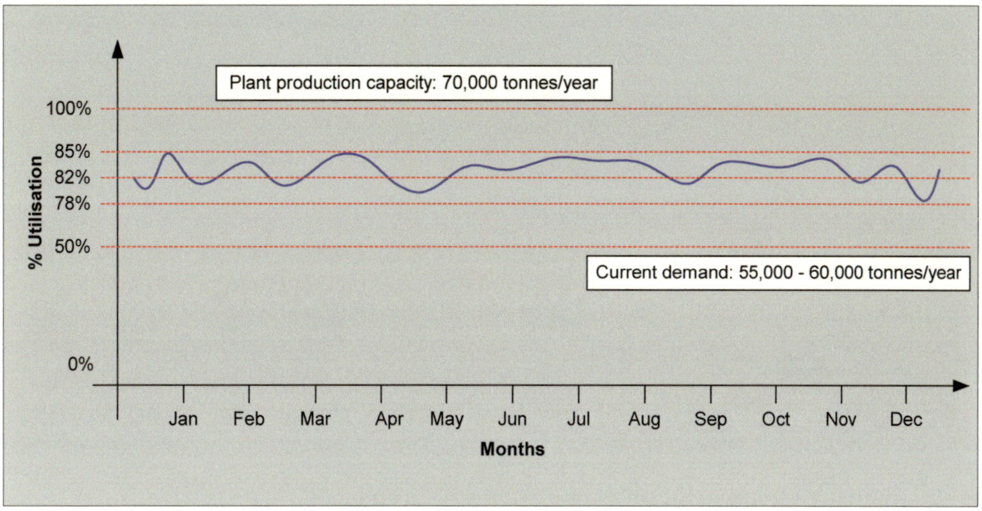

Fig. 8. Current trend of utilization of the facility

LIFE CYCLE OBJECTIVE FUNCTIONS (FINANCIAL)	TARGET
Total Life Cycle Cost (TLCC) $ in present value	A$100 million
Equity Internal Rate of Return (EIRR) %	30%
Net Present Value to Capital Investment Ratio (NPV/C)	1.50
Total Life Cycle Cost (TLCC/Po) $ p.a. to unit production output	Confidential
Cost to Worth Ratio (C/W)	Confidential
Environmental emission standard	Confidential

Table 2. Targeted LCOFs

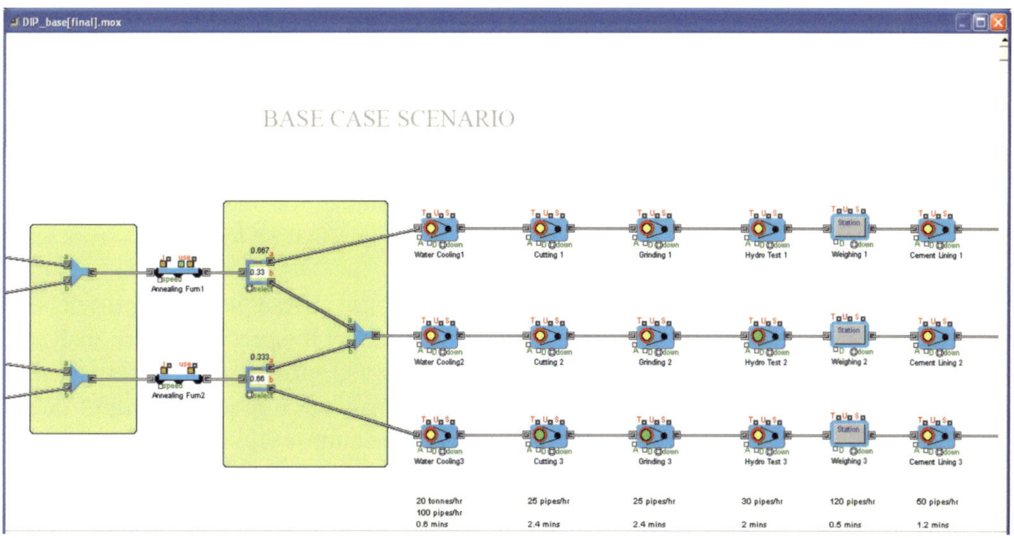

Fig. 9. Base case processes for production of ductile iron pipes

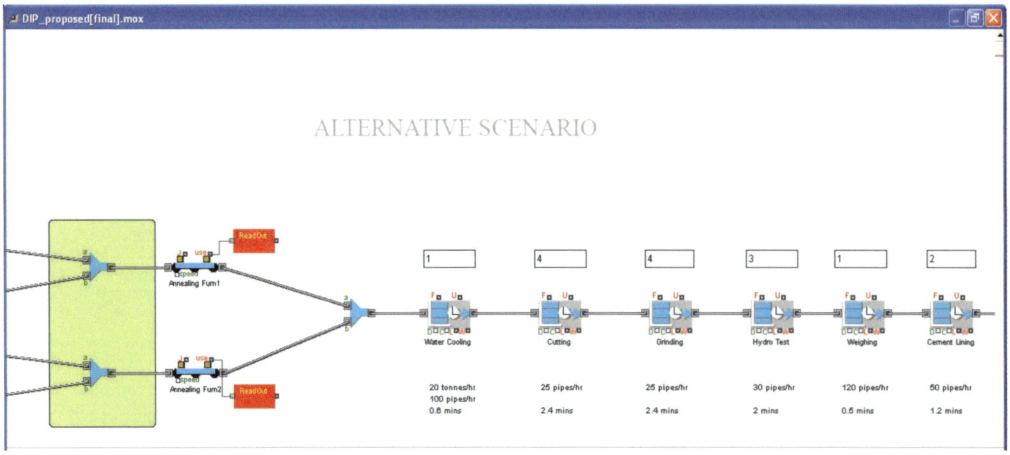

Fig. 10. Alternative scenario for production of ductile iron pipes

12.5 Simulation output and the LCOFs evaluation in scenario 1

Details of the statistical outputs from the simulation run for Scenario 1 have not been presented. The simulation model was run for 500 hrs and the average utilization of processes was found to be about 62%. In Scenario 1, seven processes: water cooling, cutting, grinding, hydraulic test, weighting, cement lining and coating processes were highlighted. It was found that while the first four processes (water cooling, cutting, grinding and hydraulic test) were utilised on average 85%, the remaining processes were utilised less than 30% on average. A severe bottleneck has been experienced near the water cooling and cutting processes.

12.6 Alternative scenario and optimization

An alternative was developed by reconfiguring some of the processes under consideration as shown in Fig.10. In this reconfiguration process, one additional cutting and grinding processes were added while weighing processes were reduced to only one and the cement lining processes were cut down to two. Cement lining processes also have been reduced from three to two as these processes were found underutilized in the base case scenario. Details of the network process diagrams have not been shown for brevity. Simulation was run for the equal time period as the base case and capacity utilization for the processes were recorded.

In order to optimise the proposed design, evolutionary optimization approach was employed on proposed scenario and impact on performance of the processes were analysed. In the optimization process, the modelling parameters were varied and best performance was monitored by defining a range of objective functions. Fig.11 shows an output of the optimiser with approximately 99% convergence for maximum output in the model. The Genetic Algorithm based optimiser produces significantly better operational performance and utilization of the proposed processes over existing situation. The optimiser includes a number of parameters such as the probabilities of crossover and mutation, the population size and the number of generations (Khral, 2002).

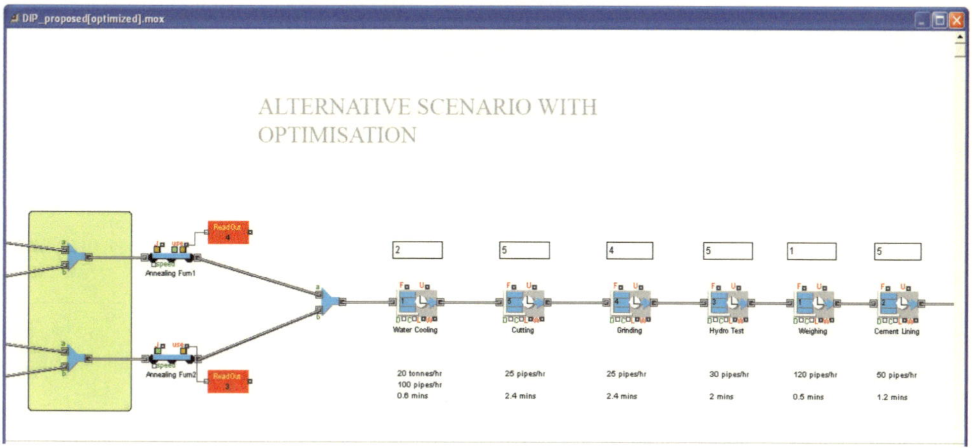

Fig. 11. Alternative case with optimization

12.7 Impact of new process configurations

Fig. 12 shows a comparative analysis of process utilization between base case, proposed and optimized scenarios. The optimized process configuration for maximum output values in

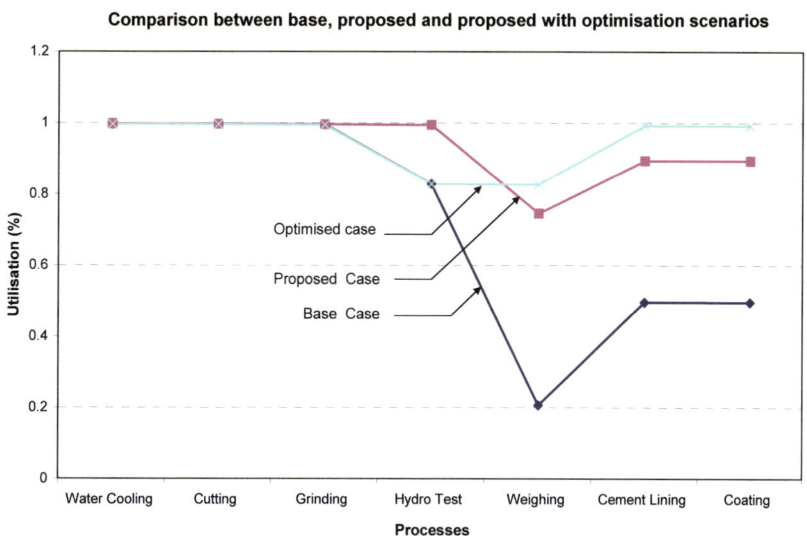

Fig. 12. Comparative impacts of new process configuration

the proposed design over the base case scenario was achieved by increasing the capacity of four processes over the proposed scenario. As seen, there is a good balance with about 95% average utilization of processes in the new optimized design. An introduction of an additional processes along with the alteration of flow sequences on processes have significant impact on overall process performances of the project. It is evident that the capability of the manufacturing facility could be enhanced by altering the current baseline operation; obviously there is a limit to what can be achieved without significant investment in new plant and facilities.

These decisions then investigated in terms of target LCOFs in the integrated framework by using the existing operations as the starting point. Management strategies and require capability are then built supporting the reengineered processes and project operation. As has been demonstrated in this example, the process simulation approach is a powerful tool in achieving this objective.

Feasible alternative scenarios (1)	% Utilization of project facility (2)	% Utilization of operational resources (3)	TLCC* (%) (4)	Strategic Business Objectives							
				Financial (5)		Facility/Asset performance (6)		Market demand (7)		Sustainability & Risk (8)	
				Unit cost (%)	ROI (%)	Waste reduction (%)	Shorter cycle, (%)	Improv e-ment (%)	New custome rs (%)	Sustain-ability (0 - 6)**	Reduced Risks (0 - 6)**
Base case	72	65	100	0	12	0	0	0	0	0	0
Proposed case	93	79	95	10	15	10	15	10	5	1	1
Optimised case	95	87	91	12	19	15	10	20	8	1	2

* TLCC is the Total Life Cycle Cost for scenario under consideration.
** Sustainability and Reduced Risks are measured on an index scale from 0 for no effect to 6 for highly effective.

Table 3. Holistic analysis of alternative project solutions

13. Life cycle decision analysis

Table 3 shows the overall evaluation matrix by integrating upstream and downstream information for optimal management decisions in the project. While all three options in column (1) are assumed feasible, each has an estimated total life cycle cost and a corresponding level of various high level criteria influencing strategic business objectives (LCOFs). Result of simulation analysis provides the input values in columns (2) and (3). Values in column (4) are the result of the traditional life cycle cost analysis using the raw data from Table 2 (Jaafari, 2000). In order to determine the optimal solution, values in columns (5) - (8) are used to see what tradeoffs are available against values in columns (2), (3) and (4). These trade-offs are then analysed using Multi-criteria Decision Modelling (MCDM) technique (Doloi, 2007) to locate the optimal solution among those which meet the target criteria. Details of MCDM techniques can be found in Doloi (2007) and Jaafari et al. (2004).

As already stated, an appropriate conceptual model to facilitate holistic evaluation and management of project's complexity is not currently available. Much work needs to be done to better understand and apply a project-based approach by integrating processes and operations in the front-end management practices. An optimization model was in dire need to evaluate a given operation to show if current processes are in balance within the expected present or future demand patterns while maintaining its business and environmental performances (Doloi, 2007; Cleland, 2004). This chapter has demonstrated an approach that sets a benchmark for an integrated framework enabling management of complex projects. It has shown a way forward in computational aspect of the project management approaches for sustainable project development and management practices.

14. Conclusion

Simulation modelling has been introduced as a decision support tool for front end planning and design analysis of projects. An integrated approach has been discussed linking project scope, end product or project facility performance and the strategic project objectives at the early stage of projects. The case study example on tram network demonstrates that application of simulation helps assessing performance of project operation and making appropriate investment decisions over life cycle of project.

Optimised design and maintenance of physical project facilities in competitive business environment triggers the strategic positioning of the project organisations over life cycle of the project. The preliminary research has identified the key roots of inefficient operations in terms of the capabilities and utilisation of the project facilities and resources and contributed in devising optimal solutions based on life cycle objective functions of the project. The framework assists organisations in their management decisions in respond to market dynamics, customer needs and organisational intents.

In developing the prototype, the process simulation approach has been used in the projects. The simulation based framework facilitates evaluating the functionality and operability of feasible project configuration for strategic implementation. Research by the author reveals that there has been little attempt to assess the link between the physical project's facility and the underlying business capability and ability to respond to market shifts in contemporary project management practices. The concept presented in this research has taken into consideration multiple views of project facility within a business operating environment. Process reengineering or investment decision on the existing facility depends on the target LCOFs of the project. Analysis of alternative project solutions (based on alternative process scope and configuration) rather than focusing on well designed activities for project implementation has significant contribution in supporting decision making and management of future project outcomes.

While for design visualisation, the simulation modelling is immensely valued, project selection and overall investment decisions are holistically evaluated incorporating strategic business objectives in the cycle project model. The simulation based framework put forward provides the engineering assistance in optimizing project's configuration, planning and design and investment decision on capital projects. The ability for quick exploration of the multiple scenarios of significant benefits and the capability incorporating results on design and engineering processes in devising the best possible solution on complex projects are the significant contributions in this chapter.

15. References

Artto, K., Lehtonen, J.M. & Saranen, J. (2001). Managing Projects front-end: incorporating a strategic early view to project management with simulation, *International Journal of Project Management*, Vol. 19, pp. 255-264.

Berk, V. (1994). *The Architect as a Project Manager,* A publication of the University of New South Wales.

Chaaya, M. & Jaafari, A. (2000). Collaboration and integration of project life cycle design information using IT systems, *Proceedings of International Conference on Construction Information Technology*, pp. 277-291, 2000.

Cleland, D.I. (2004). Strategic Management: The Project Linkages, In: *The Wiley Guide to Managing Projects*, J.K. Pinto and P.W.G. Morris (Eds.), pp. 206-222, Wiley, New York.

Cyberplaces (1998). http://www.cyberplaces.com/columns/archive/

Dikmen, I., Birgonul, M.T. & Artuk, S.U. (2005). Integrated framework to investigate value innovation, *Journal of Management in Engineering*, Vol. 21, No. 2, pp. 81-90.

Doloi, H. (2007). Developing an integrated management system for optimising project options, *Journal of Enterprise Information Management*, Vol. 20, No. 4, pp. 465-486.

Doloi, H. & Jaafari, A. (2002). Towards a dynamic simulation model for strategic decision making in life cycle project management, *Project Management Journal*, Vol. 33, No. 4, pp. 23-38.

Goldschmidt, G. (1992). Criteria for design evaluation: a process oriented paradigm, In: *Evaluating and Predicting Design Performance*, Kalay, Y.E. (Ed.), pp. 67-79, John Wiley, New York.

Heath, T., Scott, D. & Boyland, M. (1994). A prototype computer-based design management tool, *Construction Management and Economics*, Vol. 12, pp. 543-549.

Irani, Z., Hlupic, V, Baldwin, L.P. & Love, P. (2000). Re-engineering manufacturing processes through simulation modelling, *Logistics Information Management*, Vol. 13, No. 1, pp. 7-13.

Jaafari, A. (2000). Life cycle project management: a new paradigm for development and implementation of capital projects, *Project Management Journal*, Vol. 31, No. 1, pp. 44-53.

Jaafari, A. & Doloi, H. K. (2002). A simulation model for life cycle project management. *Journal of Computer-Aided Civil and Infrastructure Engineering*, Vol. 17, pp. 162-174.

Jaafari, A., Doloi, H. & Gunaratnam, D. (2004). Life cycle project management: a platform for strategic project management, *Proceedings of Innovations: Project Management Research 2004*, D. P. Slevin, D. I. Cleland, & J. K. Pinto (Eds.), pp. 141-159, Four Campus Boulevard, Newtown Square, 2004, Project Management Institute (PMI), Pennsylvania, USA.

Jamieson, A. & Morris, P. (2004). Moving from corporate strategy to project strategy, In: *The Wiley Guide to Managing Projects*, J.K. Pinto & P.W.G. Morris (Eds.), pp. 177-205, Wiley, New York.

Khral, D. (2002). The extend simulation environment, *Proceedings of the 2002 Winter Simulation Conference*, E. Yücesan, C.-H. Chen, J. L. Snowdon, & J. M. Charnes (Eds), pp. 205-213.

Kirkham, R.J. (2005). Re-engineering the whole life cycle costing process, *Construction Management & Economics*, Vol. 23, No. 1, pp. 9-14.

Laufer, A. (1999). Essentials of project planning: owner's perspective, *Journal of Management in Engineering*, Vol. 2, pp. 162-176.

Luk, M. (1990). Hong King Air Cargo Terminals to work in Synch because of simulation applications, *Industrial Engineering*, Vol. 11, pp. 42-45.

Marmon, C. (1991). Teledyne applies simulation to the design and justification of a new facility, *Industrial Engineering*, Vol. 13, pp. 29-32.

Morris, P. W. G. (1983). Managing project interfaces - key points for project success, In : *Project Management Handbook*, Cleland, D. I. & King, W. R. (Eds.), Van Nostrand Reinhold Company, New York.

Nicholson, M.P. & Naamani, Z. (1992). Managing architectural design - a recent survey, *Construction Engineering and Economics*, Vol. 10, pp. 479-487.

O'Kane, J. F. (2003). Simulation as an enabler for organisational excellence, *Measuring Business Excellence*, Vol. 7, No. 4, pp. 12-19.

Paul, S. (1982). *Managing Development Programs – The Lessons of Success*, Westview Press, Inc. USA.

Pidd, M. (1984). *Computer Simulation in Management Science*, John Wiley & Sons.

PMBOK (2004), A *Guide to the Project Management Body of Knowledge*, Project Management Institute, PA, USA.

Puthamont, S. & Charoenngam, C. (2007). Strategic project selection in public sector: construction projects of the Ministry of Defence in Thailand, *International Journal of Project Management*, Vol. 25, pp. 178-188.

Shi, J. & Abourizk, S. (1998). Continuous and combined event-process models for simulating pipeline construction, *Construction Management & Economics*, Vol. 16, No. 4, pp. 489 – 498.

Yeo, K.T. (1995). Planning and learning in major infrastructure development: systems perspectives, *International Journal of Project Management*, Vol. 13, No. 5, pp. 287-293.

9

Risk and Reliability Analysis of Flexible Construction Robotized Systems

Calin Ciufudean
Stefan cel Mare University of Suceava
Romania

1. Introduction

We consider this discussion, as there is a lot of confusion about the definition of the risk and the reliability of flexible manufacturing system analysis, both being risk analysts and decision makers. Thinking of risk and reliability analysis of flexible construction robotized systems (FCRS's) from a probabilistic perspective, we come to the conclusion that probability is a measure of expressing uncertainty about the process seen through the point of view of the assessor (i.e. the controller of a process), and based on some background information and knowledge that we have at the time we quantify our uncertainty. A sharp distinction between objective, real risk, and perceived risk cannot be made, simply because complete knowledge about the world does not exist in most cases, and the analysis provides a tool for dealing with these uncertainties based on coherence by using the rules of probabilities. If sufficient data become available, consensus in probability assignments may be achieved, but not necessarily, as there are always subjective elements involved in the assessment process. Risk is primarily a judgement, not a fact. As risk expresses uncertainty about the world, i.e. about consequences and outcomes of an activity, risk perception has a role to play to guide decision makers. It is, however, not obvious how such a thinking should be implemented in practice, in a decision making context, and different frameworks can be established. This approach emphasises the so-called observable quantities and their prediction (Aven, 2004). Examples of observable quantities are the number of facilities and production volumes. The starting point is an activity or a system that we would like to analyse now, to provide decision support for investments, design, operation etc. Therefore, the interesting quantities for risk and reliability analysis of an FCRS are the performances of that system, for example measured by production, production loss, number of fatalities, and so on. Unfortunately, in most cases, we are led to predictions of these quantities that reflect our expectations. But these predictions will normally not provide sufficient information; assessment of uncertainties is required. In order to express the uncertainties, we need a measure, and we choose the probability for measuring uncertainties.

2. Basic consideration on risk and reliability analysis

Basically, uncertainty related to any value the observable quantities will take, is expressed by probabilities. This uncertainty is a result of lack of knowledge.

The purpose of risk and reliability analysis is to provide decision support. This can be done by analysing and describing risk in a decision-making context, for example for a flexible construction robotized system (FCRS) efficient job schedule, these tasks define the area of the FCRS availability. In our view the scientific basis of risk and reliability analysis (e.g. the determination of the availability of the considered manufacturing system) can be summarized by the following points (Kask & Dechter, 1999), (Aven, 2004):

1. Knowledge about the system's performance and associated observable quantities are described by models, observed data and expert opinions;
2. Coherent and certainty assessments using the rules of probability.

Models: a number of different types of models are used in risk and availability analysis. These include both quantity-oriented and event-oriented models.

The purpose of quantity-oriented models is to predict the value of an observable quantity C by expressing knowledge of C in terms of a set of quantities $X = (X_1, X_2, \ldots, X_n)$, and the functional relationship f between C and X, i.e., $C = f(X)$. The function f is typically established on the basis of a mixture of commonly accepted, constitutive models from the field of physics, mathematics, chemistry, empirical knowledge, and more intuitive assumptions regarding the analysed system.

While the quantity-oriented models describe a functional relationship between a set of factors and the numeric value of quantity, event-oriented or logical models describe the conditions under which events occur. Such models consist of conditions and logical terms, and they usually have a binary outcome space (typically 0-1, failure-not failure, etc). In risk models logical models capture one's attention by developing events from a low initial level to large scale scenarios, threatening human lives and health, environmental and economic values. Three basic examples of logical models in risk and reliability analysis are:

$$F = X_1 - X_2 \leq 0 \tag{1}$$

$$F_s = X_1 \cup X_2 \tag{2}$$

$$F_P = X_1 \cap X_2 \tag{3}$$

Equation (1) describes a failure event F according to a simple load capacity consideration. F occurs when a load X_2 takes a higher value than the capacity value X_1. Equation (2) shows the conditions of failure of a two components system F_s, i.e., at least one of the two components X_1 or X_2 fails. Equation (3) represents the failure of a system of two components in parallel, i.e., both components fail. As a matter of fact, natural sciences provide theories and laws describing real world phenomena. These theories and laws are models of the world or they provide a basis for establishing models of the world, i.e. simplified representations of the world. For a specific risk availability analysis, application of the models may be better or worse in describing the world and useful for its purpose. Strictly speaking, a model is always wrong; otherwise it would not be a model - a simplification of the world. The accuracy of the models as representations of the world (i.e. the analysed system) needs to be addressed to. We notice that there is always a balance to be made, i.e., we need to simplify the system (i.e. the world) to obtain a tool that can be used in practice, to see the key factors; and this would result in less accuracy. It is difficult to give a detailed specification of what a satisfactory model is. How accurate must a model be in order to be considered acceptable? Well, the ultimate requirement for an availability model is that any improvement in the model to make it more accurate should not lead to a change of the conclusions that the analyst draws.

We must notice that this requirement may be difficult to verify – the best the analyst can do in most cases is to use sensitivity analysis to see how changes in the model affect the result.
A model C = f (X) is a purely deterministic representation of factors judged to be essential by the analyst. It provides a framework for mapping uncertainty about the observable quantity of interest, C, from epistemic uncertainty related to the observable quantities X, and does not introduce by itself additional uncertainty. In this approach the model is merely a tool judged to be useful for expressing the knowledge of a system. The model is a part of the background information of the probability distribution specified for C. If we change the model, we change the background information.
We must say that, when conducting a risk analysis we cannot "verify" an assigned probability, as it expresses the analyst's uncertain prior observation. What can be done is a review of the background information used as the rationale for the assignment, but in most cases it will not be possible to explicitly document all the transformation steps from this background information to the assigned probability. We do not search for the true probabilities, as such numbers do not exist, but for a consistent assessment of uncertainties steaming from the result of lack of knowledge. It seems that some probabilities are easy to assign and look sure, others are vague and it is doubtful that the single number means anything. Should the vagueness not be specified? In order to provide a basis for the reply, let us remember that probability P(A) is in fact a short version of a conditional probability of A given the background information K, i.e., P(A) = P(A/K). This means that even if we assign the same probability for two probabilities, they may be considered different, as the background information is different. For example, in some cases we may know much about the process and phenomena leading to the event A, and in other cases, very little, but still we assign a probability of 0.50 in both cases (Aven, 2004). However, by considering several similar events of the type A, i.e., we change the performance measure; the difference in the background information will often be revealed. We conclude that we always need to address the background information, as it provides a basis for the evaluation. Of course, there are a number of potential pitfalls in such judgement processes, but the risk analysts are aware of them and make use of tools in order to avoid them. Such tools include calibration procedures, the use of reference probabilities, standardisation, and more detailed modelling (Russell & Norvig, 2003). In order to exemplify the given approaches, we consider the following example, where an efficient discrete event simulation model has been proposed for reducing the risk of the availability analysis based on failure for FCRS's. The proposed model has been successfully applied and tested for the reliability value analysis of FCRS's.

3. Modelling the controllers of flexible construction robotized systems

An important characteristic of flexible construction robotized systems (FCRS's) is modularity. Since building materials transfer lines are widely used and constitute an important subclass of FCRS's, in this paper we focus on formal analysis of logic controllers for high volume transfer lines. We first propose a systematic approach to model and to verify the space size of the logic controllers using Petri nets. We consider that Petri net models of FCRS's can be decomposed in a few typical modules and the analysis of the entire system is made following a bottom-up algorithm. We have chosen to model the logic controllers with Petri nets, in order to take advantage of well developed theories for analysis and verification.

A controller is a discrete event supervisory system which controls synchronized sequences of basic operations of each data transmission cell in order to achieve the goal of the entire transmission system. Although logic controllers are very important in machining industry there is not yet a standard integrated tool, which is powerful, versatile and which can carry out formal analysis of correctness. High volume transfer lines are very usually machining systems in the FCRS's for mass building resources flow. In this paper we propose a systematic way to verify and implement the algorithms of the logic controllers for high volume transfer lines, using Petri nets. Petri nets are a well defined system modeling methodology, although the typical state-space explosion is problematic. For this reason, in this paper we describe an approach to estimate the state-space size of the Petri nets. This estimation algorithm is based on subnets and interconnections that mean a bottom-up approach. This algorithm allows typical subnets and is extendable for development of additional interconnections. We notice that the problem of state-space size estimation of PN's is being pursued in two distinct manners: top-down and bottom-up (Ferrarini, 1992), (Ferrarini, 1994). The top-down approach can analyze entire models without identifying subnets or imposing design constraints. The bottom-up approach builds the PN's model to a set of interconnected subnets. We notice that at the expense of complete generality, the bottom-up approach offers better accuracy.

4. Petri nets models for FCRS's controllers

We will assume that the reader is familiar with Petri nets theory and their application to discrete event systems (such as those involved in data transmission) or we refer the reader to (Murata 1989), (Zhon & DiCesare, 1992). In an ordinary Petri net PN = (P, T, F, M_0), where P and T are two disjointed sets of nodes named, respectively, places and transitions. $F \subseteq (PXT) \cup (TXP)$ is a set of directed arcs. $M_0: P \rightarrow N$ is the initial marking.

Two transitions t_i and t_j are said to be in conflict if they have at least one common input place. A transition t is said to be conflict free if it is not in conflict with any other transition. A transition may fire if it is enabled. A transition $t \in T$ is said to be enabled at marking M if for all $p \in {}^*t$, $M(p) \geq 1$. The SPN's considered here are ordinary Petri nets with timed transitions. Timed transitions can be in conflict therefore we say that a marking is stable if no conflict transitions are enabled. In the following lines we assume that the initial marking is a stable marking. We note by (M,T) a stable marking reachable from M by firing t. The new stable marking M* is obtained from M according to some routing probability. The basic idea is that in order to guarantee that a stable marking can be reached we must ensure that the respective circuit contains at least one timed transition. A SPN can be defined by the following elements (Park, 1999), (Tilburg & Khargonekar, 1999): T_t - Set of timed transitions; $M_s(M,t)$ - Set of stable markings reachable from M by firing transition t; $p(M^*,M,t)$ - Probability of reaching a stable marking M* from M when t fires (obviously, we have: $p(M^*,M,t) = 0$ if $M^* \notin Ms(M,t)$); $F_t(.)$ - Distribution functions of the firing time of t.

The GSMP representation of the SPN can be characterized by the following parameters:

X(t,k) - Independent random variables, where $t \in T_t$, and $k \in N$(each X(t,k) has distribution F_t and corresponds to the time of the kth firing of transition t); U(t,k) - Random variables on [0,1]. Each U(t,k) corresponds to the routing indicator at the kth completion of t. $r_n(t)$ - Remaining firing time of transition t at S_n; S(t,k) - Independent uniform random variables on [0,1], where $t \in T_t$, $k \in N$ (each U(t,k) corresponds to the routing indicator at the kth

completion of t); t_n - n^{th} completed timed transition; M_n - Stable marking reached at the firing of t_n; S_n - Completion time of t_n; τ_n - Holding time of marking M_{n-1}; V(t,n) - Number of instances of t among $t_1, \ldots, t_n$.

The dynamic behaviour of an SPN can be explained in the following way: at the initial marking M_0, set $r_n(t) = X(t,1)$, $\forall\, t \in T_t(M_0)$ and set $V(t,0) = 0$, $\forall\, t \in T_t$. All other parameters t_{n+1}, τ_{n+1}, s_{n+1}, V(t,n+1), M_{n+1}, r_{n+1} can be determined recursively as usually done in discrete event simulation. Recursive equations are given in (Zhou & Twiss 1998). The following routing mechanism is used in GSMP:

$$M_{n+1} = \varnothing(M_n, t_{n+1}, U(t_{n+1}, V(t_{n+1}, n+1))) \tag{4}$$

Where $\varnothing$ is a mapping so that $P(\varnothing(M,t,U) = M^*) = P(M^*,M,t)$.

Following the approach given in (Hopkins, 2002), we suppose that the distributions of firing times depend on a parameter Θ. In perturbation analysis the following results hold (Watson & Desrochers 1994), where the performance measures under consideration are of the form $g(M_1, t_1, \tau_1, \ldots, M_n, t_n, \tau_n)$ and a shorthand notation $g(\Theta)$ is used:

a) For each Θ, g(Θ) is a.s. continuously differentiable at Θ and the infinitesimal perturbation indicator is:

$$\frac{dg(\theta)}{d\theta} = \sum_{i=1}^{n} \frac{\partial g}{\partial \tau_i} \cdot \frac{d\tau_i}{d\theta} \tag{5}$$

b) If $d \in [g(\Theta)]/d\Theta$ exists, the following perturbation estimator is unbiased:

$$\sum_{i=1}^{n} \frac{\partial g}{\partial \tau_i} \cdot \frac{d\tau_i}{d\theta} + \sum_{k=1}^{n} f_k(h_k) \cdot G_k \tag{6}$$

$$f_k = \frac{f_{tk+1}(L_k(t_{k+1}))}{F_{tk+1}(L_k(t_{k+1})+y_k - F_{tk+1}(L_k(t_{k+1})))} \tag{7}$$

$$y_k = \min\{r_k(t) : \forall t \in T(M_k) - \{t_{k+1}\}\} \tag{8}$$

$$\tau_k = \frac{dL_k(t_{k+1})}{d\theta} - \frac{dX(t_{k+1})}{d\theta} \tag{9}$$

$L_k(t)$ is the age of time transition t at S_k; $G_k = g_{pp,k} - g_{DNP,k}$. The sample path $(M_1(\Theta), t_1(\Theta), \tau_1(\Theta), \ldots, M_n(\Theta), t_n(\Theta), \tau_n(\Theta))$ is the nominal path denoted by NP.

The $g_{DNP,k}$ is the performance measure of the kth degenerated nominal path, denoted by DNP_k. It is identical to NP except for the sojourn time of the $(k+1)^{th}$ stable marking in DNP_k. $g_{pp,k}$ is the performance measure of a so-called k^{th} perturbed path, denoted by PP_k. It is identical to DNP_k up to time s_k. At this instant the order of transition t_k and t_{k+1} is reversed, i.e., the firing of t_{k+1} completes just before that of t_k in PP_k. We notice that by definition, DNP_k and PP_k are identical up to s_k. At s_k, the events t_k and t_{k+1} occur almost simultaneously, but t_k occurs first in DNP and t_{k+1} occurs first in PP_k.

The commuting condition given in (Hopkins, 2002) guarantees that the two sample paths became identical after the firing of both t_k and t_{k+1}. Our goal is to introduce a correction mechanism in the structure of the SPN so that the transition t_k and t_{k+1} fire in the desired order, and the routing mechanism given in relation (4) is re-established. We will exemplify this approach, and we will correlate the theoretical assumption with some practical mechanisms in order to verify the approach. In a high volume transfer line (i.e., in a FCRS's, as shown above) the logic controller modules are related by synchronizations. Using these synchronizations, the Petri nets models for modules can be integrated in one Petri net for the entire logic controller (Zaitoon, 1996), (Murata, 1989). Some advantages of this module synthesis are that the structure of the entire net model is a marked graph and the synchronized transitions in the model have physical meaning.

The functional properties of the synthesized model can be analyzed using well-developed theories of marked graphs. The Petri net model of the entire system is defined as a modular logic controller.

The modules in a modular logic controller are simplified by the modified reduction rule to overcome the complexity in the Petri net model. For example, any transition which is not a synchronized transition can be rejected. Therefore, only synchronized transitions appear in the modular logic controller. Modules are connected by transitions. Each transition in a module is a synchronized transition, and appears in at least one other module. For example, in the figure 1 we have a modular logic controller which consists of three modules and three synchronized transitions. The initial place of each module has one token. The Petri net model for a logic controller is a reduced size model, which represents the specifications of the controller hierarchically. Therefore, the structure and initial marking of a modular logic controller should be live, safe, and reversible (Murata, 1989).

We notice that the logical behavior of the controller can be ensured from the functional correctness of its Petri net model. A common and convenient representation of a marked Petri net is by its state equation.

The main terms involved in the state equation of a Petri net are the incidence matrix, C and the initial marking M_0, which can be represented for the modular logic controllers, as the above given matrix, see relation (10).

Following the definition of an incidence matrix, for a Petri net with k modules and n_i number of places in the ith module, the incidence matrix of each module, C_i, where $i = 1, \ldots, k$, can be represented as a (n_i x m) matrix, where m is the number of transitions in the system. This matrix is constructed with the places of each module and the transitions of the system: $C_i(t)$.

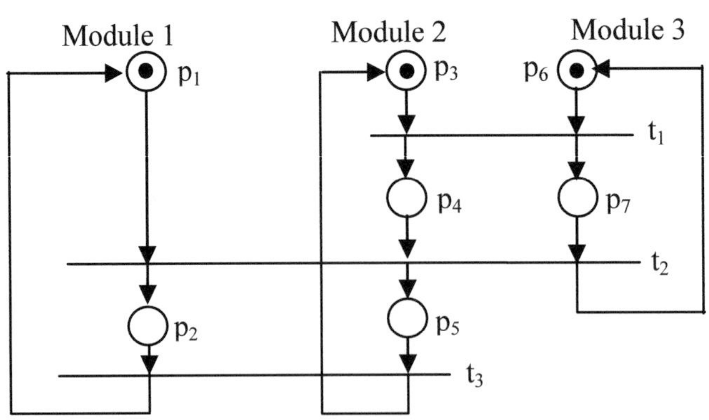

Fig. 1. An example of a modular logic controller

$$C = \begin{matrix} p_1 \\ p_2 \\ p_3 \\ p_4 \\ p_5 \\ p_6 \\ p_7 \end{matrix} \begin{bmatrix} & \begin{matrix} t_1 & t_2 & t_3 \end{matrix} \\ C_1 & \begin{bmatrix} 0 & -1 & 1 \\ 0 & 1 & -1 \end{bmatrix} \\ C_2 & \begin{bmatrix} -1 & 0 & 1 \\ 1 & -1 & 0 \\ 0 & 1 & -1 \end{bmatrix} \\ C_3 & \begin{bmatrix} -1 & 1 & 0 \\ 1 & -1 & 0 \end{bmatrix} \end{bmatrix}, M = \begin{bmatrix} 1 \\ 0 \\ 1 \\ 0 \\ 0 \\ 1 \\ 0 \end{bmatrix} \tag{10}$$

The incidence matrix of the system can be constructed using the following equation:

$$C = \begin{bmatrix} C_1^+ & - & C_1^- \\ - & - & - \\ C_2^+ & - & C_2^- \\ - & - & - \\ & \vdots & \\ - & - & - \\ C_k^+ & - & C_k^- \end{bmatrix} = \begin{bmatrix} C_1 \\ - \\ C_2 \\ - \\ \vdots \\ - \\ C_k \end{bmatrix} \tag{11}$$

Where C_i^+ and C_i^- are post and pre – incidence matrices of the ith module respectively and the incidence matrix C is a n x m matrix and $c_{ij} \in \{0,-1,1\}$. The initial places of a modular logic controller are assumed to be the first place of each module and can be represented by an n-dimensional vector. The initial marking is represented by:

$$M_0 \in \{0,1\}^n \tag{12}$$

Here 1 represents the initial places of the modules. This modular construction can be easily modified and reconfigured (i.e. it is suitable for FCRS's representation) by replacing incidence matrix of modules. The dynamic evolution of a modular logic controller can be determined by this incidence matrix and initial marking using the following relation (state equation):

$$M = M_0 + C \cdot f_C \tag{13}$$

Where, f_C is the firing count vector of the firing sequence of transition f in the net. An important parameter of the FCRS's is the resources flow volume. This is determined by the cycle time of a system in normal operation. Generally, performance analysis of event based systems is done by adding time specifications to the Petri net model. The performance analysis of timed Petri nets has been done for the evaluation of the cycle time. For strongly connected timed marked graphs, a classic method for computing the minimum cycle time C_T is given by the following relation (Park 1999), (Tilburg & Khargonekar, 1999):

$$C_T = \underbrace{\max}_{v \in \Gamma} \left\{ \frac{D(\gamma)}{N(\gamma)} \right\} \tag{14}$$

Where, Γ is the set of directed circuits of the pure Petri net; $D(\gamma) = \sum_{p_i \in \gamma} \tau_i$ is the sum of times of the places in the directed circuit γ; N(γ) is the number of tokens in the places in directed circuit γ. As pointed out in (Zhou & Twiss, 1998), the cyclic behavior of timed Petri nets is closely related to the number of tokens and to the number of states in the directed circuit which decides the cycle time C_T. As we know, model analysis and control algorithms implemented with Petri nets are based on the model state-space, and hence they are adversely affected by large state-space sizes. Thus, in the next section we'll give a bottom-up approach for the state-space size estimation of Petri nets.

5. Size estimation of modular controllers of FCRS's

In order to estimate the state space of Petri nets, they are divided into typical subnets, i.e., subnets with basic interconnections, such as: series, parallel, blocking, resource sharing, failure repair inter-connection, etc. Each subnet is associated with a state counting function (Zaitoon, 1996) (SC-function) that describes the subnet's state-space size when it contains r "flow" tokens. We notice that "flow" tokens (those that enter and leave the subnet via its entry and exit paths) are different from control tokens in a controlled Petri net. Petri nets model the execution of sequential parallel and choice operations, which are abstracted to be subnets (SN). Figure 2 illustrates two subnets in series, where tokens pass from SN_1 to SN_2. The interconnection's SC-function is given by the following relation (Watson & Desrochers, 1994).

$$S_{series}(r) = \sum_{r} S_1(r_1) \cdot S_2(r_2) = \sum_{i=0}^{r} S_1(i) \cdot S_2(r-i) \tag{15}$$

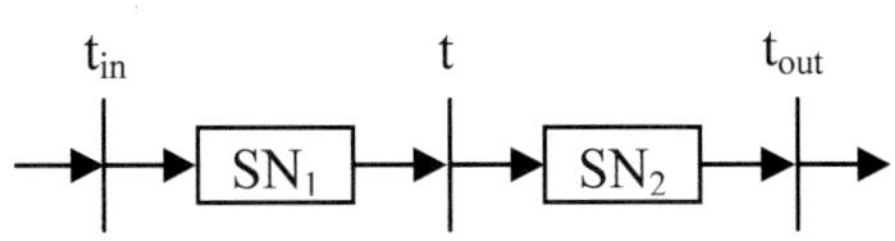

Fig. 2. Series interconnection of two Petri subnets

Analogous with the previous approaches, in the figure 3 we have the basic interconnections for parallel subnets (Fig.3.a); choice among subnets (Fig.3.b); blocking (Fig.3.c), and resource sharing (Fig.3.d).
The SC-functions (Zaitoon, 1996) for the nets in Fig.3.a, b, c, d are given by relations (16), (17), (18), (19), respectively:

$$S_{paralel}(r) = S_1(r) \cdot S_2(r) \tag{16}$$

$$S_{choice}(r) = \sum_{i=1}^{r} S_1(i) \cdot S_2(r-i) \cdot S_3(r-i-1) \tag{17}$$

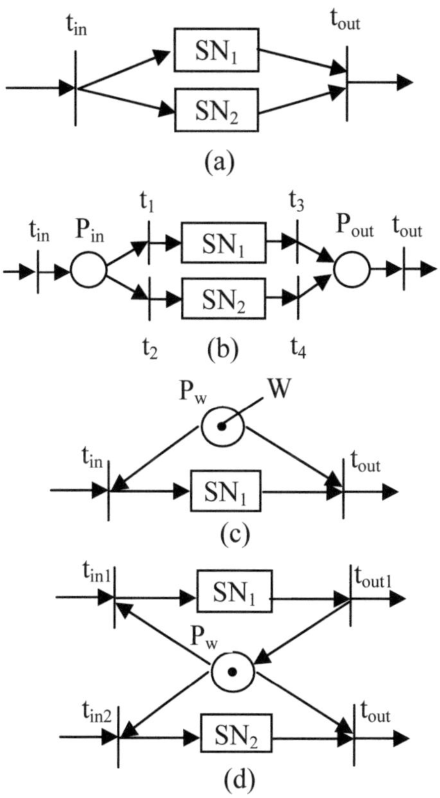

Fig. 3. Basic interconnections of Petri subnets

In relation (16) places P_{in} and P_{out} are considered as a group which forms the third subnet.

$$S_{blocking}(r) = \begin{cases} S_1(r), r \leq w \\ 0 \quad , r > w \end{cases} \tag{18}$$

$$S_{share}(r) = \begin{cases} S_1(r_1) \cdot S_2(r_2), & r_1 + r_2 \leq w \\ 0 \quad , & r_1 + r_2 > w \end{cases} \tag{19}$$

For example, in the figure 4 we have a system composed of three interconnections: the innermost is a choice between two subnets (each of the places); the middle interconnection is a resource block with queue; the outermost interconnection is a resource block. The SC-function for the inner choice is:

$$S_{in(r)} = \begin{cases} 1 \ , r = 0 \\ 4 \ , r = 1 \\ 10, r = 2 \end{cases} \tag{20}$$

The SC-function of the middle resource block is:

$$S_{mid(r)} = \begin{cases} 1 & ,r = 0 \\ 5 & ,r = 1 \\ 15 & ,r \geq 2 \end{cases} \quad (21)$$

The SC-function of the outer resource block is:

$$S_{out(r)} = \begin{cases} 1 & ,r = 0 \\ 5 & ,r = 1 \\ 15 & ,2 \leq r \leq 4 \\ 0 & ,r > 4 \end{cases} \quad (22)$$

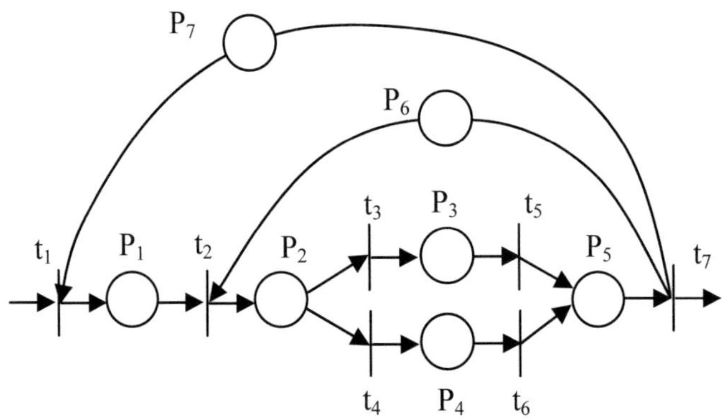

Fig. 4. Example of a multiple interconnection system

Following the above approach for calculating the size of the Petri net models of the modular controllers, we can adjust or modify the models accordingly to a reasonable size or in order to achieve the system requirements. We notice that state-space size estimation provides a tool for the model developer and the resulting data can be used to evaluate detail trade-off. As noted before, the longest directed circuits of the timed Petri net model determine the cycle time. Since for a high volume transfer line, the cycle time is determined by a directed circuit, we can use many of the known results to get more efficient algorithms for finding the critical operations of a timed modular logic controller (Murata, 1989). For example, because all transitions in the Petri net model of a modular controller are synchronized, we can assume that the sequence of transitions for the cyclic behavior is obtained by firing all transitions in the system only at once. Then the markings of the cyclic behavior of the system can be generated by the state equation (4) from the initial marking M_0.

6. The interaction Man-Machine in FCRS's

A characteristic of high level security control systems, such as those used in FCRS's is that an answer to a flaw that makes the man-machine system go to a lower level of security is considered a false answer, namely a dangerous failure, while an answer leading to a higher level of security for the man-machine system is considered an erroneous answer, namely a non-dangerous failure. That is the reason for the inclusion of some component parts with maximum failure probability towards the erroneous answer and parts with minimum failure probability towards the false answer. One must notice that the imperfect functioning states of the components of the man-machine system imply the partially correct functioning state of the FCRS. In the following lines the notion of imperfection will be named imperfect coverage, and it will be defined as the probability "c" that the system executes the task successfully when derangements of the system components arise. The imperfect reparation of a component part implies that this part will never work at the same parameters as before the derangement (Ciufudean et al., 2008). In other words, for us, the hypothesis that a component part of the man-machine system is as good as new after the reparation will be excluded. We will show the impact of the imperfect coverage on the performances of the man-machine system in railway transport, namely we will demonstrate that the availability of the system is seriously diminished even if the imperfect coverage's are a small percent of the many possible faults of the system. This aspect is generally ignored or even unknown in current managerial practice. The availability of a system is the probability that the system is operational when it is solicited. It is calculated as the sum of all the probabilities of the operational states of the system. In order to calculate the availability of a system, one must establish the acceptable functioning levels of the system states. The availability is considered to be acceptable when the production capacity of the system is ensured. Taking into account the large size of a FCRS, the interactions between the elements of the system and between the system and the environment, one must simplify the graphic representation. For this purpose the system is divided into two subsystems: the equipment subsystem and the human subsystem. The equipment subsystem is divided into several cells. A Markov chain is built for each cell i, where i=1,2,...n, in order to establish the probability that at least k_i equipments are operational at the moment t, where k_i is the least equipment in good functioning state that can maintain the cell i in an operational state. Thus, the probability of good functioning will be established by the probability that the human subsystem works between k_i operational machines in the cell i and k_{i+1} operational machines in the cell (i+1) at the moment t, where i=1,2,...n; n representing the number of cells in the equipment subsystem (Thomson & Wittaker, 1996). Assuming that the levels of the subsystems are statistically independent, the availability of the whole system is:

$$A(t) = \left[\prod_{i=1}^{n} Ai(t) \right] \cdot A_h(t) \tag{23}$$

Where: A (t) = the availability of the FCRS (e.g. the man-machine system); A_i (t) = the availability of the cell i of the equipment subsystem at the moment t; A_h (t) = the availability of the human subsystem at the moment t; n = the number of cells i in the equipment subsystem.

6.1 The equipment subsystem

The requirement for a cell i of the equipment subsystem is that the cell including N_i equipment of the type M_i ensures the functioning of at least k_i of the equipment, so that the system is operational. In order to establish the availability of the system containing imperfect coverage and deficient reparations, a state of derangement caused either by the imperfect coverage or by a technical malfunction for each cell, has been introduced. In order to explain the effect of the imperfect coverage on the system, we consider that the operation O_1 can be done by using one of the two equipments M_1 and M_2, as shown in the figure 5.

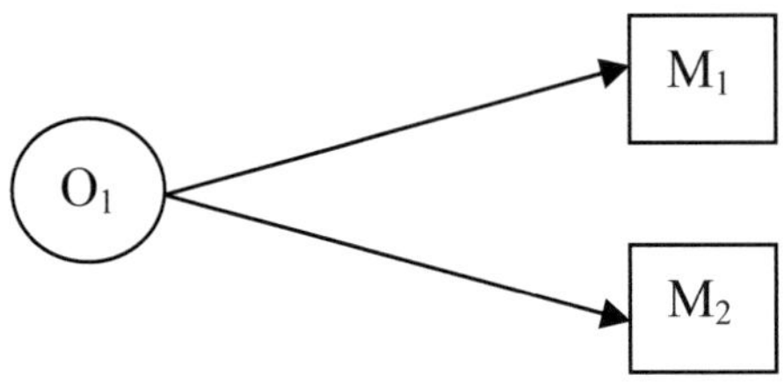

Fig. 5. A subsystem consisting of one operation and two equipments

If the coverage of the subsystem in the figure 1 is perfect, that is c =1, then the operation O_1 is fulfilled as long as at least one of the equipments is functional. If the coverage is imperfect, the operation O_1 falls with the probability 1-c if one of the equipments M_1 or M_2 goes out of order. In other words, if the operation O_1 was programmed on the equipment M_1 which is out of order, then the system in the figure 1 falls with the probability 1-c (Kask & Dechter, 1999). The Markov chain built for the cell i of the equipment subsystem is given in figure 6.

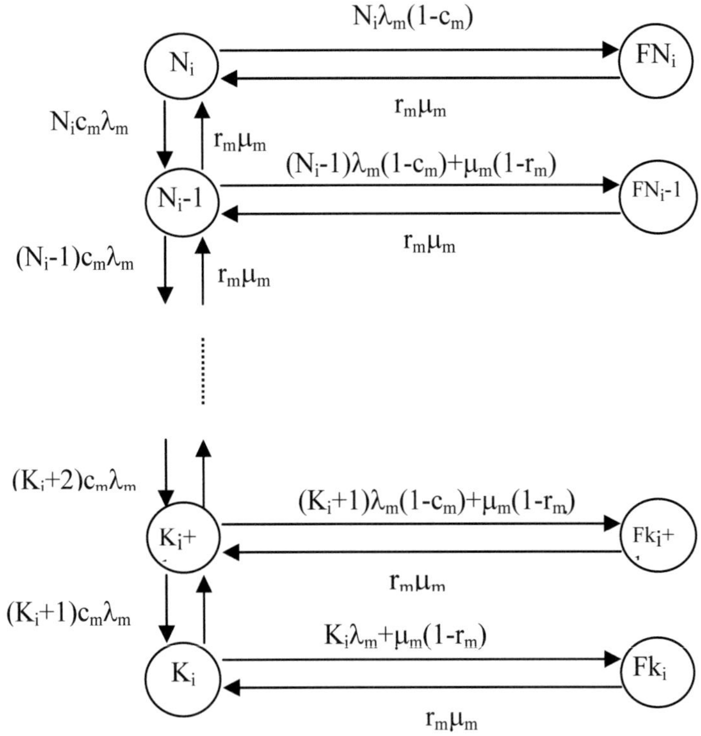

Fig. 6. The Markov model for the cell i of the equipment subsystem

The coverage factor is denoted as c_m, the failure rate of the equipment is λ_m (it is exponential), the reparation rate is μ_m (also exponential), and the successful reparation rate is r_m, where all the equipments in the cell are of the same type. In the state k_i the cell i has only k_i operational equipments. In the state N_i the cell works with all the N_i equipments. The

state of the cell i changes from the work state K_i, for $K_i \le k_i \le N_i$, to the derangement state Fk_i, either because of the imperfect coverage $(1\text{-}c_m)$ or because of a deficient reparation $(1\text{-}r_m)$. The solution of the Markov chain in the figure 6 is the probability that at least k_i equipments work in the cell i at the moment t.
The formula of this probability is:

$$A(t) = \sum_{k=ki}^{Ni} P_{ki}(t) \tag{24}$$

Where, $A_i(t)$=the availability of the cell i at the moment t; $P_{ki}(t)$=the probability that k_i operational equipments are in the cell i at the moment t, i=1,2,…,n; N_i= the total number of the M_i type equipments in the cell i; K_i=the minimum number of operational equipments in the cell i.

6.2 The human subsystem

The requirement for the human subsystem is the exploitation of the equipment subsystem in terms of efficiency and security. In order to establish the availability of the operator for doing his work at the moment t, we build the following Markov chain, which models the behaviour of the subsystem (Ciufudean et al., 2006):

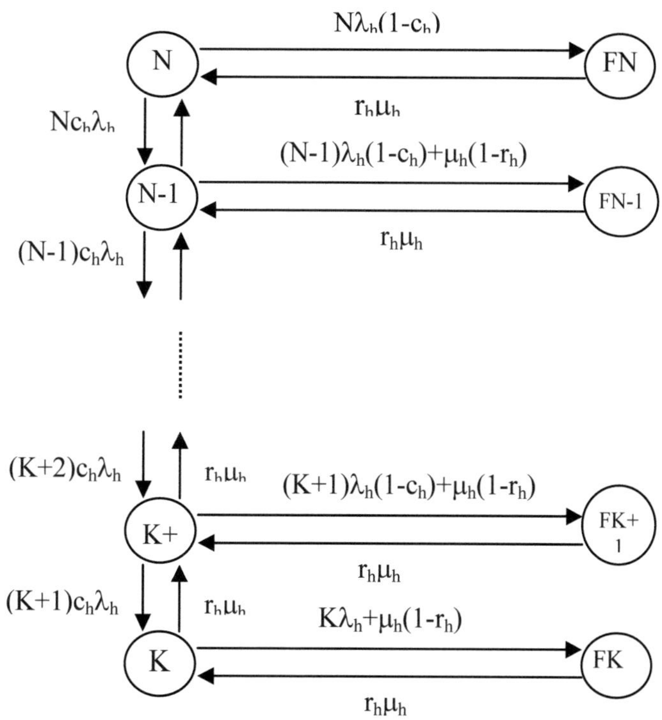

Fig. 7. The Markov chain corresponding to the human subsystem

Where, λ_h = the rate of making an incorrect decision by the operator; μ_h = the rate of making a correct decision in case of derangement; c_h = the rate of coverage for the problems caused

by incorrect decisions or by the occurrence of some unwanted events; r_h = the rate of successfully going back in case of an incorrect decision (Bucholz, 2002).
According to the figure 7, the human operator can be in one of the following states:
The state N = the normal state of work, in which all the N human factors in the system participate in the decisional process;
The state K = the work state in which k persons participate in the decisional process;
The state $F_{(k+u)}$ = the work state that comes after taking an incorrect decision or after an inappropriate repair that can lead to technological disorders with no severe impact on the traffic safety, where u=0,…N-k;
The state F_k =the state of work interdiction due to incorrect decisions with severe impact on the traffic safety.
In the figure 7, the transition between the states of the subsystem is made by the successive withdrawal of the decision right of the human factors who made the incorrect decisions.
The working availability of the human factor under normal circumstances is:

$$A_h(t) = \sum_{x=j}^{m} P_x(t) \tag{25}$$

Where, $P_x(t)$ = the probability that at the moment t the operator is in the working state X; m=the total number of working states allowed in the system; j = the minimal admitted number of working states.
Assigning new working states to the human factor increases the complexity of the calculus. Besides, although the man-machine system continues to work, some technological standards are exceeded, and that leads to a decrease in the reliability of the system.
The highlighting of new states of the human subsystem, that is the development of complex models with higher and higher precision, renders more difficult because of the increasing volume of calculus and the decreasing relevance of these models.
In order to lighten the application of complex models of Markov chains, a reduction of these models is required, until the best ratio precision/relevance is reached.
We notice that it is relatively easy to calculate the probabilities of good functioning for the machines (engines, electronic and mechanic equipments, building and transport control circuits, dispatcher installations etc.), while the reliability indicators of the decisional action of the human operator are difficult to estimate. The human operator is subjected to some detection psychological tests in which he must perceive and act according to the apparition of some random signals in the real system man-machine. However, these measurements for stereotype functions have a low accuracy level.
The man-machine interface plays a great part in the throughput increase of the FCRS's. The incorrect conception of the interface for presenting the information and the inadequate display of the commands may create malfunctions in the system.

7. An example of reliability analysis of construction robotized system

In order to illustrate the above-mentioned method, we shall consider a building site equipped with electronic and mechanic equipments consisting of three robot arms for load/unload operations and five conveyors. Two robots (e.g. robot arms) and three conveyors are necessary for the daily traffic of building materials and for the shunting

activity. That means that the electronic and mechanic equipment for two robots and three conveyors should be functional, so that the construction materials traffic is fluent.
The technician on duty has to make the technical revision for the five conveyors and for the three robots, so that at least three conveyors and two robot arms of the building site work permanently (Ciufudean et al., 2008).
On the other side, the construction engineer has to coordinate the traffic and the manoeuvres in such a manner as to keep free at least three conveyors and two robot arms, while the maintenance activities take place on the other two conveyors and one robot.
In this example the subsystem of the human factor consists of the decisional factors: the designer (i.e. architect), the construction engineer and the equipments technician (electro-mechanic). The subsystem of the equipments consists of the three robots and five conveyors (including the necessary devices). This subsystem is divided into two cells, depending on the necessary devices (e.g. electro-mechanisms and the electronic equipment for the conveyors, and respectively the electronic and mechanic equipment for the robots).
All the necessary equipments for the conveyors section are grouped together in the cell A_1, are denoted by $Ap_{1...5}$ and serve for the operation O_1 (the transport of building materials). The rest of the equipments denoted by $E_{1...3}$ are grouped together in the cell A_2 and serve for the operation O_2 (the load/unload operations of building materials by conveyors), according to the figure 8.

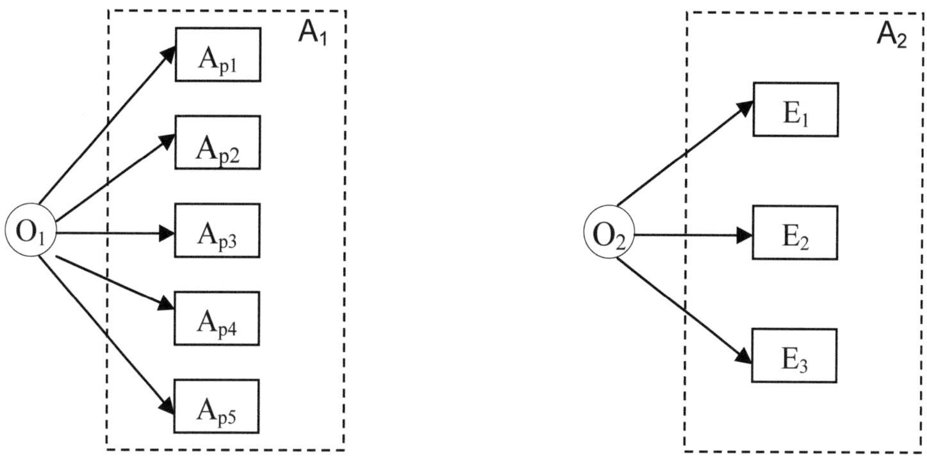

Fig. 8. The cells structure of the equipment subsystem

In the next table the rates of spoiling/repairing of the components are given.

The components of the system	C	μ	λ	r	K_i	N_i
A_{pi}	0.8	1.0	0.03	0.8	3	5
E_i	0.8	0.5	0.025	0.8	2	3
The components of the human subsystem	0.8	0.2	0.01	0.8	1	1,2,3

Table 1. The failing/repairing rates for the components of the system

$$
\begin{array}{c|cccccc|}
 & 3 & 4 & 5 & F_3 & F_4 & F_5 \\
3 & -3\lambda+\mu & 0{,}8\mu & 0 & 3\lambda+0{,}2\mu & 0 & 0 \\
4 & 3{,}2\lambda & -(4\lambda+\mu) & 0{,}8\mu & 0 & 0{,}8\lambda+0{,}2\mu & 0 \\
5 & 0 & 4\lambda & -(5\lambda) & 0 & 0 & \lambda \\
F_3 & 0{,}8\mu & 0 & 0 & -(0{,}8\lambda) & 0 & 0 \\
F_4 & 0 & 0{,}8\mu & 0 & 0 & -(0{,}8\lambda) & 0 \\
F_5 & 0 & 0 & 0{,}8\mu & 0 & 0 & -(0{,}8\lambda)
\end{array}
$$

Fig. 9. The matrix of the state probabilities for the cell A_1 from the equipment subsystem

$$
\begin{array}{c|cccc|}
 & 2 & 3 & F_2 & F_3 \\
2 & -(2\lambda+\mu) & 0{,}8\mu & 2\lambda+0{,}2\mu & 0 \\
3 & 2{,}4\lambda & -(3\lambda) & 0 & 0{,}6\lambda \\
F_2 & 0{,}8\mu & 0 & -(0{,}8\mu) & \\
F_3 & 0 & 0{,}8\mu & 0 & -(0{,}8\mu)
\end{array}
$$

Fig. 10. The matrix of the state probabilities for the cell A_2 from the equipment subsystem

For the equipment subsystem there are two Markov chains, one with six states (cell A_1) and one with four states (cell A_2); the matrix in the figure 9 corresponds to the first one and the matrix in the figure 10 corresponds to the second one. The following Markov chains correspond to the human subsystem:

- with six states (the decisions are made by the three factors: the designer, the construction engineer and the electro-mechanic);
- with four states (the decisions are made only by two of the above-mentioned factors);
- with two states (the decisions are made by only one human factor).

A matrix of the state probabilities corresponds to each Markov chain:

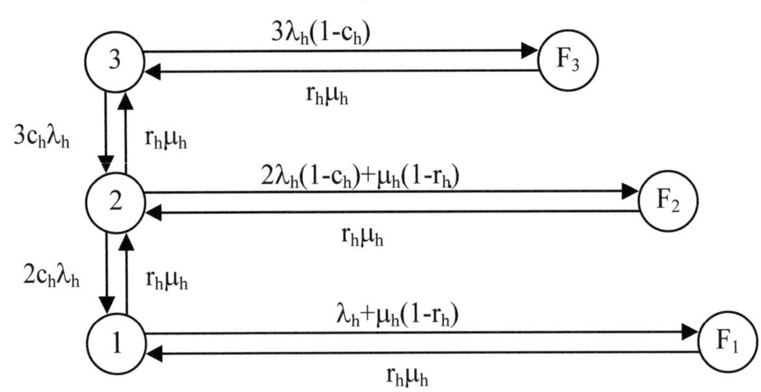

Fig. 11. The Markov chain corresponding to three of the decisional factors

$$
\begin{array}{c|cccccc|}
 & 1 & 2 & 3 & F_1 & F_2 & F_3 \\
1 & -(\lambda+\mu) & 0{,}8\mu & 0 & \lambda+2\mu & 0 & 0 \\
2 & 1{,}6\lambda & -(2\lambda+\mu) & 0{,}8\mu & 0 & 0{,}4\lambda+0{,}2\mu & 0 \\
3 & 0 & 2{,}4\lambda & -(3\lambda) & 0 & 0 & 0{,}6\lambda \\
F_1 & 0{,}8\mu & 0 & 0 & -(0{,}8\mu) & 0 & 0 \\
F & 0 & 0{,}8\mu & 0 & 0 & -(0{,}8\mu) & 0 \\
F_3 & 0 & 0 & 0{,}8 & 0 & 0 & -(0{,}8\mu)
\end{array}
$$

Fig. 12. The matrix of the state probabilities corresponding to the Markov chain in the Fig.11

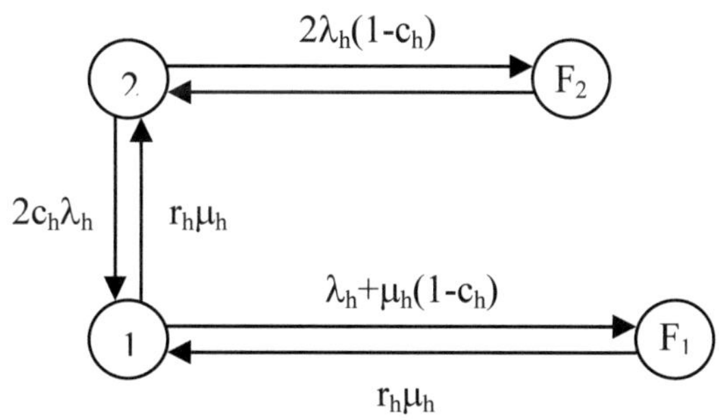

Fig. 13. The Markov chain corresponding to two decisional factors

$$
\begin{array}{c|cccc|}
 & 1 & 2 & F_1 & F_2 \\
1 & -(\lambda+\mu) & 0{,}8\mu & \lambda+0{,}2\mu & 0 \\
2 & 1{,}6\lambda & -(2\lambda) & 0 & 0{,}4\lambda \\
F_1 & 0{,}8\mu & 0 & -(0{,}8\mu) & 0 \\
F_2 & 0 & 0{,}8\mu & 0 & -(0{,}8\mu)
\end{array}
$$

Fig. 14. The matrix of the state probabilities corresponding to the Markov chain in the Fig.13

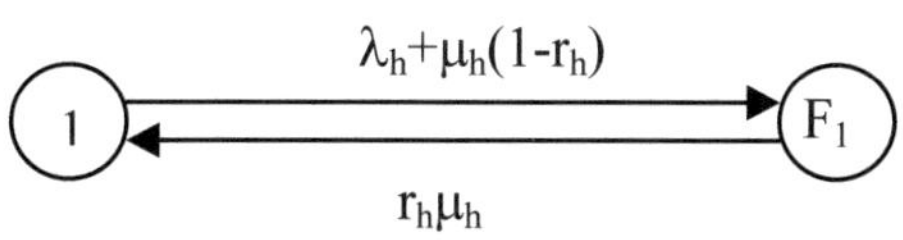

Fig. 15. The Markov chain corresponding to one decisional factor

$$
\begin{array}{c|cc|}
 & 1 & F_1 \\
1 & -(\lambda+0{,}2\mu) & \lambda+0{,}2\mu \\
F_1 & 0{,}8\mu & -(0{,}8\mu)
\end{array}
$$

Fig. 16. The matrix of the state probabilities corresponding to the Markov chain in the Fig.15

The equations given by the matrix of the state probabilities are functions of time and by solving them we obtain:

- The expressions of the availabilities for the cell A_1, and respectively A_2 from the equipment subsystem calculated with the relation (18);
- The expression of the availability of the human subsystem calculated with the relation (19);
- The expression of the availability of the whole system calculated with the relation (17).

The values of these availabilities depending on time are given in the table 2.

Time [hours]	Cell A_1	Cell A_2	The human subsystem A_h	The availability of the railway system A
0	1.000000	1.000000	1.0000000	1.00000000
1	0.980013	0.985010	0.9548293	0.92171802
4	0.947011	0.951341	0.8645392	0.77888946
8	0.933510	0.933468	0.8061449	0.70247605
12	0.933010	0.927481	0.7809707	0.67581225
16	0.933129	0.926133	0.7701171	0.66553631
20	0.933060	0.925951	0.7654364	0.66131243
24	0.932891	0.925600	0.7647893	0.65970171
28	0,932762	0,925012	0,7635876	0.65876005
32	0.932132	0.924910	0.7631243	0.65781145
36	0.931902	0.924830	0.7625786	0.65716133
40	0.931819	0.924690	0.7621289	0.65640272
44	0.931791	0.924600	0.7619786	0.65640272
48	0.931499	0.924582	0.7619456	0.65618425

Table 2. The availability values for the elements of the exemplified system

8. Conclusion

An advantage of the above-mentioned calculus method is the easy calculation of the availability of the whole system and of the elements of the system. The availabilities of the exemplified system are drawn in figure 17, depending on time and on the number of decision factors. In figure 17, the numbers x=1,2,3 show the availability of the systems corresponding to the Markov chains in figure 11, figure 13, respectively figure 15. The figure 17 shows that the best functioning of the system can be obtained by using two decisional factors: while the availability of the system in figure 15 is 65% after 12 hours of functioning, the availability of the system in figure 13 is 82%. The availability of the system decreases when the third decisional factor appears, because the diminution due to the risk of imperfect coverage or due to an incorrect decision is greater than the increase due to the excess of information.
In the figure 18 the availability of the system depending on the coverage factors (c_m), and on the successful repairing (r_m) of deficient equipment is illustrated. One may notice that the availability increases with 5 percents when the coverage is perfect (c_m=1). Moreover, when the repairing of a deficient equipment is perfect (r_m=1), the availability increases with 10 percents (we mention that the increases refer to a concrete case where c_m=0.8 and r_m=0.8). An important conclusion that we can draw is that the presumption of perfect coverage and repairing affects the accuracy of the final result. This presumption is made in the literature in the majority of the analysis models of the system availability (Hopkins, 2002).

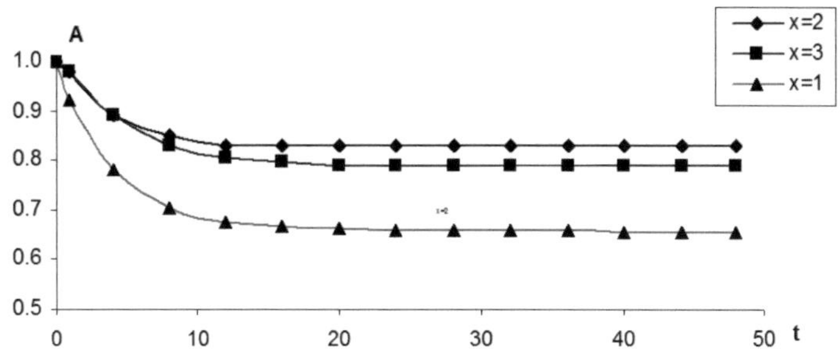

Fig. 17. The availability of the railway system depending on the number of the decisional factors

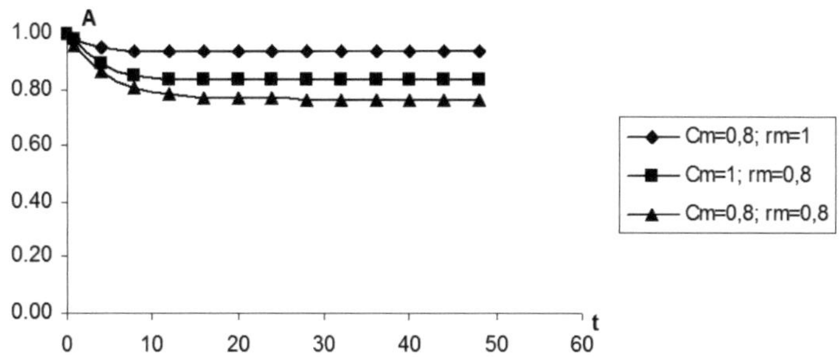

Fig. 18. The variation of the system availability depending on the factors c_m and r_m

The analysis of the availabilities of the operation O_1 and O_2 done by the cell A_1 and respectively by the cell A_2 from the equipment subsystem shows that an increase of the number of the conveyors (from N_i=5 and k_i=3 to N_i=5 and k_i=4) in the cell A_1 would lead to a decrease of the availability of the operator O_1 with 4% (as shown in the figure 19). In the case of the cell A_2, a decrease of the total number of robots (from N_i=3, k_i=2 to N_i=2, k_i=2) would lead to a decrease of the availability of the operator O_2 with 20% (as shown in the figure 20). The conclusion is that an extra robot is critical for the system, because it improves considerably the availability of O_2 and hence, the availability of the system.

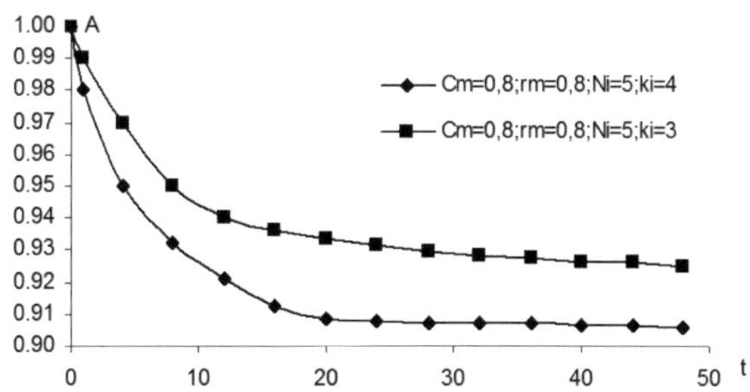

Fig. 19. The analysis of the availability of the cell A_1

The analysis of the availability allows us to establish the lapse of time when changes must be made in the structure of the system (major overhaul, the rotation of the personnel in shifts etc). For example, from the figure 17, if the availability is 70%, the human decisional factor must be replaced every 12 hours (for the system in the figure 15 that is rotating the personnel every 12 hours).

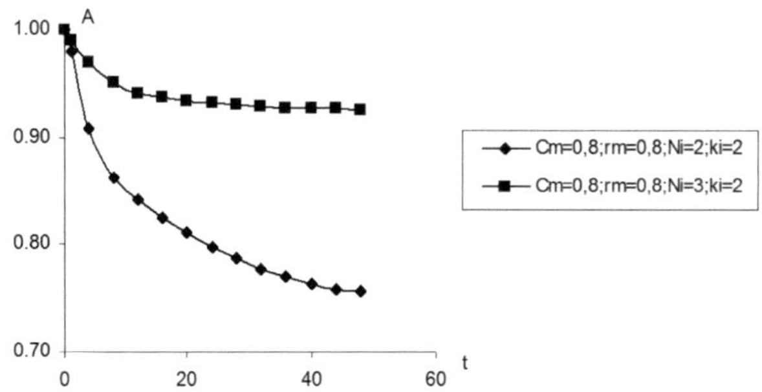

Fig. 20. The analysis of the availability of the cell A_2

9. References

Aven, T. (2004). Risk Analysis and Science, *International Journal of Reliability, Quality and Safety Engineering*, vol. 11, no. 1, pp. 1-15

Ferrarini, L. (1992). An incremental approach to logic controller design with Petri nets, *IEEE Trans. Syst. Man. Cybern.*, vol. 22, pp. 461-473

Ferrarini, (1994). L. A new approach to modular liveness analysis conceived for large logic controllers design, *IEEE Trans. Robot. Automat*, vol. 10, pp.169-184

Zaitoon, J. (1996). Specification and design of logic controllers for automated manufacturing systems, *Robot. Comput-Integr. Man.*, vol. 12, no. 4, pp. 353-366

Murata, T. (1989). Petri nets: Properties, analysis and applications, *Proc. IEEE*, pp. 541-580,

Zhon, M. C. & DiCesare, F. (1992). Design and implementation of a Petri net based supervisor for a flexible manufacturing system, *IFAC J. Automatica*, vol. 28, no. 6, pp. 1199-1208

Park, E.; Tilburg D.; Khargonekar, P. (1999). Modular logic controllers for machining systems: formal representation and performance analysis using Petri nets, *IEEE Trans. Rob. and Autom.*, vol. 15, no. 6, pp. 1046-1060

Watson J. & Desrochers, A. (1994). State space size estimation of Petri nets: a bottom-up perspective, *IEEE Trans. Rob. and Autom.*, vol. 10, no. 4, pp. 555-561

Zhou M. & Twiss, E. (1998). Design of industrial automated systems via relay ladder logic programming and Petri nets, *IEEE Trans. Man. Cyber*,vol.28,pp.137-150

a. Hopkins, M. (2002). Strategies for determining causes of events, *Technical Report R-306*, UCLA Cognitive Systems Laboratory

b. Hopkins, M. (2002). A proof of the conjuctive cause conjecture in causes and explanations, *Technical Report R-306*, UCLA Cognitive Systems Laboratory

Thomson, M. G. & Wittaker, J. A. (1996). Rare Failure State in a Markov Chain Model for Software Reliability, *IEEE Trans. Reliab. 48(2)*, pp. 107-115

Kask, K. & Dechter, R. (1999). Stochastic local search for Bayesian networks, *In Workshop on AI and Statistics 99*, pp. 113-122

Russell, S. & Norvig, P. (2003). *Artificial Intelligence: A Modern Approach*, J. Willey and Sons, N.Y.

Bucholz, P. (2002). Complexity of memory-efficient Kroneker operations with applications to the solutions of the Markov models, *Informs J. Comp.*, no. 12(3), pp. 203-222

Ciufudean, C. & Graur, A. & Filote, C. (2006). Determining the Performances of Cellular Manufacturing Systems, In *Scientific and Technical Aerospace Reports*, vol.14, Issue 6, NASA, Langley Research Center, USA

Ciufudean, C. & Filote, C. & Amarandei, D. (2008). Scheduling Availability of Discrete Event Systems, *The 14th IEEE Mediterranean Electrotechnical Conference, MELECON'2008*, Palais des Congrès François Lanzi - Ajaccio – France.

10

Precast Storage and Transportation Planning via Component Zoning Optimization

Kuo-Chuan Shih, Shu-Shun Liu and Chun-Nen Huang
National Yunlin University of Science and Technology
Kainan University
Taiwan (ROC)

1. Introduction

Industry management issues, such as enterprise resource planning (ERP) and supply chain management (SCM), are discussed and implemented successfully in many manufacturing industries but construction. No matter what the nature of construction is manufacturing buildings, risks and uncertainties make its characteristic different to other manufacturing industries. In order to reduce effects of these two scourges, precast is an evolutional method what is adopted to remove construction work environment from outdoor to indoor and make the procedure of component producing regular as an automatic factory. Thus, precast method is a construction method with its industrial characteristics being closest to those of manufacture industry. However, Practical plans and information identification in working process must be further recognized and achieved. This study proposes an optimization model which focuses on planning issues of precast manufacturing procedure.

The storage and transportation planning of a construction precast project is mainly discussed herein. Generally, whole process of a precast project can be divided into 5 stages: design, production, storage, transportation, and installation. Besides, at least 4 important roles: client, architect, subcontractor, and precast factory, are involved in these 5 stages. Relationships among these four roles depend on contracts of a project. From perspective of the precast factory, two stages are out of their control: design stage and installation stage. In design stage, the architect confirms details of all precast components, such as shape, strength and material, with the client, and then makes components exact. The precast factory receives these component details and then produces components according to architect's designs as orders. In installation stage, the subcontractor installs all completed components at where the places according to architect's design. The precast factory supplies components on time in the installation stage of most cases. It is obvious that the design stage and the installation stage involves two or more roles. Thus, production, storage, and transportation stage are more controllable than these two stages from precast factory's viewpoint. Furthermore, production stage was the issues most frequently investigated and analyzed in prior precast management related study. However, the planning of storage and transportation are still very significant to a precast factory. To complete the management mechanism of precast factory, these two stages need to be investigated.

2. Reference

There are many production stage related studies of precast related studies can be referred and enumerated. For example, Chan has studied a lot in precast production planning. In order to suit different standardization degrees of components, Chan (2002) has proposed two production planning models are comprehensive method, which the method utilizes resources regularly in component producing, and specialized method, which the method considers low standardization components, for factory business organization. Furthermore, a coordinated production scheduling and rescheduling model has proposed by Chan (2003, 2005) to deal with risks of component demand. Chan (2005) has adopted simulation to build scenarios of viaduct producing for analysis. Considering resource-constrained environment, Leu (2002) has proposed a GA base scheduling model that discussed the importance of manpower, cranes, steam curing capacity and reinforcement cage storage space. Besides, Viaduct precast considers both supply-demand matching and high productivity.
Storage and transportation planning issues of precast project have been few studied in construction field but manufacturing industry. Storage and transportation plan are two sides of a coin that are presented in many studies in manufacturing industry field. A common study type is the total cost optimization of transportation among factory, warehouse and customer. Furthermore, outsourcing of inventory and transportation may be related to time-base and quantity-base issues (Sıla, 2006). For precast project, how to store a special object such as precast component which reduce secondary movement and component finding is a key question. The Acheson & Glover Group (2006) tried to develop a precast storage system. However, ways to store component are also related to safety of labor. For construction process issues in job-site, Hu (2005) proposed a geometric reasoning method that can help to determine the component sequence of demand. Nevertheless, characteristics of construction precast component: huge; heavy; unique in design or installation procedure cause these achievements cannot easily perform in precast factory.
One of problem of precast storage and transportation is huge component. However, containerization is an issue that is valuable to be explored. Vis and Koster (2002) discussed the process of transporting containers from ship to stack in terminal. Means are used to move huge objects, including cranes, vehicles, straddle carrier… etc. Thus, a plan to simplify the process and to use these means efficiently is necessary. Avriel (1998) proposed a mathematic model that tried to deal with the storage planning problem and reduce the shifts of container in a ship. However, the size of the precast component is not the only item to be concerned of one should also be concerned about the different shapes of the component. Sadiq (1996) proposed a cluster-analysis base model the classified all the objects into several sets, then allocated all the sets to different storage zones. This model tried to link the relationship between objects to reduce secondary-movement for objects. The cluster-analysis can also work in precast project and a cluster-analysis named zoning strategy in precast project will be discussed herein.

3. Precast component storage and transportation

Optimization is a usual tool for precast factory planning in previous production stage's researches. Thus, this research follows mathematical model discussion with more concern in storage and transportation stages.

3.1 Storage stage

Generally speaking, storage stage had been considered in the component producing planning but simplified as an inventory calculation. Daily inventory is a vehicle variable between component producing and the demand. Produced components are stored in factory as inventory and this inventory add up all produced components in factory in a period. In order to match the demand, the component inventory must equal to or exceed the demand of contract at the deadline of project. Thus, a component producing plan considers daily inventory is able to create, and it is still practical for factory business mainly considering production stage. Furthermore, the cost of inventory can be also calculated through the quantity of stored components, and the inventory limits can be restrained if storage space is further concerned as constraints. Nevertheless, traditional precast factory can perform this kind of production planning formulation without considering how to store components.

Component storage must be planed from perspective of a precast factory business. No matter how a precast factory closed to a manufacturing one, the nature of product of precast factory, construction component, is very unique to other industries. In addition, there are less consistent among components when component's sharp, strength and position of a building are in consideration. A component that is unique to any other one is commonly concerned in most construction precast project. Thus, to identify each component is usually important in storage stage. Furthermore, there are several circumstances must be regarded in practical storage work: size of components; limitation in vertical loading of ground; safe distance between components and ways to store components. Briefly, the problem of component storage is a 2-dimensional or 3-dimensional spatial allocation with component identification. These considerations cause component storage complex and identification of component is necessary. It is hard to ignore the storage stage to a precast factory business.

A good storage plan is also benefit to help component delivering in right order and on time. Two sequences are accompanied with component delivering are sequence from production stage to storage stage and sequence from storage stage to installation stage. In the first case, components and molds can be grouped that had mentioned by Chen 2005. In order to produce components smoothly and continually, Components which can be grouped into a same sharp, strength and material can be produced orderly by the same mold or grouped molds which belong to the same category certainly. Under this circumstance, components can be produced as soon as possible in production stage with minimal operation change of mold. This sequence is named production sequence, PS, in this research that means components are delivered according to their group. Besides, PS always causes grouped components storage. The other sequence is named installation sequence or IS. Components must be installed at the positions where they are appointed in architect's design. Thus, components which belong to the same scope of once installation work, for example a one-floor installation, are needed in a short moment. Therefore, components are delivered according to their demand timing from factory to installation worksite. By view of these two sequences, it can be recognize that there is a conflict between PS and IS in precast factory business. Therefore, the functionality of storage plan is not only a quantity calculation of component inventory, but rearrangement of component sequence from PS to IS to resolve this conflict before their installation.

Foreign site storage that component are stored in a space out of both factory and work site is another issue in practical factory business. When the space of the precast factory is insufficient to store all components, foreign site storage is a common alternative. Extra movements of components are needed to deliver components between different sites.

Foreign site storage refers that the precast factory has to look for other storage sites to store the components that cannot be stored in the precast factory during project period. This Foreign site storage issue combines component storage and transportation, and it is complex in both storage and transportation planning. This alternative occurs in practice. Therefore, further component planning and controlling mechanism in storage stage is needed to analysis all above issues for a precast factory.

3.2 Transportation stage

Transportation stage is ignored in previous precast management researches. Components always produced with few redundancies in a construction precast project, and all of them must be transported for installation to meet project requirement. Therefore, the cost of transportation can be treated as a fixed cost in most cases without detail delivery consideration because component delivery is necessary in a project. Thus, component transportation has been a parameter of fix cost which does not need to plan. However, the component transportation still plays an important role in factory business.

There are two kinds of transportation must be recognized in factory business: component movement within a site and component transportation between two sites for a long distance. Component movement within a site means that components are moved within the factory, a storage site, or the work site in short distance. Equipments such as cranes and trams can be utilized for this case. These equipments are owned or rented for daily business by precast factory. Hence, transportation cost in this case can be neglected from single precast project or transformed onto the cost for factory or site setting cost. The other, component transportation for a long distance is performed by trucks. In practice, trucks are mostly rented case by case when components transport in sites or turn over from any site to work site are sure. Two important factors: weight of components and transported distance are commonly adopted for truck rental fee calculation. This long distance transportation is variable case by case. For example, components are delivered from the factory to a foreign site, the factory to the work site, and a foreign site to the work site.

4. Component zoning strategy

Taking to above issues as well as problems with component storage and transportation into consideration, a mechanism for precast factory planning which employed the concept of basic zoning with minimization of total cost is purposed. The related definitions and assumptions are explained in the following sections:

4.1 Component Zoning

Components are grouped into zones herein. A zone is a space that components can be stored following specific rules which assigned by planner. This component zoning is most like the behavior of the goods package in manufacturing that goods can be encased into a box by fallowing rules of what kind of box it is. By the way, a box is both storage and transportation strategy basis. However, there is no real box for precast component, but it can be instead by a specified storage space as a zone without encasement. A zone is also similar to a container to collect components. Components can be moved in respectively. Therefore, the behavior of a zone is flexible that dependent on rules what the planner made to form it as a box or container. However, the rules of zones must be clearly declared before planning.

Zones are able to be alternated to release storage space too. This study mainly focus on how zoning strategy working in storage and transportation stage to represent component zoning. To recognize zoning rules is benefit for precast factory planning and management issue. Zones try to retain the flexibility of storage that planner can declare their own rule. Whole storage and transportation process can be formed by zones and their own rules. Zoning rules of a zone basically contain what kind of component can be stored, how many components can be stored, how much space are required and other specific rules what made by planner. Zoning rules help planner to control storage and transportation process because zones can force components well-regulated in preset rules. Thus, making zoning rules appropriate is a further important issue to meet the request of factory business management. The PS, to store grouped component, and IS, to store component by installation scope, are mainly discussed in component zoning herein.

The component zoning according to grouped component occurs in PS that is the common situation in practical storage business. To store components by group has several advantages: Components can be easily found; the space utility is well in most of cases; storage space demand can be easily calculated. This is why most factories which include all kinds of industry store goods by group of goods type for warehousing, and most of precast factories are working without exception. However, this kind of zoning strategy is not always suit precast factory storage. Hundred or thousand of component groups always occur because of architect's design.

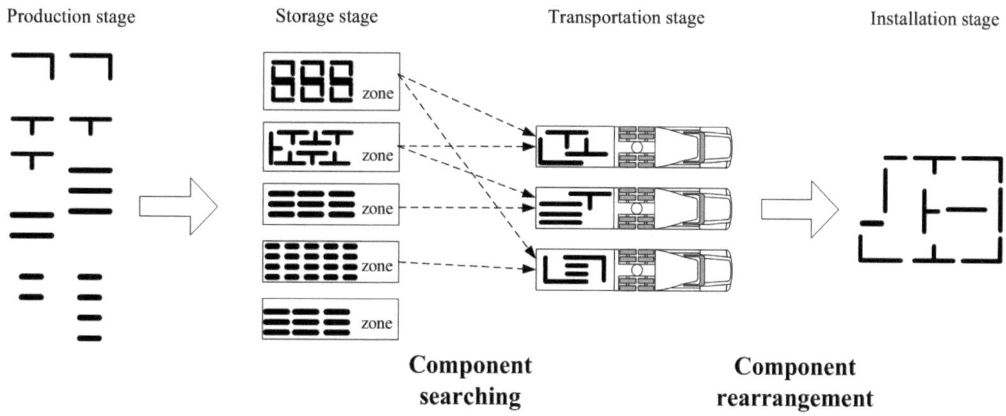

Fig. 1. Zones by following PS

Figure 1 shows a possible case of PS zoning rules and normal storage practice. Components are grouped and stored into zones in storage stage. However component searching or component rearrangement are needed before component installation because IS occurs after transportation stage. Trucks must find out required components through overall zones for component searching, or operations in work site must rearrange components before installation. Additional time and cost are caused in practical. Nevertheless, it can be also a choice in consideration.

Zoning with IS aim to overcome the conflict between PS and IS in factory. Component movement within a site is easier and cheaper than long distance transportation of trucks because cranes and trams can move and rearrange components conveniently. Thus, only component rearrangement is needed in storage stage. Figure 2 shows this situation.

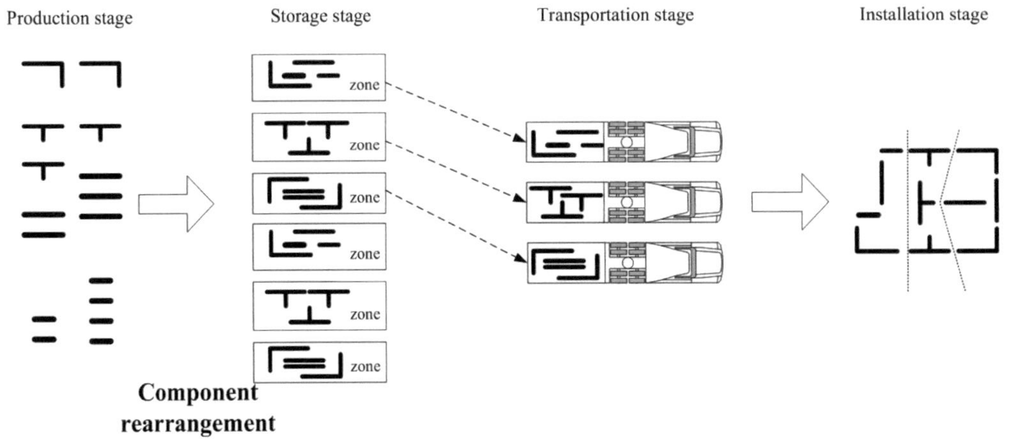

Fig. 2. Zones by following IS

4.2 Zoning strategy

A zoning strategy is composed by zones what are chose for a storage and transportation plan during a planned period. Thus, a zoning strategy can contain zones with same rules, zones with different rules, mutualism zones, and mutually exclusive zones if they are needed. All zones whether PS or IS are alternatives when planner do not really recognize what kind of zone rules are suitable before practice. Planner can create kinds of zones and manifold rules in zones if they are recognized before or during decision making procedure. An optimization model for seek out the optimal zoning strategy with minimal operation cost of precast factory in storage stage and transportation stage by zone selection and allocation is proposed as below.

4.3 Zone selection and allocation

From perspective of component storage, zones are used as basis elements for checking the component storage and utility of each storage site. In order to form an optimized zoning strategy, procedure of picking up appropriate zones, in term of zone selection, is very important. Figure 3 shows a possible situation of zone selection. First at all, components must be collected into zones fallowing the rules of each zone is a basic assumption. This assumption makes sure that whole process of component storage can be represented by zones. Beside, whole inventory space is divided into sites. Two kinds of site that are site inside factory and foreign site are involved according foreign site inventory behaviour, extra site rental fee are considered if a foreign site is adopted during the period of a project. This rental fee contains land usage fee and necessary facility fee to operate storage business. Besides, truck rental fee can also be recognized by location of a foreign site and weight of component which are planned to store in this site.

The zone selection can be explained as relationships among components and zones. Components must be stored for sure, so that at least one zone must exist whenever any components are stored in. In other word, this zone is adopted when any component is planed to be stored in. For example, zone 1, 2, 3, and 5 in figure 3 are selected to store components. On the contrary, zone 4 is not selected to store any component. Besides, component can be stored into a zone when only they are permitted by rules of this zone.

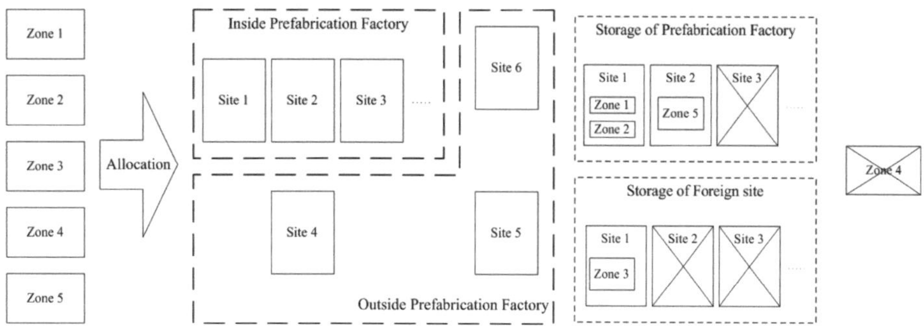

Fig. 3. Illustration of the Relationship among Zones and Storage Sites

The zone allocation can also be present as relationship among zones and sites. As the same circumstance, at least one site must be used or rented because there is at least one zone must be adopted to store components. A zone can be allocated into a site, no matter site inside factory or foreign site, when the storage space of this site is sufficient. The required storage space of this zone is according as its zone rules. Site inside factory or foreign site is allowed to allocate zones, but one zone can be only allocated once and into one site during a planed period. Figure 3 shows that site 1 and site 2 inside factory are occupied by zone1, zone 2 and zone5, and foreign site 1 is occupied by zone 3 respectively. It is allowed that two or more zones allocated in a site. Besides, whenever a foreign site is selected, the rental fee that contains land usage fee and the charge of necessary resource to operate storage business will be added into project cost for entire project period.

4.4 Transportation between sites

The whole transportation problem can be divided into 3 layers component movement according to zone allocation that mention above are: 1. Factory, in other word production stage, to sites inside factory; 2. sites inside factory to foreign site; 3. sites to work site, in other word installation stage. The route of component transportation diagram is shown in figure 4 as follows:

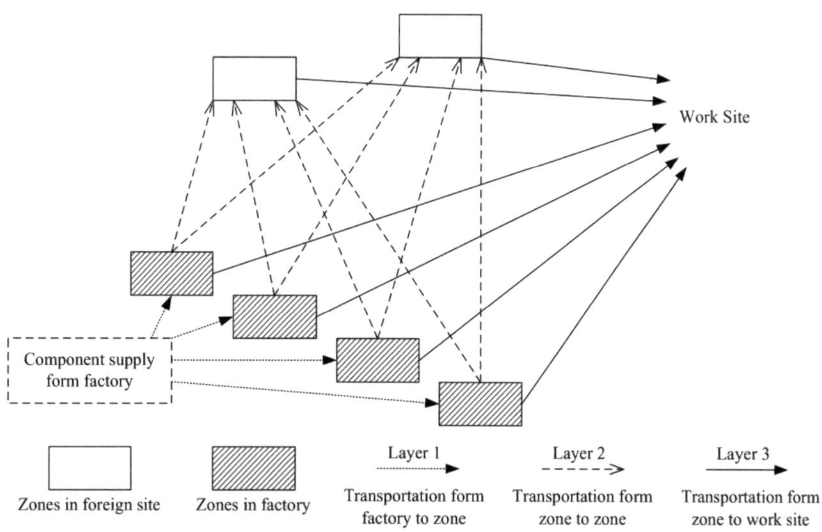

Fig. 4. Component transportation layer

Two types of transportation, within a site and long distance transportation, can be identified into these 3 movements. The case of transportation within the same site occurs when components are moved from factory to sites inside factory, layer 1, obviously. No extra transportation cost will be charged because these movements are completed by equipments belonging to factory. The other, long distance movement occurs in layer 2 and 3 and the truck rental fee according to weight of components and distance between sites will be charged.

5. Storage-transportation optimization model

5.1 Cost classification of model

Base on the proposed component zoning method, the whole procedure of component storage and transportation can be integrated and transferred into a problem of zone selection and allocation. A mathematic optimization model that belong to mix-integer planning, MIP, has been developed. The objective function of this model is to minimize total cost in whole storage and transportation stage of precast factory. Inventory cost has been further divided into two parts, inside and outside factory, that presented as function (2) are site fee of selected sites. In addition, the transportation cost has been divided into at least three parts that presented as function (3) according to different truck rental fee calculation. The structure and classification show as fig 5.

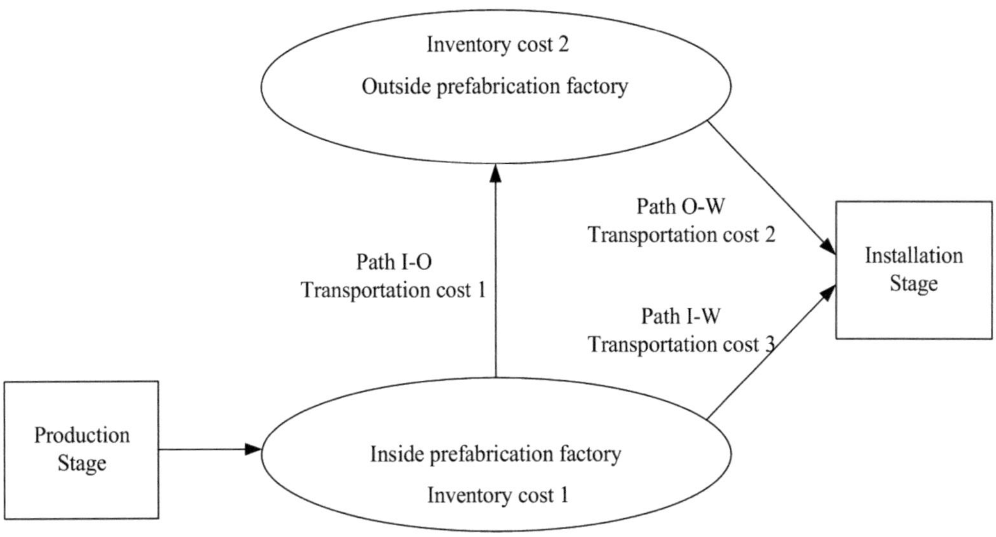

Fig. 5. Cost Classification Chart of the Model

$$\text{Objective function} \quad \text{Minimize } Total\_Cost = IC + TC \tag{1}$$

Where IC is the sum cost in storage stage and TC is the sum cost in transportation stage.

$$IC = \sum_{i}^{np} UP_i * pcs_i + \sum_{j}^{ni} UI_j * ics_j \tag{2}$$

Where i is the index of each site inside factory; np is the total number of sites inside factory; UP_i are binary variables for judgment of usage for each site i by 0 or 1; pcs_i are parameters of site maintain or holding cost of each site when UP_i are value 1; j is the index of each foreign site; ni is the total number of foreign site; UI_j are binary variables for judgment of site usage for each site j by 0 or 1; ics_j are parameters of site rental and holding cost of each site when UI_j are value 1.
pcs_i and ics_j are fixed in consideration of single precast project. Besides, foreign site must be rented until project is completed.

$$TC = \sum_{k}^{ns}\sum_{l}^{ns}\sum_{m}^{ct}\sum_{n}^{p} TSQ_{k,l,m,n} * (tc1_m + tc2_m) + \sum_{k}^{ns}\sum_{l}^{ns}\sum_{m}^{ct}\sum_{n}^{p} (d_{m,n} - TSQ_{k,l,m,n}) * tc3_m \quad (3)$$

Where k and l are both the index of zone. These two parameters present the index of zone that components are transported from and components are transported to respectively when transportation between zones occurs. These situations occur in transportation between sites from inside factory to foreign site; ns is the total number of zones; m is the index of component type considering its sharp, weight and strength; ct is the total number of component type; n is the index of project time by working days; p is total working days of whole project period; $TSQ_{k,l,m,n}$ are positive variables to calculate the quantity of component transportation between zones; $tc1_m$ are parameters of component transportation cost between zones which are calculated by distance and weight of component; $tc2_m$ are the parameters of component transportation cost from foreign site to worksite which are calculated by distance and weight of component; $d_{m,n}$ are parameter of component demand of worksite which present component type and working day as a two dimension matrix; $tc3_m$ are the parameters of component transportation cost from sites inside factory to worksite which calculation by distance and weight of component.
The demand of components is fixed after design stage of a project. In addition, there are 2 paths the components can be only transported by zones inside factory to worksite through foreign site or zone inside factory to worksite directly. The component demand, parameter $d_{m,n}$, of worksite is equal to sum of the component number which transported through these 2 paths and also equal to total sum of produced component of a project in main consideration. Beside, foreign site are rented till the end of project and cannot retain component. Thus, the quantity of transported component from site inside factory to foreign site is equal to the quantity of transported component from foreign site to worksite.
Constraints:
Function (4) - (8) present rules of zone selection and allocation in whole project period.
Judging of site usage

$$\forall i \qquad M * UP_i \geq \sum_{k}^{ns}\sum_{n}^{p} SLP_{k,i,n} \quad (4)$$

$$\forall j \qquad M * UI_j \geq \sum_{k}^{ns}\sum_{n}^{p} SLI_{k,j,n} \quad (5)$$

Where M is a parameter with infinity value named big M in IP; $SLP_{k,i,n}$ are binary variables for zones are allocated into the sites i which inside factory on day n. $SLI_{k,j,n}$are binary variables as $SLP_{k,i,n}$ for foreign site j.
Zone allocation limit 1: Each zone cannot be located in more than one site at the same time.

$$\forall k,n \qquad \sum_{i}^{np} SLP_{k,i,n} + \sum_{j}^{ni} SLI_{k,j,n} \leq 1 \tag{6}$$

Zone allocation limitation 2: No zone's movement can be made after zone allocation.

$$\forall k,i,n \qquad SLP_{k,i,n} \geq SLP_{k,i,n-1} \tag{7}$$

$$\forall k,j,n \qquad SLI_{k,j,n} \geq SLI_{k,j,n-1} \tag{8}$$

Function (9) - (10) present the space limit of each site for zone allocation. In most practical case, the unit of space is square meter.
Site space limit:

$$\forall i,n \qquad pa_i \geq \sum_{k}^{ns} (SLP_{k,i,n} * sa_k) \tag{9}$$

$$\forall j,n \qquad ia_j \geq \sum_{k}^{ns} (SLI_{k,j,n} * sa_k) \tag{10}$$

Where pa_i are parameters of space limit of site i inside factory; sa_k are parameters of space requirement of zone k; ia_j are parameters of the space limit of foreign site j.
Function (11) presents that only when the zone is inside the precast factory can acquire component from production stage.
Supply acquisition Limit:

$$\forall k,m,n \qquad \sum_{i}^{np} M \times SLP_{k,i,n} \geq SS_{k,m,n} \tag{11}$$

Where $SS_{k,m,n}$ are variables to calculate the quantity of component which acquire from production stage on day n.
Function (12) presents that only the zone inside factory can keep component at the end of project.
Final inventory control:

$$\forall k,m \qquad \sum_{i}^{np} M * SLP_{k,i,n} \geq IS_{k,m,period} \tag{12}$$

Where $IS_{k,m,period}$ are variables that the component quantity stored in each zone at the and of project; period is a parameter of index n to present the last working day of project.
Function (13) presents that all components produced in production stage should be first moved into zones which are allocated in sites inside factory.
Supply allocation:

$$\forall m,n \quad \sum_{k}^{ns} SS_{k,m,n} = sc_{m,n} \tag{13}$$

Where, $sc_{m,n}$ are paramaters for the quantity of component in type m which produced in production stage on day n.
Function (14) presents that every zone can have initial inventory at the beginning of project. This function retains the flexibility in multiple project planning and rescheduling issue.
Initial inventory of zone:

$$\forall k,m \quad IS_{k,m,0} = ois_{k,m} \tag{14}$$

Where, $ois_{k,m}$ are parameters for quantity of component m which stored in zone k at the beginning of project.
Function (15) presents that daily inventory calculation of components which stored in each zone.
Daily inventory:

$$\forall k,m,n \quad IS_{k,m,n} = IS_{k,m,n-1} + SS_{k,m,n} + \sum_{l}^{ns} TSQ_{l,k,m,n} - \sum_{l}^{ns} TSQ_{k,l,m,n} - TWQ_{k,m,n} \tag{15}$$

Where, $TWQ_{k,m,n}$ are variables for quantity of component m which stored in worksite on day n.
Function (16) presents as the base zoning rules that the quantity of component which is stored in a zone can not exceed the component quantity limit of this zone.
Zone storage limit:

$$\forall k,m,n \quad IS_{k,m,n} \le msq_{k,m} \tag{16}$$

Where, $msq_{k,m}$ are parameters for the maximum component quantity in zone m.
Functions (17)-(20) present inventory relationships in working site.
Initial inventory of worksite:

$$\forall m \quad IW_{m,n} = oiw_{m} \tag{17}$$

Where, oiw_m are parameters of initial inventory of component m which are stored in worksite at the beginning day of project.

Daily inventory of worksite:

$$\forall m,n \qquad IW_{m,n} = IW_{m,n-1} + \sum_{k}^{ns} TWQ_{k,m,n} - d_{m,n} \tag{18}$$

Limit of component in worksite storage:

$$\forall m,n \qquad IW_{m,n} \le mwq_{m} \tag{19}$$

Where, mwq_m are parameters for limit of components which the max quantity of component can store in worksite.
Final inventory of worksite:

$$\forall m \qquad IW_{m,p} = 0 \tag{20}$$

Functions (21)-(23) are detail constraints of transportation between zones to make transportation path of component fallow as fig. 4.
Zone Transportation Limit 1:

$$\forall k,m,n \qquad TSQ_{k,k,m,n} = 0 \tag{21}$$

Transportation within a site is not allowed
Zone Transportation Limit 2:

$$\forall k,n \qquad \sum_{i}^{np} SLP_{k,i,n} * M \ge \sum_{l}^{ns} \sum_{m}^{ct} TSQ_{k,l,m,n} \tag{22}$$

Zones in foreign site are not allowed to transport components to other zones.
Zone Transportation Limit 3:

$$\forall k,n \qquad \sum_{j}^{ni} SLI_{k,j,n} * M \ge \sum_{l}^{ns} \sum_{m}^{ct} TSQ_{k,l,m,n} \tag{23}$$

The storage zones inside the factory are not allowed to accept any components transported from zones.
All related variables and parameters in the purposed model are positive integers.

5.2 Solving tool

The optimization model proposed herein can be modeled and solved by most of mathematical programming tools. This study adopts ILOG OPL program that contains a mathematic optimization engine named CPLEX to solve the proposed model.

6. Experiment of model

6.1 Input data

In order to verify the accuracy of the model proposed in this study, an experiment is presented by using a small case study. Data include environmental parameters such as production supply, demand information of worksite and rules of zoning strategies, as shown in Table 3 as follows:

Symbol	Value
np	2 (sites)
ni	3 (sites)
ns	3 (zones)

Symbol	Value
ct	3 (types)
p	7 (days)

	Site 1 inside factory	Site 2 inside factory
pcs_i	18,000	30,000
pa_i	1,500	2,500

	site 1 outside factory	site 2 outside factory	site 3 outside factory
ics_j	100,000	50,000	25,000
ia_j	5,000	3,000	2,000

	Com 1	Com 2	Com 3
oiw_m	0	0	0
mwq_m	5	5	5
$tc1_m$	60	60	70
$tc2_m$	350	350	450
$tc3_m$	350	350	450

	Zone 1	Zone 2	Zone 3
sa_k	2,000	2,000	2,500

	Zone #	Com 1	Com 2	Com 3
$ois_{k,m}$	1	0	0	0
	2	0	0	0
	3	0	0	0
$msq_{k,m}$	1	0	0	25
	2	0	0	20
	3	10	10	20

	Com #	Day 1	Day 2	Day 3	Day 4	Day 5	Day 6	Day 7
$d_{m,n}$	1	0	0	0	0	0	10	0
	2	0	0	0	0	0	10	0
	3	0	0	0	0	0	0	30
$sc_{m,n}$	1	0	0	0	10	0	0	0
	2	0	0	5	0	5	0	0
	3	10	10	5	0	5	10	0

Table 1. Related Input Parameters of a Case Study

6.2 Output results

The objective value of optimal solution is 76,550 that contains the inventory cost is 55,000 and the transportation cost is 21,550. The details of the optimization solution for variables can be presented as follow:

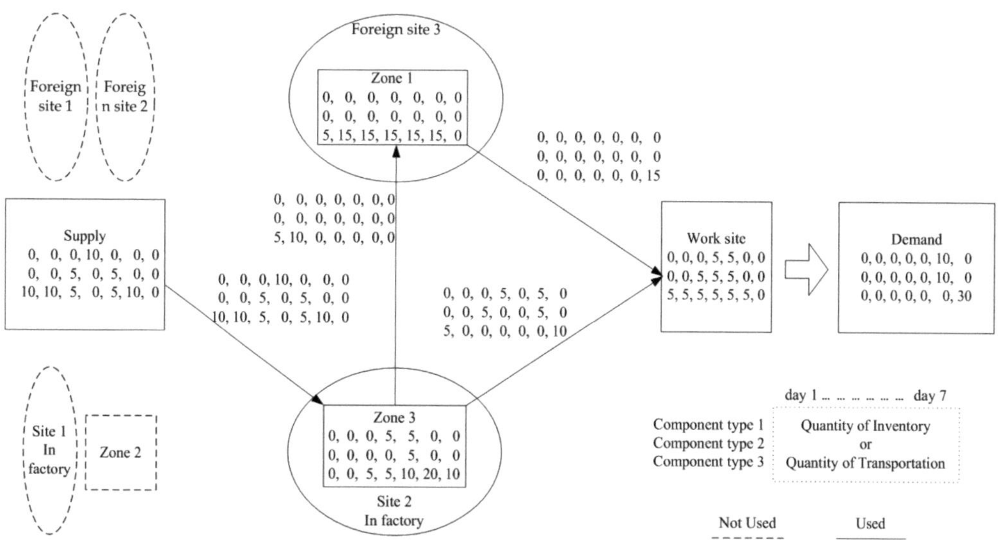

Fig. 7. Output of case experiment

This case experiment is not complex enough to match a practical situation of precast factory business. However, it is easier to explain how the proposed model works and how the zone strategy are formed according to data in table III. A supply-demand relationship that three components' demand of installation and supply of production are identified by array $d_{m,n}$ and $sc_{m,n}$. Component quantity of supply exceeds demand to form a pure case of storage and transportation problem that components is sufficient and some of them will be residual at end of considered period. Figure 7 shows whole storage and transportation information.

In order to store components, zone selection and allocation must be done before any other decisions. However, 3 zones with their rules of $msq_{k,m}$; sa_k and 5 sites that their rental fee: pcs_i ; ics_j and storage space: pa_i ; ia_j are known in consideration. Zone 1 and zone 3 are selected during optimization proceed and they are allocated into site 2 inside factory and foreign site 3 respectively to meet the objective function. By the way, zone 2, site 1 inside factory, foreign site 1 and foreign site 2 are not adopted. All this decisions assembles the best zoning strategy. It is one of important purposes of this research that planners do not fully recognize how to store all components they have, but some clues of zoning rules are known according to their experience of factory business. In this model, all kinds of zones with their own rules that can be considered as parameter or constraints, and several foreign sites that can be use to assist factory in storing components are allowed. In addition, the best zoning strategy will be picked out through this optimization model. In order to avoid unnecessary limit of zoning strategy forming, planner can consider a great quantity of zones with possible rules. However, additional zones and sites will cause more calculation efforts are needed, but a planner with experience is supposed to screen impossible zones and sites to make all parameters reasonable.

Transportation issue can be further resulted after zone allocation. Thus, vehicle path is also determined with certain transported component. This model avoids unnecessary component movement according to function 21-23. That is why zone 1 stores only 15 components although $msq_{1,3}$ is equal to 25. All other components are transported to worksite directly.

Repeat the optimization process is encouraged to refine the zone strategy because unsuitable zone may occur. The example of zone 1 that mention before implies rules of zone 1 need to be adjusted if able or other suitable zone can be argued because it is not fulfilled during the considered period. Zones of this case study are superficial. However, the output can assist planner to refine their zone strategy with their experience or the latest optimized zoning strategy.

PS and IS are not only two choices of zone types but they are practical. In this case study, zone 1 and zone 2 can be classified into PS because only component 3 can be stored in but zone 3 is relaxed. This study retains possibilities of zones in order to meet kinds of storage environment or ways to store component. Of course, PS and IS are included. However PS always occurs in production stage and IS in installation stage. No matter how varied zone strategy is formed, PS and IS still exist even the conflict between them. To a practical zone strategy, zones that contain PS and IS are encouraged.

The whole process of solution finding took 0.46 seconds only with 1.6GHz CPU and 768 MB RAM of computer. It implies that despite the complicatedness of constraints, which contained 594 constraints and 523 variables, the model could still obtain optimized results. The solving screen of the model is shown in Figure 5 as below:

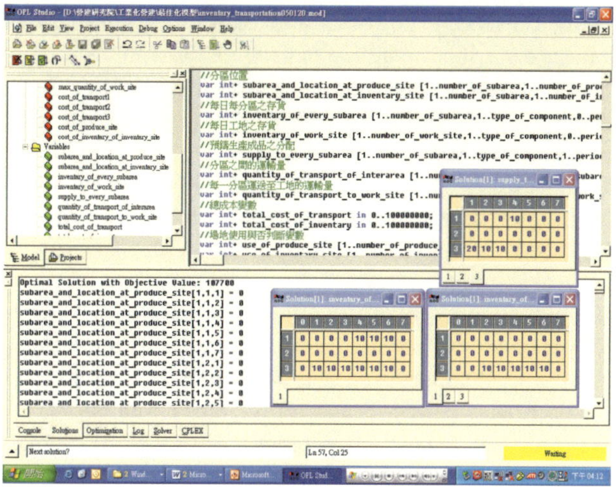

Fig. 8. Screen of Solving by ILOG OPL

7. Conclusion

Storage and transportation of precast component are quite practical issues. This study tries to point out its importance through PS and IS discussions, and an optimization model is proposed to refine a best zoning strategy through zone selection and allocation procedure. This model is proved that it can be executed and obtain solutions effectively.

To create zones with practical and significant rules is important to perform this proposed model, and then zones will be confirmed for best zone strategy. This model is used as means to assist planner approaching a beneficial decision of precast factory in storage and transportation stage. Nevertheless, to set rules of each zone based on planner's experience is still critical.

8. References

Chan, W. T. & Hao, H. (2002). Constraint programming approach to precast production scheduling, *Journal of Construction Engineering and Management,* Vol. 128, No. 6, November/December pp. 513-521,.

Sou-Sen Leu & Shao-Ting Hwang (2002). GA-based resource-constrained flow-shop scheduling model for mixed precast production. *Automation in Construction,* Vol. 11, Issue 4, pp. 439- 452.

Weng-Tat Chan & Zhen Zeng (2003). Coordinated production scheduling of prefabricated building components, *Construction Research Congress - Wind of Change: Integration and Innovation.*

W. T. Chan & Hao Hu (2002). Production scheduling for precast plants using a flow shop sequencing model, *journal of computing in civil engineering,* Vol. 16, No. 3, July 2002, pp. 165-174.

Weng-Tat Chan and Zhen Zeng (2005). Rescheduling precast production with multiobjective optimization, *Computing in Civil Engineering.*

Wenfa Hu (2005). Automatic construction process of prefabricated buildings on geometric reasoning, *Construction Research Congress 2005: Broadening Perspectives.*

Wah-Ho Chan & Ming Lu (2005). Logistics and operations simulation in precast viaduct construction: case study, *Computing in Civil Engineering*.

Luisa Equi & Giorgio Gallo & Silvia Marziale & Andres Weintraub (1997). "A combined transportation and scheduling problem", *European Journal of Operational Research*, Vol. 97, Issue 1, pp. 94-104.

Sıla Cetinkaya & Fatih Mutlu & Chung-Yee Lee (2006). A comparison of outbound dispatch policies for integrated inventory and transportation decisions, *European Journal of Operational Research*, Vol. 171, Issue 3, pp. 1094-1112.

Acheson & Glover, Precast storage system,
http://www.acheson-glover.com/constructionpro/precast/storage_systems.

New Zealand the Department of Labour (2006). Unsafe A-frames used for transportation and storage of precast concrete panels, *Accident Alert.* No. 13.

Iris F.A. Vis & Ren_e de Koster (2003). Transshipment of containers at a container terminal: An overview, *European Journal of Operational Research,* 147 1–16.

Mordecai Avriel & Michal Penn & Naomi Shpirer & Smadar Witteboon (1998). Stowage planning for container ships to reduce the number of shifts, *Annals of Operations Researc,h* 7655 - 71.

Sadiq, M. B. & Landers, T. L. & Taylor, G. D. (1996). An assignment algorithm for dynamic picking systems, IIE Transactions, Vol. 28, pp. 607-616.

11

Pose Estimation of Construction Materials by Acquisition of Multiple IDs of Devices

Tomohiro Umetani[1], Kenji Inoue[2] and Tatsuo Arai[3]
[1] *Department of Intelligence and Informatics, Konan University*
[2] *Graduate School of Science and Engineering, Yamagata University*
[3] *Graduate School of Engineering Science, Osaka University*
Japan

1. Introduction

It is a crucial issue for building the relationship between states of the construction materials at the construction site and their information such as their existence, pose (pose and orientation), forms, task information for workers or robots, and so on, for efficient construction automation (Umetani et al., 2006). One of the methods for building the relationship is using ID devices, for example, radio frequency identification devices and barcodes (Want et al., 1999) (Penttila et al., 2004). Several studies using ID devices attached to the construction materials have been introduced, for example, tracking of the construction materials at the construction site (Akinchi et al., 2002) (Jaselskis et al., 1995) (Jaselskis & El-Misalami, 2003), tracking of the tools for the workers (Goodrum et al., 2006), and scheduling of the construction site including the production of the construction materials (Yagi et al., 2005).

This chapter describes a method of the pose (position and orientation) of the construction materials using multiple ID devise to maintain the relationship between the status of the materials and their information. Object pose is important status for construction tasks; however, the object pose can be changed by the workers or robots that cannot update the data of the object. In this case, the workers and robots should measure the object pose. The proposed method estimates the object pose using at least two ID devices attached to the object. The chapter focuses on the simplification of the pose estimation and the geometrical condition of the ID devices using pose estimation. Several pose estimation method of the construction materials. Furlani has been proposed pose estimation using at least not aligned three RFID devices based on the device position with respect to the object coordination frame (Furlani & Stone, 1999). Umetani has been proposed pose estimation using at least two ID devices attached to the different sides of the object based on the device position and orientation with respect to the object coordination frame (Umetani et al., 2003). Former one is not considered to the motion planning of the ID reader because the method does not apply the orientation of the ID reader. On the other hand, the latter one uses the full orientation of the ID device. It is difficult to register the orientation of the ID devices.

This chapter shows feasibility of the proposed method through the modeling result of the pose estimation and experimental results. The paper introduces integration of the objects

such as the construction materials and their information using ID devices. We show the method for estimation of the object pose based on acquisition of the multiple ID from the ID devices attached to the object. Experimental results show feasibility and effectiveness of the method.
The organization of this chapter is the following. Section 2 introduces object-pose estimation using multiple ID devices. Section 3 describes an algorithm of estimation of pose of object using multiple ID devices. Section 4 shows experimental results by numerical simulation and pose estimation experiment using the motion capture system. Section 5 describes the discussion and the conclusion of the chapter.

2. Object-pose estimation using ID devices

This section describes a method of object-pose estimation using ID device attached to the large object such as the construction materials. First, we introduce the integration of objects at the construction site and their information using ID devices. Then, the section illustrates the object pose using ID devices to maintain the relationship between the objects at the site and their information.

2.1 Integrated management of objects and their information using ID devices

The recent advancement of information and communication technologies has brought the integrated management of objects and their information using ID devices. Figure 1 shows the concept of the intelligent environment using ID devices at the construction site for building the relationship. Using the ID devices, workers and robots can obtain the information of the construction materials by obtaining the IDs of the devices attached to the materials using their ID readers. Then, the workers or robots can achieve tasks using the information using acquisition of the ID attached to the objects.

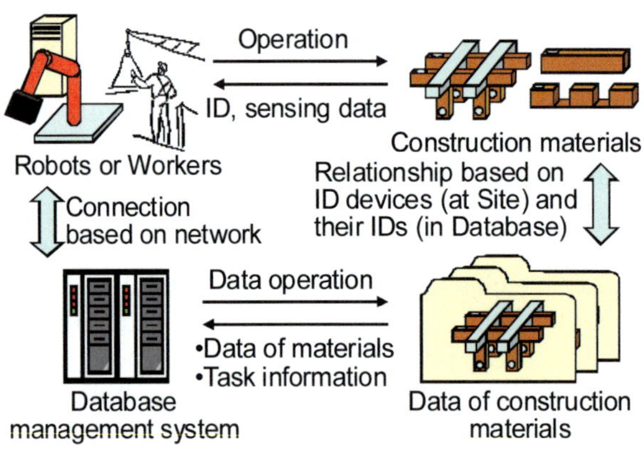

Fig. 1. Integration of objects at the construction site and their information using ID devices.

The integrated management of the status of the objects and their information is applied to various situations in robotics and automation fields, for example, navigation in the environment (Hahnel et al., 2004) (Jia & Takase, 2007), a manipulation task (Umetani et al., 2006). Using the ID devices, the workers or robots can identify the objects easily and perfectly, since the ID of the device is unique in the environment. The workers or robots updates the information stored in the database using the relation between the objects in the

site and the information in the database since the database server can search the information that should be updated easily.
It is required to maintain the relationship between the status of the materials and their information to realize integration of the construction materials with their information using ID devices. It is required that the status of the construction materials at the construction site is identical with that of the information of the construction materials in order to realize the integration. If the stored information in the database conflicts with the status of the materials at the site, the workers or robots cannot obtain the correct information using the ID devices. Therefore, the integrated management of the construction materials and their information using ID devices ends in failure. In this case, the workers and robots should acquire the status of the materials using their equipments and update the information of the materials stored in the database.

2.2 Object-pose estimation using ID devices

The pose of construction materials is important status for the workers and robots to achieve the tasks. However, the construction materials can be moved by the workers or robots that cannot update the information of the materials using the ID device. As a discussion in the former subsection, the object pose should be estimated by the workers and robots at the construction site. In addition, it is difficult to estimate the pose of the large object directly using the measurement equipment such as the fixed cameras at the site since the range of the cameras is narrow so that the large environment is required.
We propose a method of estimation of object pose (position and orientation) using multiple ID devices. The method estimates the object pose using the acquisition of the ID of the devices attached to the object and the relative movement of the ID reader. The assumptions of the method are described as follows:

- The object size is much larger than the communication area of the reader.
- Multiple ID devices are attached to each object.
- The position and orientation of the ID reader with respect to the reference coordination frame can be measured accurately
- The position and direction of each ID device with respect to the object coordination frame are registered in the database.
- The proposed method does not conflict the assumption of the conventional updating method for the information of the construction materials.

In the following sections, we define the geometrical model of the ID reader and the device when the ID reader acquires the ID of the device attached to the object. The properties of the method are described using the derived geometrical model of the ID acquisition. In addition, the paper illustrates the pose estimation experiment. Modeling results of the pose estimation and experimental results show feasibility of the proposed method.

3. Modelling of object pose estimation using multiple ID devices

This section describes a method of object-pose estimation using multiple ID devices. The method supposes that the position and orientation of the ID reader can be measured on acquisition of the ID. The section introduces the geometrical relation between the ID reader and the device. Then, the section shows that the object pose is estimated using at least two ID devices attached to the object.

3.1 Geometrical condition of ID reader

First, we define the geometrical condition of the ID reader and the device on acquisition of the ID. We suppose a commercial small RFID reader as an ID reader in the paper. The ID reader has small communication area. In addition, the object size is much larger than the size of the communication area. The properties of the ID reader are described as follows:

- The ID reader has a communication area along the axis of the reader as shown in figure 2.
- The ID reader can acquire the ID of the device if the ID device is in the communication area of the ID reader and the device directs to the reader. The ID reader has the range of the inclination of the device to the reader.
- The ID reader can acquire the ID of the device even if the device rotates about the axis of the direction of the ID reader.

Figure 2 shows the model of the ID reader and ID device on acquisition of the ID. $\Sigma_{R,D1}$, $\mathbf{p}_{c,D1}$, R_{WD1}, $\mathbf{p}_{o,D1}$ and $\mathbf{k}_{o,D1}$, indicate the coordination frame of the ID reader, position of the device with respect to the reader coordination frame, the device position with respect to the object coordination frame, rotation with respect to the reader coordination frame, the device position with respect to the object coordination frame and the device direction with respect to the object coordination frame, respectively. The size of $\mathbf{k}_{o,D1}$ is 1. The reader direction is set to the z-axis of the reader coordination frame. The range of $\mathbf{p}_{c,D1}$ and R_{WD1} are determined by the properties of the ID reader, while these parameters cannot be determined on acquisition of the ID.

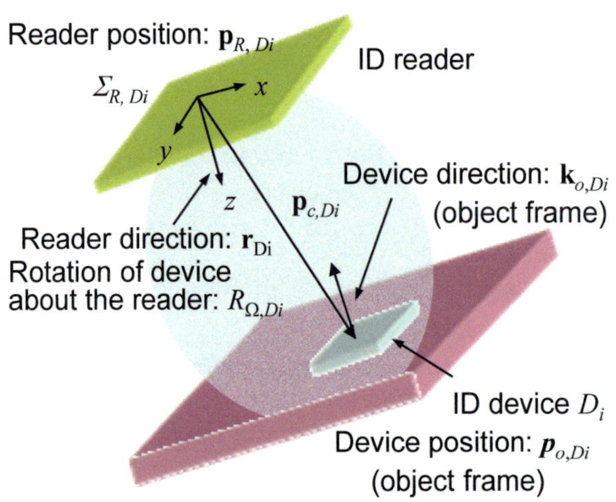

Fig. 2. Geometrical relationship between ID reader and ID device.

The geometrical relation between the ID reader and the ID device D_i on acquisition of the ID is described as follows:

$$\mathbf{p}_{R,D_i} + {}^{W}R_{R,D_i}\mathbf{p}_{o,D_i} - {}^{W}R_o\mathbf{p}_{o,D_i} - \mathbf{p}_o = \mathbf{0} \tag{1}$$

$${}^{W}R_{R,D_i}\begin{bmatrix}0\\0\\1\end{bmatrix} \cdot {}^{W}R_o R_{\Omega_{D_i}} \mathbf{k}_{o,D_i} = -1 \tag{2}$$

$$ {}^{W}R_{R,D_i}\begin{bmatrix}0\\0\\1\end{bmatrix}\times {}^{W}R_o R_{\Omega_{D_i}}\mathbf{k}_{o,D_i}=\mathbf{0}, \tag{3}$$

where ${}^{W}R_o$ and $\mathbf{p}_o$ indicate the object orientation and position, respectively. ${}^{W}R_o$ is given by a rotation matrix. The set of the object pose is described by the parameter sets of the ID reader. For each ID acquisition, the geometrical relation between the reader and the device is defined. The product set of the object poses is the estimated object pose.

3.2 Object pose estimation using multiple ID devices

We describe a model of estimation of the object pose using minimum two ID devices attached to the object based on the relative displacement between each ID device. It is difficult to estimate the relative position and orientation in the communication area of the reader when the reader acquires the ID of the device. In addition, the orientation of the device attached to the object is not determined by the definition of the reader model.

The proposed method estimates the object pose as the model estimation based on the measurement data of the ID reader and the devices. We add the assumption that the position of the device with respect to the reference coordination frame is the position of the reader, and the device direction faces the front of the ID reader. Therefore, we consider $\mathbf{p}_{c,Di}$ in equation (1) and ${}^{R}W_{Di}$ in equations (2) and (3) as the minimal displacement from the reader and initial rotation matrix, respectively.

We define the model of the pose estimation based on the measurement of the pose of the ID reader with former assumption. Equations (1), (2), and (3) are derived as follows:

$$ {}^{W}R_o\mathbf{p}_{o,D_i}+\mathbf{p}_o=\mathbf{p}_{R,D_i} \tag{4}$$

$$ \mathbf{r}_D\cdot {}^{W}R_o R_{\Omega_{D_i}}\mathbf{k}_{o,D_i}=-1 \tag{5}$$

$$ \mathbf{r}_D\times {}^{W}R_o R_{\Omega_{D_i}}\mathbf{k}_{o,D_i}=\mathbf{0}, \tag{6}$$

where $\mathbf{r}_{Di}$ indicates the direction vector of the ID reader when the reader acquires the ID of the device. The size of direction vector $\mathbf{r}_{Di}$ is 1. The estimation parameters are object pose $\mathbf{p}_o$ and ${}^{W}R_o$, we set the matrix of the object orientation ${}^{W}R_o$ as

$$ {}^{W}R_o=[\mathbf{n}_o \quad \mathbf{s}_o \quad \mathbf{a}_o]=\begin{bmatrix} n_{o,x} & s_{o,x} & a_{o,x}\\ n_{o,y} & s_{o,y} & a_{o,y}\\ n_{o,z} & s_{o,z} & a_{o,z}\end{bmatrix}.$$

We obtain seven equations by each ID acquisition based on equations (1), (2), and (3) for vector $\mathbf{x}^T=[\mathbf{p}_o^T\ \mathbf{n}_o^T\ \mathbf{s}_o^T\ \mathbf{a}_o^T]^T$. We can estimate the object pose using multiple devices since the object pose with respect to the reference coordination frame is not changed. From equations (4), (5), and (6), if there are the ID devices that the object direction is the same and opposite as that of the other device, efficient equations for pose estimation is decreased. We express the vector $\mathbf{a}_o$ by the cross product of vector $\mathbf{n}_o$ and $\mathbf{s}_o$ since the matrix ${}^{W}R_O$ is the orientation matrix.

Through the above process, the object pose is estimated using the measurement data of the ID reader when the reader acquires the ID of the device. In the case that the reader acquires the ID of the device that the device direction is the same and opposite as that of the other device, efficient equations for pose estimation is decreased, however, the equation (1) can be set independently since the position of the other device is different. As a result, at least 10 equations are defined for nine parameters of the object. Therefore, the method can estimate the object pose based on the acquisition of the ID of at least two devices as shown in Figure 3.

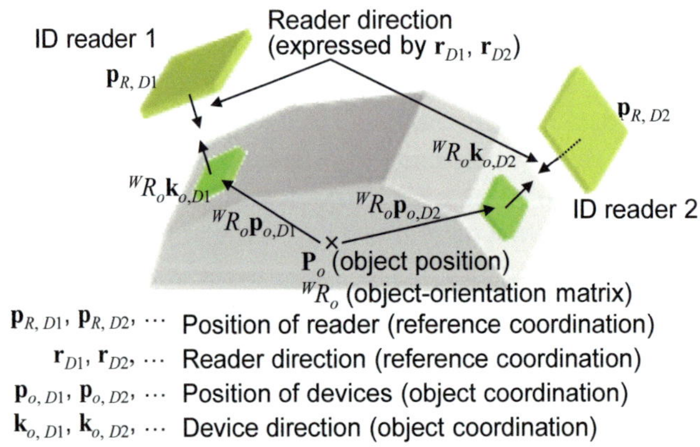

Fig. 3. Object pose estimation using two ID devices

3.3 Features of proposed method

We describe the features of the proposed pose estimation method using multiple ID devices. The method can estimate the object pose based on acquisition of the ID of at least two ID devices. In addition, the method can estimate the object pose using the acquisition of the device attached to the same side of the object. This fact makes the motion planning of the ID reader easier. In the previous pose estimation method, the device poses of the devices that are attached to the other side of the object are needed.

The method uses the direction of the ID reader with respect to the reference coordination frame. This fact is that the motion planning of the ID reader is considered. If the direction of the ID reader is not considered, the geometrical relation between the reader and the device is not obtained when the ID reader acquires the ID of the first device. The other sensing device is needed to obtain the geometrical relation between the reader and the device. By the proposed method, the object side that the ID device is attached faces the front of the ID reader. The workers or robots can move the ID reader using acquisition of the ID of the other device.

The method uses the direction of the ID devices with respect to the object coordination frame. This fact enables the motion planning of the ID reader. Moreover, the direction of the ID device is easy to register the database, since the device is attached to the object. In fact, the direction of the device can be registered as the normal vector of the object surface; the parameters are need for production of the construction materials.

The object pose estimated by the proposed method is not always satisfied that the relative position of each device $\mathbf{p}_{c,Di}$ and $^{R}W_{Di}$ from the estimated object pose can be out of the range of the ID reader. However, the ID reader acquires the ID of the device at the measured object pose. We can derive the real set of the object pose using the estimated object pose as the initial value.

4. Object-pose estimation experiment

This section describes an object-pose estimation experiment to show feasibility of the proposed method. The experiment assumes that workers obtain the ID of the device attached to the object using the small RFID reader. First, we introduce the numerical simulation including the pose error of the reader. Next, we show the pose estimation experiment using the motion-capture system.

4.1 Numerical examples

We have carried out the numerical simulation of the pose estimation including the pose error of the ID reader to show feasibility of the method. The ID reader has communication area and the reader can acquire the ID of the device that is inclined from the ID reader. The fact causes the error and variance of the pose estimation results. We have carried out the simulation on the assumption that the workers and robots acquire the ID of the device that is inclined in the communication area of the reader.

We set two conditions of the layout of the ID devices; aligned two ID devices that are same direction (condition 1), and not-aligned three ID devices that are same direction (condition 2), respectively. The registered positions of the ID devices with respect to the object coordination frame are set at random. The distance between each ID device is set 1.0 - 2.0 [m]. In condition 2, the direction of the devices is set as the normal vector of the plane whose vertices indicate the device position. That is, the condition 2 is considered as the condition that the ID devices are attached to the same side of the object. The object is set in the reference coordination frame at random.

The pose of the ID reader is set the position of the ID device with respect to the reference coordination frame for position and the opposite direction of the device with respect to the reference coordination frame, respectively. In addition, the translational and rotational displacements of the reader are set at random. The ranges of the displacements are 20 [mm] for each axis about the translational displacement and 5, 30 [deg] for each axis about the rotational displacement, respectively. The distribution of the displacement is uniform distribution.

We have carried out the pose estimation under the following conditions. We used the object pose based on Gauss - Newton Method. The initial value of the object pose was set $\mathbf{p}_o = [0, 0, 0]^T$ and the first and second column of the initial orientation matrix as $\mathbf{n}_o$ and $\mathbf{s}_o$, respectively. We set 10 poses of the object, and estimate the object pose 10 times for each object layout.

Table 1 shows the average estimation error of the object pose for each condition of the acquisition. This table shows the average error of the position and the average error of the angle for each axis under the accurate and worst accurate case. Table 2 shows the pose error of the worst accurate case with respect to the position accuracy. These tables show the result of the numerical simulation; the simulation results are qualitative result, however, the results are not analytical.

As shown in Tables 1 and 2, in condition 1, the error of the object pose increases as the orientation error of the ID reader. On the other hand, in condition 2, the error of the object pose does not increase as the orientation error of the ID reader. In the case that the ID reader acquires the ID of the collinear and same-directed devices, the estimated pose error of the object increases as the orientation error of the ID reader increases.

Conditions	Average position error (Min. - Max.) [mm]	Average error of angle for each axis (Min. - Max.) [deg]
Condition 1 (Orientation error: 5[deg])	32.70 - 74.06	0.30 - 2.734
Condition 2 (Orientation error: 5[deg])	29.90 - 64.76	0.13 - 2.464
Condition 1 (Orientation error: 30[deg])	62.00 - 406.75	3.49 - 19.59
Condition 2 (Orientation error: 30[deg])	27.55 - 64.86	0.13 - 2.04

Table 1. Average estimation errors of object pose in numerical simulation.

Conditions	Largest position error [mm]	Oriental error for each axis [deg] (Roll - Pitch - Yaw)
Condition 1 (Orientation error: 5[deg])	153.47	2.66 - 1.84 - 6.60
Condition 2 (Orientation error: 5[deg])	115.18	1.11 - 0.31 - 5.70
Condition 1 (Orientation error: 30[deg])	937.28	9.56 - 2.00 - 34.20
Condition 2 (Orientation error: 30[deg])	112.93	1.00 - 1.17 - 1.49

Table 2. Position error in the worst accurate case.

If the workers or robots estimates the object pose based on the acquisition of the ID of the collinear devices, it is difficult to estimate object pose accurately and stably because of the range of the communication area of the reader and the range of the inclination of the device to the front of the ID reader. The ID reader has the range of the inclination of the device to the front of the ID reader.

The workers or robots can estimate the object pose using the additional ID devices to improve the pose accuracy. They can obtain the position and direction of the neighbor ID device when they acquire the ID of the device attached to the object, since they can search for the information of the other device attached to the object. In addition, the workers or robots make a motion plan for acquisition of the other ID of the device if the workers or robots estimate the device pose precisely using the range of the communication area of the ID reader (Umetani et al., 2005).

4.2 Pose estimation experiment

Next, we have carried out the pose estimation experiment using the motion capture system to show feasibility of the proposed method. The experiment has been carried out on the assumption that the workers at the construction site read the ID devices attached to the object using the small ID reader such as the mobile ID reader.

Figure 4 shows the procedure of the pose estimation experiment using the motion capture system. The motion capture system obtains position and direction of the ID reader, then the experimental system estimates the object pose using the data, which are position and direction of the ID devices. On the other hand, the motion capture system obtains the position of the markers attached to the object, then the experimental system estimates the object pose using the motion capture data. The estimated object pose is compared with the object pose estimated by the acquisition of the ID from the devices attached to the object. In this experiment, the operator moves the ID reader by hand.

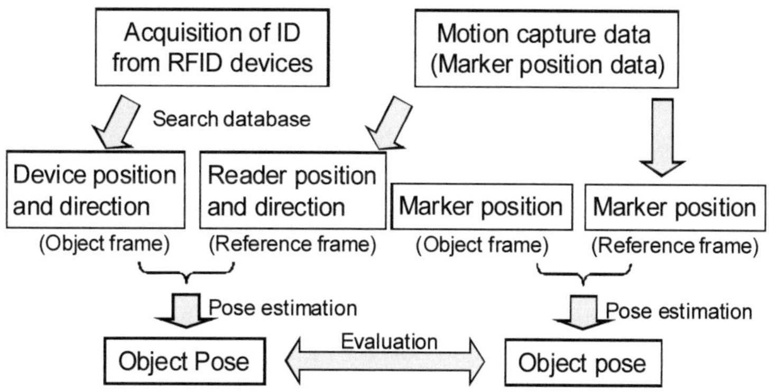

Fig. 4. Procedure of pose estimation experiment.

The registered positions of the ID devices with respect to the object coordination frame are set D_1 to D_3 as (207, 0, -290), (-193, 0, -235) and (165, 0, 320) [mm], respectively. The direction of each device $\mathbf{k}_{o,Di}$ is set (0, 1, 0). We set the object in the field of measurement of the motion capture system with propriety and estimate the object pose.

We have carried out the pose estimation under the following conditions. We used the object pose based on Gauss - Newton Method. The initial value of the object pose was set $\mathbf{p}_o = [0, 0, 0]^T$ and the first and second column of the initial orientation matrix as $\mathbf{n}_o$ and $\mathbf{s}_o$, respectively. We set nine poses of the object, and estimated the object pose using ID reader motion. The operator acquires the IDs of the devices attached to the object in the predetermined sequence, $D_1 \rightarrow D_2 \rightarrow D_3$ in each object layout. We compared the estimation results using three ID devices with those using two ID devices. The estimation results correspond with the result based on the aligned two ID devices and not-aligned three ID devices, respectively.

We have estimated the pose of the object under various conditions, then the error and standard deviation of the object pose using the not-aligned three ID devices are small. Table 3 shows the estimation error and standard deviation of the object pose when the operator estimates the same object pose 10 times. The upper row of the element in the table indicates the position error or standard deviations for each axis. The lower row of the element indicates the orientation error or standard deviations for each axis, which is Roll - Pitch - Yaw. From table 2, the workers or robots can estimate the object pose in each condition. In addition, they can estimate the object pose accurately and stably using not-aligned ID devices. The estimation results are qualitative; the quantitative analysis of the pose error in the estimation is required in order to introduce the proposed method in the real environment for construction automation.

Conditions	Average pose error Upper: (x - y - z) [mm] Lower: (R - P - Y) [deg]	Std. dev. of object pose Upper: (x - y - z) [mm] Lower: (R - P - Y) [deg]
Level object (3 devices)	9.83 - 11.00 - 19.39 0.38 - 1.83 - 12.60	2.46 - 1.89 - 2.30 0.14 - 0.41 - 0.19
Level object (2 devices)	16.12 - 4.38 - 20.67 1.43 - 2.35 - 11.62	5.48 - 4.31 - 5.07 0.42 - 0.46 - 0.63
Inclined object (3 devices)	14.25 - 5.72 - 16.56 2.50 - 0.11 - 6.82	4.02 - 1.55 - 2.29 0.83 - 0.53 - 0.52
Inclined object (2 devices)	11.30 - 26.87 - 29.71 5.11 - 1.69 - 2.77	8.31 - 6.32 - 5.24 1.24 - 1.22 - 0.84

Table 3. Estimation error in pose estimation experiment.

The registered positions of the ID devices with respect to the object coordination frame are set D_1 to D_3 as (207, 0, -290), (-193, 0, -235) and (165, 0, 320) [mm], respectively. The direction of each device $\mathbf{k}_{o,Di}$ is set (0, 1, 0). We set the object in the field of measurement of the motion capture system with propriety and estimate the object pose.

We have carried out the pose estimation under the following conditions. We used the object pose based on Gauss - Newton Method. The initial value of the object pose was set $\mathbf{p}_o = [0, 0, 0]^T$ and the first and second column of the initial orientation matrix as $\mathbf{n}_o$ and $\mathbf{s}_o$, respectively. We set nine poses of the object, and estimated the object pose using ID reader motion. The operator acquires the IDs of the devices attached to the object in the predetermined sequence, $D_1 \rightarrow D_2 \rightarrow D_3$ in each object layout. We compared the estimation results using three ID devices with those using two ID devices. The estimation results correspond with the result based on the aligned two ID devices and not-aligned three ID devices, respectively.

We have estimated the pose of the object under various conditions, then the error and standard deviation of the object pose using the not-aligned three ID devices are small. Table 3 shows the estimation error and standard deviation of the object pose when the operator estimates the same object pose 10 times. The upper row of the element in the table indicates the position error or standard deviations for each axis. The lower row of the element indicates the orientation error or standard deviations for each axis, which is Roll - Pitch - Yaw. From table 2, the workers or robots can estimate the object pose in each condition. In addition, they can estimate the object pose accurately and stably using not-aligned ID devices. The estimation results are qualitative; the quantitative analysis of the pose error in the estimation is required in order to introduce the proposed method in the real environment for construction automation.

5. Conclusions

This chapter describes a method of estimation of object pose using multiple ID devices for construction automation. We have introduced the geometrical model of the ID reader and the device in acquisition of the IDs. The proposed method can estimate the object pose using at least two ID devices attached to the object. The chapter discuss the properties of the method based on the geometrical model of the ID reader. Experimental results have shown feasibility of the proposed method.

The proposed method uses the position and direction of the ID reader and the devices attached to the object. Using the direction of the ID device makes the registration of the object pose easier. In addition, the assumption of the ID reader has feasibility of the development of the ID reader that the workers operate at the construction site. These features are more suitable for the real environment.
The quantitative analysis of the pose error in the estimation, improvement of the pose accuracy using the other sensing devices, simplification of the pose estimation, and the planning of the ID reader and the robots at the construction site are the future works for realization of the proposed method.

6. Acknowledgements

The authors would express our appreciation to Prof. Kiyoko YOKOYAMA and Mr. Tsuyoshi MATSUKAWA, Nagoya City University, for suggestive discussion in the planning of the experiment using motion capture system, and Mr. Hirofumi SAKAMOTO, Nagoya City University for cooperation in the pose estimation experiment.

7. References

Akinchi, B.; Patton, M. & Ergrn, E. (2002). Utilizing Radio Frequency Identification on Precast Concrete Components - Supplier's Perspective, *Proceedings of the 19th International Symposium on Automation and Robotics in Construction*, pp. 381 - 386, Gaithersburg, USA, Sept. 2002

Furlani, K. M. & Stone, W. C. (1999). Architecture for Discrete Construction Component Tracking, *Proceedings of the 16th IAARC/IFAC/IEEE International Symposium on Automation and Robotics in Construction*, pp. 289 - 294, Madrid, Sept. 1999

Goodrum, P. M.; McLeren, M. A. & Durfee, A. (2006). The application of active radio frequency identification technology for tool tracking on construction job sites, *Automation in Construction*, Vol. 15, No. 3, pp. 292 - 302, ISSN 0926 - 5805

Hahnel, D.; Burgard, W.; Fox, D.; Fishkin., K. & Phillipose, M. (2004). Mapping and Localization with RFID Technology, *Proceedings of the 2004 IEEE International Conference on Robotics and Automation*, pp. 1015 - 1020, New Orleans, USA, 2004, ISBN 0-7803-8232-3, IEEE, NewYork, USA

Jaselskis, E. J.; Anderson, M. R.; Jahren, C. T.; Rodriguez, Y. & Njos, S. (1995). Radio-Frequency Identification Applications in Construction Industry, *Journal of Construction Engineering and Management*, Vol. 121, No. 2, pp. 189 - 196, ISSN 0733-9364

Jaselskis, E. J. & El-Misalami, T. (2003). Implementing Radio Frequency Identification in the Construction Process, *Journal of Construction Engineering and Management*, Vol. 129, No. 6, pp. 680 - 688, ISSN 0773 - 9364

Jia, S & Takase, K. (2007). Development of Service Robot System with Multiple Human User Interface, In: *Human-Robot Interaction*, Sarkar, N. Ed., pp. 139 - 156, I-Tech Education and Publishing, ISBN 978-3-902613-4, Vienna, Austria

Penttila, K. M; Engels, D. W. & Kivikoski, M. A. (2004). Radio Frequency Identification Systems in Supply Chain Management, *International Journal of Robotics and Automation*, Vol. 19, No. 3, pp. 143 - 151, ISSN 0826 - 8185

Umetani, T.; Mae, Y.; Inoue, K.; Arai, T. & Yagi, J. (2003). Automated Handling of Construction Components Based on Parts and Packets Unification, *Proceedings of the 20th International Symposium on Automation and Robotics in Construction*, pp. 339 - 344, Eindhoven, the Netherlands, 2003, ISBN 90 - 6814 - 574 - 4, Technische Universiteit Eindhoven, Eindhoven, the Netherland

Umetani, T; Mae, Y.; Inoue, K.; Arai, T & Yagi, J. (2005). Pose estimation of objects using multiple ID devices, *Journal of the Robotics Society of Japan*, Vol. 23, No. 1, pp. 84 - 94. ISSN 0289 - 1824 (in Japanese)

Umetani, T.;Arai, T.; Mae, Y.; Inoue, K & Maeda, J. (2006). Construction Automation Based on Parts and Packets Unification, *Automation in Construction*, Vol. 15, No. 6, pp. 777 - 784, ISSN 0926 - 5805

Want, R; Fishkin, K. P.; Gujar, A & Harrison, B. L. (1999). Bridging Physical and Virtual Worlds with Electronic Tags, *Proceedings of the SIGCHI Conference on Human Factors in Computing Systems*, pp. 370 - 377, Pittsburgh, USA, 1999, ISBN 0 - 201 - 48559 - 1, ACM, New York, USA

Yagi, J.; Arai, E. & Arai T. (2005). Parts and packets unification radio frequency identification (RFID) application for construction automation, *Automation in Construction*, Vol. 14, No. 4, pp. 477 - 490, ISSN 0926 - 5805

12

High PerformanceTracking Control of Automated Slewing Cranes

Frank Palis and Stefan Palis
Otto-von-Guericke-University Magdeburg
Germany

1. Introduction

Automation of slewing cranes in handling and transport processes is getting more and more important. However, each crane operation causes swaying movements of load, which are weakly damped. Therefore active damping of swaying movements is a major task in automated crane operation. The main difficulty here, in contrast to overhead travelling cranes, is the coupling between rotary and translation movements, which calls for a special approach when designing the control scheme.

Fig. 1. Slewing crane

There has been extensive research on active damping of overhead travelling cranes. Here, in most cases linear models are used and controlled in open loop operation via time or energy optimal control laws (Buch, 1999) or via feedback control for sway angle rejection (Palis, 1989). However, application of these strategy for slewing cranes raises difficulties because of non-linearity and complexity of motion. Publications on gantry cranes and boom cranes (Arnold et al., 2007) tackle this problem but here no complete model of load dynamics is given.
The actuator system of a slewing crane consists generally of an electrical drive system. Its control system is usually designed in cascade structure and optimized using linear standard

criteria (Betragsoptimum, symmetrical optimum) with P/PI-controllers. Current limitation is realized by current reference value limitation. Because of its easy implementation this structure is highly accepted in practice. Specific technological tasks can be taken into consideration by overlaid control loops that provide the appropriate reference values for the speed controlled drive system.

To this aim, as will be shown, crane motion can be separated into a rapid and a slow component according to the principle of motion decomposition. The rapid component is controlled by the high dynamic speed or position control loop able to reproduce the desired trajectory and to compensate for existing force interactions between the motion axis. The slow component describes load dynamics governed by the trajectory of the corresponding axis. As shown for overhead travelling cranes in (Palis, 1989) this approach facilitates considerably control system design and allows the use of standard speed controlled drive systems. Moreover, the proposed motion separation offers ideal possibilities for system linearization, adaptive control and flatness based trajectory tracking.

Mathematical description of slewing crane movement differs considerably from gantry cranes and overhead travelling cranes. Crane movement is characterized by the appearance of Coriolis and centrifugal forces. Consequently, the system of governing equations is becoming significantly non-linear and the application of linear control theory methods may lead to problems. Practical and simulation investigations prove that both components may considerably influence the movement of slewing cranes and that they must by taken into account when designing and optimizing the controller system. To solve this problem, three different approaches can be utilized:

- Consideration of parameter variations and non-linear force interactions via control parameter adaptation,
- Linearization of the plant via appropriate compensating loops and linearization of plant parameters and
- Application of robust control strategies able to guaranty satisfactory system behaviour in the whole range of working even if plant parameters are changing and Coriolis and centrifugal forces appear.

It is obvious that the first mentioned two methods starts from the supposition of a known and defined plant model. Here, the main problem consists in finding the complete system of governing equations to describe the load motion trajectory. In (Sawodny et al., 2002) a solution for a boom crane with hydraulic drive system is given where centripetal and Coriolis terms are neglected and the rope length is taken constant. Consequently, the obtained results cover only a limited range of applications. To generalize these results all significant force couplings and changing rope length have to be taken into account. The following investigations will show that these effects may lead to considerable deviations between motion of simplified models where these effects are neglected and the real load. Therefore, one of the main concerns of the paper will be the derivation of an complete mathematical model of the load trajectory for slewing cranes. It will be shown that the obtained relatively complex system of governing equations for the swinging load can be considerably simplified using the above mentioned principle of motion separation.

Due to the complexity of the load dynamics and starting from the supposition that handling tasks for slewing cranes do not require extremely high position nor travel trajectory accuracy the authors proposed in (Palis & Palis, 2005) a robust control strategy. These works were aimed at finding a control strategy that features robustness in the whole range of possible parameter variations as well as Coriolis and centrifugal forces and which, in addition, guaranties necessary motion quality. To solve this task a LQ-controller was implemented without taking into consideration the mentioned nonlinear effects and force interactions. It was shown that this approach may be satisfying for many practical handling tasks. However, when more precise tracking performances are required more sophisticated control schemes are needed.

2. Nonlinear model of the slewing crane

To describe motion of the slewing crane, the equivalent kinematic scheme with concentrated masses represented in figure 2 is utilized. Hence, dynamics of the generalized rotating crane represents a multibody system with 5 independent degrees of freedom (DOF)

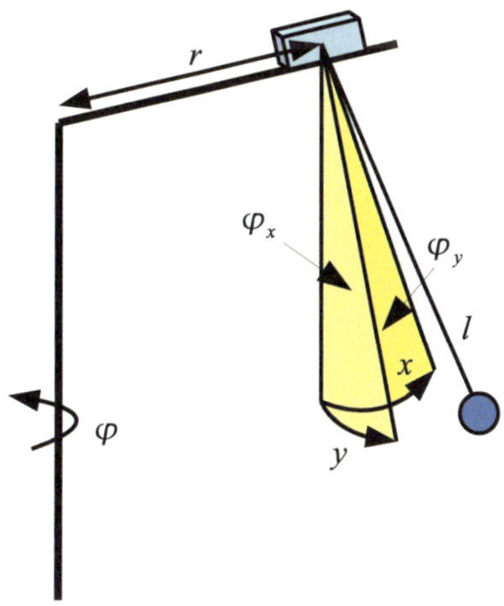

$q_1 = \varphi$ slewing gear rotation angle

$q_2 = r$ radial position of the trolley

$q_3 = l$ hoisting gear movement

$q_4 = \varphi_x$ sway angle in radial direction

$q_5 = \varphi_y$ sway angle perpendicular to the radial direction.

Fig. 2. Equivalent kinematic scheme

The task consists in load positioning along a desired trajectory with given accuracy using a position controlled drive system for the trolley, slewing and hoisting gear. Referring to the equivalent kinematic scheme the following energy balances can be established:

- the potential energy of the load,

$$V = m_L g l \left(1 - \cos\varphi_x \cos\varphi_y\right) \tag{1}$$

- the kinetic energy of the tower T_T, the boom T_B, the trolley T_R and the load T_L.

$$T_T = \frac{1}{2} J_T \dot{\varphi}^2 \tag{2}$$

$$T_B = \frac{1}{2} m_B r_B^2 \dot{\varphi}^2 \tag{3}$$

$$T_R = \frac{1}{2} m_R \left(r^2 \dot{\varphi}^2 + \dot{r}^2\right) \tag{4}$$

$$T_L = \frac{1}{2} m_L \left[\begin{array}{l} \left(\begin{array}{l} -\dot{\varphi}\sin\varphi\left(-\sin\varphi_x + r\right) + \cos\varphi\left(\dot{l}\sin\varphi_x + \dot{\varphi}_x l\cos\varphi_x + \dot{r}\right) \\ -\dot{l}\sin\varphi\sin\varphi_y - l\dot{\varphi}\cos\varphi\sin\varphi_y - l\dot{\varphi}_y\sin\varphi_y\cos\varphi_y \end{array} \right)^2 \\ + \left(\begin{array}{l} \dot{\varphi}\cos\varphi\left(l\sin\varphi_x + r\right) + \sin\varphi\left(\dot{l}\sin\varphi_x + l\dot{\varphi}_x\cos\varphi_x + \dot{r}\right) \\ +\dot{l}\cos\varphi\sin\varphi_y - l\dot{\varphi}\sin\varphi\sin\varphi_y + l\dot{\varphi}_y\cos\varphi \end{array} \right)^2 \\ + \left(\dot{l}\cos\varphi_x\cos\varphi_y - l\dot{\varphi}_x\sin\varphi_x\cos\varphi_y - l\dot{\varphi}_y\cos\varphi_x\sin\varphi_y\right)^2 \end{array} \right] \tag{5}$$

Establishing the Lagrange function

$$L = T - V \tag{6}$$

and applying the Lagrange formalism

$$\frac{d}{dt}\left(\frac{\partial L}{\partial \dot{q}_i}\right) - \frac{\partial L}{\partial q_i} = Q_i \tag{7}$$

we obtain the system of governing equations with the generalized coordinates q_i that can be written in the general form.

$$Q = D(q)\ddot{q} + C(q,\dot{q}) + G(q) \tag{8}$$

Figure 3 depicts equation (8) with

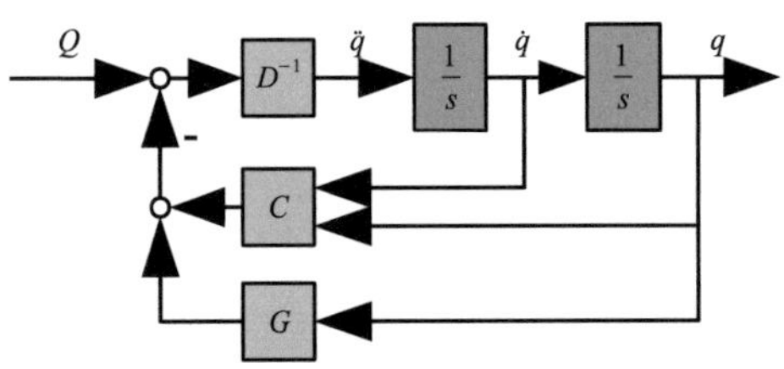

Q	generalized actor forces
q	generalized coordinates
$\mathbf{C} = \mathbf{C}_{centr} + \mathbf{C}_{cor}$	coupling matrix to calculate Coriolis and centrifugal effects respectively
G	gravitation matrix

Fig. 3. General scheme of motion

The exact calculation using eq. (5) leads to very large and complex expressions difficult to interpret physically and inconvenient to implement practically. To simplify the system of governing equations we take into account that only small sway angles are admitted. Consequently the following assumptions can be made.

$$\begin{array}{l} \sin\varphi_x \approx \varphi_x,\ \sin\varphi_y \approx \varphi_y,\ \cos\varphi_x \approx 1,\ \cos\varphi_y \approx 1 \\ \sin^2\varphi_x \approx \sin^2\varphi_y \approx \sin\varphi_x\sin\varphi_y \approx 0 \\ \varphi_y\ddot{\varphi}_x \approx \varphi_x\ddot{\varphi}_y \approx \varphi_y\ddot{\varphi}_y \approx \varphi_x\ddot{\varphi}_x \approx \dot{\varphi}_x\dot{\varphi}_y \approx \varphi_x\dot{\varphi}_y \approx \varphi_y\dot{\varphi}_x \end{array} \tag{9}$$

With these simplifying assumptions the respective terms of eq. (3) can be specified

$$\mathbf{D}(\mathbf{q})\ddot{q} = \left[\begin{array}{ccc|cc} (m_K + m_L) & -m_L l\varphi & m_L\varphi_x & m_L l & 0 \\ -m_L l\varphi_x & J_H & m_L r\varphi_y & 0 & m_L rl \\ m_L\varphi_x & 0 & m_L & 0 & 0 \\ \hline 1 & -l\varphi_y & 0 & l & 0 \\ 0 & (r + l\varphi_x) & 0 & 0 & l \end{array}\right]\begin{bmatrix} \ddot{r} \\ \ddot{\varphi} \\ \ddot{l} \\ \hline \ddot{\varphi}_x \\ \ddot{\varphi}_y \end{bmatrix} = \left[\begin{array}{c|c} \mathbf{A}_1 & \mathbf{D}_1 \\ \hline \mathbf{D}_2 & \mathbf{A}_2 \end{array}\right]\begin{bmatrix} \ddot{r} \\ \ddot{\varphi} \\ \ddot{l} \\ \hline \ddot{\varphi}_x \\ \ddot{\varphi}_y \end{bmatrix} \tag{10}$$

$$\mathbf{C}(\mathbf{q},\dot{\mathbf{q}}) = \begin{bmatrix} -m_L(r + l\varphi_x) - m_K r \\ 0 \\ -m_L r\varphi_x \\ \hline -(r + l\varphi_x) \\ -l\varphi_y \end{bmatrix}\dot{\varphi}^2 + 2\begin{bmatrix} m_L\left(-\dot{l}\dot{\varphi}_x + \varphi_y\dot{l}\dot{\varphi} + l\dot{\varphi}\dot{\varphi}_y\right) \\ m_L(r\dot{l}\dot{\varphi}_y + \varphi_x r\dot{\varphi}\dot{l} + rl\dot{\varphi}\dot{\varphi}_x + r\dot{r}\dot{\varphi} + l\varphi_x\dot{\varphi}\dot{r}) + m_K r\dot{r}\dot{\varphi} \\ m_L\varphi_y\dot{r}\dot{\varphi} \\ \hline \dot{l}\dot{\varphi} - l\dot{\varphi}\dot{\varphi}_y - \varphi_y\dot{l}\dot{\varphi} \\ l\dot{\varphi}\dot{\varphi}_x + \dot{\varphi}\dot{r} + \dot{l}\dot{\varphi}_y + \varphi_x\dot{l}\dot{\varphi} \end{bmatrix} \tag{11}$$

$$\mathbf{C}(\mathbf{q},\dot{\mathbf{q}}) = \begin{bmatrix} \mathbf{C}_{\mathbf{cent,1}} \\ \hline \mathbf{C}_{\mathbf{centr,2}} \end{bmatrix}\dot{\varphi}^2 + \begin{bmatrix} \mathbf{C}_{\mathbf{cor,1}} \\ \hline \mathbf{C}_{\mathbf{cor,2}} \end{bmatrix} \tag{12}$$

with

$$J_H = m_K r^2 + m_L\left(r + l\varphi_x\right)^2 + m_T r_a^2 + J_T \tag{10}$$

$$\mathbf{Q} = \left[F_K \quad M \quad F_H \mid 0 \quad 0\right]^T = \begin{bmatrix} \mathbf{Q}_1 \\ \hline \mathbf{0} \end{bmatrix} \tag{14}$$

$$\mathbf{G} = \left[0 \quad 0 \quad g \mid -g\varphi_x \quad -g\varphi_y\right]^T = \begin{bmatrix} \mathbf{G}_1 \\ \hline \mathbf{G}_2 \end{bmatrix} \tag{15}$$

3. System decomposition

3.1 Fast subsystem (drive control system) and slow subsystem (swinging load)

From equations (10)-(15) can be drawn the conclusion that the overall system can be formally decomposed into 2 subsystems governing the DOF $\mathbf{q}_1 = \left[r \quad \varphi \quad l\right]^T$ and $\mathbf{q}_2 = \left[\varphi_x \quad \varphi_y\right]^T$ described by the following equations.

$$\begin{bmatrix} \mathbf{Q}_1 \\ \hline \mathbf{Q}_2 \end{bmatrix} = \begin{bmatrix} \mathbf{A}_1\ddot{\mathbf{q}}_1 + D_1\ddot{\mathbf{q}}_2 \\ \hline \mathbf{D}_2\ddot{\mathbf{q}}_1 + \mathbf{A}_2\ddot{\mathbf{q}}_2 \end{bmatrix} + \begin{bmatrix} \mathbf{C}_{\mathbf{ce1}} \\ \hline \mathbf{C}_{\mathbf{ce2}} \end{bmatrix}\dot{\varphi}^2 + \begin{bmatrix} \mathbf{C}_{\mathbf{co1}} \\ \hline \mathbf{C}_{\mathbf{co2}} \end{bmatrix} + \begin{bmatrix} \mathbf{G}_1\mathbf{q}_1 \\ \hline \mathbf{G}_2\mathbf{q}_2 \end{bmatrix} \tag{16}$$

Both subsystems are graphically represented in fig. 4.

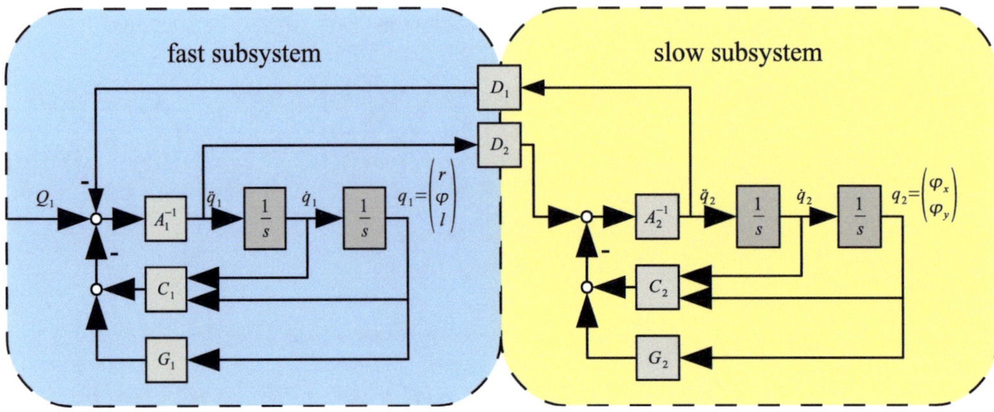

Fig. 4. System decomposition

On the supposition that a speed or position controlled electric drive system is utilized the fast subsystem in fig. 4 must be extended by introducing inner current and speed and outer position control loops. Fig. 5 depicts the drive related fast subsystem for the coordinates $\mathbf{q}_1 = \begin{bmatrix} r & \varphi & l \end{bmatrix}^T$.

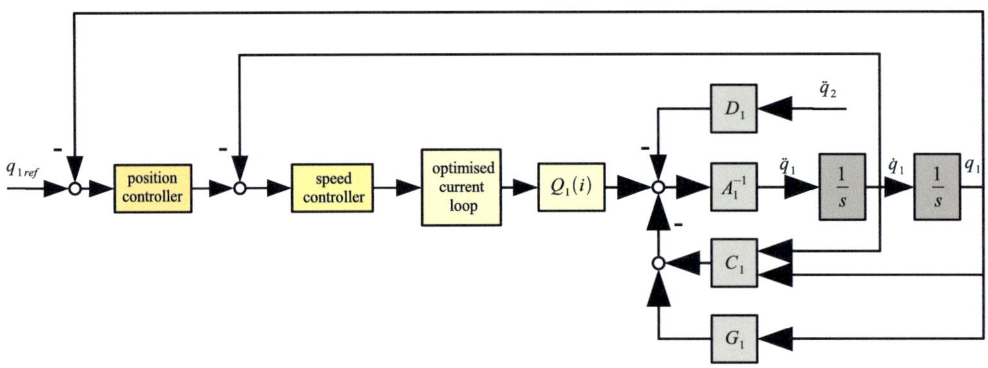

Fig. 5. Speed and position controlled electric drive system

In accordance with the rules of cascade structure systems commonly used in drive systems the time constant of the inner current loop determines the dynamics of the overall system. Assuming a time constant of $(1...2)ms$ for the current loop, speed and position control reactions are much more faster than load dynamics. Therefore all force couplings affecting the drive system are compensated by the speed controller. To improve dynamics of the speed control system related to these couplings well known disturbance observers or decoupling networks can be used. Assuming the above discussed structural conditions system dynamics can be decomposed into the two subsystems as assumed above. The first subsystem is characterized by the fast drive control system with the output variables $\mathbf{q}_1$, whereas the second subsystem reflects the dynamics of the relatively slow load swinging governed by the output of the fast subsystem $\ddot{\mathbf{q}}_1$. Due to their difference in dynamics both systems can be considered separately what significantly facilitates control design. Strictly speaking, the matrix $\mathbf{A}_1$ is affected by the output of the slow subsystem $\mathbf{q}_2 = \begin{bmatrix} \varphi_x & \varphi_y \end{bmatrix}^T$. However, due to small sway angles and its slow changes matrix $\mathbf{A}_1$ can be taken constant.

3.2 Transformation of the slow subsystem and damping strategy

It is obvious that the slow system depicted in fig. 4 is governed by the acceleration of the fast position controlled drive system. Hence, any control signal to damp the swinging load must be transformed into the corresponding position signal. To this aim the governing equation of the slow subsystem in equation (16) must be integrated. The velocity governed subsystem yields

$$\int \mathbf{D}_2 \ddot{\mathbf{q}}_1 dt + \int \mathbf{A}_2 \ddot{\mathbf{q}}_2 dt + \int \mathbf{C}_2 dt - \int g \mathbf{q}_2 dt = \mathbf{0} \tag{11}$$

and the position governed subsystem is described by

$$\iint \mathbf{D}_2 \ddot{\mathbf{q}}_1 dt^2 + \iint \mathbf{A}_2 \ddot{\mathbf{q}}_2 dt^2 + \iint \mathbf{C}_2 dt^2 - \iint g \mathbf{q}_2 dt^2 = \mathbf{0} \tag{12}$$

Neglecting the time dependence of $\mathbf{D}_2$ and $\mathbf{A}_2$ leads to a simplified system of equations with a given error $\mathbf{\Delta}$ for the velocity governed subsystem

$$\mathbf{D}_2 \dot{\mathbf{q}}_1 + \mathbf{A}_2 \dot{\mathbf{q}}_2 + \int \mathbf{C}_2 dt - g \int \mathbf{q}_2 dt + \mathbf{\Delta}_{\dot{q}} = \mathbf{0} \tag{13}$$

and the position governed subsystem

$$\mathbf{D}_2 \mathbf{q}_1 + \mathbf{A}_2 \mathbf{q}_2 + \iint \mathbf{C}_2 dt^2 - g \iint \mathbf{q}_2 dt^2 + \mathbf{\Delta}_q = \mathbf{0} \tag{14}$$

Equation (20) is represented in fig. 6 and leads to the conclusion that to damp this system we have to apply a negative feedback $\int \mathbf{q}_2 dt$ to the position loop of the fast system.

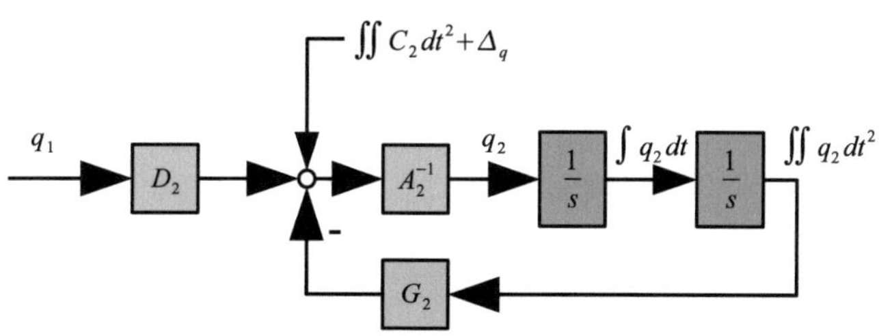

Fig. 6. Position controlled swinging load

Figure 7 illustrates how this feedback signal can be generated from the velocity governed model using the well known observer principle. Here we start from the supposition that the acceleration governed model of the slow system can be easily obtained taking into account all existing force couplings according to fig. 4.

This model is taken as reference model and its sway angles serve as fix point to adapt the simplified velocity governed model (19) where the coupling forces and the error due to the non-exact integration are taken into account via additive model correction. Hence, the simplified velocity governed model reflects all variables with high accuracy.

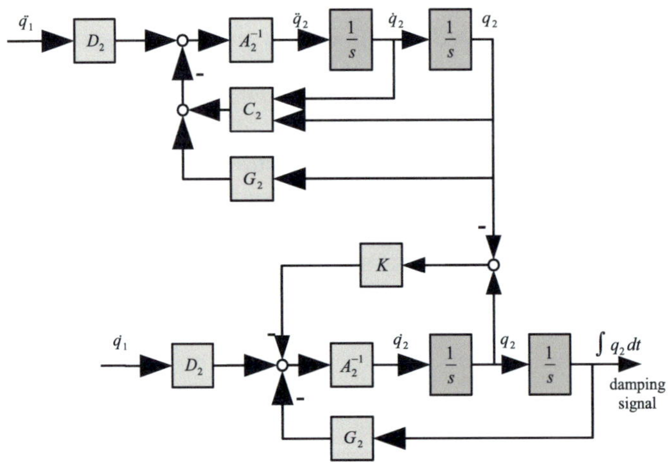

Fig. 7. Observer structure for deriving the damping signal

4. Strategy of sway angle rejection

Figure 8 depicts the general scheme for the proposed damping strategy. The sway angle is modelled based on the acceleration governed model of the slow system taking into consideration all significant couplings, force interactions and changing rope length. Using

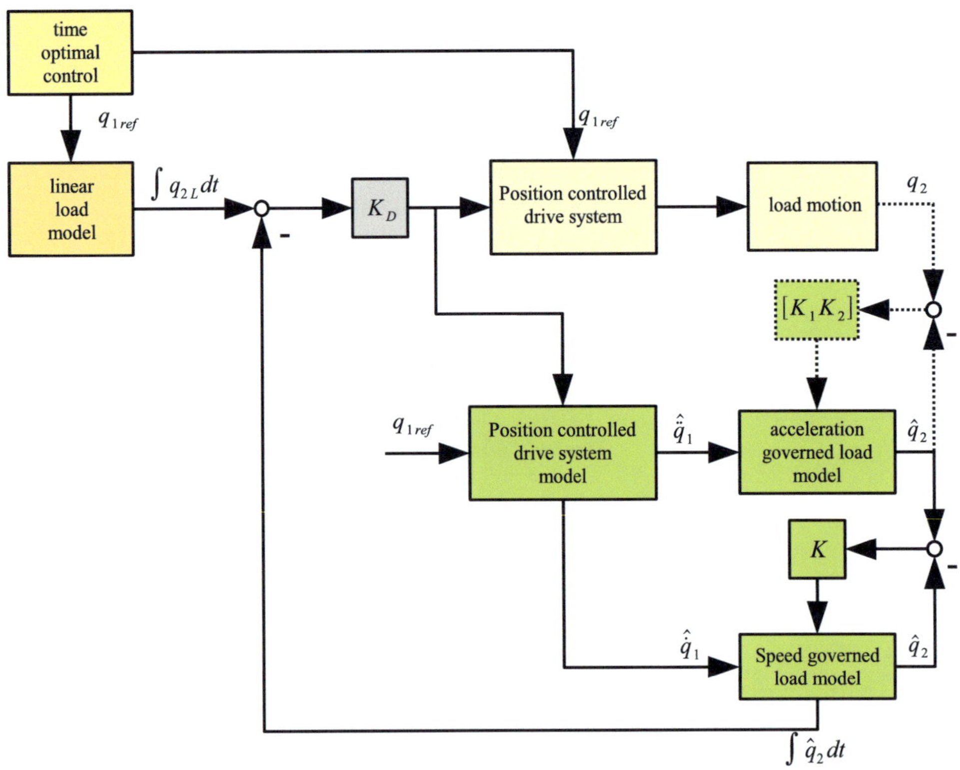

Fig. 8. General control scheme

state observer structure the acceleration governed load model is transformed into the velocity governed load model in order to determine the damping signal $\int \hat{\mathbf{q}}_2 dt$, that is applied to the position controlled drive system. The advantage of the model based approach consists in the fact that all effects like changing rope length, Coriolis and centrifugal forces can be taken into consideration. Hence, in some cases no measurement of the sway angles is needed and the system can work in open loop operation. However, it is obvious that in open loop operation external forces like e.g. wind forces and non-zero initial conditions of the angles are not reflected by the model. Here measurement is needed to correct the model variables as shown in fig. 8 by the dotted line. But even in these cases the model approach is necessary to generate the damping signal avoiding open integration of $\hat{\mathbf{q}}_2$. Moreover the model possesses good smoothing capacities so measurement disturbances can be filtered.
As known, damping leads to a prolongation of the acceleration and breaking phase of the positioning. To overcome this shortage a time or energy optimal control law for the linear load model with constant rope length is calculated (Buch, 1999) and applied to the velocity governed linear load model to determine its damping signal $\int \hat{\mathbf{q}}_{2L} dt$. The difference between this signal and the damping signal of the nonlinear load model $\int \hat{\mathbf{q}}_2 dt$ gives an error, which is applied through the scaling parameter K_D to the position controlled drive system. Hence, the presented damping strategy has an effect only on the deviation between the real load trajectory or trajectory of the nonlinear model and the optimal linear load trajectory. Consequently, a given load sway necessary to bring the load into the target position without final swaying in an optimal way is deducted from the damping algorithm.

5. Simulation results

5.1 Open loop system simulation

Figure 9 depicts the load trajectory in bird's eye view in open loop operation when a time optimal control law designed for the linear load model is applied. It is obvious that in this

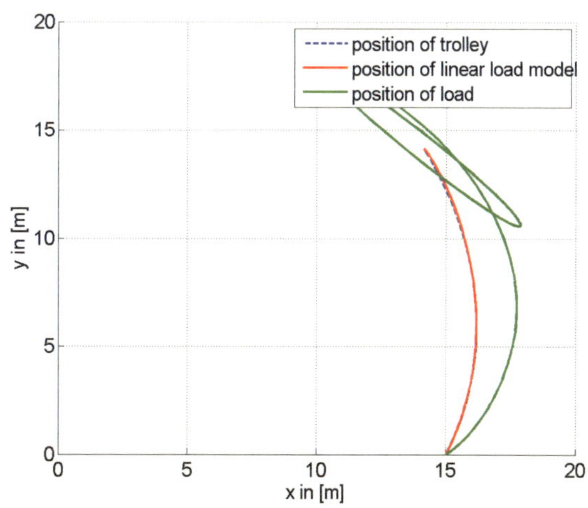

Fig. 9. Bird's eye view of load motion in open loop operation

case the load reaches its final position without swaying (red line). In contrast to this coupling forces and nonlinear system dynamics of the slewing crane lead to considerable deviations from the time optimal trajectory and to significant oscillations around the target position (green line). Figure 10 shows the motion as function of time for x and y in the global coordinate frame. Figure 11 depicts the same motion in the local coordinate frame attached to the trolley (fig. 2).

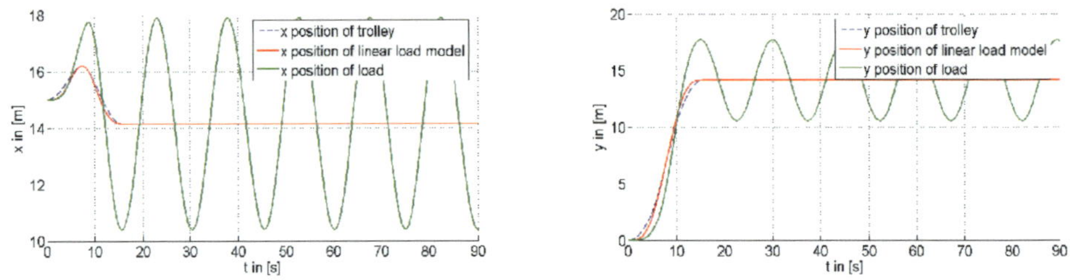

Fig. 10. Load motion in open loop operation as function of time in the global coordinate frame

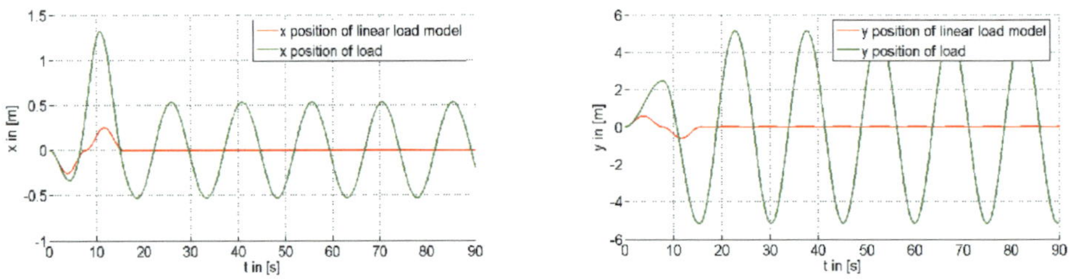

Fig. 11. Load motion in open loop operation as function of time in the local coordinate frame attached to the trolley

4.1 Closed loop system simulation

Figure 12 shows the load trajectory in bird's eye view in closed loop operation when the time optimal control is superposed by the proposed damping strategy. As can be seen

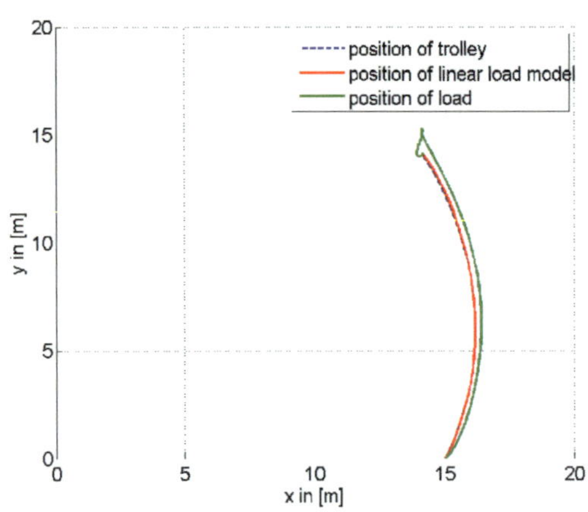

Fig. 12. Bird's eye view of load motion in closed loop operation

deviations from the time optimal trajectory are considerably reduced and swaying around the target position is completely rejected (green line). Figure 13 illustrates the motion in the global coordinate frame x and y over time. Figure 4 depicts the same motion in the local coordinate frame.

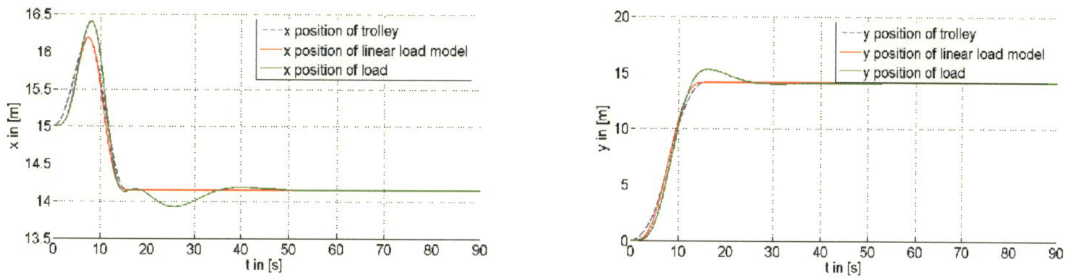

Fig. 13. Load motion in closed loop operation as function of time in the global coordinate frame

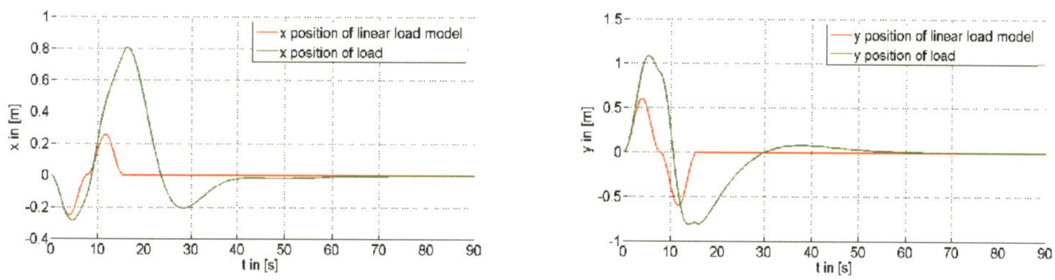

Fig. 14. Load motion in closed loop operation as function of time in the local coordinate frame attached to the trolley

5. Conclusion

Slewing crane motion is affected by nonlinear effects as Coriolis and centrifugal forces and couplings. A mathematical model containing all relevant effects and couplings under mild assumptions has been derived. Due to the difference in dynamics the general motion can be decomposed into a slow and fast subsystem, which can be optimized independently. The slow system reflects load dynamics and serves as basis for designing the proposed damping strategy. Promising simulation results have been obtained. Future investigations will focus on deriving optimal control laws for the nonlinear model and implementation and practical verification of the presented approach on a slewing crane.

6. References

Arnold, E.; Neupert, J. & Sawodny, O. (2007). Trajectory Tracking for Boom Cranes Based on Nonlinear Control and Optimal Trajectory Generation, *Proceedings of Control Applications*, pp.1444-1449, ISBN: 1085-1992, Singapore, Oct. 2007, Singapore

Sawodny, O.; Aschemann, H.; Kumpel, J.; Tarin, C. & Schneider, K. (2002). Anti-sway control for boom cranes, *Proceedings of American Control Conference*, pp. 244-249 vol. 1, ISBN: 0-7803-7298-0, Alaska, May 2002, Anchorage

Kim, Y. S.; Hong, K. S. & Sul, S.K., R. (2001). Anti-Sway Control of Container Cranes: Inclinometer, Observer and State Feedback. *International Journal of Control, Automation and Systems,* Vol. 2., No. 4, (December 2004) pp. 435-449

Palis, S.; Palis, F.; Lehnert M. (2005). Anti-Sway System for Slewing Cranes. *Proceedings of 22nd International Symposium on Automation and Robotics in Construction ISARC 2005,* Italy, Sept. 2005, Ferrara

Buch, A. (1999). *Optimale Bewegungssteuerung von schwingungsfähigen mechatronischen Systemen mit zwei Freiheitsgraden am Beispiel eines Krans mit pendelnder Last und elastischer Mechanik.* University of Magdeburg, Faculty of Electrical Engineering. Dissertation.

Palis, F. (1989). *Prozessangepasste Steuerung von elektrischen Kranantrieben mit Mikrorechnern.* Universität Magdeburg, Fakultät für technische Wissenschaften. Habilitationsschrift

13

Enhancing Facility Management Using RFID and Web Technology in Construction

Yu-Cheng Lin
National Taipei University of Technology/ Civil Engineering
Taiwan

1. Introduction

Facility management (FM) has become an important subject for research and academic study in real estate and construction. To benefit most, organizations need to understand that they must be informed clients in managing their facilities and properties. Managing facility management effectively are extremely difficult owing to various environments. With the advent of the Internet, web-based information management solutions enable information dissemination and information sharing among related participants. Generally, facility managers and staffs require access to the facility location to handle inspection and maintenance work at any facility locations. However, current desktops and notebooks are not suitable for using on facility locations because of problems in transportability. Facility managers and staffs generally handle various types of information, including specification, checklists and reports. Facility managers and staffs generally use sheets of paper and/or field notes. Consequently, there is serious rework progress regarding the data capture and input in inspection and maintenance progress.

Managing the facility maintenance information play an important role in the view of facility management. Therefore, controlling and managing facilities effectively are extremely difficult owing to the environment of various facilities conditions. With the advent of the Internet, web-based information management solutions enable information dissemination and information sharing among involved facility management participants. Furthermore, integrating promising information technologies such as Personal Digital Assistants (PDA), Radio Frequency Identification (RFID) scanning and data entry mechanisms, can help improve the effectiveness and convenience of information flow in the tracing management. This study presents a novel system called Mobile RFID-based Facility Management (M-RFIDFM) System for the acquisition and tracing of facility maintains information on locations and providing a facility maintains information sharing platform among all participants using web technology and RFID enabled PDA.

2. Problem statement

Facility management and control performance can be enhanced by enabling participants to share information with each other. However, two major key aspects of information sharing are information acquisition and information communication. Information acquisition problems in facility management follow from most of the data and information being gathered from the facility location, which is an extension of the facility management. The effectiveness of information and data acquisition influences the information flow between the office and the facility locations. However, facility managers and facility staff members generally use sheets of paper and/or field notes for maintenance progress. Restated, existing means of processing information and accumulating data are not only time-consuming and expensive, but also compromise maintenance management in information acquisition. Such means of communicating information between facility locations and offices, and among all participants, are ineffective and inconvenient. According to the questionnaire survey from 3 facility manager and 10 facility staff members, the primary problems in inspection and maintenance regarding to data capture and sharing are as follows: (1) the efficiency and quality are low, especially in the inspection and maintenance progress in facility management though document-based media, and (2) there are serious rework progress regarding the data capture and input in inspection and maintenance progress. However, few suitable platforms are developed to assist facility staff members with capturing and sharing the inspection and maintenance information when facility staff members need to handle inspection and maintenance work. Therefore, to capture data effective and enhance information communication in facility management will be primary and significant challenge in the study.

3. Research objectives

This study utilizes the RFID and web technology to enhance the maintenance progress and effectiveness in facility management service. This portal is controlled by the management division, and provides facility managers, facility staff members and suppliers with real-time facility-related information-sharing services, enabling them to dynamically respond to the entire facility management network. This study develops Mobile RFID-based Facility management (M-RFIDFM) system to improve efficiency and cost-effectiveness of facility management, improve practical communication among participants, and increase flexibility in terms of service delivery and response times. M-RFIDFM system is a web-based system for effectively integrating facility managers, facility staff members and suppliers, to distribute facility services in the appropriate information, to the right locations and at the right times. PDAs can extend M-RFIDFM systems from offices to facility locations. Data collection efficiency can also be enhanced using RFID-enabled PDAs to enter and edit data on the facility location. By using web technology and mobile devices, the M-RFIDFM system for the management division has tremendous potential to increase the efficiency and effectiveness of information flow, thus streamlining services processes with other participants.

Facility managers and staff members frequently waste time by travelling to obtain information in the absence of other efficient means of communication. The portal and PDAs enable facility location engineers to update data from the facility location and immediately upload it to the system portal; suppliers can receive maintenance information and make better decisions regarding future facility management and control.

The main purposes of this study include (1) developing a framework for a mobile facility management system for facility management division; (2) applying such a system that integrates RFID technology with PDA technology to increase the efficiency of facility inspection and maintenance data collection; (3) designing a web-based portal for facility management and control, providing real-time information and wireless communication between offices and facility locations, and suppliers, and (4) apply RFID technology to carry brief maintenance data in the facility itself. Figure 1 illustrates solutions used in a real case utilized M-RFIDFM system in a factory building in Taiwan. With appropriate modifications, the M-RFIDFM system can be utilized at any facility inspection and maintenance service for facility management divisions or suppliers in support of the M-RFIDFM system.

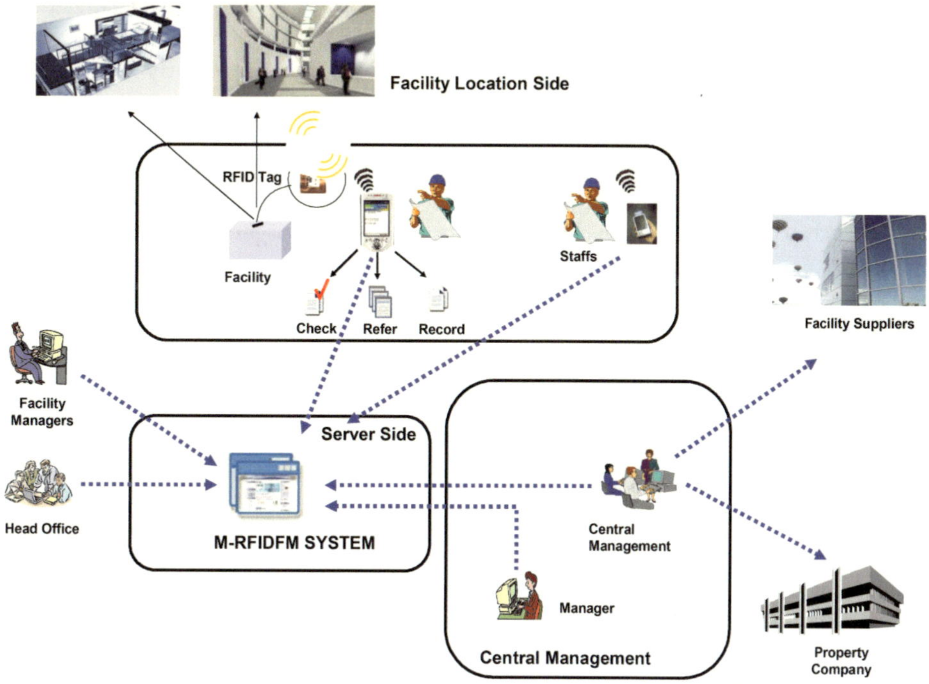

Fig. 1. M-RFIDTM System Framework Overview

4. Background research

RFID is an automatic identification solution that streamlines identification and data acquisition, operating similarly to bar codes. Automatic identification procedures have recently become very popular in numerous service industries for purchasing and distribution logistics, and in manufacturing companies and material flow systems. Jaselskis & Anderson (1995) investigated the applications and limitations of RFID technology in the construction industry, and attached read/write RFID tags to the surfaces of concrete test that were cast from the job site to test lab. This RFID technology has been widely applied in many areas in the construction industries for the following reasons: (1) to provide owners and contractors with information to enhance operation using RFID technology (Jaselskis and Tarek, 2003); (2) to propose a novel concept of "parts and packets unified architecture" in order to handle data or information related to a product carried by product itself by utilizing RFID technology (Yagi et al., 2005); (3) to apply RFID technology as a solution to problems in pipe spools, and identify potential economic benefits from adopting RFID technology in automated tracking (Song et al., 2006); (4) to apply RFID combined with GIS technology in order to locate precast concrete components with minimal worker input in the storage yard (Ergen et al., 2006); (5) to improve the efficiency of tracing tools and tool availability using RFID (Goodrum et al., 2006); (6) to develop mobile construction supply chain system integrated with RFID technology (Wang et al., 2006); (7) to describe a prototype of an advanced tower crane equipped with wireless video control and RFID technology (Lee et al., 2006), and (8) to improve tracing of material on construction using materials tagged with RFID tags (Song et al., 2006).

The use of technology to improve delivery process control is not a novel concept. Many industries have applied barcodes to track materials for many years. Construction companies began to examine the use of barcodes for tool management in the early 1990s. Although barcode is an established and affordable technology, it has presented problems in the construction industry due to the short read range and poor durability of barcodes — a barcode requires a line of sight, and becomes unreadable when scratched or dirty.

An RFID system is composed of an RFID tag and an RFID reader. The RFID tag comprises a small microchip and an antenna. Data are stored in the tag, generally as a unique serial number. The RFID tags can be either passive (no battery) or active (battery present). Active tags are more expensive than passive tags and have a read range of 10–100 meters. Passive tags have a read range of 10mm to approximately 5m (Manish & Shahram, 2005). The vast majority of RFID tags applied in the construction industries are passive.

The RFID reader functions as a transmitter/receiver. The reader transmits an electromagnetic field that "wakes up" the tag and provides the power required for it to operate (Lahiri, 2005). The tag then transfers data to the reader via the antenna. This data are then read by the RFID reader, and transferred to a Pocket PC or computer. Unlike barcodes, RFID tags do not require line-of-sight to be read; they only need to be within the reader's radio range. Additionally, RFID tags, unlike barcodes, can be read through most materials. RFID tags are shrinking, with some measuring only 0.33mm across. Although RFID systems can apply different frequencies, the most common frequencies are low (125KHz), high (13.56MHz) and ultra-high (UHF) (850–900MHz) (Lahiri, 2005).

Notably, RFID systems are one of the most anticipated technologies that will potentially transform processes in the engineering and construction industries. In the construction industry, RFID technology can be utilized with PDAs, thereby allowing staff members to integrate seamlessly work processes at labs and sites, due to the ability to capture and carry data. With a RFID scanner plugged into a PDA, the RFID-enabled PDA is a powerful portable data collection tool. Additionally, RFID readings increase the accuracy and speed of information communication, indirectly enhancing performance and productivity. Figure 2 shows an application of an RFID-enabled PDA used in facility management.

The advantages of using mobile devices in the construction industry are well documented (Baldwin et al., 1994; Fayek et al., 1998; McCullough, 1997). Moreover, mobile devices have been applied in numerous construction industries, to provide the following support: (1) providing wearable field inspection systems (Sunkpho & Garrett, 2003); (2) supporting pen-based computer data acquisition for recording construction surveys (Elzarka & Bell, 1997); (3) supporting collaborative and information-sharing platforms (Pena-Mora & Dwivedi, 2002); (4) using mobile computers to capture data for piling work (Ward et al., 2003), and (5) utilizing mobile devices in construction supply chain management systems (Tserng et al., 2005).

A portal is a web site that collects information concerning specific themes or topics, and facilitates access to related services and information sources. When using a portal, all case-related information is centralized in a case database that can only be obtained via a web interface. The portal also has authentication and access control mechanisms that enable participants to access information according to user privileges. However, exchanging information among participants is more difficult in practice than it seems at first glance. For example, several different systems and standards are used to exchange information; peer-to-peer relationships among companies in the network are generally extremely difficult to manage, and most systems do not facilitate seamless information exchange with other systems. Furthermore, most participants are extremely reluctant to share information with other parties because paper-based information is difficult and inconvenient to share each other. Portals can resolve these problems. Standardized interactions with a single portal are easier to manage than numerous peer-to-peer relationships.

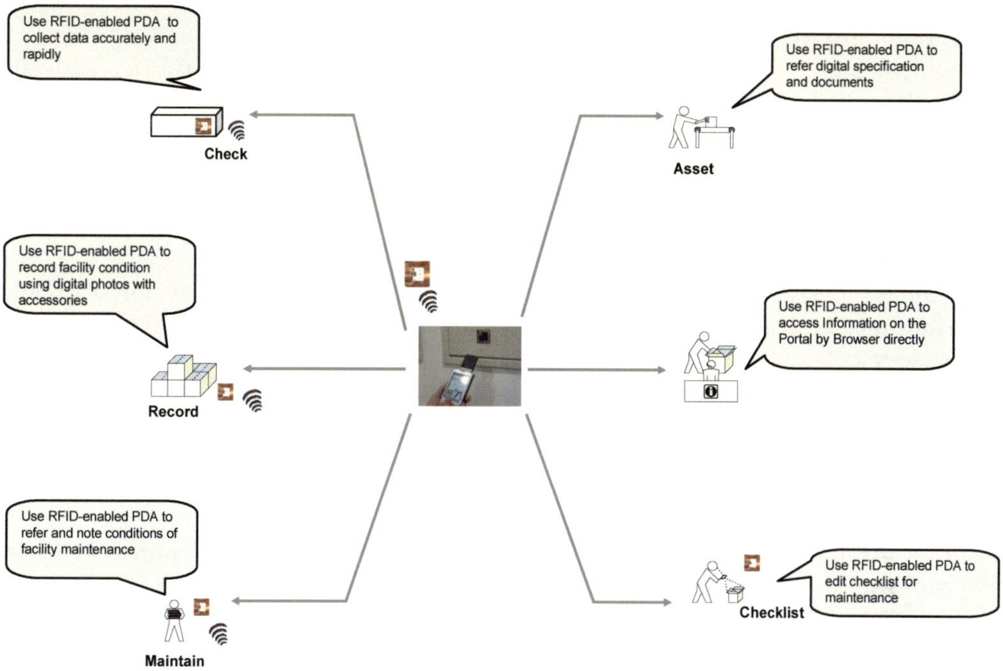

Fig. 2. Application of RFID-enabled PDA Used in Facility Management

5. System implementation

5.1 System architecture

The M-RFIDFM system has three main components, a PDA, RFID and a portal. Significantly, both the PDA and RFID components are located on the client side, while the portal is on the server side. All facility-related information acquired by facility staff members within the M-RFIDFM system is recorded in a centralized FM system database. All participants can access required information via the portal based on their access privileges. Moreover, the portal is limited by design to thirty persons logging in when all participants acquire the same case information at same time. The M-RFIDFM system extends the RFID-based facility management system from control offices to facility locations to assist with inspection and maintenance services, while the HubFM portal primarily deals with data transactions in all departments or systems integration within a facility management system.

When the data are updated on the server side, e-mails are automatically sent from the server to the facility managers of the management division and to participants involved in the relevant activity.
The M-RFIDFM system consists of an inspection and management portal integrated with mobile devices and RFID technology (RFID-enabled PDA). Each module is briefly described below.

Mobile Device (PDA) Module of M-RFIDFM System

The M-RFIDFM system is operated on Windows CE. The programming language and tools used in module development are Visual Basic and eMbedded Visual Tools. SQL Server for Windows CE serves as the PDA database for the Windows CE-based PDA. All data files in the PDA module are first stored in the PDA database, before being transmitted to the server through the Internet.

RFID Module of M-RFIDFM System

The Windows CE mobile device platform was chosen as the RFID-enabled PDA hardware system. The M-RFIDFM system adopts a Pocket PC with an MPR-1230 RFID Reader. The RFID technology can be either a passive or active system. The major difference between an active and a passive RFID system is that an active tag contains a battery, and can transmit information to the reader without the reader generating an electromagnetic field. The case study uses passive RFID technology due to budget restrictions.

Web Portal Module of M-RFIDFM System - HubFM Portal

The HubFM portal is an information hub in the M-RFIDFM system for a facility management. The HubFM portal enables all participants to log onto a single portal, and immediately obtain information required for planning. The portal provides suppliers and customers with information about the inventory levels of other portal users. The users can access different information and services via a single front-end on the Internet. For example, a customer can log onto the portal, enter an assigned security password, and access real-time inspection schedule information. A general contractor can check the test or inspection status, availability of reports and various other case-related data. The HubFM portal is based on the Microsoft Windows 2003 operating system with Internet Information Server (IIS) as the web server. The prototype was developed using Java Server Pages (JSP), which are easily combined with HTML and JavaScript technologies to transform an Internet browser into a user-friendly interface. The HubFM Portal provides a solution involving a single, unified database linked to all functional systems with different levels of access to information, based on user role, both within an organization and across organizations and other participants.

5.2 Modules of system functions

This section describes the implementation of each module in the RFID-enabled PDA system.

Test Report Module:

The report module provides facility staff members with a complete record of inspection and maintenance performed in the facility management. Facility staff members can record data for dates, conditions, inspection result, descriptions of problems and suggestions that have arisen during maintenance.

Inspection and Maintenance Module:

Facility staff members can download the most up-to-date maintenance results from the Internet, and enter maintenance results directly via a PDA. Additionally, PDAs display the checklist for every task. Facility staff members can also mark unacceptable positions, and select relevant items from lists in the PDA. The module has the benefit that facility staff members can enter/edit inspection and maintenance test results, and all test records can be transferred between the PDA and portal by real-time synchronization, eliminating the need to enter the same data repeatedly.

Progress Monitor Module:

This module is designed to enable facility staff members to monitor the progress of inspections and tests. Additionally, managers, facility staff members and participants can access the progress or inspection condition of critical work. The progress monitor module provides an easily accessed and portable environment where facility staff members can trace and record all information regarding the status of inspections delivered to the maintenance or scheduled for repair.

E-specification Module:

This module allows facility staff members to download specifications in advance, and reference them during inspection. This module also has a search function that enables the information to be found and retrieved easily, which is a valuable feature in dynamic environments. Moreover, facility staff members who do not need paper-based specifications can download e-specifications and access them directly using their PDAs.

6. Case study

This study uses a factory building in Taiwan as the basis for the case study. This study utilizes an M-RFIDFM system in the facility management in Taiwan. Existing approaches for tracking and managing the inspection in facilities adopt manually updated paper-based records. However, information collected using such labor-intensive methods is unreliable and ineffective for tacking and managing maintenance results. The results of the inspection and maintenance results are generally obtained by telephone or fax. The inspection process is currently not easy to trace or monitor. The RFID-based facility inspection and management system is applied to enhance the effective management of facilities inspection and maintenance.

In the case study, facility management division and facility staff members utilized the M-RFIDFM system to enhance facility inspection and maintenance management. Passive read/write RFID tags (radio frequency of 13.56MHz) are used in the case study. Passwords are utilized to protect the data in the RFID tags. After the critical facilities are selected, an RFID tag for the facility is made, and the relevant information is entered into the system. The primary description in the case study is presented to elucidate the application of the maintenance using the RFID-enabled PDA and web portal.

In the initial phase, the facility staff members first discuss which facility will be maintained during special period under the facility management division. Following the facility is designed to be monitored for maintenance, the facility is scanned with a RFID tag to enter the portal information.

During the setup phase, the RFID will be prepared and entered the basic information brief regarding facility basic information and stores before it is delivered to the final location. Moreover, the facility staff member uses the PDA to scan the facility and enters the basic data of the facility. The data in the PDA is used to synchronously update the data/information to the portal, and the manager may understand that the facility has already been setup.

The staffs use RFID-enabled PDA to scan RFID tags and select the status when the facility is ready for use to the facility location. The data in the PDA synchronously update the data/information to the portal, and the facility management division is notified to update the status of the facility for starting use.
When this facility has been used in the facility location, the facility staff member scans the RFID tag to update the inspection/maintenance status. PDA displays the basic component and checklist information for each item. Facility staff member enters the result of inspection, edit the description in the PDA and provide the updated information to the portal. Meanwhile, the system automatically sends an alarm message to the facility manager in the facility management office to enter the portal and check the updated information.
Following the facility is inspection/maintenance in the facility location, the facility staff member must check facility quantity and quality, and then record the result via PDA. Finally, the facility tracking process will update the status to pass the test synchronously in the portal, and allows the facility manager or the authorized suppliers to check the process. Furthermore, the facility staff member utilizes PDA directly to write test data in to RFID tag. Once the facility component is ready to install following the maintained process, the facility with the RFID tag must be scanned again by facility staff member to update the information during the maintenance phase. Additionally, the updated information is updated and announced synchronously via the portal.

7. Field tests and results

Overall, the field test results indicate that passive RFID tags are effective tools for facility maintenance management. All tags survived use in the facility environment over one year testing period. The number of facilities for inspection and maintenance progress in field trials was about sixty two (two floors). The M-RFIDFM system was installed on main server in the facility management division. A user's guide of the M-RFIDFM system was briefly explained to staff involved in facility maintenance management progress.
During the field trials, verification and validation tests were performed to evaluate the system. The verification aims to evaluate whether the system operates correctly according to the design and specification; and validation evaluates the usefulness of the system. The verification test was carried out by checking whether the M-RFIDFM system can perform tasks as specified in the system analysis and design. The validation test was undertaken by asking selected case participators to use the system, and provide feedback by answering a questionnaire. The case participators consisted of two facility managers with 6 years of experience; four facility staff members with 4 years of experience, and three repair engineers with one year or less of experience in the case study. To evaluate system function and the level of system capability satisfaction, we distributed questionnaires, and the users of the system were asked to grade the conditions of system testing, system function, and system capability separately, compared with the typical paper-based maintenance method, on the five Likert scale. Some comments for future improvements of M-RFIDFM system were also obtained from the case participators through user satisfaction survey. Table 1 shows a comparison of the approximate time required for a typical facility maintenance service using a traditional paper-based inspection approach and the proposed system. The next section presents the detailed results of the performance evaluation and the user survey conducted during the field trials.

Item	Paper-based Approach		Proposed Approach	
	Method	Average Time (Min)	Method	Average Time (Min)
Find inspected related Information	Referring to inspection menu	1.9	Automatic selection	1.2
Input defect description	Referring to Inspected item and checklist	2.5	Use electronic forms	1.1
Check history record	Paper forms	2.2	Read information from RFID tag	0.1
Entry the problems	Describing and Sketching	3.2	Entry the PDA and store in RFID and system	0.1
Archive data	Re-entry at the office	8.1	Real-time Update database	0.2
Sharing defect information	Send the e-mail or by Fax	2.6	Accee the system directly and share information	0.8

Table 1. System Evaluation Result

The 88% obtained from user satisfaction survey indicates that the M-RFIDFM system is quite adaptable to the current facilities maintenance management practices, and is attractive to users. This result implies that the M-RFIDFM system was well designed, and could enhance the current time-consuming facilities maintenance process.

The 85 % obtained from facility staff members satisfaction survey indicates that the system automatically generated all documentation, and accumulated the related historical data in the central database server. The facility staff members could thus collect maintenance data, and send them electronically to the M-RFIDFM system. No additional work was required for any documentation or maintenance analysis after the data collection.

The advantages and disadvantages of M-RFIDFM system identified from the real case studies application are identified. However, 84% of users obtained from facility staff members satisfaction survey agree that the M-RFIDFM system is useful for improving the efficiency and effectiveness of automated data acquisition and information sharing in facility maintenance service, thus assisting facility managers and facility staff members in managing and monitoring the maintenance progress of facilities in the building. HF Passive tags are less expensive that active tags. Thus, HF passive tags are suited to facility maintenance management.

The use of RFID and web technology to collect and capture information significantly enhanced the efficiency of inspection and maintenance processes of facilities. The case study produced an overall read rate of 91% for the tags on the facility. RFID readers and tags are widely thought likely to improve in the future, thus bringing the read rate to an industrially acceptable level, and significantly improving the operation efficiency.

The tags adopted in this study cost under $0.5 US dollars each in 2007. The cost of these tags is decreasing every year. The total cost of the equipment applied in this study was $2650 US dollars (including RFID-enabled PDA reader and one server personal computer). Experimental results demonstrate that M-RFIDFM system can significantly enhance the maintenance progresses. The use of RFID reduces the amount of manual checking activities in the location area and the management area, thus significantly decreasing the overall maintenance operation time.

8. Conclusions

This study presents a mobile RFID-based facility management system that incorporates wireless technology and mobile devices to improve the efficiency and effectiveness of on facility location data acquisition and information sharing among participants to assist managers in maintenance service. The M-RFIDFM system not only improves the acquisition of data on facility maintenance efficiency using RFID-enabled PDA, but also provides a monitor to service the facility maintenance progress. On the client side, Facility staff members use RFID-enabled PDAs to overcome time and space constraints, enabling them to seamlessly maintenance work processes at facility locations, owing to its accuracy and ability to capture data effectively. Plugging a RFID scanner into a PDA creates a powerful portable data collection tool. Additionally, RFID readings increase the accuracy and speed of information communication, indirectly enhancing performance and productivity. Meanwhile, on the server side, the HubFM portal offers a hub center to provide facility suppliers and facility management division with real-time updated property-related information and to monitor the service progress. In a case study, the application of the M-RFIDFM system helps to improve the process of inspection and maintenance work for the factory facilities in Taiwan. Based on experimental result, this study demonstrated that passive RFID technology has significant potential to enhance inspection and maintenance work in facility management. The integration of real-time maintenance information from facility supplier helps managers to track and control the whole facility management in inspection and maintenance progress. Furthermore, the facility supplier can update the service progress in real time to arrange all services. Real-time feedback on the status of maintenance progress in facility management is provided to management division so process steps can be re-sequenced as required. Compared with current methods, communication of information/data on the facility management can be enhanced by using RFID and Web technologies, and the information sharing among participants can be accelerated and made more efficient via the portal technology.

9. Recommendations

The case study findings indicate that the RFID technology and mobile devices are useful tools to enhance the progress of inspection and maintenance works. Recommendations for implementing the proposed system in the future are given below.

- Cost is a currently significant factor limiting the widespread use of RFID tags in the construction industry. Passive tags are cheaper that active tags. Additionally, most passive tags have small data storage capabilities, thereby decreasing the cost of chips and motherboards. Therefore, passive tags are suited to the facility management.
- The high data storage capabilities will be considered and suggested for use in facility management in the future to carry complete the maintenance information in the facility.
- If the RFID tag needs to place the interface of the metal facilities, the RFID tag should be using the special designed tag that can avoid influence by metal facilities.
- It is necessary to consider the usage time and reading distance of RFID tags. Currently, the average of longest time regarding to RFID tags is ten year. Therefore, if the facility need to track over ten years then the RFID tag should be attached to replace easily and workable.
- This study has indicated that RFID devices are more useful than barcodes, because they can carry information. However, most of tags can't carry all maintenance data because of the capacity of tag data.

10. References

Baldwin, A. N., Thorpe, A. and Alkaabi, J. A. (1994), "Improved material management through bar-code: results and implications of a feasibility study," Proceedings of the institution of Civil Engineers, Civil Engineering, 102(6), 156-162.

Elzarka, H. M. and Bell, L. C. (1997), "Development of Pen-Based Computer Field Application," Journal of Computing in Civil Engineering, ASCE, 11(2), 140-143.

Esin Ergen et al. (2006) "Tracking and locating components in a precast storage yard utilizing radio frequency identification technology and GPS," International Journal of Automation in Construction, doi:10.1016/j.autcon.2006.07.004.

Fayek, A., AbouRizk, S. and Boyd, B. (1998), "Implementation of automated site data collection with a medium-size contractor," in Proc. ASCE Computing in Civil Engineering, Boston, MA, 454-6.

Jaselskis, E. J. and Anderson, M. R. (1995). "Radio-Frequency Identification Applications in Construction Industry," Journal of Construction Engineering and Management, 121(2), 189-196.

Jaselskis, E. J. and El-Misalami, Tarek (2003). "Implementing Radio Frequency Identification in the Construction Process," Journal of Construction Engineering and Management, 129(6), 680-688.

Johnson, D. and Broida, R. (2000), How to Do Everything with Your Palm Handheld, Osborne/McGraw-Hill, NY.

Lahiri, Sandip (2005), RFID Sourcebook, Prentice Hall PTR.

Lee, Ung-Kyun, Kang, Kyung-In, and Kim, Gwang-Hee (2006). "Improving Tower Crane Productivity Using Wireless Technology." Journal of Computer-Aided Civil and Infrastructure Engineering, Vol. 21, pp.594-604.

Manish Bhuptani and Shahram Moradpour (2005), RFID Field Guide: Deploying Radio Frequency Identification Systems, Prentice Hall PTR.

McCullouch, B. G. (1997), "Automating field data collection in construction organizations," in Proc. ASCE Construction Congress V, Minneapolis, MN, 957-63

McPherson, F. (2000), How to Do Everything with Your PocketPC & Handheld PC, Osborne/McGraw-Hill, NY.

Paul M. Goodrum Matt A. McLaren, and Adam Durfee. (2006) "The application of active radio frequency identification technology for tool tracking on construction job sites," International Journal of Automation in Construction, 15(3), 292-302.

Pena-Mora, F. and Dwivedi, G. D. (2002), "Multiple Device Collaborative and Real Time Analysis System for Project Management in Civil Engineering," Journal of Computing in Civil Engineering, ASCE, 16(1), 23-38.

Song, J., Haas, C. T. and Caldas, C. (2006). "Tracking the Location of Materials on Construction Job Sites," Journal of Construction Engineering and Management, 132(9), 680-688.

Song, J., Haas, C. T., Caldas, C., Ergen, Esin, and Akinci, B. (2006). "Automating the task of tracking the delivery and receipt of fabricated pipe spools in industrial projects," International Journal of Automation in Construction, 15(2), 166-177.

Sunkpho, Jirapon and Garrett, J. H., Jr. (2003), "Java Inspection Framework: Developing Field Inspection Support System for Civil Systems Inspection," Journal of Computing in Civil Engineering, ASCE, 17(4), 209-218.

Tserng, H. P., Dzeng, R. J., Lin, Y. C. and Lin, S. T. (2005). "Mobile Construction Supply Chain Management Using PDA and Bar Codes." Journal of Computer-Aided Civil and Infrastructure Engineering, Vol. 20, pp.242-264.

Wang, L. C., Lin, Y. C. and Lin, P. H. (2006). "Dynamic Mobile RFID-based Supply Chain Control and Management System in Construction." International Journal of Advanced Engineering Informatics - Special Issue on RFID Applications in Engineering, Vol. 21 (4), pp.377-390.

Ward, M. J., Thorpe, A. and Price, A. D. F. (2003), "SHERPA: mobile wireless data capture for piling works, "Computer-Aided Civil and Infrastructure Engineering, 18, 299-314.

Yagi, Junichi, Arai, Eiji and Arai, Tatsuo (2005). "Construction automation based on parts and packets unification," International Journal of Automation in Construction, 12(1), 477-490.

14

Improving Human-Machine Interfaces for Construction Equipment Operations with Mixed and Augmented Reality

Xiangyu Wang
The University of Sydney
Australia

1. Introduction

Construction equipment operation and management is a process that operates the machine in different working conditions, monitors the operation condition of equipment, maintains and repairs equipment components by referring to technical specifications, and generates more information for future access (Wang & Dunston 2006a). These construction operations are being carried out on different platforms. This makes it difficult to visualize and process critical information at any time and place. Many evolving information technology (IT) systems bring solutions into the construction field data management and sharing (Yang et al. 2000). Based on the data/information managed and stored in such IT tools, innovative and efficient human-machine interfaces that could visually merge these data into the current workspace of crew can be developed to increase the operators' situational awareness and achieve less information access cost of operators and maintenance crew.

Mixed and Augmented Reality can fulfill this vision by creating an augmented workspace through inserting digital contents into the physical space where operators work. Such augmented workspace is realized by integrating the power and flexibility of computing environments with the comfort and familiarity of the traditional workspace (Wang & Dunston 2006a). Augmented Reality (AR) opens a promising gate for integrating designs into their to-be-built real-world context. For example, AR could assist in quality assurance by comparing as-built facility with corresponding design as well as constructability review prior to the actual construction. Outdoor AR systems could assist in demonstrating architects/designs/constructors/owners what a new structure/building would look like at its final setting for evaluating functions and esthetics of a design. For example, an urban designer could show clients/politicians what a new building would look like as they walk around the neighborhood, to better understand how a tall building might affect surrounding residents.

Outdoor and mobile AR systems have just begun to become feasible due to advances in tracking and computing. Traditional tracking systems used for AR registration are intended for use in controlled indoor spaces and are unsuitable for unprepared outdoor environments such as those found on typical construction sites (Behzadan Kamat 2006). Except wide applications in tabletop and indoor AR systems, fiducial markers could also be an easy way for outdoor AR. For example, Klinker et al. (2001) accurately erected a virtual wall at a real construction site by using fiducial markers as tracking technology. Shen et al. (2001) developed an Augmented Reality system for urban planning and environmental assessment based on the approach of affine representation. As an image-based method, in addition to saving more time, it also presents more realistic and objective results than traditional graphics-based Virtual Reality (VR). With this system, users only need a digital camera to realize a more realistic VR by fusing the virtual buildings of different sizes or sites into a real 3D scene. Kamat and El-Tawil (2005) conceptualized a system which could allow on-site damage inspectors to retrieve previously stored building information, superimpose that information onto a real building in augmented reality, and evaluate damage, structural integrity, and safety by measuring and interpreting key differences between a baseline image and the real facility view. Thomas et al. (1999) developed an Augmented Reality system TINMITH2 to visualize a simple extension to a building in spatial context of its outdoor final physical surroundings.

Unlike indoor AR where the user is limited to navigating in a restricted space, users with outdoor AR systems needs the ability to navigate freely with minimum constraints. Behzadan and Kamat (2006) developed an Augmented Reality prototype system UM-AR-GPR-ROVER. They used a Delorme Earthmate WAAS enabled Global Positioning System (GPS) receiver and an InterSense InterTrax2 head tracking system to determine the user's position and orientation. These tracking data could achieve generating an accurate representation of augmented space in a real time manner so that the user could experience the world where virtual excavator stay fixed to its real-world location. The tracked information is then reconciled with the known global position and orientation of computer aided design (CAD) objects which are to be registered into the user's real world view. From this computation, the relative translation and axial rotations between the user's eyes and the CAD objects are calculated at each time frame. The relative geometric transformations are applied to the CAD objects in order to generate an augmented outdoor scene.

Observation and monitoring of change and attempts at the creation of more realistic simulations of the evolving environment have also become key research fields. Ghadirian and Bishop (2002) developed an interactive visualization system which combines Geographical Information System (GIS)-based environmental process modeling with use of AR technology to present environmental change and then monitor people's responses to and behavior in real world situations in a 3D immersive environment.

Besides the application of AR to facilitate construction processes, there are other research projects where AR technology is used to visualize hidden information (e.g., wires, pipes, and beams in a wall), non-graphical information (e.g., procedural information, maintenance schedules and records), and invisible information (e.g., heat and pressure of pipes) in assembly, maintenance, and repair of complex facility and machinery. It has been suggested that an AR system might give users "X-ray vision" inside a building, allowing them to see where the pipes, electric ducting, and structural supports are inside walls and above ceilings (Webster et al. 1996). Examples include the Augmented Reality systems by Roberts et al. (2002) and Lawson & Pretlove (1998). In all of these applications the main thrust of the research is to use AR techniques to augment physical objects in the environment with additional information or views of their hidden structure. Maintainer and repairer could use AR as a visualization aid for maintenance and repair. These datasets could then be rendered and combined in real time with a view of the real structure. This would be very useful for minimally-excavation/invasive operation, which reduces the opening of a structure/object by using small cuts or no cuts at all. AR technology could provide an internal view without the need for larger openings. This might also guide precision tasks, such as displaying where to drill a hole into the pipe for testing. The information from the non-invasive sensors would be directly displayed on the object/structure, showing exactly where to perform the operation. Non-graphical representation can also be used as in the Augmented Reality-based equipment management system (AR EMS) by Wang & Dunston (2006). Invisible information can also be indicated by AR, for instance, a red-to-blue coloring scheme was used by Klinker et al. (2001) to represent temperature variations and distributions within heating pipes.

This chapter introduces information on Mixed Reality (MR) and Augmented Reality (AR), and their existing research applications in the area of construction equipment operations. This chapter also elaborates two systems in the area of construction equipment operations: the first system is a Mixed Reality-based interface for backhoe excavations in dangerous working environments; and the other system is an Augmented Reality-based information management system for maintenance, repair and inventory checking of heavy construction equipment. Conceptual designs for these two systems are developed and presented with certain scenarios identified. Discussion of limitations of Mixed and Augmented Reality technologies in these systems are also discussed.

2. Mixed and augmented reality immersive technologies

Virtual Reality generates a complete virtual environment (VE), simulated or completely synthetic, that surrounds or immerses the users. Mixed Reality, defined by Milgram et al. (1994, 1999), is a special class of Virtual Reality – related technologies for creating environments wherein real world and virtual world objects are presented together in a single display. Augmented Reality that is a sub-mode of MR can create an augmented workspace by inserting the virtual space in which we store and interact with digital contents into the physical space where we work. Mixed and Augmented Reality differs from immersive VR in that it does not try to block the surrounding real environment from user.

Mixed and Augmented Reality systems use more varied media representations to augment the real environment than VR and two or more types of media representation can be presented one single system termed as hybrid representation. Based on the level of fidelity, the identified major classes of media representations are texts, indicators, platform (tablet and screen), 2D image, 3D wireframe, 3D data, and 3D object (Wang & Dunston 2007). High-fidelity representation conveys more detailed and richer information that can assist human's comprehension ability thus augmenting the cognitive process and activity. However, such high-fidelity setting may indirectly trigger usability problems associated with lag and low frame rate simply because computing power is limited (Wang & Dunston 2007). Input device is typically used to manipulate digital information displayed over the real environment. Most of the input devices used in virtual environments can also be applied in AR systems. More information regarding the input device can be found in (Gabbard 1997). Voice input, a more direct, natural form of interaction, can be achieved as an extra input to increase input capability of the whole system. Also, haptic devices can also be input module through devices like Phantom (Massie & Salisbury 1994). The display can be generally classified as visual display (e.g., head-mounted displays), acoustic display (e.g., 3D localized sound systems), and tactile display (e.g., force feedback devices). Visual displays are the most popular one used in AR systems with other types of displays as supplement. An elaboration on advantages and disadvantages of video-based see-through mixing and optical see through combination can be found in (Azuma 1997). Trackers are used to measure the motion of the user's head or hands, and sometimes eyes. Accurate registration and positioning of virtual objects in the real environment requires accurate tracking of the user's head and sensing the locations of other objects in the environment. Most tracking technologies are context-aware approaches and different technologies are available for tracking depending upon the application. Major trackers include mechanical, inertial, magnetic, ultrasonic, optical, and infrared tracking systems. For detailed comparison of tracking technologies, readers are referred to the surveys in (Ferrin 1991; Meyer 1992; Applewhite 1991; Holloway & Lastra 1995). For mixed environments that require large user-roaming areas, sophisticated ultrasonic tracking systems may be used to increase user range. Future tracking systems that can meet the stringent requirements of AR will probably be hybrid systems, such as a combination of inertial and optical technologies (Azuma 1997).

3. Application in remote excavation of dangerous materials

The existing dangers for human life in construction sites with dangerous materials beneath extort the needs to elaborate the remote and automatic systems for universally used machines and vehicles to make them safer for operators. Even though in this section we will discuss a civil-related system example proposed to use for demonstrating the concept, the concept can comprehensively facilitate the state-of-the-art of the whole teleoperation and telerobotics. This new Mixed Reality-based human-machine interface provides a more immersive, and compelling way to view video from real world scenes than the traditional method with limited field of view. The cameras mounted on the remote vehicle could acquire images of the explored environment. The camera parameters can be tracked by a combination between GPS and mechanical tracker, which are then sent to vision system for rendering the mixed scene as shown in Figure 1. These camera parameters and video

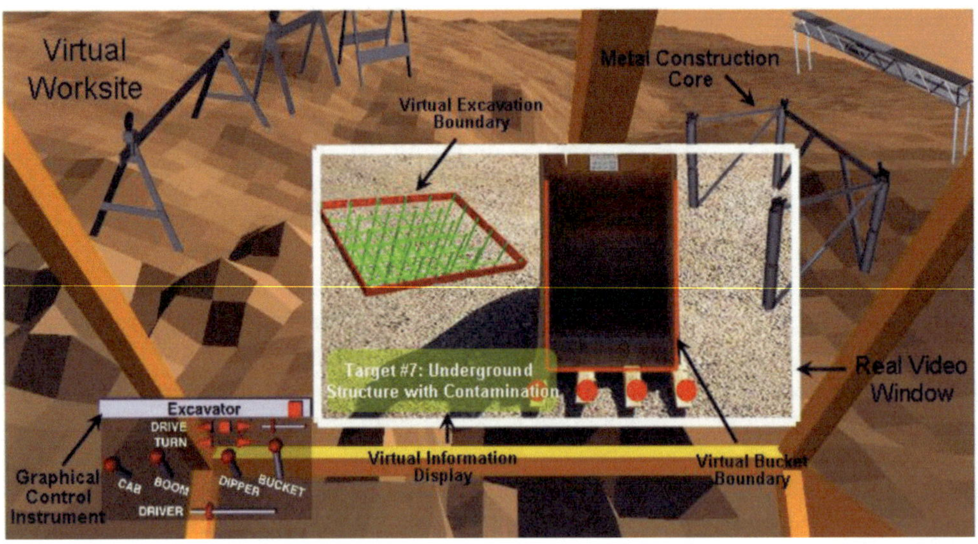

Fig. 1. A Sample Mixed Reality Display Screen for Teleoperation (Wang and Dunston 2006b)

streams are automatically acquired and synchronized in a real time manner. The extent of seamlessness of border between the video and the outer virtual environment depend on the following factors: calibration, the accuracy of modeling the real surroundings, lighting and the angle the camera makes with the objects surface (Wang & Dunston 2006b)

3.1 Current issues in remote excavation

The practical example application area that we are studying is the remote inspection and excavation of accidentally spilled or released hazardous materials in unstructured environments, which is a hazardous and, indeed, largely inaccessible environment. Currently, remote inspection and excavation are performed using a single camera mounted on a remote-controlled vehicle/robot that can transmit monoscopic color video information from the field to the human operator. The operator, using the visual feedback of the video, controls the vehicle and the camera head using certain control device. The inspection and excavation are assessed through the data from sensors, operator's knowledge and experience, inspection conditions, and camera characteristics. The accurate detection and excavation of hazard materials in a large unstructured environment using current methods is therefore a demanding task for the operator. Accurate assessments may not be achieved in practice as the nature of current teleoperation system is often too limited in visual interface/aids, placing high demands upon the skill and concentration of the operator. For example, in traditional CCTV teleoperation systems, the use of monoscopic video displays eliminates all binocular depth cues. The loss of these important cues results in situations wherein the location and size of an object in the remote scene is ambiguous. An additional failing as noted earlier is that it is very difficult to keep a satisfactory level of situational awareness for safe exploration due to inevitable restrictions in the field of view of the remote camera(s) and in the quality of the visual displays fed back from the remote worksite (Wang & Dunston 2006b). To address the above issues, a MR-enhanced visual interface was conceptually designed for remote excavation. The mechanisms and strategies of the teleoperation system are formulized in the next section.

3.2 Architecture of MR-enhanced visual interface for remote excavation

The framework of the teleoperation system with visual interface enhanced by Mixed Reality concept is conceptually designed for remote excavation of dangerous materials in an unstructured environment. Here, the technical components for the application of the MR-enhanced visual interface instead of other automotive and robotic components (e.g., how robots are controlled) are explored, verified and presented. Basically, the entire visual system consists of two basic modules: observation set and operator position. The developed conceptual scheme of elaborated visual system is presented in the Figure 2. In this system, three channels radio link has been applied. Two one-direction channels serve to send television signal while the two-direction channel serves to send steering and manipulation signals. A tethered vehicle is proposed to be built so that, in as many ways as possible, it mimics the basic design of commercial remote excavation platforms. Our goal is to improve the interface between the remote environment and the operator using MR and stereoscopic vision.

Observation Module

Observation set consists of a steering camera-head mounted symmetrically on a rotary beam. Observation module serves to observe the explore environment or chosen objects with cameras. The remote machine is able to roam around a real space while sending video updates to a virtual world model of the real space. Its construction allows to observe chosen objects with both cameras (carrying out stereovision process) or observation the objects located in different directions. Stereoscopic cameras allow achieving relatively high spatial vision and in comparison with one eye techniques (one-camera) it leads to quantity measurements, more direct and unambiguous. Stereo cameras are mounted on the vehicle via a pan and tilt robotic head system (Pretlove, 1996). The stereo device is capable of independent vergence and common pan and tilt of two remote-head color CCD cameras with independent control of all four degrees of freedom. The change of the axis of rotary beam rotation is done by its steering head. The typical operation of the head is that the pan and tilt axes are slaved to the motion of a human operator's head using two degree of freedom (DOF) of a three-DOF head-tracker mounted on a head mounted display (HMD). The system should allow an operator to intuitively control the gaze of the cameras and provides additional stereoscopic depth cues to enhance their perception of the remote scene. The visual display proposed to be used is HMD through which the operator is immersed into a mixed scene environment disregard the surrounding working environment.

Tracking Module

As elaborated in the last section, the graphics methodology for overlaying two virtual cameras as the same position to the real camera requires the parameters of the real observation cameras. Then the virtual machine might be properly situated in the virtual workspace, that is, its base reference frame can then be positioned correctly with respect to the workspace frame. The accuracy of this virtual/real correspondence is quite critical. In principle, the system must always know the position of the camera frame with respect to the mobile base, any changes in vehicle position should be updated by certain tracker. The following types of trackers are envisaged to be combined to produce the data that can be readily used by graphics module for rendering the virtual entities.

- *Position of observation cameras*: In mobile robotics applications, it is a common practice to use external sensing systems, such as radio frequency beacons, optical tracking, GPS, etc., to report the robot position in the workspace. GPS system mounted to observation camera reads their azimuth and specifies: set location in geographical coordinates, and sea level. The data about camera parameters and the location and orientation of observation set are sent to controller and moreover might by displayed on the monitor screen. Also, the data from GPS might be plotted on digital map which would allow drawing e.g., the route of a vehicle drive (with mounted vision system). Then the route by which the robots have traveled can be displayed in virtual environment as a reference.
- *Orientation of observation cameras*: The orientation of the camera can be tracked by a mechanical tracker mounted onto them, which can reach is high precision in a limited physical volume.
- *Machine states measurements*: Another source of data fusion comes from the remote machines, which typically have a variety of internal sensors that are used to measure and report current states. These states are typically positions, joint angles, tilt angles, velocities, accelerations and forces. These signals are relayed to the system as a real-time data stream.

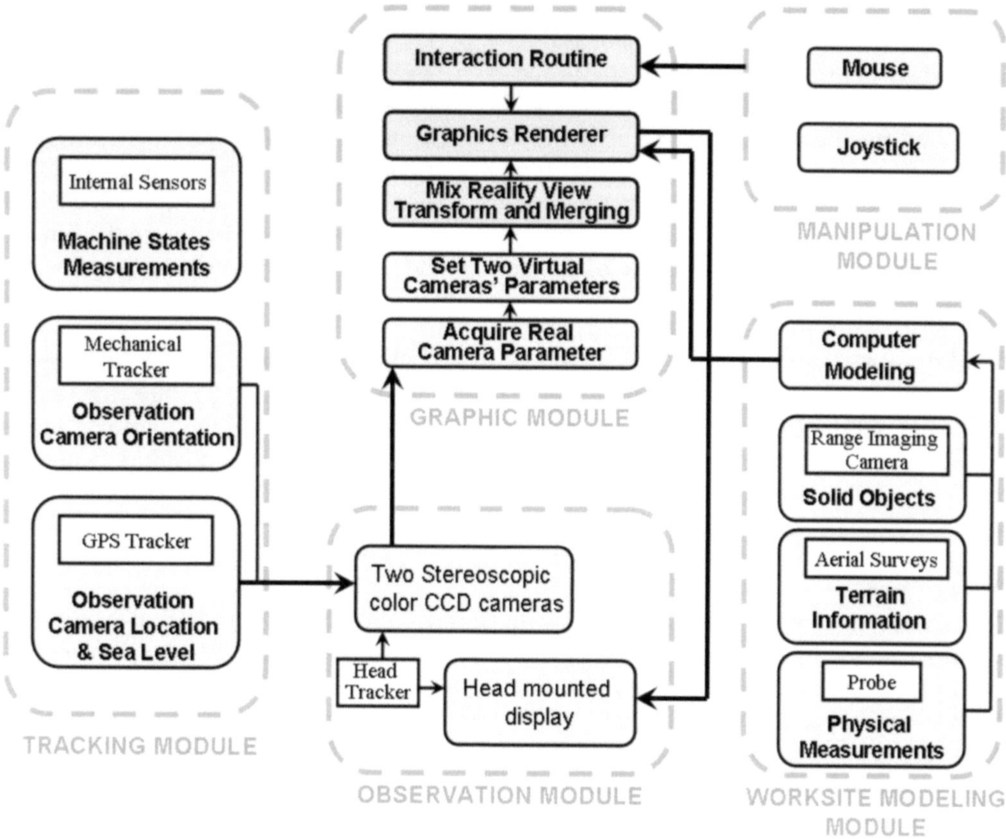

Fig. 2. System Architecture of MR-Enhanced Visual Interface for Remote Excavation

Worksite Modeling Module

The construction of a simplified virtual counterpart of the real environment that is not captured by the observation cameras, relies heavily on sensors and their ability to generate models of the world. The proper method should be exploited based on the features of working space and task requirement. The following sensing methods are proposed to be used together in constructing a virtual working environment.

- *Remote range imaging camera*: The range imaging data can be acquired by 3D laser range scanning system (Greenspan et al 1995). Data sets produced by this system can be displayed as surface representation, and are usually readily interpreted as images by the human operator, especially if the viewpoint is dynamically adjustable. It provides a normal vector at every surface location, thereby enabling artificial lighting models. Range sensors are used to sense solid objects in the working site such as obstacles, humans, other equipment and materials, and working units.
- *Aerial surveys*: This method is able to cover large working area and thus tend to involve very large files due to maintaining tolerable resolution of the data sets. The digital terrain map of working site can be collected and then integrated with other sources of data sets to construct a nearly complete virtual representation of real working environment.

- *Probe*: This can be used to collect readings made at several placements and orientations of a collimated sensor probe. For example, by knowing the probe characteristics, and the position and direction of the probe with respect to the surface that contains the radioactive material, an estimate of the radioactivity level at the surface of the objects being monitored may be calculated. Other measurements which may be treated this way include eddy-current signatures for metal surface inspection, temperature readings and coating thickness. Usually, these physical measurements do not explicitly include those particular locations data in 3D space. These must instead be inferred from the known or estimated positions of the sensor and other local objects at the time the sensors are sampled.

Manipulation Module

In addition to standard mouse-activated pull-down menus, buttons, sliders, and pop-up forms, two 2-axis joysticks can be integrated with force reflection capability, for remote excavator control. In the "dig" mode, joysticks can be used to control the four axes of the excavator arm (e.g., swing, boom, stick, shovel, etc.). The force reflection feature is used to indicate to the operator any collision with the virtual barriers mentioned above. Alternatively, in the "travel" mode, the force reflection feature is used to indicate "movement" by simply vibrating the stick.

Graphics Module

The major task of graphics module is to receive data from tracking and worksite constructing modules and then build the whole mixed environment scene following the methodology presented in last section. For virtual entities in AR portion, the graphics module can import and display 3D CAD models of objects such as scene props, virtual barriers and 3D stencils onto the real video stream. Also, the 3D models and surfaced range images can be used as geometric templates to support the display of other data being measured, such as temperature, pressure, etc.

3.3 Challenges and issues

While providing human viewers with extra-sensory information has a wide range of applications, adding visual information also presents the viewer with several perceptual and engineering hurdles. Some important issues for the development of the teleoperation system, though not covered in depth here, include: managing and controlling the quantity of extra-sensory information displayed, integrating multimodal cues (haptic and auditory sensations) from obscured locations, and interacting with information across multiple depth planes. The author believes that the largest and most challenging class of issues confronted in the MR-enhanced operator interface is depth ambiguity, more specifically relative distance judgment, while visualizing obscured information. The MR-enhanced operator interface can enhance the operator's perception of the real environment by showing information the operator cannot directly sense when unaided. The operator can be endowed with "X-ray vision," enabling him/her to see through the ground surface to view the buried hazardous materials underneath. However, displaying such hidden virtual objects in a manner that a user intuitively understands is not always trivial. It is essential to perceptually convey the difference between what is obscured (not visible) and what would be in plain sight (visible). This is especially problematic for the visual displays that are most often a planar projection (2D) view of a Mixed Reality 3D environment. The position of virtual buried objects in depth can become ambiguous when 3D information is presented on a 2D planar surface (i.e., information about depth becomes lost and thus is ambiguous). In such a planar view, the exact location (especially in depth) of virtual objects is inherently ambiguous. Thus, in an extended vision environment, where a virtual hazardous structure could be buried 2 feet or 4 feet underground, it is important to properly convey the depth information of the target. The AR display must make this relationship intuitively obvious or it may be more confusing than it is helpful. Approaches of solving this technical issue for effective visualizations of occluded information in the MR-enhanced operator interface can be driven by the application of perceptual psychology and cognitive science principles. In the development of the entire MR interface, one of the proposed future works will focus on developing approaches to accurately visualize the obscured virtual objects relative to real objects such as ground earth and excavation equipment end effecters (e.g., bucket). Alterative approaches will also be validated through experimental techniques.

4. Application in construction equipment repair and maintenance

Everett (1994) revealed that automation and machinery are suitable for physical-intensive tasks, which can be done by machine, and humans are still more cost effective at information-intensive tasks that require judgment, sensing, and adaptability. An interesting finding is that compared to automation and machinery appropriate for the physical-intensive tasks that convert information from the paper to the real work with the physical input, Mixed and Augmented Reality-based technology is suitable for information-intensive tasks, which deal with the information communication between the paper and the work. Virtually all construction activities require performance of some information-intensive tasks, which may also be embedded into physical-intensive tasks. AR can augment human's ability of accessing information and documentation, which can enhance the decision-making abilities. For example, MR system can supply the human operator with related necessary information in order to facilitate decision-making such as when and where to start and stop the equipment, how fast to proceed, and how hard to push or dig. The primary objectives in development of this second AR system are:

- To design and prototype a wearable system that could allow a mobile user to access digital information in the form of virtual media objects and real-time data visualizations that are linked to specific locations (equipment, or parts of equipment).
- To develop methods for annotating information that can be displayed to remote experts or other on-site crew.
- To explore the concept of "context visualization", that is develop design guidelines on how to make explicit organization and display of the layers of information and digital data that are attached to objects, site, equipment, as well as information about the relationships between them for a mobile user.

4.1 Application scenario

As the equipment maintenance crew with the AR system moves around, his display is continuously updated such that the tags drawn on the display stick to the real world elements. He can click on any tag to retrieve its maintenance schedule, maintenance history, the manufacturer's contact information, and component properties etc. He can also view text or hear audio instructions for maintenance procedure. Images or pictures may also be displayed as a reference. He also can view the maintenance record to see when a particular component was last replaced. He can make any annotations for that component, which would be saved. He or other authorized personnel can log in his maintenance work he carries out any time later in the future. The expert can handle several maintenance jobs at the same time. The reason why the expert is there is not only expertise sharing but also he has large amounts of maintenance and equipment technical specification in his computer or at hand. He can either instruct the crew via audio communication or send the updated data (pictures, video, and parameters) to the crew's local database through wireless network. Then such information can be displayed into the crew's view in a real time manner as reference. For example, from the detailed technical specification/drawings at hand, the expert can determine how the wiring is running behind the equipment and find out any other devices, e.g., a component, which the component might be connected to. Figure 3 illustrate a sequential maintenance scenarios that can be implemented this system where the maintenance crew could test hydraulic oil sampled from S.O.S. sampling valves for preventative maintenance. If the user clicks the repair service mode, the AR system can provide an associated pull down list of small tools, consumables, equipment and materials needed to effectively and safely accomplish a repair work. Also, the user can have easy access to training videos for repair procedure for a specific component.

(a) Identification of engine access panel

(b) Recognition of open panel causes identification of components and instruction to remove cap

(c) Recognition of cap removal causes new instruction to press test button.

(d) Additional information about position of occluded filter bypass structure in response to user's query

Fig. 3. Hypothetical illustration of sequence of automatic instructions and information for maintenance task (Wang et al. 2004)

4.2 Conceptual design of system architecture

As a test bed to evaluate the above concept and scenario, an initial technology platform has been conceptually developed that consists of a lightweight stereoscopic HMD system and camera with wireless audio/video transmission that is mounted on a remote user's head and body. Additional subsystems (modules) are added for tracking the user's location and head orientation, interacting with digital contents, accessing data about the equipment and generating data (annotating) to be displayed etc. A total of seven separate modules were configured and Fig. 4 shows the flow diagram for the entire AR system. More details can be found in (Wang et al, 2004). The following sections discuss certain technological extensions which have not been reported in (Wang et al, 2004). They are in the representation module, identification module and input module as shown in Figure 4.

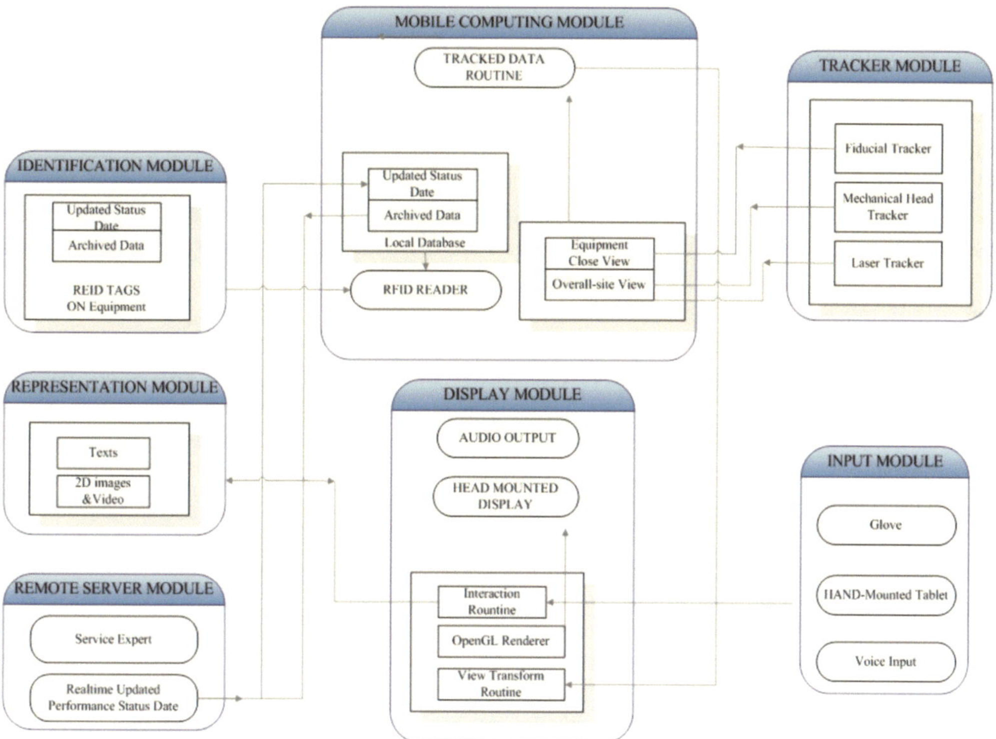

Fig. 4. Generic system architecture and data flow of the Augmented Reality system (adapted from Wang et al. 2004)

Most management information is typically stored in the format of descriptive texts and sometimes images/photos. Simple, informative media representation that actually facilitate real work may provide much more utility for serving the purpose of certain descriptive and procedural applications where users needs much explanatory information from database rather than the geometrical information that high fidelity, true-to-life virtual media can represent. The media representation should be textural description and/or 2D-image/video computationally-intensive (see Figure 4). These representations impose much less intensiveness onto the computing resources of the system.

This AR system can also be enhanced by adding a hypermedia system that provides contextual information in the form of labels and links. Like the use of HTML webpage links, more information can be accessed from clicking the hypertext which is linked to the database. The AR system allows users to experience hypermedia for themselves and manipulate complex information spaces with familiar interaction techniques.
Radio-frequency identification (RFID) technology could be used as the major identification technology to identify equipment (see Figure 4). RFID refers to a branch of automatic identification (auto ID) technologies in which radio frequencies are used to capture and transmit data. Automatic Identification Manufacturers, Inc., Pittsburgh describes RFID systems as those that read data from radio-frequency tags that are within range of the RG-reading equipment. The range is set as 50 yards to the navigating user and outlined working condition information of equipment can be queried from the RFID tags and/or from central server and displayed right beside the equipment via certain tracking technology as DGPS. A RFID tag is mounted onto equipment and signals can be received by readers on the mobile computing unit via antenna. The tags we use is read/write (R/W) tags in which, additional data can be written by overriding or extending the existing data stored in the tags. This enables the whole RFID to be able to update. Jaselskis et al. (1995) provided information on RFID and also discussed potential construction applications and conceptual design systems including concrete processing and handling, cost coding for labor and equipment, and materials control. These potential applications are envisioned to be integrated into the AR system to achieve more application arenas.
The input module stipulates the selection and manipulation of the digital contents. For simple command input, speech recognition (e.g., Microsoft speech recognition software) is sophisticated enough to realize this functionality. Data glove is used to intuitively interact with digital entities (for example, click a linked hypertext) and does not occupy user's hands like other unintuitive devices (e.g., keyboard, mouse, and joystick). A 3D virtual cursor can be controlled via data glove. Hand-mounted tablet can be used to quickly annotate information into the database after daily regular inspection.

5. Summary

Mixed and Augmented Reality is a promising advanced technology for the construction industry that can be integrated into heavy construction operations and equipment management. These technologies can reduce the cost, time, and hazards levels by augmenting relevant activities with digital contents. They are also promising advanced technologies for the construction industry by being integrated into visual interface of teleoperation and telerobotics. This chapter presented two Mixed and Augmented Reality systems and their application scenarios. This chapter also formulated mechanisms and strategies of these two systems.

6. References

Applewhite, H. (1991) "Position Tracking in Virtual Reality." *Proceedings of Virtual Reality '93*, Beyond the Vision: The Technology, Research, and Business of Virtual Reality, Westport, CT, 1-8.

ARToolKit (2004) Download website: http://www.hitl.washington.edu/artoolkit/

Azuma, R. T. (1997). "A Survey of Augmented Reality." *Presence: Teleoperators and Virtual Environments,* MIT Press, 6 (4), 355-385.

Behzadan, A.H. and Kamat, V.R. (2006). "Animation of construction activities in outdoor Augmented Reality", *Proceedings of Joint International Conference on Computing and Decision Making in Civil and Building Engineering,* June 14-16, Montreal, Canada, 1135-1143.

Everett, J. G., Slocum, A. H. (1994). "Automation and Robotics Opportunities: Construction Versus Manufacturing." *Journal of Construction Engineering and Management,* 120 (2), 443-452.

Ferrin, F. J. (1991) "Survey of Helmet Tracking Technologies." *SPIE Proceedings* Vol. 1456 Large-Screen Projection, Avionic, and Helmet-Mounted Displays, 86-94.

Gabbard, J. L. (1997). "A Taxonomy of Usability Characteristics in Virtual Environments." *MS Thesis for Computer Science and Applications,* December, Virginia Polytechnic Institute and State University.

Ghadirian, P. and Bishop, I.D. (2002). "Composition of Augmented Reality and GIS to Visualize Environmental changes", *Joint AURISA and Institution of Surveyors Conference,* Adelaide, South Australia, 25–30 November, 10 pages.

Greenspan, M., Lipsett, L., Ballantyne, J. (1995). "Laser Range Vision for Tele-excavation", *Robotics & Knowledge-Based Systems Workshop,* Montréal.

Holloway, R. and Lastra, A. (1995). "Virtual Environments: A Survey of the Technology." *SIGGRAPH'95 Course,* 8 (A), 1-40.

Jaselskis, E. J., Anderson, M. R., Jahren, C. T., Rodriguez, Y., Njos S. (1995). "Radio-frequency Identification Applications in Construction Industry", *Journal of Construction Engineering and Management,* 121 (2), 189-196.

Kamat, V.R. and El-Tawil, S. (2005). "Rapid Post-Disaster Evaluation of Building Damage Using Augmented Situational Visualization", *Proceedings of the 2005 Construction Research Congress, American Society of Civil Engineers,* Reston, VA, 10 pages.

Klinker, G., Stricker, D. and Reiners, D. (2001). "Augmented Reality for exterior construction applications", *Fundamentals of Wearable Computers and Augmented Reality,* 379-427.

Lawson, S.W. and Pretlove, J.R. (1998). "Augmented Reality for Underground Pipe Inspection and Maintenance", *SPIE Telemanipulator and Telepresence Technologies V,* 3524, 98-104.

Massie, T. H. and Salisbury, J. K. (1994). "The PHANTOM Haptic Interface: A Device for Probing Virtual Objects." *Proceedings of the ASME Winter Annual Meeting, Symposium on Haptic Interfaces for Virtual Environment and Teleoperator Systems,* Chicago, IL, November.

Meyer, K., Applewhite, H. L., and Biocca, F. A. (1992). "A Survey of Position-Trackers." *Presence: Teleoperators and Virtual Environments,* MIT Press, 1(2), 173-200.

Milgram, P. and H. Colquhoun (1999). A Taxonomy of Real and Virtual World Display Integration, *Mixed Reality - Merging Real and Virtual Worlds.* Y. O. a. H. Tamura, Ohmsha(Tokyo) & Springer Verlag(Berlin): 1-16.

Milgram, P. and F. Kishino (1994). "A Taxonomy of Mixed Reality Visual Displays", *IEICE Trans, on Information and Sytems (Special Issue on Networked Reality),* E77-D (12): 1321-1329.

Pretlove, J. R. G. (1996). "Get A Head in Telepresence: Active Vision For Remote Intervention." *Proceedings of the International Confeference on Remote Techniques for Hazardous Environments*, 264–269.

Roberts, G. W., Evans, A., Dodson, A., Denby, B., Cooper, S., and Hollands, R. (2002). "Look Beneath the Surface with Augmented Reality." *GPS World*, February, Internet Article, 14-20.

Shen, J., Wu, Y., and Liu, H. (2001). "Urban Planning Using Augmented Reality." *Journal of Urban Planning and Development*, 127 (3), 118-125.

Thomas, B., Piekarski, W., Gunther, B. (1999). "Using Augmented Reality to Visualize Architecture Designs in an Outdoor Environment.", *Proceedings of DCNET1999 - Design Computing on the Net*, November 30-December 3, Sydney, NSW, Australia.

Wang, X., Dunston, P. S. (2006a). "Mobile Augmented Reality for Support of Procedural Tasks." *Proceedings of Joint International Conference on Computing and Decision Making in Civil and Building Engineering*, June 14-16, Montreal, Canada, 1807-1813.

Wang, X., Dunston, P. S. (2006b). "Mixed Reality-Enhanced Operator Interface for Teleoperation Systems in Unstructured Environment." *CD Proceedings of the 10th Biennial ASCE Aerospace Division International Conference on Engineering, Construction and Operations in Challenging Environments (Earth and Space 2006)*, American Society of Civil Engineers (ASCE), March 5-8, League City/Houston, Texas, 8 pages.

Wang, X., Dunston, P. S. (2007) "Design, Strategies, and Issues towards an Augmented Reality-based Construction Training Platform." *Journal of Information Technology in Construction*, International Council for Research and Innovation in Building and Construction (CIB), Rotterdam, Netherlands, 12, 363-380.

Wang, X., Dunston, P. S., Skiniewski, M. (2004). "Mixed Reality Technology Applications in Construction Equipment Operator Training." *Proceedings of the 21st International Symposium on Automation and Robotics in Construction (ISARC 2004)*, September 21-25, Jeju, Korea, 393-400.

Webster, A., Feiner, S., MacIntyre B., Massie, W., Krueger, T. (1996). "Augmented Reality in Architectural Construction, Inspection, and Renovation." *Proceedings of ASCE Computing in Civil Engineering*, Anaheim, California, June 17-19, 913-919.

Yang, C., Hsu, Y., Lin, C. (2000). "Site Engineer Digital Assistant ---- the Integration of Information Technology in Order to Assist Site Engineers in Processing Information at All Times and Places." *Proceedings of 17th International Symposium on Automation and Robotics in Construction*, Taiwan, 18-20 September, 771-776.

15

An Active Technology for Improving the Sound Transmission Loss of Glazed Facades

Berardo Naticchia, Alessandro Carbonari and Sara Spadoni
Polytechnic University of Marche
Department of Architecture Construction and Structures
Italy

1. Introduction

The 89/106/CEE European Directive and the related national legislation of the European member countries have made noise protection a compulsory building requirement. From an acoustic point of view, building envelopes have the task of reducing external noise to an acceptable level inside the same building.

In fact, glazed facades are the preferred paths followed by disturbing noise from the outside to the inside and, in spite of the adoption of very expensive passive means, they are often unable to respect the strict limits required by standards and regulations. The two main types of passive means used are: laminated glass technology or double glazing. Both of these can be useful for reducing noise transmission at high frequencies: in particular the laminated solution is used to shift the coincidence effect at frequencies higher than the audible range, in fact improving STL values in the range of acoustic waves higher than 1500 Hz, while determining no improvements at lower frequencies. On the other hand, in double glazing, the coupling between the glass panels and the air layer adds another resonance effect at frequencies lower than 500 Hz, while improving their STL in the range of frequencies limited between the resonance and coincidence effects. As a consequence, the adoption of an active control system is needed to solve the acoustic performances drop of glass panels at low frequencies.

This chapter will show how an active system would be capable of controlling the vibrations of glazed panels and of reducing sound radiated as a consequence, avoiding the use of very expensive glazed facades equipped with massive passive means and effective only at high frequencies, whose adoption would also raise economic and technologic problems for their installation on buildings. In particular, the ASAC (Active Structural Acoustic Control) approach will be analyzed and tested, also demonstrating why it is better than the other Active Noise Control approach, commonly known as ANC. Pursuing this aim requires tackling a series of complex problems such as choosing reference and error sensors, modelling the system controller, designing the system with well-suited actuators, integrating the active system into typical glazed façades while minimizing visual impact and maximizing the controllability of structural vibrations. It will be shown that remarkable sound transmission loss (STL) improvements can be obtained and that this approach has great exploitability potential in other specialized products like light opaque building partition walls, railway and traffic noise shielding, temporary environmental noise barriers or even real-time controlled reflecting panels for the acoustic adjustment of concert halls.

2. State of the art and the innovation provided by ASAC systems for glass panels

2.1 Early Active Noise Control applications

The concept of active noise control was first introduced by Leug (Leug, 1936), who presented a patent for a system implementing feed-forward active control of sound in a duct. According to the patent, the sound field is detected using a microphone, whose signal is passed through an electronic controller, such that it adjusts the loudspeaker to produce a cancelling wave in a duct, resulting in destructive interference of the primary or noise source wave. A feed-back arrangement was produced by Olson and May (Olson et al., 1953), who developed an active noise control system that is based upon detecting the offending sound with a microphone and feeding the signal back through a controller to a control loudspeaker located close to the microphone. They also demonstrated good local reduction at the microphone over a range of frequencies from 20 to 300 Hz.

Conover (Conover, 1955) introduced the concept of an error sensor in which the radiated sound field of the transformer was monitored and used to adjust the controller in order to minimize the radiated sound. However, he experienced difficulties due to the use of analogue systems, particularly when the physical system was changing relatively rapidly.

Hence, in the early 1970s both (Kido, 1975) and (Chaplin et al., 1976) began exploring the application of digital signal processing techniques, demonstrating the feasible use of their digital active noise control approaches on a number of important applications, such as car exhaust noise, ship exhaust stack noise and active engine mounts.

2.2 Active control of structural vibrations

The introduction of digital techniques and piezoelectric actuators to vibration control by Bailey & Hubbard (1985) considered feedback modal control of large structures. Balas & Canavin (1977) discussed feedback damping control of large spacecraft structures. Balas (1978) applied theoretical modal control using velocity feedback to a simple beam. Meirovitch & Öz (1980), with later work by Meirovitch and others (e.g. Meirovitch et al., 1983), expanded the modal control method, finally arriving at a method they described as Independent Modal-Space Control (IMSC), where a coordinate transformation was used to decouple a complicated system into a set of independent second order systems in terms of modal coordinates.

Baz & Poh (1988) made modifications to the IMSC method to minimize the effect of control spillover (inadvertent localized increases in vibration levels due to constructive interference between disturbance and control fields). Modern applications include precision optical components in noisy environments, pilot seat vibrations in helicopters, sonar masking in torpedoes and submarines.

2.3 Active Structural Acoustic Control

Advances in active control of structural vibrations have been applied to develop an alternative control method for enclosed noise fields, known as active structural acoustic control (ASAC). The ASAC approach uses structurally-based actuators to exert control forces on the radiating structure itself in order to minimize radiated sound (Ruckman & Fuller, 1995; Nelson & Elliott, 1995; Fuller et al., 1996). The actuators are vibration sources (shakers, piezoceramic patches, etc.) which modify how a structure vibrates, thereby altering the way it radiates noise.

In 1990 Jones and Fuller proposed an initial work where electromagnetic shakers were used to provide point force control inputs to simplified cylindrical test sections. The controller exploited error sensor microphones placed in the cylinder interior, such that interior sound levels were minimized by the control of structural vibration. This work demonstrated that, in general, fewer control actuators are required by the ASAC approach as compared to ANC (Active Noise Control) techniques. Also, control spillover in the interior acoustic space was reduced in the ASAC experiments. In such cases, "modal suppression" is usually suggested, where only the few dominant radiating modes are reduced in amplitude. Other structural modes that are poorly coupled with the noise field may be left unchanged by the control system (Fuller et al., 1996).

Simpson, et al. (1991) performed control experiments in a test section comprising the aft portion of a furnished DC-9 aircraft. Using a typical feed-forward controller arrangement, the researchers achieved global sound level reductions of up to 9 dB using various configurations of 7 error microphones.

Because point forces are spectrally white in a spatial sense, their use as controlling forces in ASAC work can lead to undesired spillover in many structural vibration modes. In seeking a control actuator with more distributed forcing properties, many researchers recently investigated the use of piezoceramic materials for applying bending moments or in-plane strains to structures (Clark et al., 1991). In particular, lead zirconium titanate (PZT) materials have been widely used in ASAC work, providing sufficient forcing capabilities with the benefit of greatly reduced mass and space requirements as compared to electromagnetic shaker devices.

In 1992 Fuller et al. used an aluminium cylindrical test section with a removable floor structure to simulate an aircraft fuselage environment. Piezoelectric patches were bonded directly to the cylinder surface and the system was acoustically excited with an exterior loudspeaker noise source. Using two microphones as error sensors, the ASAC system provided global attenuations on the order of 10 dB in the cylinder interior.

Recently, ASAC tests were performed by Fuller et al. (1994) in the cabin of a Cessna fuselage, a typical mid-sized business jet. In the acoustic resonance case, noise reductions of 20 dB or more were achieved at the error sensors, but an average global increase of several dB was measured at 7 additional microphones. Control performance in the off-resonance case was significantly reduced, with reductions of 2-10 dB at the error sensors.

Some past work has addressed actuator position optimization via analytical approaches (Clark et al., 1992; Wang et al., 1994; Burdisso et al., 1994). Additionally, one work with a finite element model has shown the importance of position optimization to global ASAC performance, made by (Yang et al., 1995). Spatially averaged reductions of acoustic potential energy of up to 14 dB have been predicted using 16 optimally-located point forces, while poor control force placement may result in overall global sound increases.

2.4 Active Noise Control in glazed enclosures

To the authors' knowledge, no ASAC application has been performed on glazed structures. Other works by (Kaiser et al., 2003) suggest an ANC approach through the positioning of both loudspeakers and sensors inside air cavities; and (Zhu et al., 2004) place monitoring sensors outside windows. Instead, most of the recent research was developed to verify the performances of the control provided by PZT patch actuators on opaque surfaces, both for building walls and for helicopter and airplane envelopes: in the contributions of (Kaiser et al., 2003) and (Bao and Pan, 1997) it is shown that using ANC control systems for double walls needs cavities wider than 0.1 m thick, which is not usual for standard double glazed panels. Preliminary results in Wernli (2001) show that PZT laminated (multi-layer) patches can indeed actuate the heavy panes sufficiently. However, since these patches interfere substantially with the primal function of the window, namely to look through it, they may not be considered an option.

2.5 The original contribution provided by the proposed ASAC system for glass panels

The research described in this chapter proposes an original approach for applying ASAC on glazed structures, which takes into account the problem of using piezo stack actuators, whose visual impact must be managed, even if considerably reduced compared to the piezo-laminated patch actuators.

The application of ASAC systems on glazed panels is particularly critical, as there are no means of efficiently increasing their STL over all the acoustic spectrum, as will be explained in paragraph 3.1. For example, acoustic interlayers provide the most significant noise reduction in the frequency range of 1500 to 6000 Hz, while in the range of 100 to 500 Hz the Sound Transmission Loss remains significantly low. Furthermore, the double glazing approach provides appreciable noise reduction just in the frequency range above 500 Hz.

Hence the use of the ASAC technology seems to be the only one capable of increasing glazed facades' STL over all the acoustic spectrum. In this way, it would be possible to respect the strict European and national legislation on expected acoustic performance for external facades, even in very harsh areas, such as in proximity of airports, railways or high traffic roads. Additionally, the problems connected with the application of actuators on transparent facades will be tackled for the first time.

3. Adoption of active means for controlling acoustic and vibration behaviour of glass panels

3.1 Adoption of active means for STL increase

Considering that for buildings, the main type of disturbing noise is produced by road and railway traffic, panels excited by plane waves have been studied. Analytical solutions for forecasting the behaviour of unbounded non flexible panels are known (Fahy, 1985), and they can be easily extended to building envelopes. The STL of glass alone at low frequencies is determined by the panel's static stiffness. Since glass reacts best to excitation frequencies that match its natural frequencies, the low internal damping of tempered glass produces resonances that dramatically decrease its STL. Above the resonant frequencies, sound transmission follows the mass law of acoustics and is dictated by the mass or surface density of glass. However, glass has a specific coincidence frequency at which the speed of incident acoustical waves matches that of the glass bending wave and acoustic waves are especially effective at causing glass to vibrate, which makes the vibrating glass an effective sound radiator at that specific frequency and above. As a result, the first critical point for a single glass panel is in correspondence of the resonance effect (f_R), registered at very low acoustic frequencies, as in Fig. 1-a (Spagnolo, 2001). Above this value its STL increases with frequency, until the critical effect causes another drop of STL, due to the matching between the wavelength of flexural vibrations propagating through the glass and the projection of the disturbing noise wavelength which dramatically increases the overall radiation efficiency of the glass panel. The latter effect may be solved through the adoption of laminated glass panels with PVB layers, which decrease the overall flexural stiffness of laminated panels, shifting up the critical frequency (f_C), out of the audible range. But no remarkable improvements are obtained at the resonance frequency. Even if double glazing provides an average increment of STL (moreover in the middle frequency range), however it generates another resonance frequency, due to the coupling between the two glass panels and the air cavity, whose position depends on the thickness and typology of the panels (Harris, 1984). An illustrative example from measurements is presented in Fig. 1-b, based on (Harris, 1994): three double glass panels are compared in order to point out the importance of the air layer thickness with respect to the final STL. Starting from a good double glazing (two 0.006 m thick glass panels with a 0.013 m air layer interposed) it is demonstrated that even if the air layer is further widened, the low frequency STL is still low due to resonance dip which persists even if a much higher air thickness equal to 0.05 m is provided.

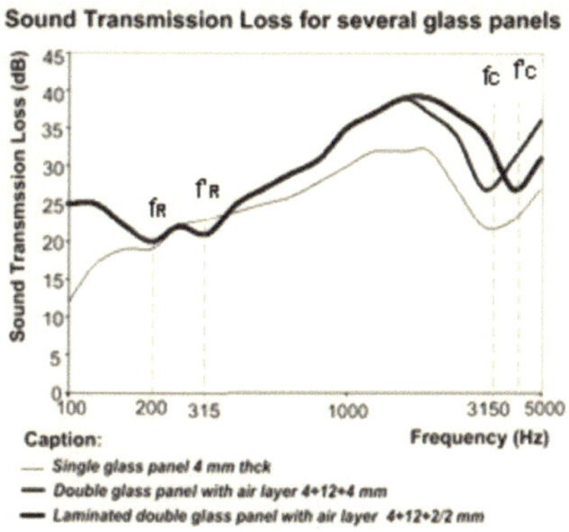

a)

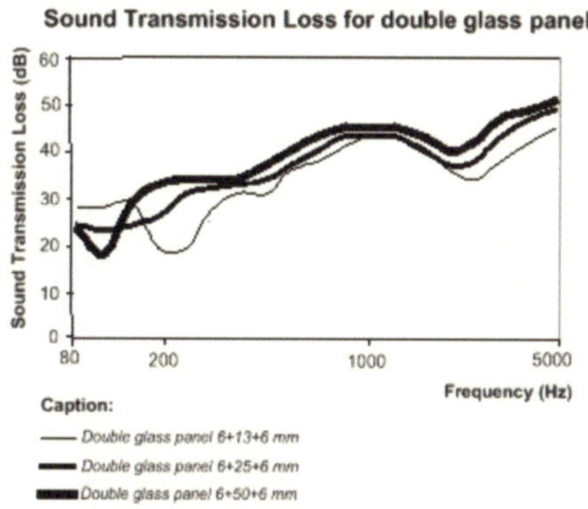

b)

Fig. 1. Sound Transmission Loss curves for single, double and laminated glass panels (a); the same for double glass panels (b).

Moreover, considering that traffic noise levels are usually very high in the low frequency range (IMAGINE, 2003), it follows that other solutions are required for guaranteeing acoustic comfort in buildings (as illustrated in Fig. 2).

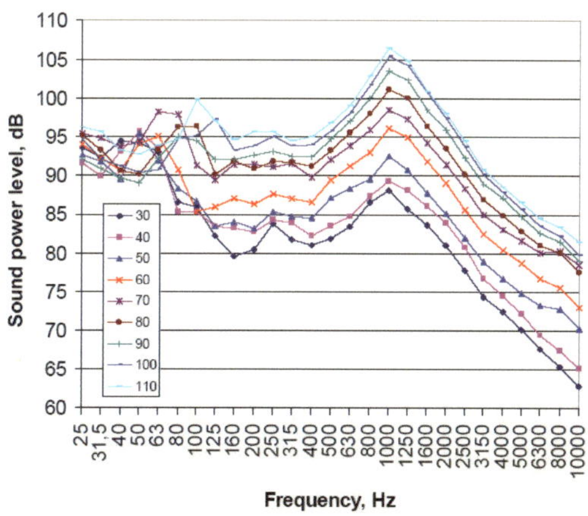

Fig. 2. Equivalent omni-directional sound power level of road traffic noise as function of a vehicle's speed.

3.2 Analytic model for the simulation of vibrating glass panels

As flexural waves in structures are the main responsible agents for sound emission (Hall, 1987) and they are induced in plates by external disturbances, it is opportune to analytically describe the flexural response of a simply supported plate to a harmonic excitation of arbitrary distribution like f(x,y,t) = F(x,y) $e^{i\omega t}$ (where $\omega = 2\pi\nu$, ν is the excitation frequency, F(x,y) is the excitation amplitude), obtaining the equilibrium model introduced by Kirchhoff-Love (Timoshenko and Woinolowsky-Krieger, 1959):

$$EI\left(\frac{\partial^4 w}{\partial x^4}+2\frac{\partial^4 w}{\partial x^2 \partial y^2}+\frac{\partial^4 w}{\partial y^4}\right)+\rho h\frac{\partial^2 w}{\partial t^2}=-F(x,y)e^{i\omega t} \tag{1}$$

being *i* the complex unit, *w* the transversal displacement, *E*, *I*, *ρ* and *h* the elasticity modulus, moment of inertia, density and thickness, respectively, of the plate. One of the possible solutions of eq. (1) for the formalization of the transverse displacement field is obtained describing it as the superposition of infinite modes of the free response of the plate, which is vibrating at the forcing frequency:

$$w(x,y)=\sum_{m=0}^{\infty}\sum_{n=0}^{\infty}W_{m,n}\Psi_{m,n}(x,y)e^{i\omega t} \tag{2}$$

where $W_{m,n}$ are the modal amplitudes, which are the maximum displacements at the free natural mode *(m, n)*. $\Psi_{m,n}$ are the modal shapes, which take into account the shape of the mode *(m,n)* on the plate, while the exponential term is aimed at describing the dependence with the time and the influence of disturbing frequency. In order to study the transversal displacement motion of a plate whose side lengths are *a* and *b*, it is necessary to compute the displacement distribution field given by an incident plane wave. Eq. (2) is generally approximated in another form, which considers only the superposition of a finite number of vibrating modes, given by the combination of two finite numbers *m* and *n* (the validity of the approximation is dependent on the choice of these two values). Thus the displacement distribution is given by (Fuller et al., 1997):

$$w(x,y,t)=\sum_{m=1}^{M}\sum_{n=1}^{N}W_{m,n}\operatorname{sen}(k_m x)\cdot\operatorname{sen}(k_n y)e^{i\omega t} \tag{3}$$

where the eigenvalues are:

$$k_m=\frac{m\pi}{a},\ k_n=\frac{n\pi}{b} \tag{4}$$

Substituting eq. (3) in eq. (1) we can find the solution as:

$$W_{m,n}=\frac{P_{m,n}}{\rho h\left(\omega_{m,n}^2-\omega^2\right)} \tag{5}$$

being $P_{m,n}$ the modal pressure, that is, the action given separately by each mode of vibration, $\omega_{m,n}$ the natural frequencies of the glass plate as reported in (Fahy, 1985):

$$\omega_{mn} = \sqrt{\frac{EI}{\rho h}}\left[\left(\frac{m\pi}{a}\right)^2 + \left(\frac{n\pi}{b}\right)^2\right] \quad (6)$$

In (Roussos, 1985) an analytical solution for the generic case of a plane wave incident on a simply supported plate, like in Fig. 3, is detailed.

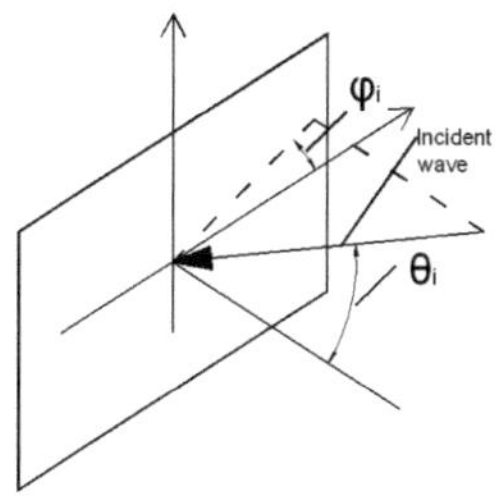

Fig. 3. Angles individuating the direction of incidence of an acoustic plane wave into a glazed plate.

4. The ASAC active control system adopted

4.1 Active Structural Acoustic Control

The approaches by Leug's patent in 1936 and (Olson & May, 1953) are the first examples of ANC feed-forward control, when prior knowledge of the noise is obtained with an upstream microphone, and ANC feedback arrangement, where the detection microphone is close to the active secondary source respectively. Both approaches relate to the main concept of ANC, and hold its drawbacks: when applied to buildings' glazed facades, they need an external microphone for disturbance monitoring, and internal error sensors and loudspeakers for control purposes. For this reason some attempts at inserting all the system components inside the air cavity of double walls were pursued but this task became much too difficult in the case of windows, due to their narrow air layers.

Therefore, the ASAC system should be preferred for this kind of application, because both reference sensors and actuators are placed on the same glazed panel, as required for the implementation of a feedback controller. Fig. 4-a and Fig. 4-b respectively depict the feed-forward and feed-back arrangements of an ASAC configuration for glazed facades' control. In both cases, actuators are positioned on the vibrating surface, which is the source of disturbing noise in the receiving room and whose vibrations are reduced to rise its final STL. ASAC does not require the use of loudspeakers or error microphones in the receiving environment, however optimum positioning of sensors and PZT actuators on the vibrating surface must be pursued. Generally, various modes of vibration have different radiation efficiencies (Fahy, 1985), and some are better coupled to the radiation field than others. This suggests two conclusions: only the most efficient modes need to be controlled, rather than the whole response; in some cases the relative phases and amplitudes of multi-modal response can be adjusted so that their radiated fields interfere destructively.

The feed-forward control of Fig. 4-a requires knowledge of the primary disturbance, which is derived by the use of a reference microphone: in the case of buildings, this seems unpractical, because it would require the installation of a microphone on the exterior of the window which, for functional and aesthetic issues, is not feasible. Therefore, the feedback type controller depicted in Fig. 4-b seems to be opportune and is detailed subsequently.

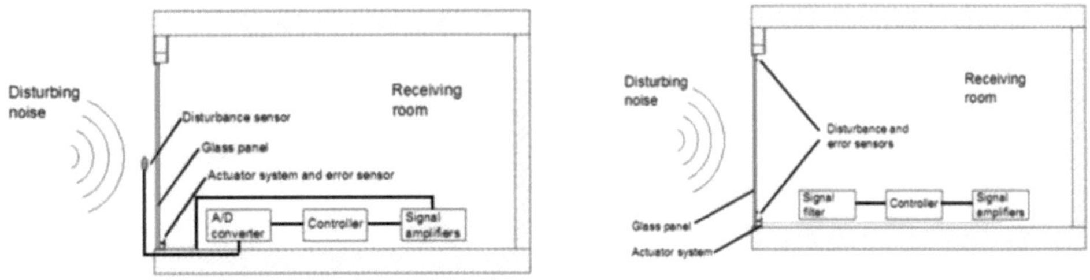

a) b)

Fig. 4. Feed-forward (case a) and feedback (case b) ASAC systems for glazed facades.

4.2 Optimal control theory

The ASAC approach is based on the reduction of structural vibrations and radiated sound (transmitted and reflected) as a consequence, by means of actuators bonded on the vibrating surfaces. ASAC can be easily integrated into buildings, as it does not require the use of loudspeakers or error microphones in the receiving environment.

As shown in Fig. 5, once the signal comes from the sensors, it must be elaborated by charge amplifiers (converting voltage signals into physical variables like displacements, velocity and accelerations), and electronic filters that have a double task:

- to separate the contribution to the total vibration field due to the primary disturbance from that due to secondary sources;
- to compute the radiated field in some positions of the receiving room, to calculate the error (difference between reference value and actual value of the radiated noise).

The controller, starting from the error signals, computes the opportune voltage to be supplied to the secondary sources, whose electric power is provided by the amplifiers. At this level an algorithm calculates the control voltage to be supplied to the secondary sources. In general, it has the task of changing the distributed mass, stiffness and damping of the plate in order to reduce the radiated noise in the internal environment, determined mainly by flexural waves which need to be decreased by the control plant for this reason. The complexity of the algorithm derives from the fact that reference and control sensors are positioned at the same place, therefore electronic filters are necessary. The controller must opportunely tune its gain coefficients to adjust the properties of the system, so that the radiated noise will be minimized. A number of papers, such as (Baumann et al., 1992), (Cunnefare, 1991), (Baumann et al., 1991) and (Pan et al., 1992), showed that optimum control is the most suitable for ASAC systems: it has the advantage that the choice in the prescribed change of the dynamic properties is motivated by its aim to reduce structural vibrations or radiated noise.

Optimal control system procedures offer the possibility of designing the controller parameters directly minimizing a cost function of performance which is proportional to the required measure of the system's response. For the sake of computational efficiency, it is convenient to define a cost function that is quadratically dependent on the response, simplifying the optimization problem. One of the possible solutions can be the one proposed by (Fuller et al., 1997), where the cost function is given by:

$$J = \frac{1}{2\rho c} \sum_{i=1}^{N} \left| p_t^i \right|^2 \Delta S_i \tag{7}$$

being J the total radiated acoustic power, ρ and c the density and sound velocity of wave in air; p_i the sound pressure values measured at some prescribed measurements points and ΔS_i the surfaces relative to each measurement point. Pressure values are not directly measured, but computed from the signals deriving from the reference sensors positioned on the glazed panel (through the use of filters). Thus, the structural-acoustic coupling is inherent in the definition of the cost function. It was demonstrated by (Fuller et al.,1997) and (Nelson & Elliott, 1992) that substituting the opportune expression of radiated sound pressure, given

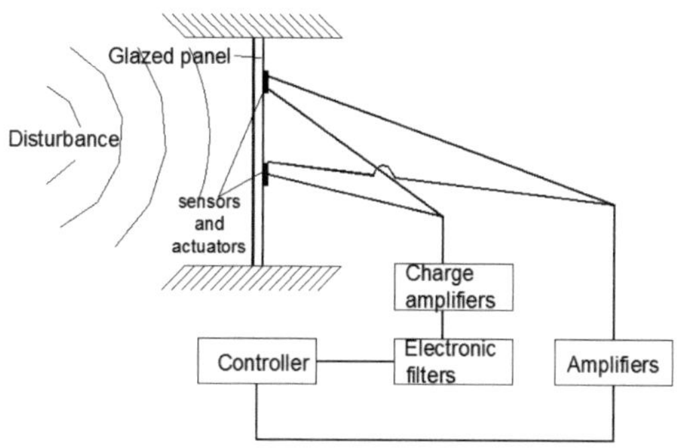

Fig. 5. Scheme of an ASAC system for glazed panels integrated in buildings

by the superposition of the two contributions of both the disturbance and the control actuators, the cost function is scalar. It can be converted into a quadratic expression of complex control voltages, and it was demonstrated that this function has a unique minimum. The general form of the equation to be minimized becomes:

$$J = \underline{h}^T \underline{q} + \underline{c}^T \underline{v}_i \tag{8}$$

where q is the vector of complex input disturbances, h is the vector of transfer functions associated with these disturbances; c is the control transfer function vector and v is the unknown vector. In this way a vector of voltages v_i is computed, minimizing the total radiated field.

Once the transfer functions between the reference signal deriving from reference sensors and the acoustic radiated noise is known for a given system, the control plant will automatically execute all these steps, minimizing the radiated noise even if glazed panels are subject to time-dependent input disturbances, giving back an automated glazed facade, that actively changes its properties according to the disturbance. Before implementing this control system, it is necessary to calculate the control transfer functions, which requires as a preliminary stage, the choice of the opportune kind of secondary sources, carried out in the next section.

However, the analytical model can be implemented only following a series of simplifications, which appear difficult to apply in terms of the actual situations that one can come across in the building field:

- simple support boundary constraints, whereas in fact, constraint situations are more complex and more similar to a semi- fixed or yielding joint;
- applications of only point forces, without the association of mass as occurs in the real case when control is effected through the use of actuators contrasted by stiffening structures.

Given the above considerations, it has been established that the numeric model based on the theory of Kirchhoff-Love, will be substituted by a model built using finite element software programs (ANSYS™, LMS VIRTUAL LAB™), which allows overcoming the simplifications tied to the analytic model.

4.3 Piezoelectric actuators

Two main types of actuators, suitable for glazed facades, are presently marketed (Fig. 6):

- Piezoelectric (PZT) patch actuators providing bending actions to excite structures;
- PZT stack actuators providing point forces to excite structures.

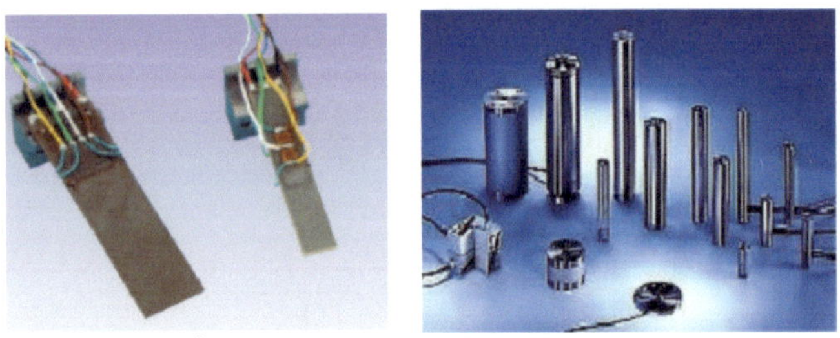

a) b)

Fig. 6. PZT patch (a) and stack actuators for glazed facades (b).

The first type is usually bonded to a surface while the second needs a stiffening structure to fix it and make it transfer forces to a surface for controlling purposes. These actuators are available in a wide range of sizes (from few centimetres to various decimetres) and are capable of generating high forces (with reduced displacements) inside a wide range of frequencies (Dimitriadis, Fuller, Rogers, 1991). Even if they were shown to work properly for many applications, however they have not been tested in applications on glazed facades, and most of the experiments were carried out in the automotive and aeronautic fields of research. As far as concerns the choice of actuators, the first rectangular shaped patch may interfere with visibility (Fig. 7-a); the stack one instead is very small but needs a stiffener in order to work properly (Fig. 7-b).

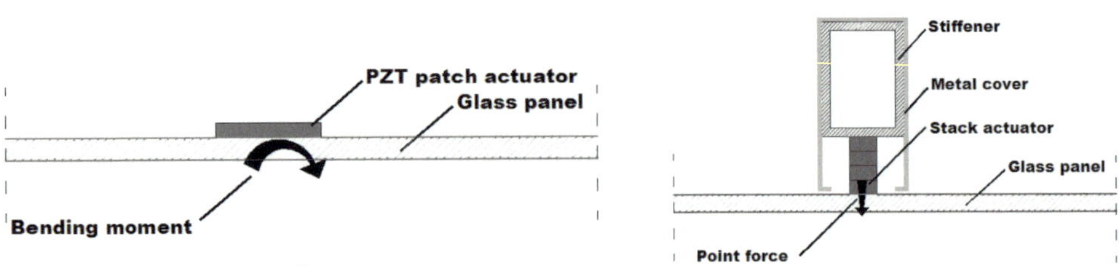

a) b)

Fig. 7. PZT patches (a) and PZT stack actuators (b), as applied on a glass panel.

In the asymmetric disposal of Fig. 7-a, the PZT patch excites the 2D structure with pure bending, that can be simulated with the numerical model developed in (Dimitriadis, Fuller, Rogers, 1991). It is assumed that the strain slope is continuous through the thickness of the glass plate and of the PZT patch, but different along the directions parallel to the plate sides, which in turn are assumed parallel to the coordinate axes (the strain slopes are billed C_x and C_y). The mathematical relation between strain and z-coordinate is:

$$\varepsilon_x = C_x \cdot z \text{ and } \varepsilon_y = C_y \cdot z \tag{9}$$

being the origin of the z-axis in the middle of the plate thickness and ε the strain. The unconstrained strain of the actuator (ε_{pe}) along plate axes is dependent to the voltage applied (V), the actuator thickness (h_a) and the PZT strain constant along x or y directions ($d_x = d_y$):

$$\varepsilon_{pe} = \frac{d_x V}{h_a} \tag{10}$$

Considering that the plate is subject to pure bending, no longitudinal waves will be excited, and by applying the moment equilibrium condition about the centre of the plate along x and y directions as in (Fuller, Elliott, Nelson, 1997), assuming that the plate thickness is $2h_b$, the plate elastic modulus is E_p, the actuator elastic modulus is E_{pe}, and ν_p and ν_{pe} are the Poisson coefficients of the plate and actuators respectively; also assuming that moments induced in the x and y directions (billed with m_x and m_y) are present only under the PZT patch, and assuming that it is located between the points of coordinates (x_1,y_1) and (x_2,y_2), in (Dimitriadis, Fuller, Rogers, 1991) it is shown that:

$$m_x = m_y = C\varepsilon_{pe}[H(x-x_1)-H(x-x_2)][H(y-y_1)-H(y-y_2)] \tag{11}$$

being $H(x)$ the Heaviside function and $C=EIK_f$, where I is the moment of inertia of the plate; then the equation of motion for plates subject to flexural waves can be written:

$$EI\left(\frac{\partial^4 w}{\partial x^4} + \frac{\partial^4 w}{\partial y^4}\right) + \rho h \frac{\partial^2 w}{\partial t^2} = -p(x,y) \tag{12}$$

where p is an external uniform pressure applied on the plate. Eq. (11), if written with the actuator induced moment, becomes:

$$\frac{\partial^2[M_x(x)-m_x(x)]}{\partial x^2} + \frac{\partial^2[M_y(y)-m_y(y)]}{\partial y^2} - \omega^2 \rho S w = 0 \tag{13}$$

where M is the internal plate moment and m is the actuator induced bending moment; ρ and S are density and surface of the plate; w is the displacement and ω is the wave phase change. Assuming that the actuator is perfectly bonded on the glass plate and substituting (11) inside (13), the solution of (12) can be calculated by using the modal expansion of (3), which gives back:

$$W_{mn} = \frac{4C_0\varepsilon_{pe}}{\rho h m n \pi^2\left(\omega_{mn}^2 - \omega^2\right)}\left(k_m^2 + k_n^2\right)p_1 p_2 \tag{14}$$

where: $p_1 = cos(kmx_1) - cos(kmx_2)$, $p_2 = cos(kny_1) - cos(kny_2)$.

Equation (14) can be written in terms of (3) and (5), defining the variable:

$$P_{mn} = \frac{4C_0 \varepsilon_{pe}}{mn\,\pi^2}\left(k_m^2 + k_n^2\right)p_1 p_2 \tag{15}$$

Thus, given the properties of the PZT patches under use and the ones of the plate, (14) together with (5) and (3) gives back the transversal displacement function on the 2D plate caused by PZT patch actuators with respect to x and y coordinates. In the case shown in Fig. 7-b, the stack actuator has the task of providing a punctual force, instead of a bending moment. Following a procedure similar to the one explained above, it is possible to calculate a numerical model that describes the vibration field in terms of (3) and (5) exploiting the following relation:

$$P_{mn} = \frac{4F_a}{ab}\sin k_m x_f \sin k_n y_f \tag{16}$$

where a and b are the side lengths of the plate; x_f and y_f are the coordinate of the point where the force F_a is applied, that is the action provided by the stack actuator, which is dependent to the reaction system stiffness. Assuming d_z the strain constant of the actuator along the z-direction, its unconstrained displacement will be computed by:

$$w_a = \frac{d_z V}{L_a} \tag{17}$$

where L_a is its height. In fact the real displacement of the stack is lower than (16) because the reaction system has finite stiffness K, and the force effectively exerted by the stack along the z-direction is:

$$F_a = \frac{d_z V K}{1 + \frac{K}{K_a}} \tag{18}$$

being ka the actuator stiffness. As in the previous case, the transverse vibration displacement of a 2D plate can be calculated by (14) with (5) and (3).

In the following numerical simulations, performed according to the model described above, the disturbance is assumed to be a wave with frequency near the frequency of the mode of vibration (2,2) of a typical building façade's panel, whose effect is compared with the one given by the use of the two aforementioned kinds of actuators. The glazed panel is supposed to be simply supported along the edges. The two configurations of Fig. 7 are studied analytically. The properties of the glazed plate used for these simulations are listed in Tab. 1, while for PZT patches in Tab. 2. For the simply supported plates of Tab. 1, natural frequencies of vibration are given by (6), whose results are listed in Tab. 3 for the smallest modes; so the frequency of the disturbance was chosen equal to 78 Hz. In the first case of Fig. 7-a, the behaviour of the panel of Tab. 1 is simulated when equipped with two dispositions of PZT patches:

- 8 patches equally distributed 0.05 *m* far from the panel edges;
- 26 patches equally distributed 0.05 *m* far from the panel edges.

Each rectangular shaped patch measures (0.05 x 0.04) *m*. Fig. 8 shows the distribution of the maximum amplitude vibration field along the middle axis of the plate, computed along the y=l/2. One of the diagrams is referred to the effect due to the disturbance wave at frequency v = 78 Hz and intensity 100 dB. For a voltage of 150 V (that is the highest limit for low-voltage actuators) PZT patches can generate vibration fields far lower than the one generated by the disturbance.

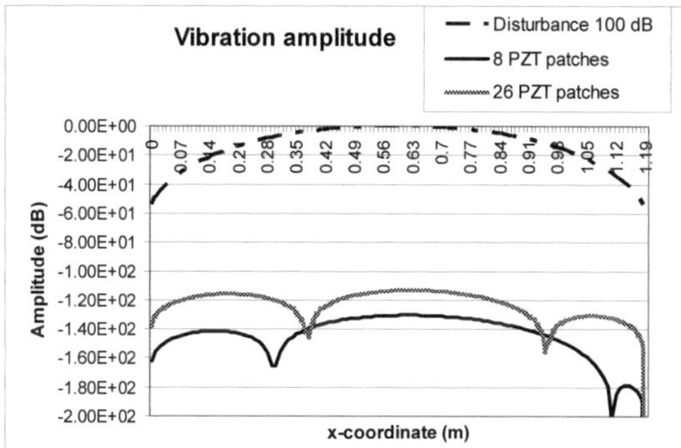

Fig. 8. Amplitude displacement along the y=l/2 axis due to the positioning of PZT patches actuators, normalized with respect to the maximum disturbance value.

In the second case vibration amplitudes are computed for the stack configuration shown in Fig. 7-b. In Fig. 9 such vibration amplitudes are drawn with dependence to the voltage provided to stack actuators. It is assumed that the panel is equipped with 3 actuators (0.02 *m* long with $7.8 \cdot 10^{-5}$ m^2 cross sectional area) per each side, equally spaced and at a 0.03 *m* distance from the two edges; the stiffness of the reaction system is assumed equal to 200 *N/μm*. Fig. 9 shows that, regardless of the small rigidity of the reaction system, the stack actuators can produce a vibration amplitude comparable with the one due to the disturbance with only a voltage of 100 V.

Symbol	QUANTITY	Units of measurement	Value
E_p	Modulus of elasticity	Pa	$6..9 \cdot 10^{10}$
ν_p	Poisson coefficient	-	0.23
ρ_p	density	Kg/m^3	2457
h_p	thickness	m	0.006
l_p	Side length	m	1.2

Tab. 1. Glazed plate's properties.

Symbol	QUANTITY	Units of measurem.	Value
E_{pe}	Modulus of elasticity	Pa	$6.3 \cdot 10^{10}$
ν_{pe}	Poisson coefficient		0.3
ρ_{pe}	density	Kg/m^3	7650
h_{pe}	thickness	m	0.0002
d_{31}	Expansion constant	m/V	-0.000000000166

Tab. 2. PZT patch's properties.

Mode	FREQUENCY (HZ)	Mode	Frequency (Hz)
(1,1)	20.6	(2,2)	82.4
(1,2)	51.5	(2,3)	133.8
(1,3)	102.9	(3,3)	185.2

Tab. 3. Natural frequencies of vibration.

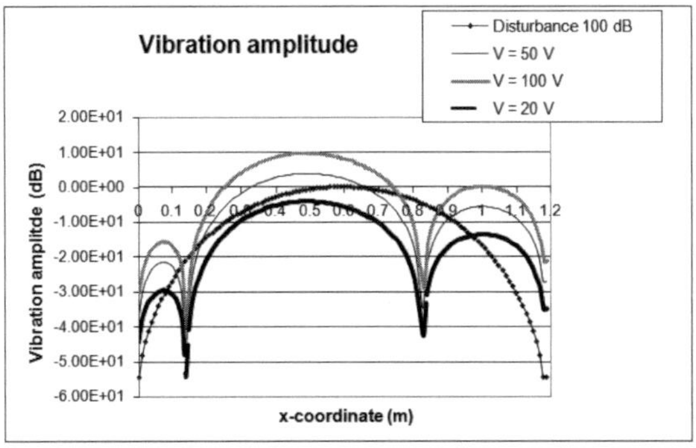

Fig. 9. Amplitude displacement along the y=l/2 axis due to the positioning of stack stiffened actuators, normalized with respect to the maximum disturbance value.

Therefore, given the high controllability provided by stack actuators, they have been considered suitable for controlling glazed facades and they have been object of the experimental campaign and technologic development carried out in this research.

5. The case study: An Active Structural Acoustic control for a window pane

5.1 The components of ASAC System for glazed facades

In paragraph 4.2 the two basic arrangements for an ASAC system configuration have been introduced, that are feed-forward and feed-back types. As already discussed, the first one requires the knowledge of the primary disturbance, which implies the use of a reference microphone. This solution seems to be unpractical for the suggested application, requiring the installation of a microphone on the exterior of the window, unfeasible for functional and aesthetical issues. Hence, the feedback arrangement is preferred by the authors and detailed in the following pages.

The components of a feedback ASAC system for glazed facades are (Fig. 10):

- sensors for detecting vibration (e.g. strain gauges);
- electronic filters for analyzing signals from sensors in order to check the vibration field induced by disturbance;
- an electronic controller for manipulating signals from the sensors and compute the most efficient control configuration at the actuators level;
- charge amplifiers for driving secondary actuators on glazed panels according to the outputs sent by the controller;
- actuators for controlling the vibration field of glazed panels.

As seen in paragraph 4.3 two different kinds of actuators are available, patch and stack actuators. For building applications, feasibility and aesthetical considerations suggest that stack actuators are preferred, as their smaller size interferes less with visibility and transparency and allow them to be easily mounted and dismantled from glass surface.

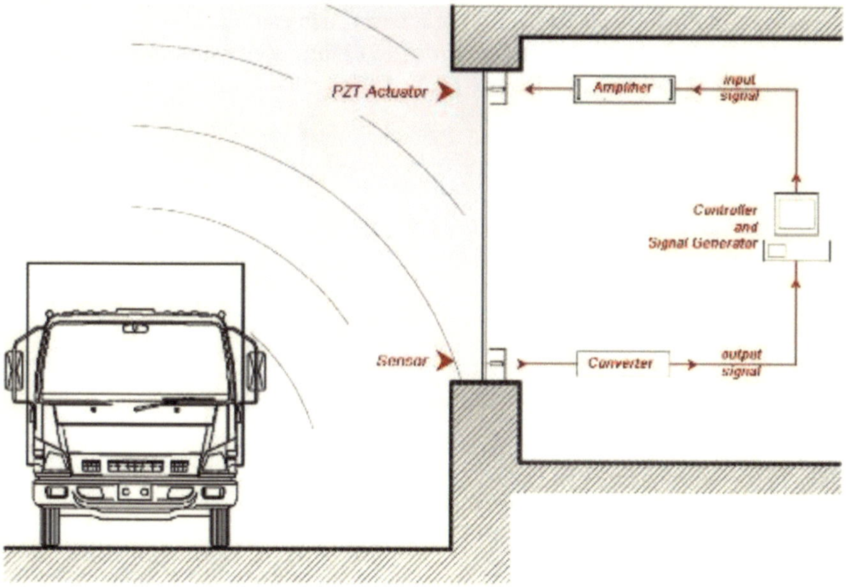

Fig. 10. Layout of the ASAC control system for glazed facades

5.2 The functioning of ASAC System for glazed facades

Signal coming from the sensors is elaborated by charge amplifiers, that convert voltage signals into physical variables like displacements, velocity and accelerations, and by electronic filters, that separate the total vibration field into one due to the primary disturbance from the other connected with the action of secondary sources. The electronic controller, starting from the error signal, estimates the radiated field in some positions of the receiving room and then computes the opportune voltage to be supplied to the actuators in order to reduce the panel's acoustic efficiency. Signal amplifiers provide for necessary electric power.

The optimization of the actuator's actions, in order to minimize the number and the size of the employed sensors and actuators, is derived from opportune algorithms implemented in the controller, like the one presented in (Clark & Fuller, 1992), based on the quadratic linear optimum control theory (see paragraph 4.2). It consists of two parts, the first dedicated to the determination of actuator size and location and the second to sensors. In both parts, the core algorithm computes the voltage to be supplied to the actuators in order to reduce glass vibrations, while the rest of the procedure defines the best actuators' configuration, upon determination of constraints relative to plate's geometry and design choices.

5.3 The technological solution developed as test-case

Stack actuators, as compared to laminated actuators, need a stiffener in order to work properly, hence a technological solution to realize this stiffener has to be designed. The presence of the stiffener, according to its position on the glass surface, may also determine interference problems with the aesthetical appearance of the glass panel which cannot be disregarded. First of all, in order to minimize the radiation efficiency of the vibrating glass surface, the correct positioning of stack actuators has to be studied. Two are the possible ways:

- by decreasing the vibration amplitude of flexural waves (Fig. 11-a);
- By changing the original vibration in order to obtain a vibration field where only even modes dominate (Fig. 11-b).

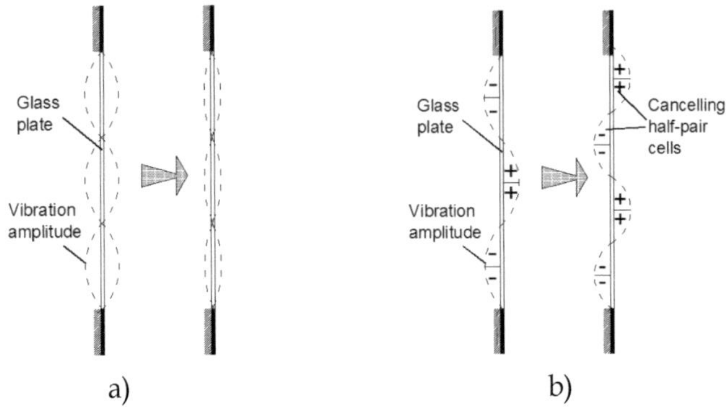

Fig. 11. Reduction of the overall acoustic radiation efficiency

In the first case actuators should act in order to reduce vibration amplitudes, while in the second one they should generate a vibration field with less radiation efficiency. To each of the alternatives listed corresponds a different positioning of actuators: in the first case they have to be installed in the points where maximum vibration amplitudes are monitored, while, in the second one, they have to be moved along the border lines, with less interference in glass panel's appearance. Starting from these considerations, in Fig. 12 three possible technological solutions are depicted (Naticchia and Carbonari, 2007):

a. stack actuators positioned close to the central axis, usually characterized by maximum amplitude vibrations, and stiffened by a metal profile (approach 1);
b. stack actuators installed along one border of the panel and stiffened by an angular profile (approach 2);
c. stack actuators placed close to the borders and stiffened with point reaction systems (approach 3).

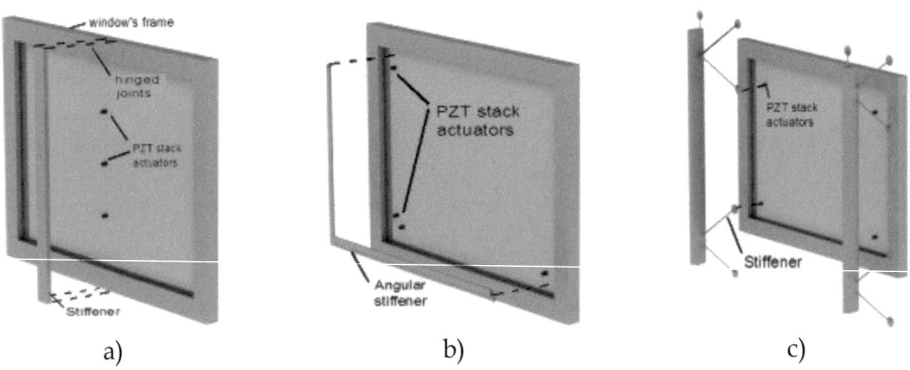

Fig. 12. Technologic solutions suggested for the installation of actuators.

Further proposals for technological solutions have been advanced, where the actuator is contrasted by a point reaction system directly attached to the glass panel's surface. For this purpose, the use of two different kinds of metallic profiles have been hypothesized: in Fig. 13-a a circular-shaped profile contrasting a stack actuator is depicted in a 3-D view and a cross-section view, while Fig. 13-b represents a similar solution realized with a z-shaped profile. Both hypotheses seem to be advantageous from an aesthetical point of view, showing little interference with visibility through the glass, and should be studied relative to profile characteristics and to the stress induced in correspondence of the connection point between the same profile and the glass panel.

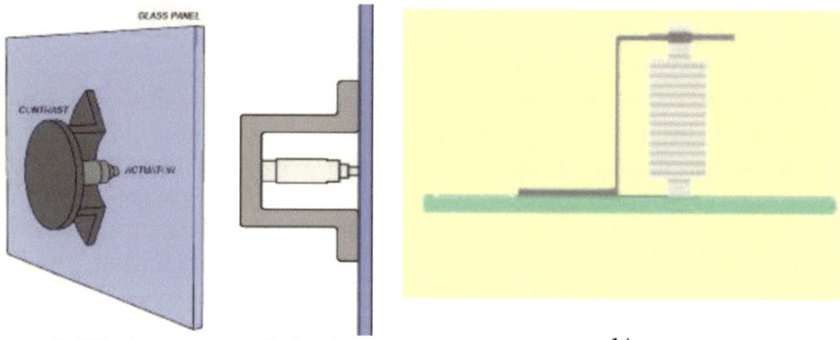

a-1: 3D view a-2: Section b)

Fig. 13. Further hypotheses of point reaction systems: circular-shaped profile (a-1;a-2); Z-shaped profile (b).

For the acoustic simulations carried out and discussed in this chapter, in order to evaluate the effectiveness of the purposed technology over the limits imposed by the choice of one solution with respect to another, an experimental solution has been developed, employing a stack actuator, stiffened by a mass, realized with a cylinder of metallic material overlapped and connected to the free extreme of the actuator, as will be detailed in paragraph 6.2.

6. Experimental analysis

In the following paragraphs, the results of experimental and numerical analyses carried out to evaluate acoustic improvements deriving from the application of the suggested active control technology will be presented (Carbonari and Spadoni, 2007). For this purpose, a finite element model and an experimental prototype were built: in both models the stiffener has been simulated with a 0.177 *Kg* weighted mass contrasting the free extreme of the actuator (Fig. 15-e and 15-f).

6.1 The building of the experimental prototype

Experimental simulations were performed on a prototype, realized by assembling a (1.00x1.40) *m* sized glazed pane with an aluminium profile frame. The main problem regarding the realization of the prototype was the simulation of a simply supporting boundary constraint: it was pursued with the interposition of two cylindrical Teflon bars between the glass panel and the two window frame profiles, as can be seen in Fig. 14-a. Every screw fixing the glass panel in the window frame was subjected to the same torque (through the use of a dynamometric spanner) equal to 0.1 *N·m*, in order to guarantee uniform contact between the glass and the Teflon bars. The whole system, as shown in Figure 14-b, was placed over dumping supports in correspondence of each panel edge, to avoid the influence of external actions on the glass's vibrations, establishing the simplest boundary conditions. A seventy-seven point grid was defined on the panel, in order to identify measurement marks.

6.2 The modal analysis performed on the prototype

The purpose of the experimental analysis is to collect data in order to evaluate the reliability of the finite element model, on which the acoustic simulations will be performed. First of all, a modal analysis was carried out on the prototype in order to determine its natural frequencies. The experimental apparatus employed for the measurements consisted in:

- a transducer for exciting the system (Fig. 15-a);
- an accelerometer for checking the vibration field (Fig. 15-b);
- a PXI platform for collecting data (Fig. 15-c).

National Instruments PXI is a rugged PC-based platform for measurements and automation systems, provided by the Mechanical Measurement Laboratory of the Polytechnic University of Marche (Castellini, Revel, Tommasini, 1998; Castellini, Paone, Tommasini, 1996), whose staff contributed to these experimental tests. PXI is a deployment platform, serving applications like manufacturing test, aerospace and military, machine monitoring, automotive and industrial tests. It is composed of three basic components: chassis, system controller and peripheral modules. PXI can be remotely controlled by PC or laptop computers, but it can also provide for embedded controllers, which eliminates the need for an external controller. In the case of the performed tests, the PXI was connected to a PC monitor in order to display the data collected from measurements on the experimental prototype used to perform its two modal analyses (see Fig. 17-b).

Experimental tests were carried out in the Advanced Robotics Laboratory of the Department of Software, Management and Automation Engineering-DIIGA ("Dipartimento di Ingegneria Informatica, Gestionale e dell'Automazione") of the Polytechnic University of Marche (Antonini, Ippoliti, Longhi, 2006; Armesto, Ippoliti, Longhi, Tornero, 2008), which is equipped with:

- one Wave Generator Hameg Instruments mod. Hm 8030-3 (see paragraph 6.3);
- one Tektronix TDS 220 oscilloscope;
- one National Instruments acquisition card mod. NI USB6009 (see paragraph 6.3).

a) b) c)

Fig. 14. The prototype used to realize simply supporting constrains (a), the window frame prototype on the dumping supports (b), Seventy-seven point grid marked on the glazed pane (c).

Modal analysis was first performed on the prototype as depicted in Fig. 14-b, that is on the prototype without any control system component in order to evaluate its natural frequencies. Subsequently, the same tests were repeated on the prototype equipped with the Device Kit, consisting in:

- one actuator acting as control system (in this first stage of the tests, the actuator was inactive to study the system's free vibration);
- one load cell for recording the values of the forces provided by the actuator;

- one stiffening mass for simulating the presence of the stiffener (total weight of the stack actuator device kit was 0.177 *Kg*).

The elements were assembled as shown in Fig. 15-e and 15-f: the stack actuator device kit was positioned along the main axis of the prototype, at a distance of 0.24 m from the edge and fixed to the glass panel with resin. Measurements were carried out keeping the position of the accelerometer unchanged and exciting each one of the seventy-seven grid point with the transducer. The data collected were processed with appropriate software in order to restore the glass panel's modal forms.

From the comparison of the natural frequencies recorded for the two, different, tested systems, summarized in Tab. 4, a maximum percentage error greater than 10% was checked, so that it had been possible to conclude that the presence of the Device Kit, with its volume and its total weight of 0.177 *Kg*, can not be omitted for the development of a correct finite element model.

m,n MODES	1.1	2.1	1.2	3.1	2.2	3.2	4.1	1.3
Prototype + Dev. Kit(Hz)	25.5	47	66.5	85	89	116.5	133.5	141
Not Controlled Prototype (Hz)	25.50	47.50	67.00	85.50	89.00	129.50	134.50	141
ABS. ERROR %	**0.00**	**1.05**	**0.75**	**0.58**	**0.00**	**10.04**	**0.74**	**0.00**

Tab. 4. Comparison between natural frequencies values in the case of non controlled glass panel and of the glass panel with the stack actuator device kit

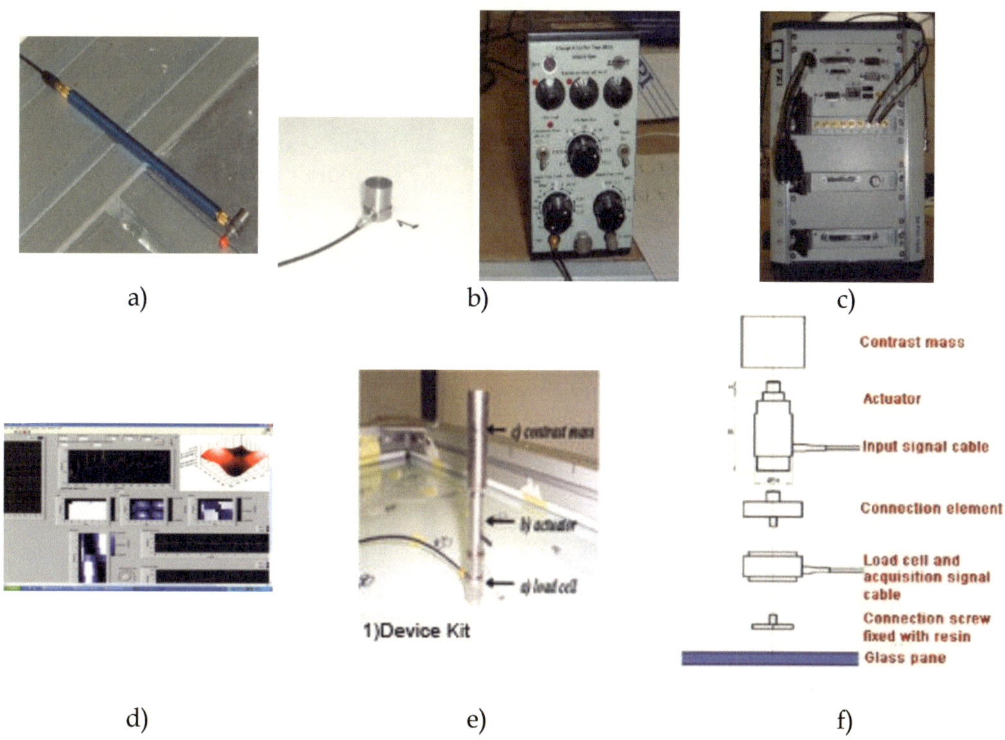

Fig. 15. Transducer (a), accelerometer B&K with its amplifier (b), PXI platform (c), modal analysis processing software (d), the stack actuator device kit (e,f).

6.3 The harmonic analysis performed on the prototype

In the second stage of the experimental measurements, harmonic analyses were performed on the prototype, in order to determine the structural response of a window pane, when excited by harmonic force. For this purpose, two frequencies were selected, 81 Hz and 142 Hz, which are very close to the panel's natural frequencies, previously defined for modes (3,1) and (1,3). This choice was influenced by the two following considerations:

- maximum structural response is recorded when a system is excited close to its natural frequencies;
- for the assumed control theory, maximum efficiency is obtained controlling modes with the maximum acoustic efficiency. From previous studies (Naticchia and Carbonari, 2006), it is possible to establish that they are coincident with the glass panel's natural frequencies, with particular reference to (3,1) and (1,3) modes.

For experimental measurements the same apparatus described in paragraph 6.2 was employed, with exception of PXI platform, replaced by the National Instruments Acquisition Card depicted in Fig. 16-a.

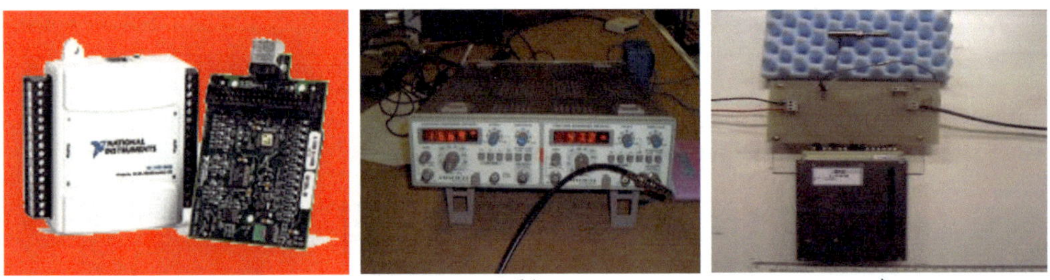

a) b) c)

Fig. 16. National Instruments acquisition card mod. NI USB6009 (a), Wave Generator Hameg Instruments mod. Hm 8030-3 (b) and E-610.00 PI amplifier employed for experimental measures (c).

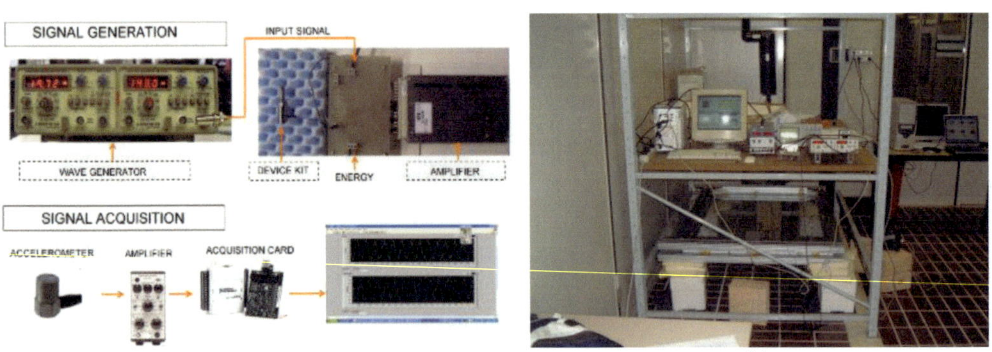

a) b)

Fig. 17. Functioning scheme of the performed tests (a), Experimental apparatus installed in the Advanced Robotics Laboratory of DIIGA of the Polytechnic University of Marche (b).

Differently from the modal analysis, the harmonic analyses were directly performed on the prototype equipped with the Device Kit. The test functioning scheme is represented in Fig.

17-a: harmonic signals exciting the prototype were generated by an analogue Wave Generator (Fig. 16-b) and sent to the Device Kit, passing through the amplifier depicted in Fig. 16-c. The measurements were carried out moving the accelerometer from one point to another of the seventy-seven point grid defined on the glass panel; signals coming from the accelerometer were collected with NI acquisition card and elaborated, with the application of opportune filtering executed using appropriate software. Applying the harmonic motion equation:

$$W_{max} = \frac{a_{max}}{(2\pi f)^2} \tag{19}$$

at every point it was possible to compute displacements along the main axis and along the axis passing through the actuator: displacements diagrams are represented in paragraph 7.3, where they will be used to validate the finite element model.

7. Numeric analysis

7.1 The modal and harmonic analyses performed on the finite element model

The finite element theory was employed for building the numerical model of a window subject to acoustic simulations for the evaluation of the real effectiveness of the technology suggested.

The same characteristics of the experimental prototype, in terms of geometry, material properties and boundary conditions, were reproduced in the finite element model. To this purpose, two different models were implemented in ANSYS 8.0™ environment: the first one represents a rectangular (1.40 x 1.00) *m* large glass plate, simply supported along the whole board (Fig. 18-a). In order to reach a high accuracy level, the plate was subdivided into square shaped finite elements of 0.02 *m* per side. The following parameters for glass material were inputted:

- elasticity Modulus $E= 6.9 \times 10^{10}$Pa;
- Poisson Coefficient $v=0.23$;
- density $\rho=2457$ Kg/m^3.

The second model was realized, adding to the first a (0.02x0.02x0.02) *m* sized parallelepiped volume to simulate the Device Kit (Fig. 18-b). In the positioning of the volume on the glass plate the same conditions as the experimental tests were respected and steel-like characteristics were assigned to it:

- elasticity Modulus $E= 2.1 \cdot 10^5$ MPa;
- Poisson Coefficient $v=0.33$;
- density $\rho=22158$ Kg/m^3 (density value was computed according to the real weight of the Device Kit).

The nomenclature of the glass natural modes was chosen according to the number of troughs along the major and secondary axes of the plate respectively.

Modal analyses were performed on both models and the results were compared, confirming that the presence of the Device Kit cannot be neglected when realizing a proper finite element model: in fact, the comparison of the model forms revealed deviation between the two models, increasing for frequencies higher than 100 Hz. Diagrams and natural frequencies values recorded for the Device Kit equipped model are represented in Fig. 19.

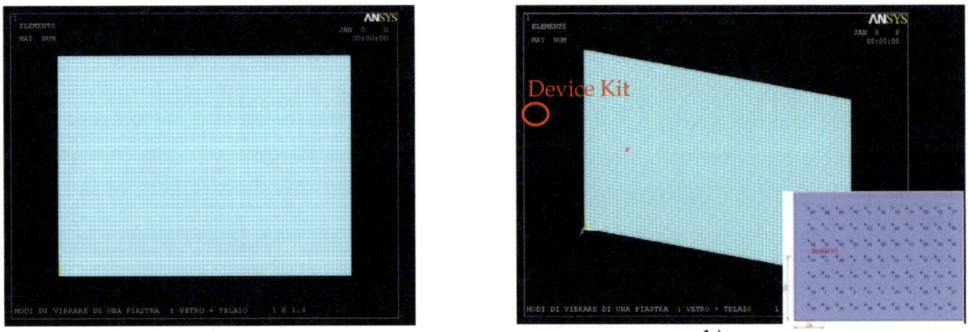

a) b)

Fig. 18. Finite elements model of the glass panel (a), Finite elements model of the glass panel with the Device Kit (b).

According to the results of the modal analysis, harmonic analysis was performed exclusively on the Device Kit equipped pane model. To this aim a point force was applied on the Device Kit volume, with an intensity of 0.17 *N* (the same value recorded by the load cell during experimental tests) at the two different frequencies of 81 Hz and 142 Hz. Results will be discussed in the following paragraph.

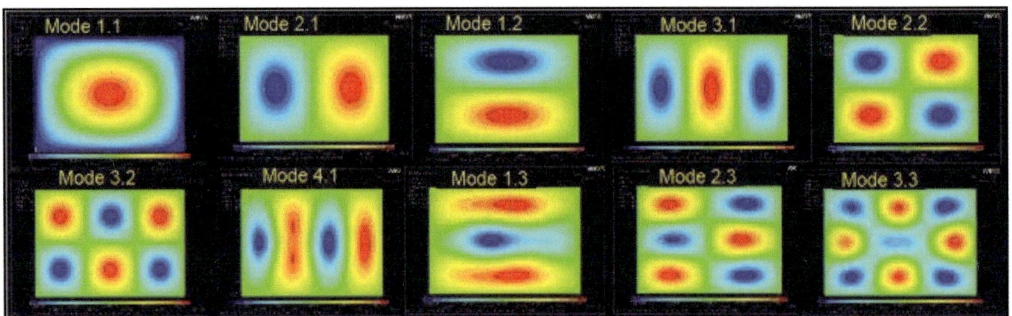

Modes m,n	1	2	3	4	5
1	22.291	66.913	140.72	245.25	379.4
2	44.498	89.503	161.75	267.52	401.57
3	81.515	127.18	197.72	304.69	438.55
4	133.94	180.08	250.92	357.16	490.38
5	203.7	248.23	319.96	424.45	557.12

Fig. 19. Modal shapes and the corresponding natural frequencies of the Device Kit equipped model.

7.3 The validation of the finite element model

Reliability of the finite element model was demonstrated through the comparison between the experimental and the numerical results. First of all, the results of the modal analysis were compared, revealing a good agreement between the values of natural frequencies for the experimental and the finite elements model: actually, a maximum percentage error of 3% was recorded.

Subsequently, diagrams relative to displacements recorded for the numerical and experimental model, due to the harmonic analysis at 81 and 142 Hz were superimposed, as represented in Fig. 20.

It can be noticed that there is a good superposition between the two models: in fact a maximum percentage difference of about 4,5% at 81 Hz frequency and of about 15% at 141 Hz frequency were registered, with an average difference of about 10%. According to these acceptable deviations, also ascribable to local effects not contemplated by the numerical model, it was considered reliable and was used for performing the acoustic simulations.

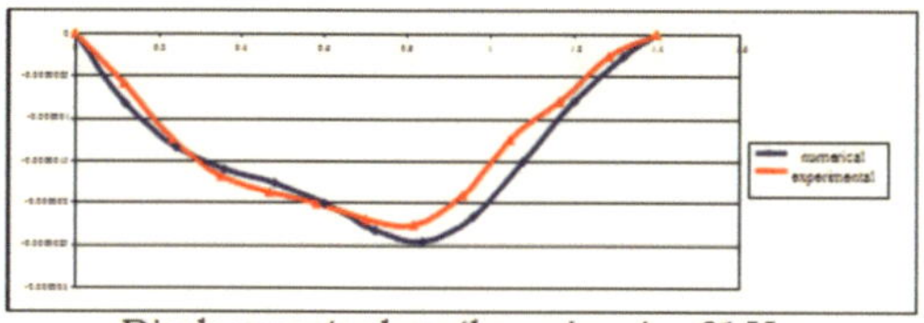

Displacements along the main axis - 81 Hz

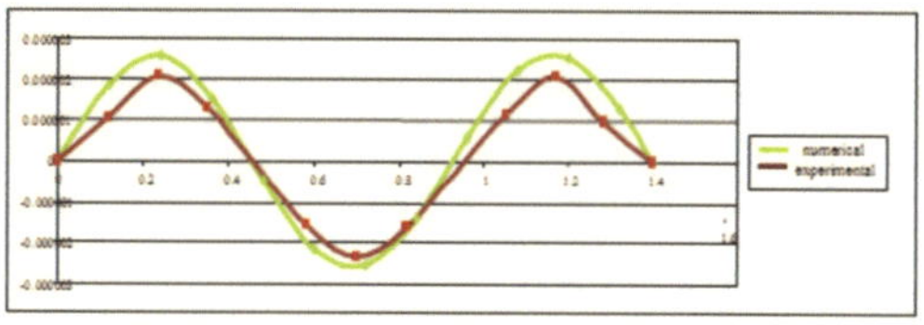

Displacements along the main axis - 142 Hz

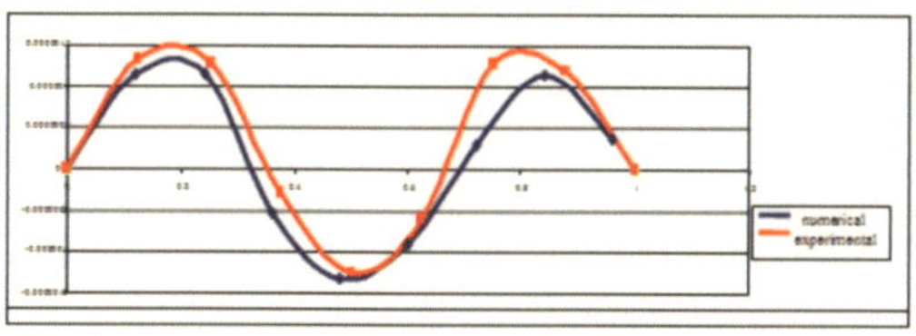

Displacements along the axis passing through actuator - 81 Hz

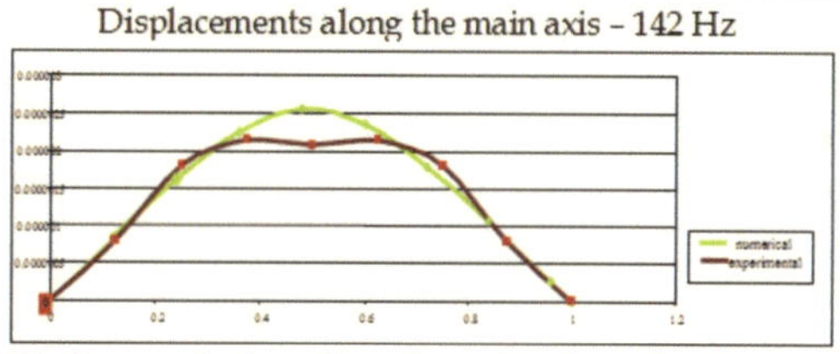

Displacements along the axis passing through actuator - 142 Hz

Fig. 20. Comparison between displacements diagrams for experimental and numerical harmonic analysis.

7.4 The evaluation of sound transmission loss improvements due to the ASAC system

For an acoustic evaluation of the suggested technology, the finite element model, developed in ANSYS 8.0TM environment, was imported in LMS VIRTUAL LABTM environment which is another finite element software, containing two dedicated sections named *noise and vibration* and *acoustics* (the first section was used to perform modal analysis and the second for the acoustic evaluations). Simulations were carried out in order to have numerical results concerning the real effectiveness of the suggested ASAC control system.

It is well-known that one of the most recurring and irritating noise sources is represented by urban traffic, especially connected with heavy vehicles, such as lorries. A research have demonstrated that a lorry, travelling a low distance and at a speed of 70 Km/h produces a noise level equal to 85 dB (Fig. 2), within a range of frequencies in which the dominant one can be identified at 140 Hz, corresponding to the glass panel's natural vibration mode (1,3). According to these assumptions, a test room measuring (2.40x2.50x2.80) *m* was developed for simulations, including within one of the walls, the validated glass panel (please refer to Fig. 21-a).

In previous research activities simulations have been led to evaluate achievable noise level reduction by the application of the ASAC technique, without considering the influence ascribable to the presence of a stiffener or a stiffening mass for the correct functioning of the actuator. In the above case, a reduction of about 15 dB in the disturbance pressure noise was estimated (Naticchia & Carbonari, 2006). Results of simulations presented in this paragraph instead, can be considered more realistic, as it also takes into account the presence of the stiffener. The developed test room was analyzed in two different configurations:

1. *anechoic room*: every wall of the room was made up of totally absorbent panels;
2. *reverberant room*: acoustic properties were associated to every wall of a typical building material like plaster for the ceiling, wallpaper for vertical walls and carpet for the floor.

In both cases, the disturbing wave incident on the glazed panel was assimilated to a uniform constant pressure on the panel equal to 0,3556 Pa, applied at the frequency of 140 Hz: this condition seems to be realistic, considering the distance that usually separates the windows of a building from the street, source of the noise. The acoustic pressure level within the room was evaluated for the following conditions:

1. acting disturbance;
2. acting actuator;
3. simultaneous acting of disturbance and actuator.

The results of the simulations are represented in Fig. 21-b with coloured diagrams relative to the anechoic room and to the reverberant room, with reference to the acoustic pressure recorded on the walls of the room with and without the application of the proposed ASAC system: the numerical values of the recorded pressure levels for both cases are synthesized in Tab. 5.

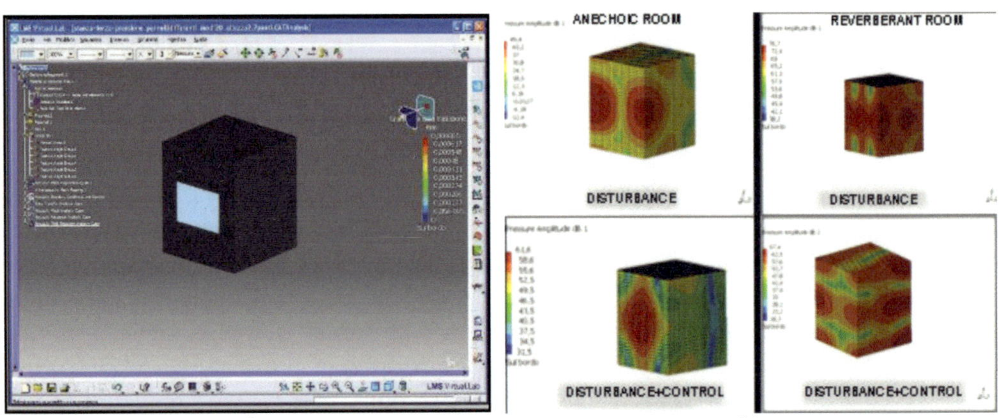

a) b)

Fig. 21. The test room model used for the acoustic simulations in LMS™ environment (a), diagrams of the noise level recorded in the anechoic room (left side) and in a real room (right side) with and without the application of ASAC system(b).

According to these results, the first observation that can be derived is that, even if the reduction achievable in the disturbance pressure level is greater in the case of the anechoic room than it is in the reverberant room, as would be expected, nevertheless in the real case, there is a sensitive drop of about 10 dB in the noise level transmitted within the room.
Besides this, it can be stated that the presence of the stiffener cannot be omitted for a real acoustic evaluation of the technology suggested.

Average interior noise level recorded with the acoustic simulations (dB)		
	Reverberant Room	Anechoic Room
Disturbance (0,3556 Pa;140Hz)	76,7	61,6
Disturbance + Acting Actuator	67,4	49,4
Maximum recorded decrease	9,3	12,2

Tab. 5. Final results of the acoustic simulations performed in LMS Virtual Lab ™ environment

8. Conclusions

In these pages it was demonstrated that the Active Structural Acoustic Control can be successfully applied in the building field, in order to provide a major improvement in glass panels' sound transmission loss in the low frequencies range: the employment of just one actuator causes a sensitive drop in the noise transmitted from the exterior to the interior, allowing the achievement of the restrictive requirements imposed by European and Italian standards. At the same time, initial considerations were presented in order to investigate the feasibility of a technological solution based on the active control of vibrations.

In order to prepare this technology for use in buildings, further efforts should be directed to facing the following two different aspects:

1. experiments using more than one actuator, to control some of the most efficient modes should be carried out, in order to determine the effects due to the interaction of different actuators and then to evaluate the final noise drop transmitted from the exterior to the interior;
2. simulations and laboratory experiments should be directed to develop a technological solution allowing the proper integration of the Active Structural Acoustic Control system with a glazed panel's structural frame.

In addition, the same technique could be adopted for applications on other specialized products such as light opaque building partition walls, railway and traffic noise shielding, temporary environmental noise barriers or even real-time controlled reflecting panels for the acoustic adjustment of concert halls

9. References

Antonini, P., Ippoliti, G., Longhi, S., (2006), Learning control of mobile robots using a multiprocessor system, *Control Engineering Practice*, vol. 14, pp. 1279-1295, ISSN: 0967-0661.

Armesto, L., Ippoliti , G., Longhi, S., Tornero, J., (In press June 2008), An asynchronous multi-rate approach to probabilistic self-localisation and mapping. *Ieee Robotics And Automation Magazine*. ISSN: 1070-9932.

Bao, C. and Pan, J., (1997), Experimental study of different approaches for active control of sound transmission through double walls, *Journal of the Acoustical Society of America*, vol. 102(3), pp. 1664-1670.

Baumann, W. T., Ho, F. S., Robertshaw, H. H., (1992), Active structural acoustic control of broadband disturbances, *Journal of the Acoustic Society of America*, 92(4).

Baumann, W. T., Saunders, W. R., Robertshaw, H. H., (1991), Active suppression of acoustic radiation from impulsively excited structures, *Journal of the Acoustic Society of America*, 90(6).

Burdisso, R.A. and Fuller, C.R., (1994), Design of active structural acoustic control systems by eigenproperty assignment, *Journal of the Acoustical Society of America*, Vol. 96, No. 3, pp. 1582-1591.

Carbonari, A. and Spadoni, S., (2007), An Active technology to increase sound transmission loss of glazed facades, *Proceedings of ARTEC Conference "The Building Envelope – a complex design"*, pp. 199-206, ISBN 978-88-6055-223-5, Ancona, November 2007, Alinea, Firenze

Castellini, P., Paone, N., Tomasini, E. P., (1996), The Laser Doppler Vibrometer as an Instrument for Non-intrusive Diagnostic of Works of Art: Application to Fresco Paintings, *Optics and Lasers in Engineering*, Vol. 25, pp. 227-246, Elsevier Science Ltd., Northern-Ireland.

Castellini, P., Revel, G. M., Tomasini, E.P., (1998), Laser Doppler Vibrometry: a Review of Advances and Applications, *The Shock and Vibration Digest*, vol. 30(6), pp. 443-456, Sage Science Press, Thousand-Oaks, CA

Chaplin, G.B.B. and Smith, R.A., (1976), Active Methods of Cancelling Repetitive Vibrations, *U.K. Patent 1,971,7176.*

Clark, R.L. and Fuller, C.R., (1992), Optimal placement of piezoelectric actuators and polyvinylidene fluoride error sensors in active structural acoustic control approaches, *Journal of the Acoustical Society of America*, Vol. 92, No. 3, pp. 1521-1533.

Clark, R.L., Fuller, C.R., Wicks, A., (1991), Characterization of multiple piezoelectric actuators for structural excitation, *Journal of the Acoustical Society of America*, Vol. 90, No. 1, pp. 346-357.

Conover, W.B., (1955), Recent contributions to transformer audible noise control, *AIEE Transactions, Applications and Industry*, Vol. 74, Part D, pp. 77-86.

Cunnefare, K. A., (1991), The minimum multimodal radiation efficiency of baffled finite beams, *Journal of the Acoustic Society of America*, 90(5).

Dimitriadis, E.K., Fuller, C. R., Rogers, C.A., (1991), Piezoelectric actuators for distributed vibration excitation of thin plates, *American Society of Mechanical Engineers Journal of Vibration and Acoustics*, 113, pp. 100-107.

Fahy, F., (1985), *Sound and structural vibration – radiation, transmission and response,* Academic Press, London - New York – Boston – Sydney – Tokyo – Toronto, par. 4.3.

Fuller, C. R., Elliott, S. J., Nelson, P. A., (1997), *Active Control of Vibration,* 2nd Edition, Academic Press, San Diego-London- Boston-New York, par. 2.10 – 8.4 – 8.5.

Hall, D. E., (1987), *Basic Acoustics*, John Wiley and sons, New York - Chichester - Brisbane - Toronto - Singapore, pp. 122-124.

Harris, C.M., (1984), *Handbook of Acoustical Measurements and Noise Control*, 3rd edition, McGraw Hill, New York.

Harris, C.M., (1994) *Noise Control in buildings*, McGraw Hill ed., New York.

IMAGINE, (2003), EC Project funded Technical Report WP 1.1: "Source modeling of road vehicles" , IMAGINE - *Improved Methods for Assessment of the Generic Impact of Noise in the Environment.*

Kido, K., (1975), Reduction of Noise by use of Additional Sound Sources, *Proceedings of Inter-Noise 75*, Sendai, Japan, pp. 647-650.

Kaiser, O. E., Pietrzko, S. J., Morari, M., (2003), Feedback control of sound transmission through a double glazed window, *Journal of Sound and Vibration*, vol. 263, pp. 775-795.

Lueg, P., (1936), Process of silencing sound oscillations, *US Patent No. 2,043,416.*

Meirovitch, L., Baruh, H. Öz, H., (1983), A comparison of control techniques for large flexible systems, *Journal of Guidance, Control, and Dynamics*, 6(4), 302-310.

Naticchia, B. and Carbonari, A., (2007): Feasibility analysis of an active technology to improve acoustic comfort in buildings, *Building and Environment*, Vol. 42, Issue 7,pp 2785-2796.

Naticchia, B. and Carbonari, A., (2006), First numerical and experimental results on active controlled glazed facades, *Proceedings of the 23rd ISARC Conference*, ISBN:4-9902717-1-8, Tokyo, October 2006

Nelson, P. A. and Elliott, S. J., (1995), *Active Control of Sound*, Academic Press Limited, London. 3rd printing.

Olson, H.L., and May, E.G., (1953), Electronic sound absorber, *Journal of the Acoustical Society of America*, Vol. 25, No. 6, pp. 1130-1136.

Pan, J., Snyder, S. D., Hansen, C. H., Fuller, C., R., (1992), Active control of far-field sound radiated by a rectangular panel - A general analysis, *Journal of the Acoustic Society of America*, 91(4).

Roussos L. A., (1985), *Noise transmission loss of a rectangular plate in an infinite baffle*, NASA, TR 2398, Washington, DC.

Ruckman, C. E. and Fuller, C. R. (1995), Optimizing actuator locations in active noise control systems using subset selection, *Journal of Sound and Vibration*, 186(3), 395 - 406.

Spagnolo, R. (2001), *Manuale di acustica*, UTET Libreria editions.

Timoshenko, S, Woinolowsky-Krieger, *Theory Of Plates And Shells*, Mcgraw-Hill, 1959

Wang, B., Burdisso, R.A., and Fuller, C.R., (1994), Optimal placement of piezoelectric actuators for active structural acoustic control, *Journal of Intelligent Material Systems and Structures*, Vol. 5, Jan., pp. 67-77.

Yang, D., and Fuller, C.R., (1995), Numerical simulation of active control of interior noise in a business jet with point force actuators - optimization of transducers, *Proceedings of Inter-noise 95.*

Zhu, H., Rajamani, R., Stelson, K. A., (2004), Active control of glass panels for reduction of sound transmission through windows, *Mechatronics*, vol. 14, pp. 805-819.

16

Development of Adaptive Construction Structure by Variable Geometry Truss

Fumihiro Inoue
Technical Research Institute, Obayashi Corporation
Japan

1. Introduction

Recent years have seen an increasing variety of attractive structures and building with movable functions worldwide (Ishii, 1995). Typical examples are bridges that open to allow ships to pass, revolving restaurants on tops of buildings, sliding roofs of baseball and soccer dome stadiums, and artistic monuments. When we focus on the moving behaviour of these structures, they simply move on rails or turn around a hinge, but do not change in structural shape. That is, their behaviours are not flexible, but monotonous according to a decided patern. However, in the near future, it may become possible to devise structures with a lively motion, freedom as well as intelligence. In particular, they will respond to voice commands with instant and adaptive shape change harmonising with their environment.

One mechanism that enables more complicated movement is the Variable Geometry Truss, we call VGT in the chapter. The VGT is a very simple truss structure composed of extendable members, fixed members and hinges, as shown in Fig.1. By controlling the lengths of the extendable members, it is possible to create various truss shapes. The VGT was originally developed as a movable actuator for a spread-type universe construction in space, and it was equipped with a small motor to perform various tasks (Natori & Miura, 1994). Thus, it is considered to be a useful structural tool in various fields as a redundant intelligent structure.

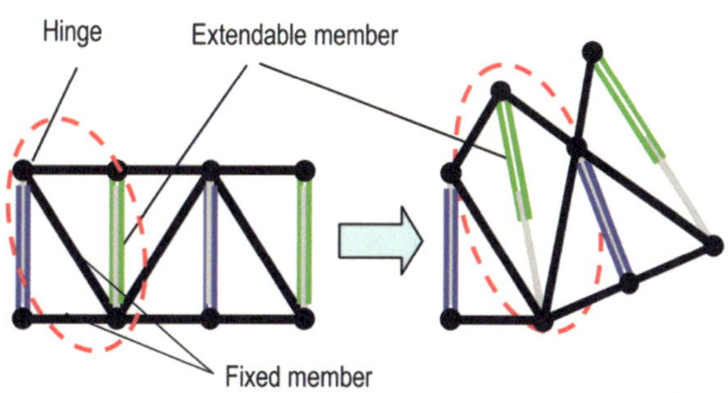

Fig. 1. Basic Mechanism of VGT and Its Shape Change

An example of the shape changes of a simple beam that combines two-dimensional VGTs is shown in Fig.2. When extendable members are extended simultaneously, the truss beam changes like a spring stretching from (a) to (b). When extendable members are extended alternately, the truss beam changes to a circular shape (c). Moreover, when they are extended optionally and their lengths are controlled, the truss beam can be changed into any intended shape (d). Based on the shape change of such a basic beam, the feasibility of the shape change in an actual building is examined.

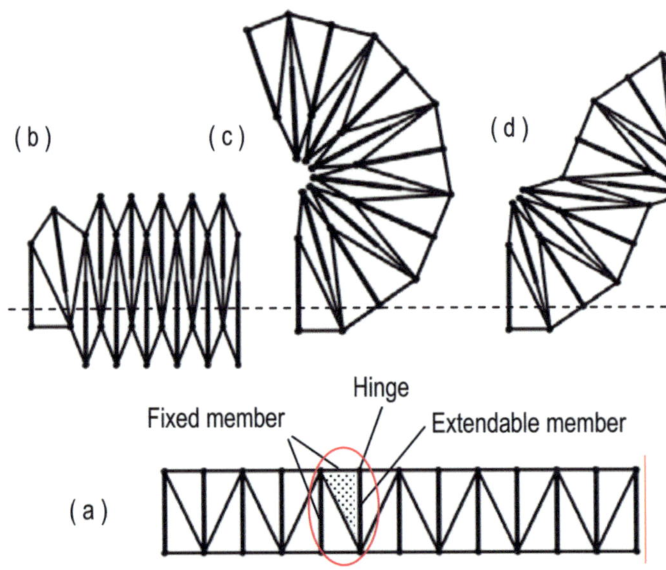

Fig. 2. Transformation of Beam Shape Using VGT

The purpose of this study was to apply VGT tevhnology to ground construction structure and to examine the development of the element technique and its applicability to moveable structures and building by numerical and experimental analysis. In this chapter, the basic characteristics and anlysis of VGT mechanism and two practical examples of adaptive construction structure using VGT are introduced. One example is a unique proposition of the elementary design of a semi-empirical dome with an adaptive roof (Inoue &Kurita, 2003), and the other is a new develpment of a cantilever-type movable monument exhibited at the 2005 International Expo in Aichi Japan (Inoue & Kurita, 2006). From thses examples, the attraction and efficiency of VGT mechanism are varified in details.

2. Characteristics and analysis of VGT mechanism

2.1 Shape analysis of VGT mechanism

In the shape analysis of a VGT mechanism, direct kinematics and inverse kinematics analyses are generally used. For the former, the shape is solved by kinematics analysis, and for the latter; numerical analysis is needed because the shape creates a very highly redundant structure.

2.1.1 Direct motion analysis

The structural units of the two-dimensional VGTs and the whole structure are shown in Fig 3. Each VGT unit is composed of two sets of fixed member and extensible actuators. Then, the whole of the structure can replace a robot manipulator combining two fixed members in series. The tip of the x, y co-ordinates of the structure combined with n ($n > 2$) VGT sets is given by equations (1) and (2) using each hinge angle θ_j .The length of the extendable member is easily found in giving θ_j and the shape of structure can be uniquely fixed

$$q_{x,n} = l_0 \cdot \sum_{k=0}^{n} \cos\left[\sum_{j=0}^{k} \theta_j\right] \tag{1}$$

$$q_{y,n} = l_0 \cdot \sum_{k=0}^{n} \sin\left[\sum_{j=0}^{k} \theta_j\right] \tag{2}$$

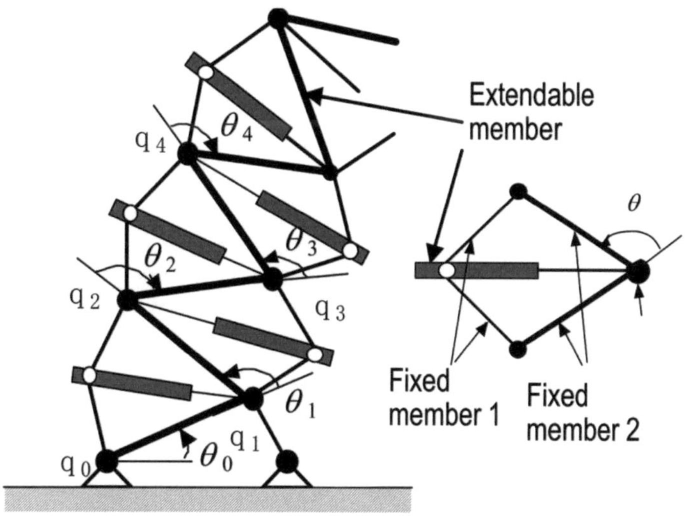

Fig. 3. Two- dimensional VGT Unit and Analysis Model

2.1.2 Inverse motion analysis

It is quite difficult to control the extensible length of each VGT to change it to the intended structural shape. In considering the temporal change of the whole of the structure, equation (3) is obtained.

$$\dot{q} = J(\theta) \cdot \dot{\theta} \tag{3}$$

Where, J indicates the Jacobean Matrix (2 × n). In this case, an inverse matrix isn't necessarily decided because J is not a regular system in n≠2. Here, it finds θ by numerical simulation from q using the pseudoinvesre matrix $J^{\#}$ shown by equation (4)

$$J^{\#} = J^T \left(J \cdot J^T\right)^{-1} \tag{4}$$

2.2 Dynamic analysis of VGT mechanism

To actually design a structure using VGTs, a motion dynamic analysis must be carried out. This is the external force acting on the structure and the torque power for its shape change.

In the analysis, a model replacing the VGT with a multi-joint manipulator and with a onedimensional tree shape structure was proposed. Using Kane's method and Lagrange's method, the final dynamical equation is as showing the equation (5).

$$M\ddot{q} + h(q, \dot{q}) + c(q) = \tau \tag{5}$$

where $\boldsymbol{M}$ is the inertia matrix, $\boldsymbol{h}(q, q)$ is the vector of q centrifugal force and Coriolis force, $\boldsymbol{c}(q)$ is the vector of external force as gravity and τ is the vector of driving torque. By solving the equation (5), the motion of the whole structure can be described.

2.3 Typical structures changes using VGT mechanism

Two basic structures changes of the cantilecer type and arch type is introduced by numerical simulation.

2.3.1 Cantilever structure

Cantilever structure is supported at a single position and a top of the structure is unrestrained from external force. The movement of the structure is thought to be that of a multi-joint robot arms. Fig.4 shows the shape changes of the cantilever structure solved by inverse kinematics. Specifying a top position moving as vector q, the length of each extensible member is found from equation (3) and the shape of the structure can be fixed at any time. It is easy to create several shapes.

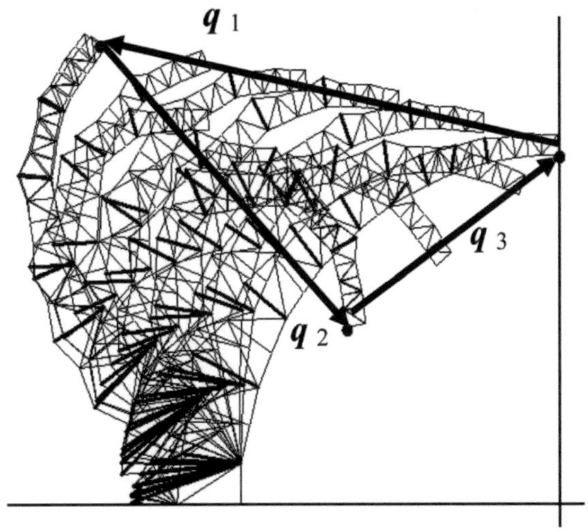

Fig. 4. Shape Change of Cantilever Structure Applying VGT Mechanism

2.3.2 Arch structure

In the arch structure, the two edges of the structure are supported at each hinge. It is very difficult to determine the shape of the structure because these edges are absolutely suited at the hinge position. Fig. 5-(a) shows the shape variations of the arch structure like a big wave change with the lengths of upper side members fixed. In this case, by slightly changing the length of a lower side member near the edge, the lengths of the lower side members are respectively analyzed by numerical simulation to correspond to the end of the structure in the hinge position. Similarly, Fig. 5-(b) shows volume changes with the extensible members sat on both lower and upper sides. A variable shape structure could be simulated numerically by inverse analysis, although both edges were fixed. In a simulation of volume change, as the lengths of the upper and lower chord members were changed, a flexible member was needed for external finishing (Inoue, 2007).

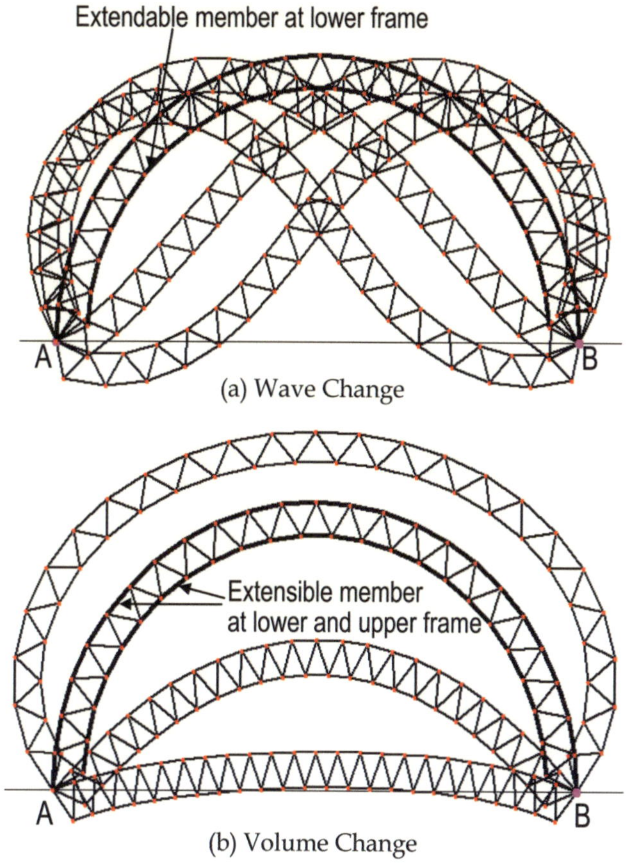

Fig. 5. Shape Change of Arch Structure Applying VGT Mechanism

2.4 Proposition of practical examples using VGT mechanism

The VGT can be used for stress control of a structure in addition to shape control. Thus, it has a wide application. The following are some possible application examples:

1. Facility equipment and temporary structures with moving parts:
   Variable-shaped work gondolas for building walls and flexible roofs in music halls to suit stage contents.
2. Shape control of the structural dome described in paragraph 3 and the pavilion changing the roof shape according to inner environment conditions.
3. Artistic moveable monuments equipped with artificial intelligence harmonizing with the surrounding environment in paragraph 4.
4. Actuators that control the stress and vibration of structure.

3. Proposition of a Hemispherical dome with an adaptive roofl

3.1 Outline of adaptive roofs dome

To obtain basis data of the VGT structure for design and manufacture, we tried to apply to a adaptive roof dome. This was classified as a cantilever-type VGT. The roofs were opened in floral manner and each roof could change into an optional shape, what we call a "Flowering Dome" as shown in Fig.6. A scale model of an adaptive roof dome was made, and a design technique for the structure, finishing materials and motion control was tested. The doom was composed of a partial roof of 10 sectors. VGT actuators were set in series at the main truss of each partial roof. To describe the roof shape and driving range of the actuator, a motion simulation of the roof was carried out by the advanced method of 2.1 and the results were indicated by a computer graphic. The shape change of from opening to closing of the roofs of the Flowering Dome is shown in Fig. 6. From the simulation, various roof shapes could be created by controlling the extensible length of each VGT.

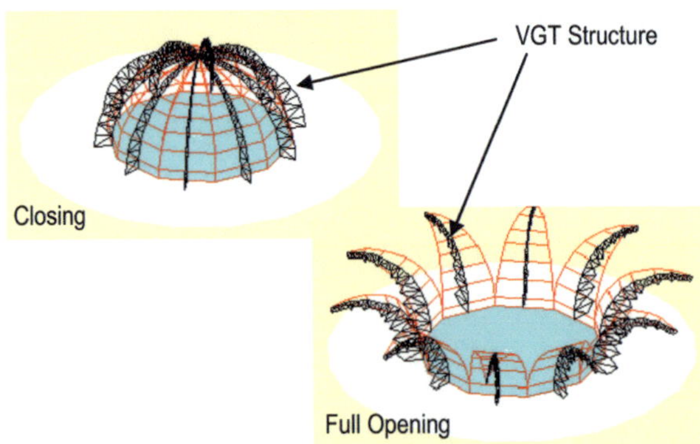

Fig. 6. Computer Simulations of Adaptive Roof Dome

3.2 Outline of adaptive roofs dome

The roof structure was composed between the moving parts containing the VGT actuator and the fixed part of the finishing materials, and these two parts were arranged mutually. The pier type and the chord type were considered for the VGT type, as shown Fig.7. The former used the extensible member for the pier parts (Fig.7-(a)). The roof opened by increasing the length of the extensible member. The latter used the extensible member for the chord parts (Fig.7-(a)). The roof was closed by increasing the length of extensible member. The characteristics of each VGT type were examined by static analysis of several items.

3.2.1 Axial force acted on the extensible member

When changing the roof to all the shapes under self-weight, the relation between the range of the axial force and the length of the extensible member is shown in Fig.8. A large tension force acted on the pier-type extensible member as the roof moved to full opening, whereas lower compression force acted on the chord-type member in the model roof. For full scale, we considered six times the model size for the Flowering Doom, and the range of compression force was much larger. Thus, the force should not act on the extensible member because of the need to avoid extensible member buckling. Therefore, the pier type VGT was more suitable for the dome roof.

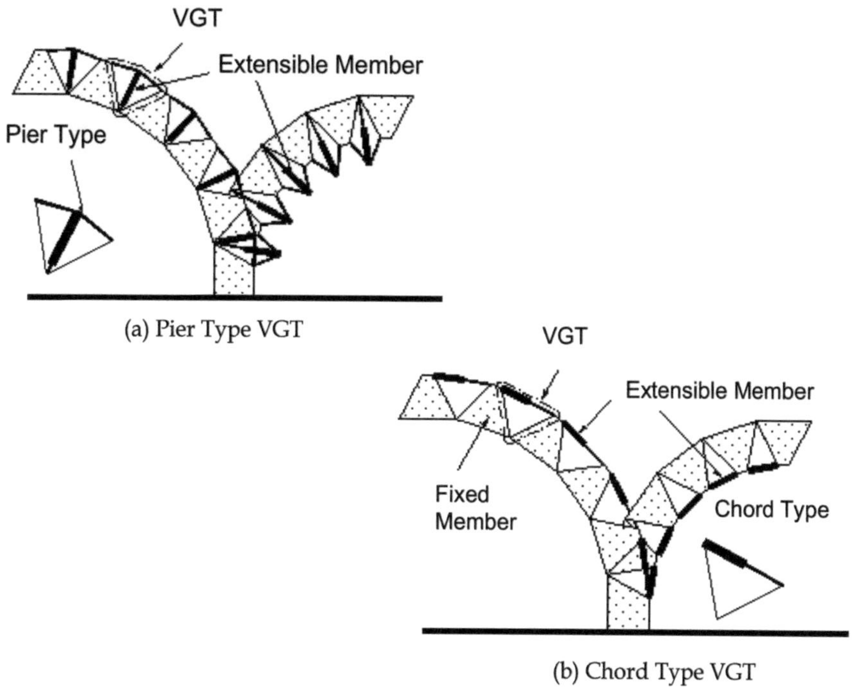

Fig. 7. Composition Type of VGT Arrangment

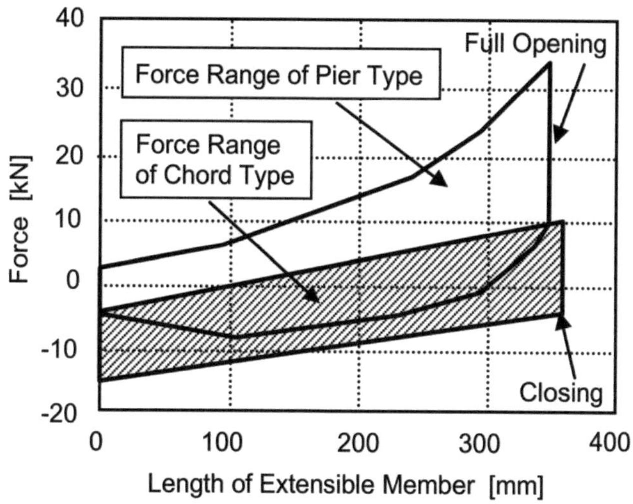

Fig. 8. Comparison of Axial Force Acted on Extensible Member

3.2.2 Stiffness of roof and precision of control length

When the roof was moved from closed to full open, the tip displacement stiffness was very important from the viewpoint of structural design. If the pier-type roof shape was similar to the chord-type roof shape, the pier-type stiffness increased with the roof opening and the chord-type stiffness was opposite. However, considering the characteristics of the hydraulic actuator, the pier-type stiffness was showed larger than the chord type at complete closure.

A few errors occurred in the control length of the extensible actuator. For the roof, accumulation error occurred in the position of the tip. The chord-type error was approximately constant regardless of shape change. However, the pier-type error was very small at complete closure. This detail was shown in a previous report (Kurita &Inoue, 2001). From these results, it was decided to use the pier-type VGT because it was superior in static characteristics at completed closure.

3.3 Dynamical characteristics acted on VGT roof

As the roof rotated around the VGT hinge, the force and rotation moment, i.e., centrifugal force and Coriolis force acted on the whole roof. In particular, by increasing the length rate, a large dynamic force was activated near the base and the lower extensible member of the roof. In this analysis, an exclusive dynamic software was applied under several conditions with a model of an extensible member as shown Fig.9. Fig.10 shows the force acting on the extensible member under a normal length rate of 5 mm/s. This force was equal to that of the static data, but the dynamic force was influenced near full opening. Fig.11 shows the force acting on the lower extensible member at several length rates. For length rates of 5 through 100 mm/s, the dynamic item wasn't considered. For length rates of more than 200 mm/s, a very heavy force was activated on the extensible member. At full scale, the dynamic force greatly increased. In this case, the extensible member as well as the material design was obliged to resume. Dynamic analysis is indispensable for solving these complicated movements of adaptive structures.

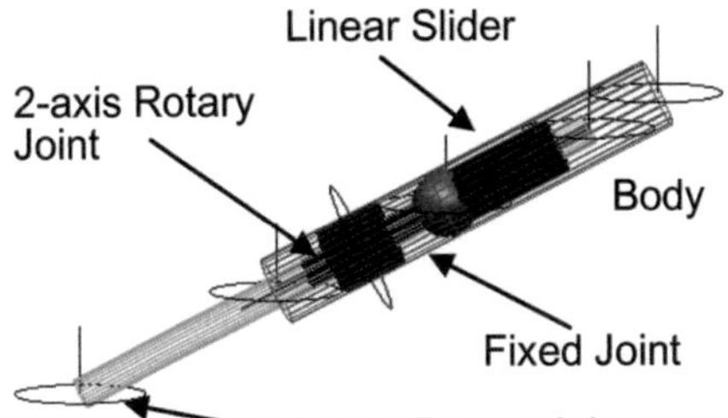

Fig. 9. A Model of an Extensible Member in Dynamic Software

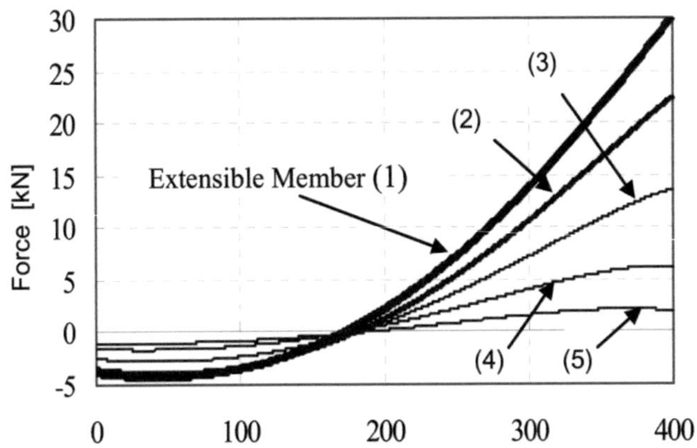

Fig. 10. Dynamical Force Acted on Extensible Member

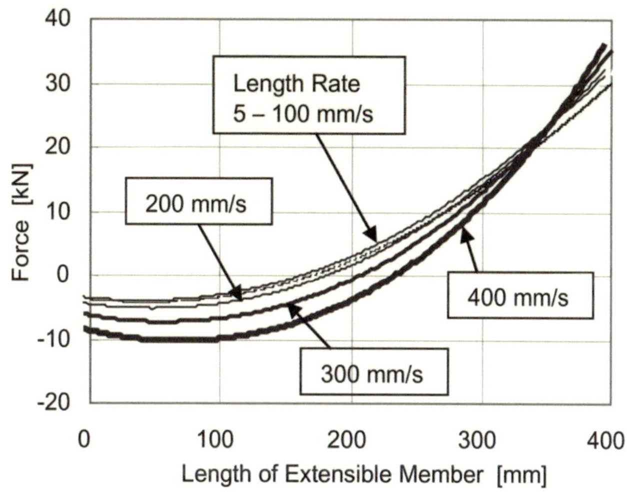

Fig. 11. Dynamical Force Acted on Extensible Member to Length Rate

3.4 Experimental evaluation by a reduced dome model

The composition of the partial roof for the Flowering Dome and the whole of the model roof are shown in Fig.12 and Fig.13 respectively. Five sets of pier-type VGT were arranged in series at the centre of each partial roof. A solid truss was installed on either side of the VGT and the roof surface was formed. The side truss was installed through the hinge outside the solid truss and each solid truss was joined continuously. The acrylic board, the finishing material, was fixed on the underside of the solid truss. When the roof was closed, the end of the roof on both sides overlapped and was waterproofed. The extensible member was comprised a hydraulic actuator, a trunnion-type special device equipped with a pump and an oil tank.

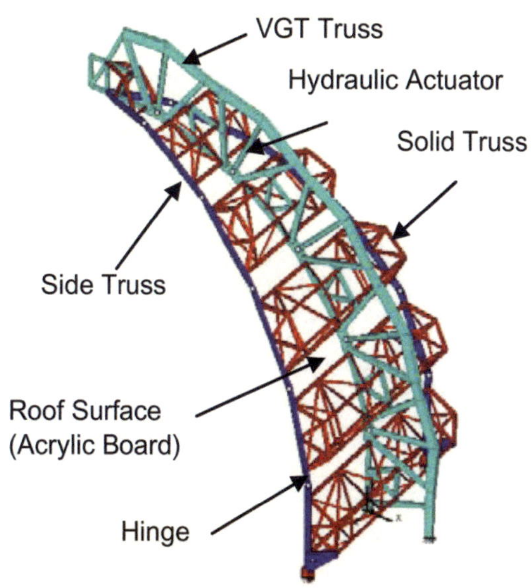

Fig. 12. Roof Structure Installed on VGT Mechanism

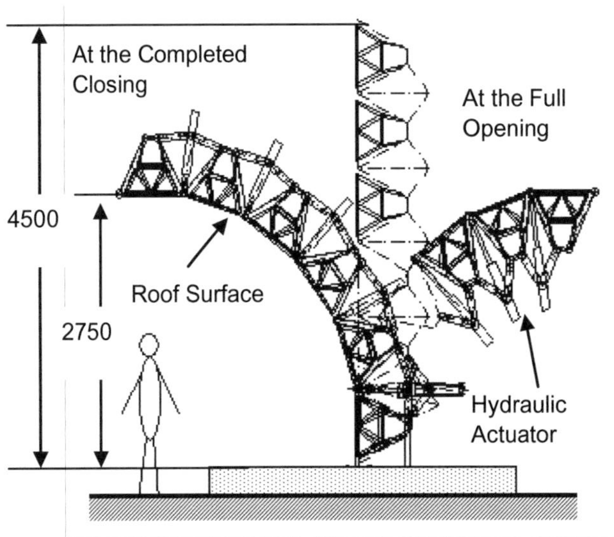

Fig. 13. Schematic Structure of a Reduced Model of Flowering Dome

The extensible length and the hydraulic pressure were measured by a stroke sensor and a pressure gage, and they were controlled by an integrated management system. The model was about 1 / 6 scale and two roof sheets divided into 10 sheets were manufactured.
Fig.14 shows the operating conditions of the moving model roof of a completed Flowering Dome. The length rate of the extensible member was set constant at 5 mm/s. The roof's movement was smooth and it took about 3 minutes to move the roof from fully closed to fully open. The variable structure roof could be set as required by controlling the length of the extensible member. Furthermore, the positioning precision of the tip of each roof was good at closing and it was precise enough to overlap each roof. The scale model verified that the actual adaptive roof dome was possible.

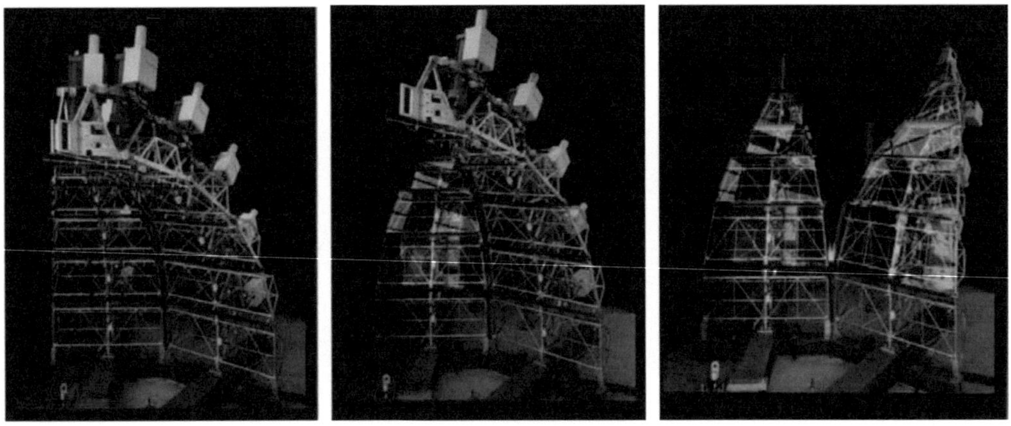

(a) Completed Closing (b) One of Roof was Closing the Other was Opening (3) Moving Roofs to Several and Shape

Fig. 14. Operation Condition of a Reduced Model of Flowering Dome

4. Develpment of a movable monument applying VGT mechanism exhibited in Expo 2005

AS for the other practical example applying VGT mechanism, Development of a movable monument is mentioned in this paragraph.

4.1 Outline of a movable monument

4.1.1 Background and outline of the development

Expo 2005 in Aichi, Japan ended successfully about 3 year ago, with more than two million people attending from around the world. There were various exhibitions, entertainments, attractive buildings, new technology and events based on the theme of "Nature's wisdom". One remarkable technology displayed at the Expo was the application and performance of automated robots. For example, music-playing robots in a group, automatic cleaning robots, guard robots, etc. These scenes might be imagined a the future and such technologies continue to be developed.

In the Japanese zone of the Aichi prefecture pavilion near the center of the exhibition as showing in Fig.15, there was a large monument in the shape of a traditional Karakuri doll beckoning visitors inside. The exhibit here was called "Dancing Tower with Karakuri doll Performance", and was exhibited as a symbol of Aichi district's culture, harmonizing traditional technology with revolutionary new ones.

Fig. 15. Overall Map of Expo Site and Nagakute Aichi Prefecture Pavilion exhibited Movable Monument (From the Expo 2005 Pamphlet of Aichi Prefecture in Japan)

For the Nagakute Aichi prefecture pavilion, the Expo organizers requested a design for a monument comprising a symbol tower to attract visitors to the pavilion. We proposed a movable monument whose shape could be changed variably and irregularly. A large movable monument using VGT was selected as a very unique and attractive monument.
A picture of the Expo site and the Nagakute Aichi Prefecture pavilion where the movable monument was set up is shown in Fig.15, and the movable monument is outlined in Table 1. The pavilion was in a picturesque position in the Japanese zone at the center of the Expo site. It faced a Japanese garden and the Kaede pond. It was located in front of the west gate and beside a global loop, making visitor access easy. The pavilion consisted of a festival plaza, a large-scale theater and a stage, a large area of the Cyube exchange pavilion, and an administration building. The exhibit contained the movable monument was built on the roof of the house.The entire exhibit was composed of movable towers and an annexed device of a Karakuri doll. Each exhibit was united by an internal control signal, and various performances by both the monument and the Karakuri doll were planned

Construction Location	Nagakute Aichi Prefecture pavilion in Expo 2005
Main Use	Movable Monument
Structure Classification	Steel structure containing adaptive Actuator
Base Condition	Combination through steel base on the roof top
Maximum Height	Ground Height :26. 6 m , Monument Height; 18.6m
Construction Period	From 2004. 11.1 to 2005 3.20
Operation Period	From 2005. 3.25 to 2005 9.25

Table 1. Outline of Movable Monument

4.1.2 Design of movable monument

Fig.16 shows schematic pictures of the entire movable monument and its base structure. The whole was composed of three movable iron towers of the same specification, and were spaced at 120-degree intervals around the circumference. Each tower comprised four truss members combined by VGT at joints. Each frame comprised a solid truss structure. The outside of the frame was equipped with a hinge, the inside combined slide actuator, and the shape of towers was changed in proportion to actuator length. Moreover, head illumination was provided at the point of the monument, and a artistic lightning rod was set up. A Karakuri device covered with a large lantern case was constructed at the center of the monument. The towers were seated on a base plate and were combined with long steelpaling that penetrated through the inside of the administration building to an anchor.

4.1.3 Structure design

In the movable structure's design, it was necessary to ensure adequate security. Then, the evaluation of the structure's design and its performance was acquired from the designated organizations. For design, the earthquake force for a projection on the rooftop and the wind loading based on the regional average wind velocity were used. A section shape was selected to ensure security and the specification of the VGT actuator was decided. Moreover, the control system for the movable mechanism, the safety mechanism, the management and the operation system were examined and approved.

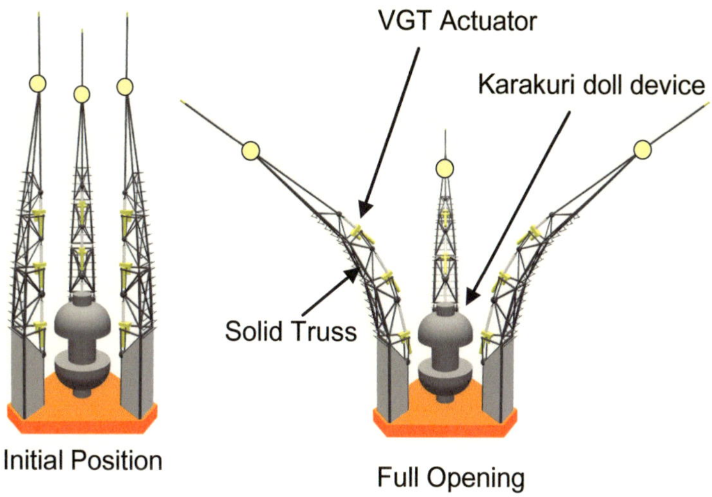

Fig. 16. Schematic Pictures of a Movable Monument and Base Structure

4.2 Composition of movable monument and control system

4.2.1 VGT actuator and monument structure

Fig.17 shows the arrangement of the VGT actuator and the movable range of the tower. Three different sized VGT actuators were set up in each tower and were controlled independently. There were two VGT mechanism arrangements: pier type and chord type. In this tower, the chord type arrangement was adopted because it had advantages of higher rigidity, higher accuracy and lower actuator load. The rotation angle of the VGT mechanism was from 2.5 degrees inside to 18 degrees outside. Thus, the total maximum rotation angle of a tower equipped with three VGT mechanisms was from 7.5 degrees inside to 54 degrees outside.

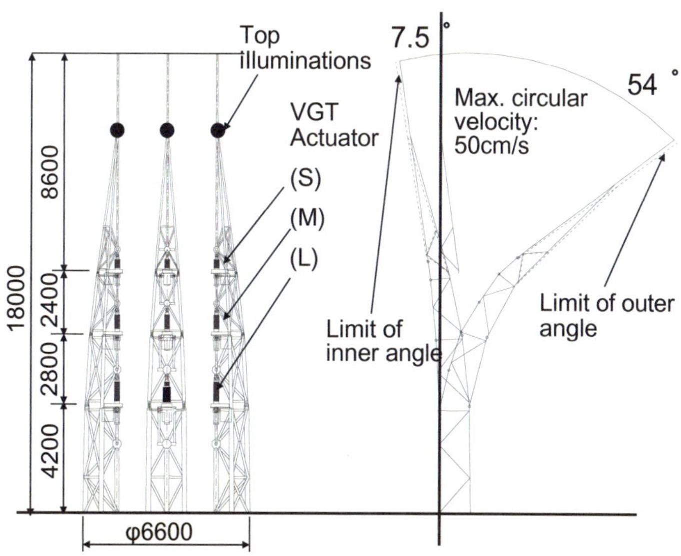

Fig. 17. Arrangement of VGT Actuator and the Movement Range of Tower

Fig.18 shows a picture of the inner structure of an extensible actuator used in a VGT Mechanism. The actuator was of the electronic type in which a screw rod was geared to a servomotor through a ball screw and a wheel gear. The top of the screw rod was linked with a truss node and the body of the actuator was carried by trunnion joints. The support bars on the node were moved in the outer stopper. Even if the screw rod broke, the tower's safety could be maintained by the support bars. In the servomotor, a magnetic brake and an encoder detect the rotating angle.

On the rod cover, both top and end limit sensors were installed. The motor was covered with waterproof covers. A cooling fan was maintained a suitable motor temperature. The electronic actuator had the advantage of high performance and energy conservation.

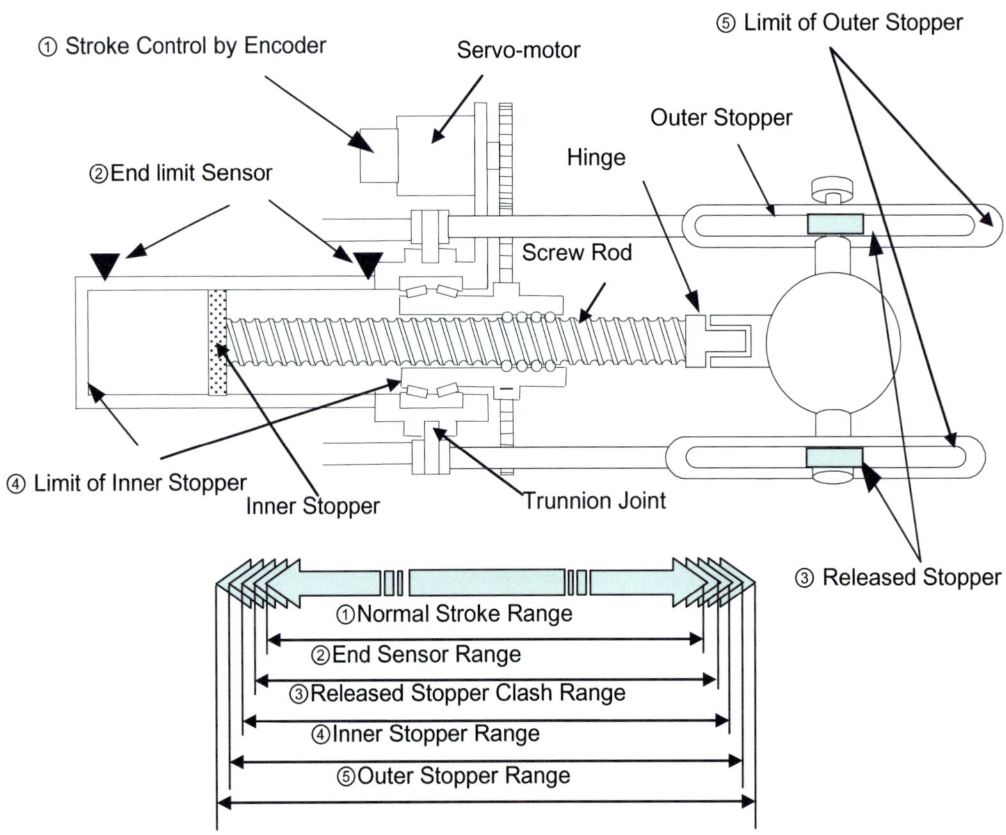

Fig. 18. Inner Structure of the Extensible Actuator of VGT and Safe Mechanisms

The relation between the load acting on the rod and the angle of each VGT mechanism are indicated in Fig.19 when the tower's movement analyzed by numerical simulation at the same angle. The load values were almost proportional to the angle, and the tension force range was wide and high. The rod's peak velocity was 20 mm/s, the tower tip rotation velocity became 500 mm/s or more when three VGT were operating at the same time. The tower movement could be expressed in an extremely dynamic and massive way in comparison with a conventional monument.

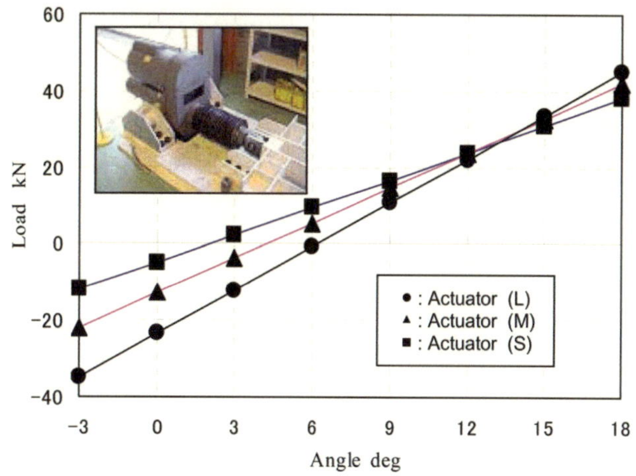

Fig. 19. Relation between Load Acting on Rod and Angle of each VGT Mechanism

4.2.2 Safety mechanism and control system

The movable monument used at the Expo had to operate continuously, so a safe structure and control system had to be developed. The actuator rod stroke was detected by the servomotor encoder data and the actuator condition was continually monitored. Various accidents to the monuments were assumed, and the check points and safety mechanisms indicated were introduced. For an accident concerning rod stroke, a five-step safety mechanism was introduced.

Fig.20 shows a chart of the operation system and plural fail-safe system. Monument operation was automated, except the initial process, and the operator mainly observed the system's safety confirmation and maintenance control. The plural fail-safe system that maintained monument safety was developed to avoid accidents. Furthermore, an emergency device; an automatic stop and warning device for earthquakes, thunderstorms, strong winds and heavy rain; and a backup device for power failure were installed.

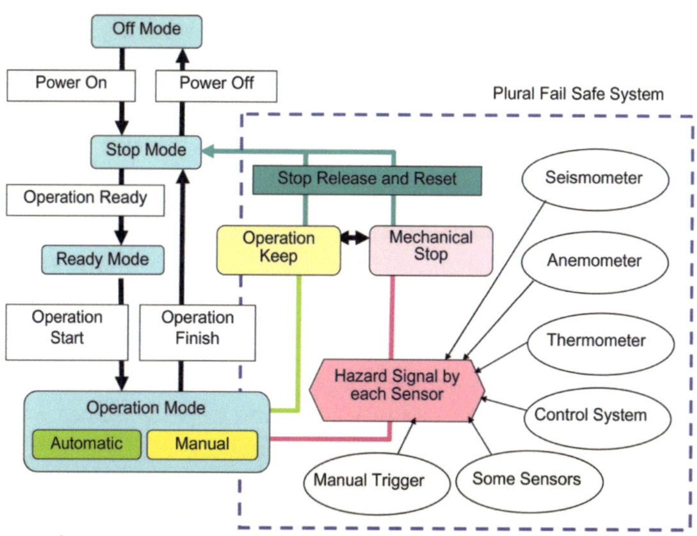

Fig. 20. Flowchart of Operation System and Monument Safety System

4.3 Performance and operation conditions of movable monument

An overview of the movable monument at Expo is shown in Fig.21. This picture expresses the coordinated performance of the three towers of the monument when fully opened (Fig.21-(a)), and Karakuri doll dancing in the center (Fig.21-(b)). A lot of visitors gathered around the monument, and they enjoyed the performance of the two exhibitions. Further, the monument was illuminated at night and its fantastic movements could be observed in the dark (Fig.21-(c))

(a) Performance of the Three Towers of the Monument When Fully Opened

(b) Karakuri Doll Performance (c)Lightening up the Monumnt

Fig. 21. Overview of the Movable Monument at EXPO

4.3.1 Performance patterns and shape change

Fig.22 shows the monument's shape changes according to performance patterns. One loop of the total performance was composed of two patterns every 30 minutes, that is, only the monument was moved for the 25 minutes of the first part, and the Karakuri doll danced with the monument for the 5 minutes of the second part. This performance loop was continuously repeated

For the monument's performance, there were two program modes. In the normal mode, the velocity and the stopping time of the actuator rod were decided by measuring and indicated in Fig.22-(a). There was a little case in which the shape of three towers reappeared at the same time. In a special mode, the monument was moved at high speed to accompany analyzing the state of the natural data (wind velocity, temperature, time, day and so on). As a result, the entire monument was moved to produce very irregular shape changes, as a preinstall program. This mode, being outside the performance loop, started suddenly, so nobody was expecting it. By selecting such modes, it was possible to express very interesting movements and monument shapes that changed slowly but dynamically.

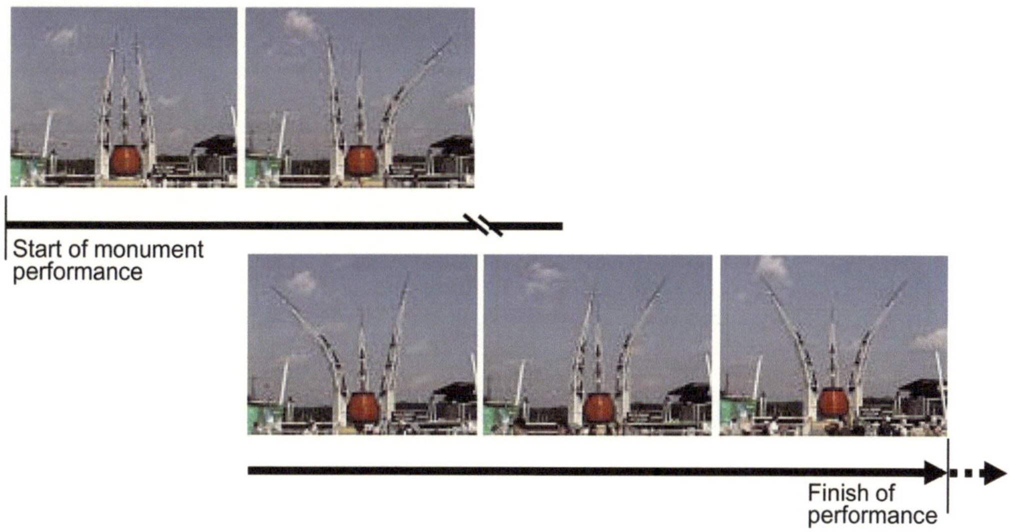

(a) Random Shape Performance of the Monument by Natural Data (25 min.)

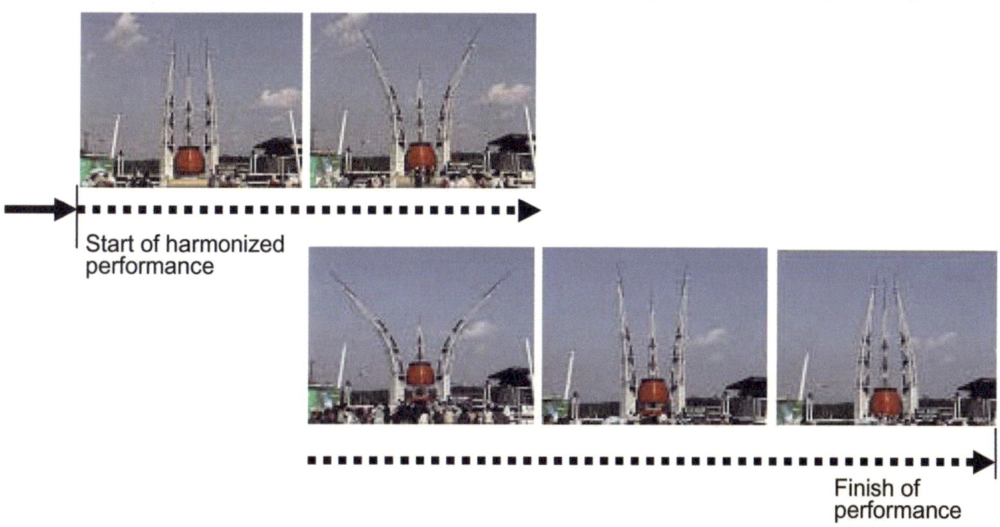

(b) Harmonized Performance of the Karakuri and the Monument (5 min.)

Fig. 22. Shape Changes of Monument According to Performance Patterns

On the other hand, in the coordinated performance with the Karakuri doll, the monument was opened and closed powerfully, synchronizing with the Karakuri doll's performance, as indicated in Fig.2-(b). In this case, the Karakuri doll performed a variable dance and somersaults with sound and illumination effect. A very traditional but innovative performance was thus created. In this performance, the Karakuri doll was the main player and the monument was a supporting player.

4.3.2 Monument operation

During the Expo, the monument was operated continuously for about 13 hours a day. However, its operation was modified every day and at times when there were unexpected special events. Fig.23 shows the record of operation frequency each day and their accumulations. When the shape of the monument was changed to open and to close, the operation was counted as one.

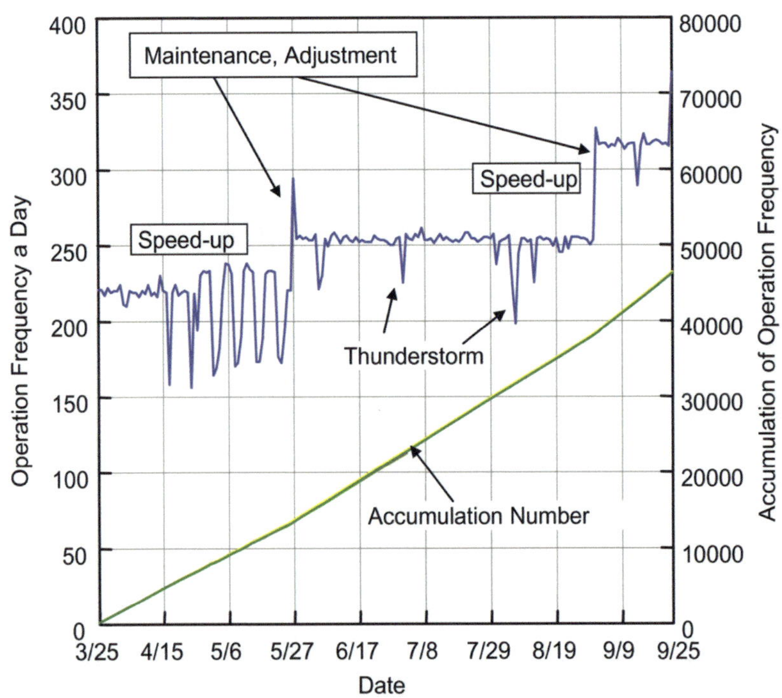

Fig. 23. Record of Monument's Operation Frequency during the Expo

At the beginning of the Expo, the average speed at which the monument moved was set at a low level, and the speed was changed depending on the day of the week. Two months after opening, the performance was switched to a random mode program. The operation frequency was observed to be almost constant. During the last month, the average speed approached the maximum level corresponding to the upsurge in attendance at the Expo site. During the Expo, the monument was operated continuously for 185 days, except during maintenance or thunderstorms, and there were neither breakdowns nor accidents. By the end of the Expo, there had been 50,000 operations, thus confirming that the monument wasoperated within the range of the initial plan.

After the Expo ended, the monument and other devices were temporarily removed from the site. In response to demands for its reconstruction, it was reconstructed in the field of the company that manufactured the VGT actuator, and is now open to the public as a memorial tower to the Expo. It may continue operating forever.

5. Conclusion

As for one movable mechanism that enables to make a future adaptive structure, we focused the VGT, and examined the development of element technolgies and its applicability to moveable structures. VGT was equipped with flexible and intelligent functions, and various shapes could be created freely by contriving its arrangement and control. The structure was considered to have a very wide application to construction sturucture. This paper has proposed an example of an adaptive structure applying the VGT. The efficiency and the characteristics of the VGT could be grasped under the several conditions by a scale model of a Flowering Dome and a movable monument exhibited in EXPO. Application of the movable monumnet was the first big project since the VGT technology had been developed in the construction field. In the development of the monument, we considered quality and security of construction. As a result, safe and excellent continuous performance was achieved, and the monument received high praise from promoters and many visitors. The VGT was shown to be a very useful technology for such movable structures whose shape can be changed variably. In the future, with progressing and spreading of VGT technology, we will propose various applications.

Finally, the author thanks all who supported the development and application of the VGT structure and a movable monument at Expo 2005.

6. References

Ishii, K. (1995). Moving Architectures. *Journal of Architecture and Building Science*, Vol. 110, No. 3, pp. 3-44

Natori, M.C., & Miura, K. (1994). Development of truss concept in Space Technology. *International. Symposium on Membrane Structure and Space Frame*, pp. 45-56

Kurita, K., Inoue, F., Natori, M. C. et al (2001). Development of Adaptive Roof Structure by Variable Geometry Truss. *Proceeding of 18th International, Symposium on Automation and Robotics in Construction* , pp.63-68, Sep. 2001, Krakow, Poland

Inoue, F., Kurita, K.et al. (2003). Application of Adaptive Structure And Control by Variable Geometry Truss. *Proceeding of The CIB 2003 International Conference on Smart and Sustainable Built Environment*, pp.59, Nov. 2003, Brisbane, Australia

Inoue, F., Kurita,K. et al (2006). Development of Adaptive Structure by Variable Geometry Truss. *Proceeding of 23th International Symposium on Automation and Robotics in Construction*, pp.704-709, Oct. 2006, Tokyo, Japan

Inoue, F. (2007). A Study of Movable Arch Strycture and its External Panel Mechanism by Variable Geometry Truss. *International Conference of Shell and Spatial Structures,* pp.704-709, Dec. 2007, Venice Italy

17

Robotic Excavation

Ahmad Hemami
McGill University
Canada

1. Intoduction

In this chapter robotic excavation or automated excavation is discussed. Excavation as performed by an excavator is the process of moving bulk material by digging, cutting and scooping, or a combination of them. The term robotic excavation implies the application of state of the art in robot motion control for automating excavating machinery, so that such a machine can accomplish excavation tasks by itself and without the continuous supervision and intervention of an operator. This type of application is still in its state of infancy and much more remains to be developed yet.

1.1 Introduction

Excavation in general term is the process of removing soil, ore or any bulk material from its original place, by digging out or digging away, and loading it (say, onto a vehicle for hauling). In this sense, it covers a variety of methods that may or may not include all the different functions of loosening (or cutting) the material, digging it and finally loading it. Also, the equipment used for this purpose are different, based on the geometry and physical properties of the environment they excavate, and the way the three basic functions (loosening, digging and loading) are executed (sequentially or combined together). There are two types of machines, in general, rotary and cyclic. The work here is intended for automation of cyclic excavating machines.

Automation of excavation process becomes required in a number of applications. For instance, in mining for more productivity, safety of workers and more efficiency an automated system is desirable; in remote places like moon an automated excavating device is required to get samples. Hazardous contaminated soil removing and mining radioactive materials are other examples.

Attention must be paid to the difference between an excavation operation and an excavation process. The latter is the function performed by an excavator. Thus, excavation process comprises a combination of cutting, digging and scooping. Automation of the process, in fact, implies the automation of the machine that performs the task. For automation, an analysis of the process and what is involved in it is necessary.

It is very important to note that the physical properties of the material to be excavated have a direct effect on the excavation method and the properties and design of the equipment. In fact they significantly determine the forces required on the cutting tool, and hence the power requirement of the machine. For simplicity of expression, hereafter, the material to be excavated will be referred to as "bulk media" or "soil". Also, instead of excavation automaton, the term "robotic excavation" is often interchangeably used.

1.2 Chapter contents

A general but brief review of the various excavation processes and the sort of machines in use is given in section 2. This will indicate the broad scope of the scenarios on excavation and its automation. The emphasis of the chapter is to make a distinction between what is involved in an excavation operation and the process of excavation, itself. In section 3 the question of automating the excavation operation is investigated. An example illustrates all the various tasks to be automated in an excavation operation by a particular class of an excavating machine. In this work, we narrow down our concern only to the issues relevant directly to automating an excavation process. Section 4 covers the analysis of the forces involved in an excavation process. These force components vary during the excavation and depend on the machine used. These are the forces that the cutting tool of an excavator must overcome. Knowledge about the force of excavation is quite important to automation of the task. Section 5 covers the analysis of robotic excavation and the necessary steps to be followed for robotic modelling of an excavating machine. The discussion is enhanced by some detailed analysis of a front-end loader and its counter part in underground mining (Hemami, 1992). This example is used for all the other discussions throughout the chapter. Section 6 is devoted to modelling of a front-end-loader type excavator as a robot manipulator. Section 7 considers the nature of what happens in an excavation process and how it can be modelled. The criteria to be used for automating such a process are also discussed. Section 8 elaborates on the conclusion of what must be carried out towards a workable automated machine.

2. Excavation methods and equipment

2.1 General excavation methods

Although not a sharp distinction line could, in general, be drawn between the various types of excavation, but because of dependency of the scope of operation on the factors such as: the quantity of the soil to be moved, the location of the excavation site, the relative width, breadth and depth, the type of soil and the purpose of excavation, excavation falls into the following basic types:

1. bulk-pit excavation
2. bulk wide area excavation
3. loose bulk excavation
4. limited-area vertical excavation
5. trench excavation
6. tunnel excavation
7. dredging

Bulk-pit excavation is excavation of considerable depth as well as considerable volume. The equipment must work against the face of nearly vertical walls from inside the pit, and the soil must be hauled away.

Bulk wide-area excavation is like the bulk-pit excavation, but shallower in depth and larger in area, and the site is accessible from many directions. The excavated soil is hauled a shorter distance.

Loose bulk excavation is like excavation of canals where the soil is not hauled away but cast into a new position. Moreover, the operation is usually performed from the surrounding ground rather than from inside the pit.

Limited-area vertical excavation is where the soil must, out of necessity, be lifted out vertically; the method is used for loose and wet soil, and the banks must be supported by shoring or sheathing.

Trench excavation sometimes falls into the category of the limited- area vertical excavation; generally, the width and depth of operation are limited.

Tunnel excavation is completely underground; the width and depth (or height) are limited. (Tunnel excavation can be divided into underground excavation and tunnelling.

Dredging is the removal of soil from underwater.

In mining, particularly considering the various equipments, excavation may be categorized as:

- Open-pit excavation (in surface mining)
- Underground excavation
- Tunnelling
- Underwater excavation

In **open-pit excavation** the various types of excavation methods, as mentioned in numbers 1 to 5 above, are used. For **underground operations**, because of space and other limitations, the type of the equipment that can be used are quite different from size, capacity and manoeuvrability points of view, as will be discussed later. Special tunnelling machines are employed in general **tunnelling** operations (which are usually different from ore mining operations). The environmental conditions for **underwater excavations** are quite different from others, as the name implies; however, depending on each particular case some of the equipment for other types of excavation might be utilized for excavating under the water, or other special equipment would be necessary. Tunnelling and underwater equipment are not discussed further in more detail in this study.

2.2 Excavating machines

In this section an overview of the various equipments that are currently in use by mining and construction industries are presented, without getting into their detailed description.. These equipments can be first categorized into **surface mining** or **open-pit mining excavating** equipment, and **underground mining** excavating equipment. Further in each category, they can be divided into **non-continuous (or Cyclic)** and **continuous** machines.

2.2.1 Open-pit mining (surface mining) cyclic machines

Five different types of cyclical excavating machines can be identified; these are:

- Power Shovel
- Dragline
- Scraper
- Clamshell
- Backhoe (hoe)

Figure 1 shows the schematics of these machines illustrating the basic difference between their structures. The **power shovel** is the most efficient of the various machines for excavating and loading large quantities of soil. The **dragline** is the unit of excavating equipment most ideally suited to handle loose bulk excavation. It operates most efficiently from an elevation higher than the soil being dug, thus very suitable for original surface; it is seldom used for excavating at or above its travel level.

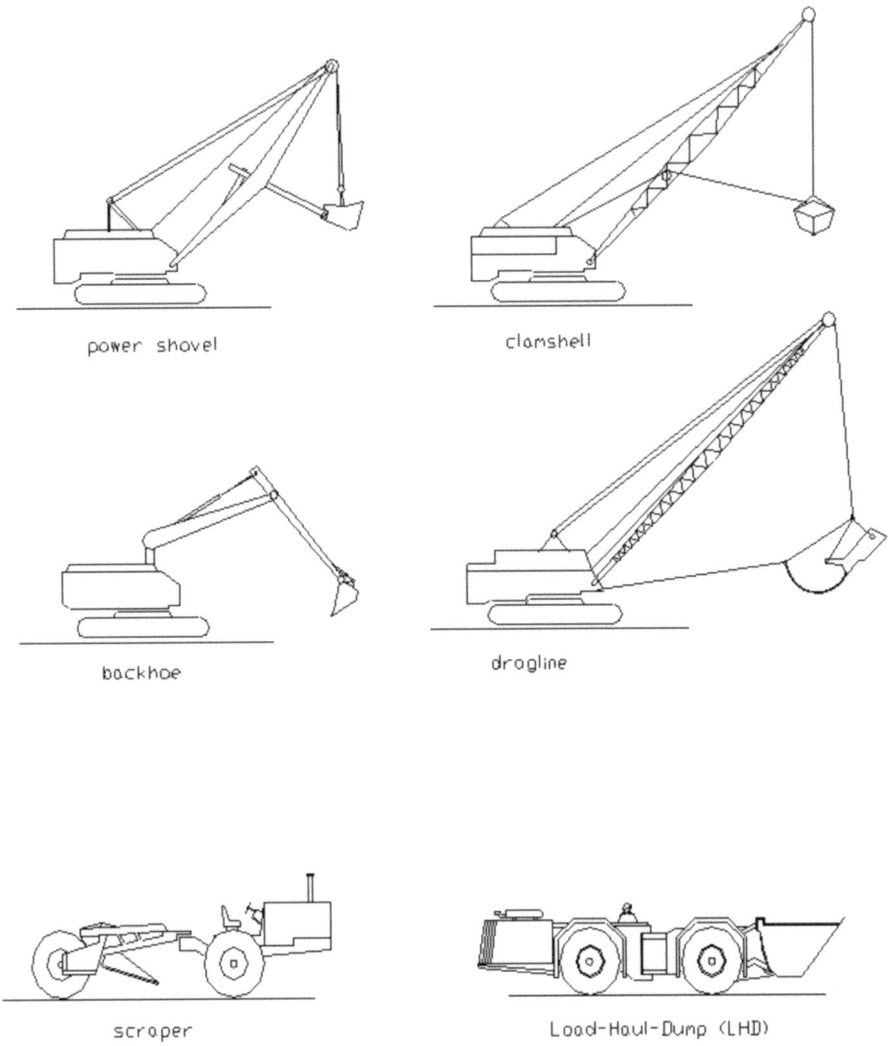

Fig. 1. More frequently used excavating machines

Unlike the power shovel none of its power is available for direct pressure on the soil. The bucket is filled by pulling it toward the dragline after being penetrated into the soil only by the force of its weight. The dragline performs many kinds of loose bulk excavation well; it has an extensive use in overburden stripping and surface mining. Draglines are mounted on crawlers which enable them to work on tight areas.

The **scraper** is designed for loading a thin layer of soil over a large area. It has the advantage of being able to haul and unload the soil to the desired destination. Thus it is very appropriate for bulk wide-area excavation. This machine is mostly pushed or/and pulled by tractors mounted on rubber tires; the cutting edge of the bowl (its bucket) penetrates 4-6 in. into the soil, depending on the density of the soil formation. Difficulty is encountered in loading loose dry sand and rock and, also, in unloading wet, sticky soil. Their greatest use is found in unconsolidated soil that requires little or no loosening; but they are finding increasing applications in loading and hauling in open-pit mining.

The **clamshell** is most suitable for limited-area vertical excavation, like foundation excavations; for this reason, most generally it is used as a secondary unit to muck out in the rear of the more productive machines. For various jobs that call for removal of the soil from below the level where the machine rests, or require moving the soil above the machine, in particular where the soil being dug is loose, soft or wet, the clamshell is ideal. The clamshell consists of a bucket hung from the boom of a crane that can be either crawler or wheel mounted. The two halves of the bucket are dropped onto the soil to be excavated; then the bucket is closed, encompassing soil between the two halves.
The **backhoe** is customarily a secondary tool in surface mining. Contrary to the shovel and dragline where the general concern is the volume excavation, the backhoe is convenient for scraping off and cleaning adhering overlay soil from surface, and also for trenching and digging ditches.

2.2.2 Underground mining cyclic machines

The choice of equipment employed for underground mining primarily is enforced by the properties of the ore to be excavated. The physical size of the ore and its geometry, that determine the method of mining, and the cost are secondary parameters. For instance, continuous mining machines can be used for soft and semi hard soil, but do not have much success with hard rock that must be fragmented by blasting (or alternative method) before excavation.
The machines for underground mining are much smaller in size and the capacity in comparison with the equipment for open-pit mining. Two most used underground mine cyclic excavators are:

- Overshoot loader
- Load-Haul-Dump (LHD) unit

Overshoot Loader is a device which picks up the blasted soil from the front and without turning discharges it to the rear in a truck to do the haulage, or to a conveyor. It may be powered by compressed air, electricity or diesel engine. The mounting can be crawlers, wheels or it is bound to move on rails. The operator stands by the side of the machine, or he can operate from a few meters away by a remote system, if equipped. Overshoot loaders are usually small in size to be able to work in small drifts.
In larger mines a **Loader** (front-end loader**)** can be used to excavate and transfer the soil to a **mining truck** (dump-truck). A loader is extensively used in construction and road work, and to a less extent in surface mining; in the same way it can be used for loading the ore. However, the size of loader, and the space it requires for operation make it less desirable for underground, unless for the wide and high stopes, where the high mobility and rapid loading overweighs the shortcoming.
A **Load-Haul-Dump** machine, or simply **LHD**, as the name implies (Figure 2), can combine the work of a loader and a dump-truck. In this way, one operator works instead of two in the former case. LHD's are specially designed for working in mines; their physical structure, and size thus enable them to operate and move without difficulty in narrower areas and with smaller height. Moreover, they are made in two parts hinged together in order to facilitate their turning through curves (Figure 3). LHD's are mounted on rubber tires which give them more mobility, and they are widely used in underground mining work. Some LHD loaders are equipped with two buckets; in this way their transportation capacity is increased.

Fig. 2. A Load-Haul-Dump unit

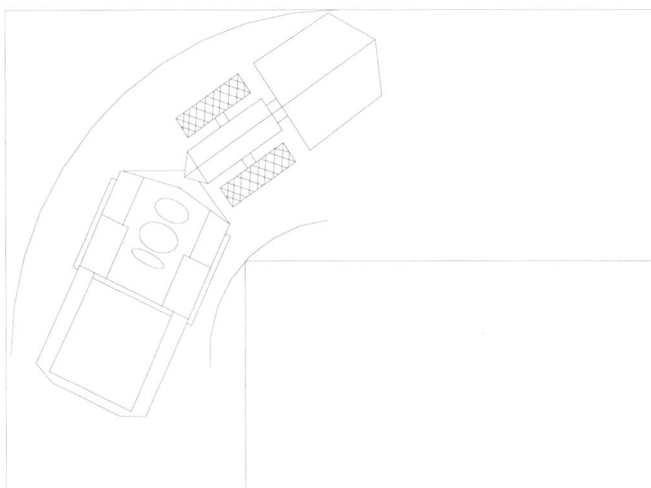

Fig. 3. LHD is designed to negotiate curves better

3. Automation of excavating operation

3.1 Autonomous excavation

In reality what is expected from automation, when it will become possible, is autonomous excavation. In fact auto-loading, which is the whole focus of this chapter, is only a necessary part of a bigger scenario of being able to perform the tasks of an operator by an automated excavation machine. When the outcome of research on excavation automation leads to technical capability for successful auto-loading, this capability must be integrated with other automated tasks, in order to be utilized. Such a scenario is shown in figure 4. In this sense, for an automated loading operation the same sort of actions that an operator does must be automatically performed. This requires the following three processes and integrating them:

a. Automation of the loading function
b. Automation of navigation
c. Automation of obstacle detection and avoidance, for movement of both vehicle and bucket during dumping

A preliminary analysis shows that in order to carry out the above tasks an operator uses his power and intelligence for the following:

1. Determining in what part of muck pile to start loading,
2. Lowering the bucket and running the vehicle forward,
3. Determining the starting point of loading action,
4. Executing the motion for the bucket,
5. Sensing whether motion is taking place in normal way,
6. Sensing if the vehicle advances or tires are slipping,
7. Deciding the final loading point,
8. Moving back and lowering the bucket for haulage,
9. Monitoring the performance of the vehicle,
10. Selecting the delivery point,
11. Deciding on the route to the delivery point,
12. Navigating to that point while watching for avoiding ground obstacles on the way,
13. Raising the bucket while watching for hitting nothing at the delivery point,
14. Dumping the loaded media.

The above list shows that there are various decision making and actions to be executed. All of these must be performed without fault for successful automation.

Fig. 4. Various tasks in an excavation scenario

3.2 Auto-loading

Auto-loading concerns only the items 4 to 8 on the list of the subtasks in section 3.1 Considering, again, in more detail, what an operator does in terms of loading function (this depends on the experience and judgement of an operator, so the exact details change from person to person), he drives the lowered bucket to the muck pile, then starts raising and simultaneously tilting the bucket, until it is almost clear from the pile. In this moment the bucket is supposed to be filled, thus the operator pulls back while lowering the bucket, for haulage. Any part of this cycle will be corrected, repeated and/or adjusted if the result is other than expected. Knowing this is what needs to be performed automatically, the following primary steps summarize what can be done to lead to the required results:

- an appropriate trajectory be defined for the bucket,
- the bucket is driven such that it follows the trajectory,
- the operation is watched to be carried out properly.

Performing the above requires that:

- the essential force is provided and adjusted when necessary,
- this provided force must be used for loading the bucket, rather than wasteful pushing,
- the bucket is getting filled in the process of following the trajectory,
- the machine parts are not damaged.

The rest of this chapter is devoted to the auto-loading only. For this, furthermore, we consider automation of LHD only. As, was said before, an LHD and front loader are similar, for this purpose. Most of the results, nevertheless, are general and applicable to all excavating machines. We are seeking the way the loading function of an LHD can be automated. That is, a machine can load itself without the interference of an operator.

4. Excavation force analysis

In the process of loading, a bucket is subject to resisting forces and torques which, in order for the bucket motion to continue, they must essentially be overcome. Also, the necessary force for moving the bulk of the loaded rock must be supplied by the active forces moving the bucket. There are four resisting forces, denoted by r_1 through r_4 in figure 5, at each instant, which must be overcome. In addition to counteracting these forces, the actuating mechanism must provide r_5 corresponding to the inertia force for motion of the load (not shown). The addressed forces are:

r_1 : weight of the loaded soil and that of above the bucket.
r_2 : force of compacting the soil by bucket.
r_3 : friction forces between bucket and the soil.
r_4 : digging resistance of the soil.
r_5 : the necessary force to move the soil in and above the bucket.

The forces to be exerted by the loader bucket at this moment are in the opposite direction of those shown.

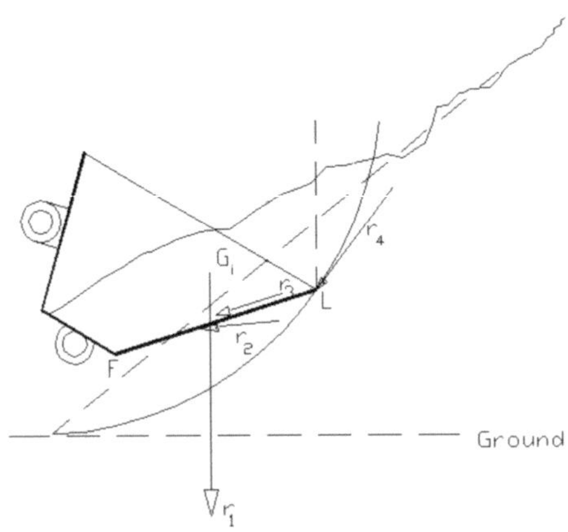

Fig. 5. Medium force components on a bucket

The above defined force elements depend on a number of factors, such as the type of soil, its properties, and the shape and material of the bucket. They have a stochastic nature and can vary considerably from point to point in the mass of a heap of bulk medium. The factors affecting their magnitude can be seen to be of four categories:

1. The tool (Bucket): The shape, size, geometry and material of the cutting device (Also, teeth on the cutting edge)
2. Environment: Temperature, gravity, terrain slope.
3. Excavator: The excavation process, itself, is not unique in various machines
4. Medium: The soil property variation is quite enormous, dictated by :
   - type of soil (its mechanical properties, like hardness and cohesion)
   - uniformity (mixture of various soils)
   - water content
   - temperature
   - particle size
   - compactness
   - adhesion (Also a function of tool material)

Certain references given at the end of this chapter indicate some previous work to formulate the forces encountered by a cutting tool. Because of the great number of the factors (Up to 32 have been reported, some of them less important, but at least 20 of them are quite important), it is quite difficult, if not impossible, to determine the force values based on given properties. Only approximate values of the force can be expected from any formulation.

5. Process analysis

The question of how to automate the operation of a loader, backhoe, Load-Haul-Dump unit, and so on leads to the fact that because motion control is involved, a robot control scheme can be employed. For this reason, the term robotic excavation is often used when referring to excavation automation. In the case of a front-end loader, LHD and power shovel, auto-loading is an analogous term. With regards to the control of industrial robots a good deal of knowledge and technology has been developed for their control. As mentioned before, in what follows all the analysis is performed for a typical LHD, which represents all the excavators of the same category.

There are two major differences between a robotic arm and an LHD (or similar excavating machine). The first difference is the relatively large tool and payload in the case of an excavator compared to usually small tool and payload for a robot manipulator. The tool for an excavator is its cutting element or bucket. This implies a significantly large interaction between the tool and its environment. The second difference is that the base of a robot manipulator is fixed, whereas for an LHD there is no fixed base. As a result of this, the final results can be different from expected, as it is depicted in figure 6, which illustrates a deformation of the vehicle structure leading to a change in the trajectory.

Control of a robot arm is based on the definition of a trajectory and then motion control to lead the tool through the defined trajectory. Based on the similarity to a robot arm, there are then two primary steps: modelling the excavator as a robot arm and defining a trajectory for the bucket to follow. Since a bucket is a rigid body defining a trajectory for it implies defining a trajectory for three of its points. If for simplification we assume the motion of the bucket during scooping to be a planar motion, then the trajectories of two points in the plane of motion are sufficient to describe the plane trajectory of the bucket. One appropriate choice for motion study is the cutting edge, because it includes the points of action of the cutting forces. Selection of a proper trajectory must be based on a sound analysis rather than repeating the same kind of trajectory as practiced by an operator. Obviously, for any given task the trajectory definition depends on a number of parameters corresponding to the bucket size and the medium to be loaded. Figure 7 illustrates some typical bucket tip trajectories for the same excavator and the same bucket, for example. AN_2 is for a heavier material compared to AN_1 (point L in the figure represents the cutting edge of the bucket; see section 6). It is necessary to point out, however, that in the process of following the trajectory is not the crucial criterion; the main objective is to fill the bucket, preferably with minimum energy.

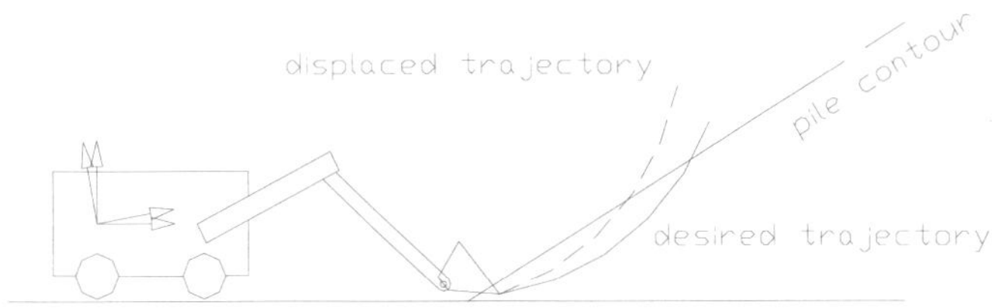

Fig. 6. Deformation of excavator changes the trajectory

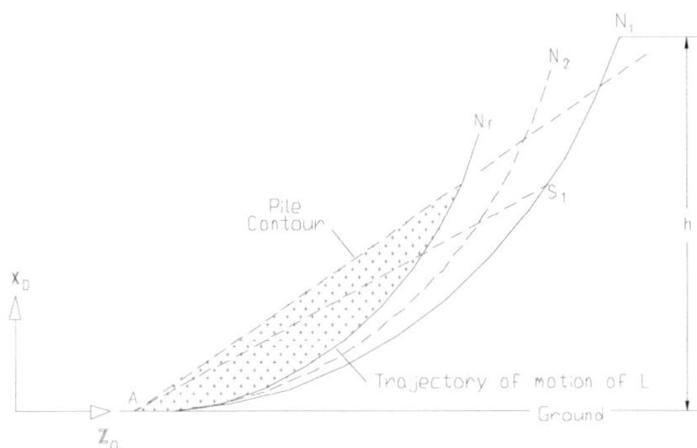

Fig. 7. Trajectories can be defined base on medium property (density, for instance)

In scooping there are a number of forces involved, as discussed before; if the active force on the bucket at each instant while it moves along a trajectory does not accord with the resistive forces, then some undesirable outcome, like an empty bucket or wasteful effort of pushing the vehicle with tires slipping instead of advancing, will result. In any of these cases the action must be stopped. That is to say, if not enough material is loaded in the bucket, the motion must not be continued; also, if a situation like that shown in figure 8 is encountered or the bucket has reached a dead stop, increasing the active force is not the right thing to do, since it is equivalent to wasting energy. This does not mean that in order to properly control the scooping operation one must have the knowledge about the exact magnitudes of the involved forces at each instant (in fact, this force has a stochastic nature, so it is impossible to define its certain magnitude). But, it implies that the force requirement must continuously be monitored, so that none of the two undesirable cases happens.

Monitoring the resistive forces is contrary to considering them as a disturbance to the problem of position control for trajectory following, which one could think of as an alternate approach. The latter is the way an industrial robot is controlled to follow a path. This is because, as pointed out earlier, the primary goal is to fill the bucket, not to just follow a trajectory (possibly with a half filled bucket). In this sense, it becomes essential to determine the approximate variation of the resultant resisting force on the bucket along the trajectory. For the sake of clarity two possible variation patterns are shown in figure 9. These are fictitious patterns, but show that at any point along the trajectory what the maximum and minimum expected values for resistive force are. Any time that the measured force is outside of the range it indicates that either the bucket is not filled enough or it has hit a sizable restriction.

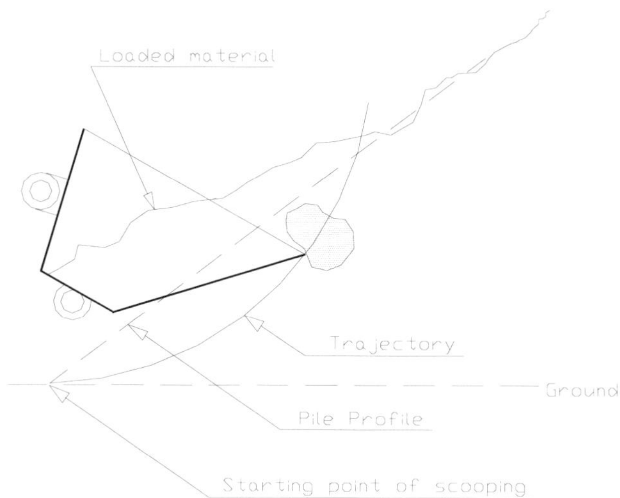

Fig. 8. A possible scenario in practice

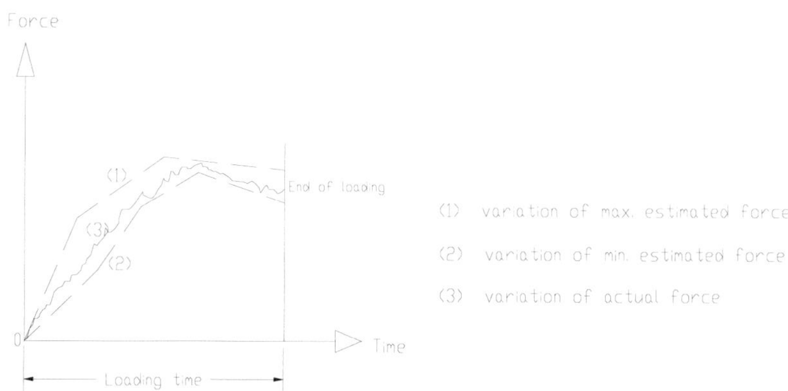

Fig. 9. A range of magnitudes can be found for the resistive forces along the trajectory

6. Kinematic modelling

In this section kinematic modelling of an LHD as a robot manipulator is carried out. A similar approach can be employed for other excavating machines (Ostoja-Starzewski and

Skibinievski, 1989, Vaehae et al, 1991), with the incorporation of more details for additional segments such as the mechanism for manipulating a backhoe bucket (Hemami, 1995). The latter can be slightly different from one manufacturer to other, but the general approach is the same. Section 6.1 describes the kinematic model and definitions, and section 6.2 presents the forward and inverse kinematic solutions for the model. Finally the relationship between the actuators efforts and resistive force at the bucket are expressed by the formulation of Jacobians for the robotic model in section 6.3

6.1 Kinematic model

An LHD consists of a driving unit to which the loading gear is attached through a pivot connection; the reason for pivoting is to make it possible for the longer vehicles to negotiate curves in narrower underground passages. Figure 10 shows the schematic of the loading gear model of a typical LHD-unit. This mechanism can be assumed to consist of a platform to which two sets of linear actuators are attached. The platform is free to move forward (and backward) with respect to the ground (this is symbolically indicated by the rollers in figure 10). There are three distinct actuations in the scooping function of this machine: a push forward by the driving vehicle, a pushing/pulling action of usually two parallel cylinders CE, raising and lowering the supporting arm BEH, and pushing/pulling action by cylinder AD. Observation of a loading action (also in dumping) reveals that the motion of the bucket provided by the three forces involved, the push of the vehicle and the forces of the hydraulic cylinders, can be assumed to be a plane motion. This makes the analysis much simpler. In this sense, the cutting edge of bucket is represented by a point L. The coordinate system $x_4y_4z_4$ is attached to the bucket at this point. The three loading forces provided by the actuators give rise to two force components along x_4 and y_4 and a torque about z_4 , at this edge of the bucket. Figure 10 shows also the definition of various dimensions in the mechanism.

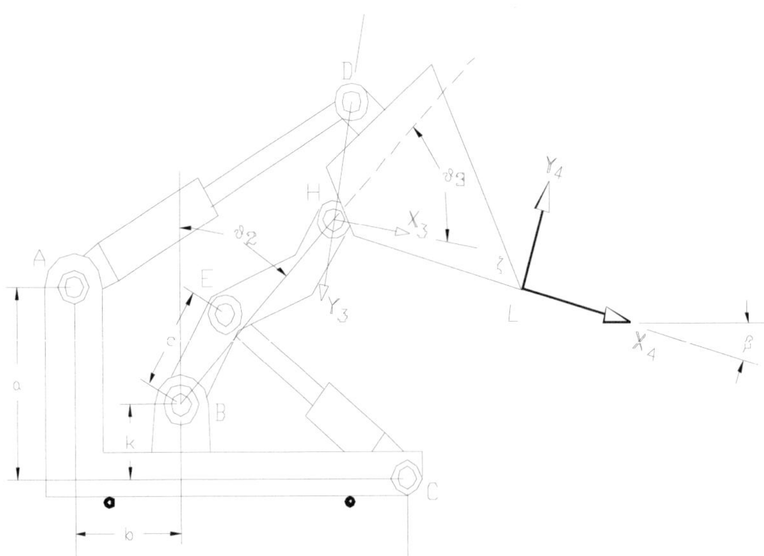

Fig. 10. Model of the loading gear in a loader or LHD

The function of the cylinder CE is to rotate the supporting link BEH about point B; and the function of cylinder AD is to give a rotation to the bucket about point H. Taking into account the fact that the direction of the forward force from the vehicle is almost unchanged during loading, it can be seen that the bucket is manipulated by one force and two torques. It can, thus, be regarded as being actuated by a prismatic joint followed by two revolute joints. Thus, the loading mechanism is modelled as a robotic arm with three degrees of freedom. Point L is taken as the tool point. Figure 11 shows the coordinate system chosen for the three joints based on (Denavit and Hartenberg, 1955) and according to (Lee, 1982).
The frame $x_0y_0z_0$ is attached to the vehicle at any arbitrary point. It serves as the reference coordinate system at any instant before a loading action is started. It serves also to define the direction of the force and movement of the prismatic actuator, both along the z_0 axis. $x_1y_1z_1$ is attached to joint 2 at point B, its direction is such that θ_2, the second joint variable, has a positive sign when clockwise, as shown. Similarly $x_2y_2z_2$ is attached to joint 3 at point H and it has the same sense of direction as joint 2. Finally the bucket attached coordinates $x_3y_3z_3$ is chosen such that the axis y_3 passes through point D (which is a distinct point on the bucket), and θ_3 is also positive clockwise. This choice of coordinate system slightly simplifies the analysis. $x_4y_4z_4$ coordinate system, which will be used to define the forces on the bucket at the tool point, however, has a different sense of direction. Table 1 shows the associated parameters for the three actuators.

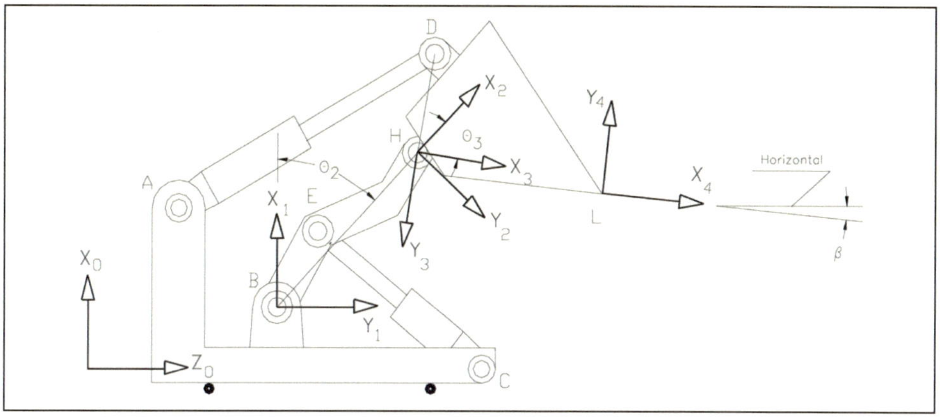

Fig. 11. Definition of coordinate systems

JOINT	α	a	d	θ	TYPE
1	90°	k	var.	0	prismatic
2	0	r	0	var.	revolute
3	0	0	0	var.	revolute

Table 1. Definition of Denavit-Hartenberg parameters

The relationship between the two coordinate systems $x_3y_3z_3$ and $x_4y_4z_4$ is constant and is defined by the transformation matrix:

$$A_4 = \begin{bmatrix} \cos\psi & \sin\psi & 0 & q_1 \\ \sin\psi & -\cos\psi & 0 & q_2 \\ 0 & 0 & -1 & 0 \\ 0 & 0 & 0 & 1 \end{bmatrix} \tag{1}$$

where constants q_1, q_2 are the coordinates of point L in frame 3 and ψ is the angle between x_3 and x_4 (see figure 10). q_1, q_2 and ψ depend on the bucket and the attachment dimensions. The transformation matrices between frame 0 to frame 3 are accordingly given by

$$A_1 = \begin{bmatrix} 1 & 0 & 0 & k \\ 0 & 0 & -1 & 0 \\ 0 & 1 & 0 & b+d_1 \\ 0 & 0 & 0 & 1 \end{bmatrix} \tag{2}$$

$$A_2 = \begin{bmatrix} C_2 & -S_2 & 0 & rC_2 \\ S_2 & C_2 & 0 & rS_2 \\ 0 & 0 & 1 & 0 \\ 0 & 0 & 0 & 1 \end{bmatrix} \tag{3}$$

and

$$A_3 = \begin{bmatrix} C_3 & -S_3 & 0 & 0 \\ S_3 & C_3 & 0 & 0 \\ 0 & 0 & 1 & 0 \\ 0 & 0 & 0 & 1 \end{bmatrix} \tag{4}$$

In equations (3) and (4), C_i and S_i stand for cos θ_i and sin θ_i, respectively. Note that for better clarity, in equation (2) the constant b (See figure 10) is separately added to the offset of joint 1; this is because $x_0y_0z_0$ is shown outside the moving platform.

6.2 Kinematic solutions

In the operation of the loader, we are concerned about the position of point L and the bucket orientation with respect to ground. The position of the loading point L is how high and how much in front of its starting position point L has moved (or must move) at any instant during scooping motion. These two values are the magnitudes of x_0 and z_0, respectively. As a measure of the bucket angle, a convenient choice is the angle β between z_0 and x_4, thus when x_4 is parallel to ground this angle is zero. The position and orientation of the bucket (position of point L and the orientation of the front side) is determined from the transformation matrix:

$$T = A_1A_2A_3A_4 = \begin{bmatrix} \cos(\theta_2+\theta_3+\psi) & \sin(\theta_2+\theta_3+\psi) & 0 & q_1\cos(\theta_2+\theta_3)-q_2\sin(\theta_2+\theta_3)+r\cos\theta_2+k \\ 0 & 0 & 1 & 0 \\ \sin(\theta_2+\theta_3+\psi) & -\cos(\theta_2+\theta_3+\psi) & 0 & q_1\sin(\theta_2+\theta_3)+q_2\cos(\theta_2+\theta_3)+r\sin\theta_2+b+d_1 \\ 0 & 0 & 0 & 1 \end{bmatrix} \tag{5}$$

Thus, the upward position of point L, as a function of θ_2 and θ_3 is defined by the first element of the 4th column of matrix T, and the forward displacement is given by the third element of the same column. Also, angle β can be seen to be given by:

$$\beta = \frac{\pi}{2} - (\theta_1 + \theta_2 + \psi) \tag{6}$$

The values of the joint angles θ_2 and θ_3 can be directly measured by means of encoders attached to the joints at B and H. The forward displacement (d_1), when necessary, can also

be determined from the forward motion of the vehicle by using an appropriate sensor and simple calculations.
The inverse kinematic problem for this three degree of freedom manipulator is quite straightforward. If at any instant of time three values are given for β, bucket orientation, h, the height, and l, the forward distance, then equation (5) and (6) can be used to determine d_1, θ_2 and θ_3 as follows. From the given value of β and equation (6) the sum of $\theta_2 + \theta_3$ can be immediately found. From the calculated value for $\theta_2 + \theta_3$ and the value for h, two values for θ_2 can be obtained; the difference between the two is a ± sign which results from using a cos function. However, more likely the negative answer is rejected because of the physical range of admissible values. Finally the value for d_1 may be determined from the result of previous calculation of θ_2 and θ_3 and the given value for l.

6.3 Jacobian relationships (relating active and resistive forces)

Jacobian matrices of the mechanism can be used for both velocity relationships and force relationships. That is, defining the bucket velocities (Two translations and a rotation, for planar motion) in terms of the actuator velocities, and the forces at the cutting edge of the bucket (two forces and one torque, associated with planar motion) in terms of actuator efforts. For the force vector at the bucket cutting edge, it is convenient to assume that the bucket is momentarily stationary. This assump-tion is not unreasonable because of the slow motion of the bucket. Because there are only three degrees of freedom which give a plane motion to this manipulator, only a 3x3 reduced Jacobian matrix is sufficient. Moreover, because weight is always a significant part of the load, then the Jacobian matrix of the tool point with reference to the world coordinates $x_0y_0z_0$ is more appropriate. This matrix is analytically determined in equation (8). The relationship between the force/torque vector in the joints coordinates and that of the load in the reference coordinates is

$$\tau = J^T F \tag{7}$$

where with the choice of J as defined above, that is $J = J^L_0$,

$$J^L_o = \begin{bmatrix} 0 & -(q_1S_{23} + q_2C_{23} + rS_2) & -(q_1S_{23} + q_2C_{23}) \\ 0 & -1 & -1 \\ 1 & (q_1C_{23} - q_2S_{23} + rC_2) & q_1C_{23} - q_2S_{23} \end{bmatrix} \tag{8}$$

$$F = \begin{bmatrix} \textit{vertical component of force at } L \\ \textit{horizontal component of force at } L \\ \textit{torque at } L \end{bmatrix} \tag{9}$$

$$\tau = \begin{bmatrix} \textit{Pushing (pulling) force of vehicle} \\ \textit{Torque at joint } B \\ \textit{Torque at joint } H \end{bmatrix} = \begin{bmatrix} f_1 \\ \tau_2 \\ \tau_3 \end{bmatrix} \tag{10}$$

and T represents transposition.
It is, however, necessary to find the relationships between f_2, the force in the hydraulic cylinder CE, and τ_2 and between f_3, the force of cylinder AD and τ_3. These relationships are functions of θ_2 and θ_3 and can be found from geometric dimensions and angle relationships. Thus:

$$\tau_2 = -ef_2 \sin(\pi - \alpha - \gamma - \varepsilon) = -ef_2 \sin(\alpha + \gamma + \varepsilon) \tag{11}$$

where α is the angle of cylinder CE with the base, γ is that of the link BE with its vertical support and ε = angle EBH = constant, all shown in figure 10. But

$$\alpha + \theta_2 = \frac{\pi}{2} \tag{12}$$

and the angles α and γ are related according to

$$\tan\gamma = \frac{k + e\sin(\alpha + \varepsilon)}{c - b - e\cos(\alpha + \varepsilon)} \tag{13}$$

In view of the equations (11), (12) and (13), therefore:

$$\tau_2 = -ef_2 \sin\left(\frac{\pi}{2} - \theta_2 + \varepsilon + \gamma\right) = -ef_2 \cos\left(\theta_2 - \varepsilon - \tan^{-1}\frac{k + e\cos(\theta_2 - \varepsilon)}{c - b - e\sin(\theta_2 - \varepsilon)}\right) \tag{14}$$

Moreover, it can be seen that

$$\tau_3 = g f_3 \sin\eta \tag{15}$$

where g = HD = constant and η = Angle HDA. Angle HDA is more conveniently, and without ambiguity, calculated from the dot product of vectors AD and HD. In this respect, the relative coordinates of points A, H and D (regardless of d_1) are determined as

$$A: \begin{bmatrix} a \\ 0 \\ 0 \end{bmatrix}$$

$$H: \begin{bmatrix} rC_2 + k \\ 0 \\ rS_2 + b \end{bmatrix} \qquad \textit{H is the origin of frame } x_2y_2z_2, \textit{ thus fourth column of } A_1A_2$$

and

$$D: A_1A_2A_3 \begin{bmatrix} 0 \\ -g \\ 0 \\ 1 \end{bmatrix} \rightarrow \begin{bmatrix} rC_2 + k + gS_{23} \\ 0 \\ rS_2 + b - gC_{23} \end{bmatrix}$$

which lead to:

$$\overline{AD} = \begin{bmatrix} rC_2 + gS_{23} + k - a \\ 0 \\ rS_2 - gC_{23} + b \end{bmatrix} \tag{16}$$

and

$$\overline{HD} = \begin{bmatrix} gS_{23} \\ 0 \\ -gC_{23} \end{bmatrix} \tag{17}$$

It follows from equations (16) and (17), after simplification, that:

$$\eta = \cos^{-1} \frac{r\,S_3 + (k-a)\,S_{23} - b\,C_{23} + g}{\sqrt{b^2 + g^2 + r^2 + (k-a)^2 + 2gr\,S_3 + 2(k-a)(r\,C_2 + g\,S_{23}) + 2b(r\,S_2 - gC_{23})}} \tag{18}$$

Since η cannot become negative there is no sign ambiguity in the determination of η from equation (18). If cos η is called w then in light of equations (15) and (18)

$$\tau_3 = g\,f_3\sqrt{1 - w^2} \tag{19}$$

that is, computation of the angle η, itself, is not necessary.
Equations (7), (14) and (19) can be used to find the three forces f_1, f_2 and f_3 that are required to give rise to a force F at the tip of the bucket. Conversely, they can be measured to monitor the resisting force/torque at the cutting edge of the bucket.

7. Motion analysis and control

A robotics approach to formulate the problem of auto-loading required the preliminary steps of modelling and analysis, as discussed in the previous sections. In this section the real control problem is touched and discussed. From a control viewpoint this question is raised: For the control of the excavation process (auto-loading) what must be measured and what must be controlled?

7.1 Motion analysis

Considering the operation of an LHD loader by an operator, initially he lowers the bucket until its front edge touches the ground; that is, a zero orientation of bucket with the terrain. He forwards the vehicle towards the pile of soil/fragmented rock, and continues until certain resistive force is sensed. He then operates the hydraulic cylinders to simultaneously tilt and raise the bucket. The entire motion is made around these actions, repeated only once or more, based on each operator's personal experience and specific method. This approach to control the motion of bucket, that is, mimicking the operator’s actions to move the bucket, has been practically implemented (Steel, et al, 1991), later blended with some induced vibration. Without the induced vibration the operation has not been successful. This mimicking the operator's action, however, is not necessarily the best way to load the bucket in terms of the energy used, or its effect on the tires and the other machine components.

A systematic approach suggests that a trajectory with minimum energy be defined for the bucket to follow. Studies for this minimum energy trajectory have given some results (Hemami, 1993). Other studies for an appropriate trajectory can be found in (Sarata et al, 2003, 2005). As mentioned earlier (see figure 6) a fundamental problem that arises is that the defined trajectory is relative to a fixed world coordinates, whereas due to deformation of the loading mechanism the trajectory can be easily displaced or distorted from its original form. For this reason some of the works done on the subject suggest using a camera to monitor the loading action, based on the load inside the bucket (Sing, 1991, Stentz et al, 1999). In reality, this approach does not define how to adjust the motion, particularly if a situation as depicted in figure 8 arises, where the camera is not able to detect).

7.2 Motion control

Various motion control schemes have been suggested. The application has been particularly for a backhoe operation in digging a trench (Seward, et al, 1988, 1992). This is based on the assumption that the machine works always with the maximum power and that the trajectory is a straight line (the bucket moves horizontally to cut a layer in the trench). The actuation power must provide the required forces to maintain the desired velocity. To achieve this, a velocity control scheme may be implemented, while the variation in the resisting forces is regarded as a disturbance to the motion. A fuzzy logic formulation has followed (Shi et al, 1995, 1996).

In a small scale, the excavation process is like when we take sugar with spoon from a pot. If at each instance the power behind the spoon is superior to the resistance faced by the spoon (and the spoon is usually strong enough not to break), the spoon will be filled. One exception is when a big chunk exists, for which we either put it inside or outside the spoon (that is the challenge for an automated system). A review of the reaction of a medium to a cutting force reveals that the medium-tool interface can be modelled as shown in figure 12. This figure shows the situation for a one dimensional tool. In a loader or excavator, this is the case for each actuator. F_A is the actuator force and F_R is the resistive force of the medium. The tool is represented by a mass-spring-damper, which transfers the active force and can deform based on the magnitude of the resistive force. The medium is also modelled as a mass-spring-damper, though the mass and damping are relatively much larger than those of the tool, whereas the stiffness is much smaller. Although this model suggests a linear behaviour, but the process itself is not linear in the sense that if the active force is higher than the resistive force, then there is a motion; otherwise, no motion takes place.

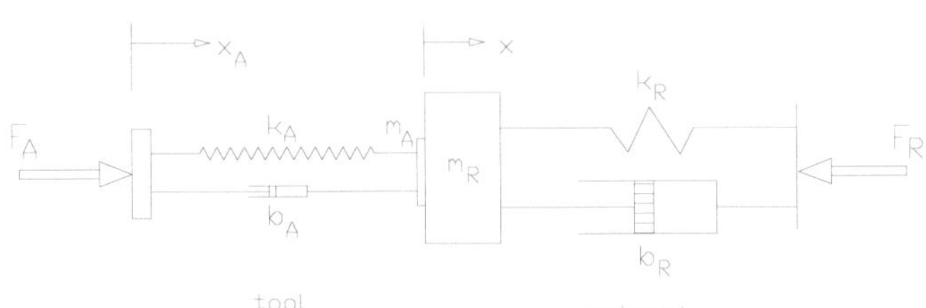

Fig. 12. modelling the interaction of too and medium

7.2 Control strategy

Figure 12 helps in comprehending what happens in a real situation in case of an excavating machine. In an excavator this can happen for each of the actuators, that, if the actuator effort is larger than the effect of the medium resistance, then there will be acceleration, or an already started motion will continue. Otherwise, the motion in that actuator stops and a deformation follows. It also helps to understand the validity of a control strategy, suggesting that a two-level control scheme is necessary (Bullock et al, 1990, Hemami, 1995). At one level motion control is performed based on a feedback from the measurement of actuator positions and velocities. This is a secondary level and the sampling of the variables must be performed at a higher frequency. A primary level control is performed with a lower frequency, which in fact is a corrective monitoring of the defined trajectory. This primary level control is based on a force feedback. That is, sampled at a lower rate the forces at the actuators are measured. If any of these forces is outside of its defined approximate range, then a correction to the trajectory is carried out and the previous trajectory is replaced by a revised one. The "defined approximate range" is based on all the information and the necessary calculations as discussed in section 4 and 6 of this chapter. Figure 13 helps to show this philosophy.

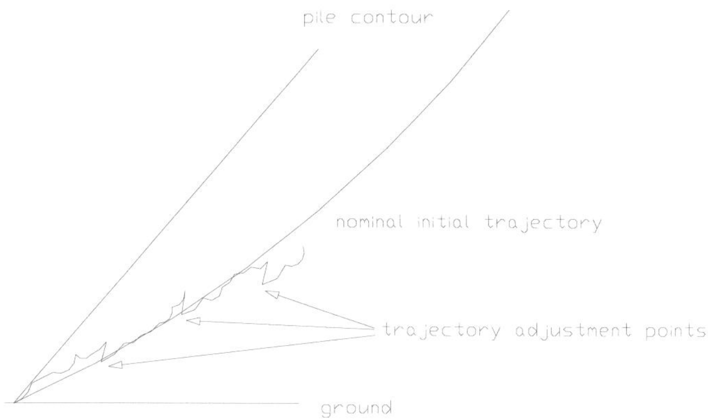

Fig. 13. Illustration of the effect of the conceptual control strategy

8. Concluding remarks

Although the advantages of auto-loading or automated excavation are obvious, and despite considerable work on the subject, a successful machine that can be readily purchased and put to work has not been reported. This implies the uncertainty about any method that has been tried at the industrial level, or that a method has not yet been tried. Alternatively this can be because of the priorities in industry and lack of interest, knowledge or both.

As far as the two-level control is concerned, an immediate conclusion at this point is the necessity of theoretical and experimental work towards investigation and formulation of the five different forces acting on the bucket at each point on the trajectory. This serves for the definition and approximation of a lower and higher range for force variation along the trajectory, in order to be employed for feedback reference values. This involves the studies for values and the variation of the five force components together with the point of action of each force. Out of these force elements the cutting force is the dominant member. Some of the works performed on this issue can be found in (Fabrichnyi, 1975, Balovnev, 1983,

Zelenine et al, 1985, Fielk & Riley, 1991, Hemami, 1994, Takahashi, et al, 1999). In fact plenty of research has been reported on the analysis of the forces in excavation and tillage. A review of all the work can be found in (Blouin et al, 2001).

Based on the conclusions in (Blouin et al, 2001), no uniform and well accepted definition, method or means to verify the results exists for formulation of the cutting force or other forces. A primary essential task is, then, a normalization and comparison of the previous work and comparison and validation of their results.

By the same token, an appropriate tool to verify the results of any control scheme for robotic excavation becomes desirable, in order to avoid the high cost of implementation of various control methods. As part of this, simulation software has been developed (Hemami & Hassani, 2007) for incorporating the behaviour of a medium during loading by a bucket. This software determines the resistive forces that work on a bucket when it moves through the medium when loading.

9. Summary

In this chapter the fundamentals of robotic excavation has been presented. This implies the modelling of an excavating machine as a robot and finding ways to control its motion, so that it can be programmed to automatically carry out an excavation task. The practice of modelling, kinematics, inverse kinematics, Jacobian relationships, etc. is the necessary work that must be repeated for each type of excavator. The present works in this chapter covers the matter for a front-end-loader or its underground counter part, a Load-Haul-Dump unit. The chapter points out the proposed methods for control of the process, although no final results are presented, since the matter can be still seen to be in its infancy stage. Analysis of the forces that a cutting tool or a bucket faces and must overcome during excavation has been done to some reasonable extent. But, this issue is by itself is very involved. An interested reader should consult the references given here for more detail.

10. References

Balovnev, V.I. (1983), *New Method for Calculating Resistance to Cutting of Soil*, Amerind Publishing Co., New Delhi.

Bernold, L. E. (1991), "Experimental Studies on Mechanics of Lunar Excavation", *Journal of Aerospace Engineering*, Vol. 4, No. 1 pp 9-23.

Blouin, S, Hemami, A and M. Lipsett (2001), Review of Resistive Force Models for Earthmoving Processes, *Journal of Aerospace Engineering*, Vol. 14, No. 3, pp 102-111.

Bullock, D. M., Apte, S. M. and Oppenheim, I. J. (1990), Force and Geometry Constraints in Robot Excavation, *Proc. Space 90 Conf.* Part 2, Albuquerque, pp 960-969.

Denavit, J., and Hartenberg, R.S. (1955), A Kinematic Notation for Lower-Pair Mechanism Based on Matrices, *J. Applied Mechanics*, pp 215-221.

Fabrichnyi, Y. F. (1975), Calculating the Resistance of Blasted Rock to Scooping by a Bucket,*Soviet Mining Review*, Vol 11, No. 4, pp 438-441.

Fielke J. M. and Riley, T. W. (1991), The Universal Earthmoving Equation Applied to Chisel Plough Wings, *Journal of Terramechanics*, Vol 28, No. 1, pp 11-19.

Hemami, A. (1992), Modelling Analysis and Preliminary Studies for Automatic Scooping/Loading in a Mechanical Loader, *Int. Journal of Surface Mining and Reclamation*, No 6, pp 151-159.

Hemami, A. (1993), A Study of the Bucket Motion Trajectory for Automatic Scooping in LHD Loaders, *Trans. of the Institution of Mining and Metallurgy*, Vol. 102, A1-70, pp A37-A42

Hemami, A. (1994), Study of Forces in the Scooping Operation of a Mechanical Loader, *Transactions of the CSME* (Canadian Society of Mechanical Engineers), Vol. 18, No. 3, pp 191-205.

Hemami, A. (1995), A Fundamental Analysis of Robotic Excavation, *J. of Aerospace Engineering*, Vol. 8, No. 4, pp 783-790.

Hemami A. , Seward D.W. and Quayle S. (1999), Some experimental Force Analysis for Automation of Excavation by a Backhoe., *Proc. 16th Int Conf. on Automation and Robotics in Construction*, Madrid, Spain, pp 503-508

Hemami, A and F. Hassani (2007), Simulation of the Resistance Forces Bulk Media to Bucket in a Loading Process, *Proc. 24th International Symposium on Automation and Robotics in Construction, ISARC 24*, India, pp 163-168

Lee, C.S.G. (1982), "Robot Arm Kinematics, Dynamics, and Control", *IEEE Computer*, Dec. 1982, pp. 62-80.

Michirev, P.A. (1986), Design of Automated Loading Buckets, *Soviet Mining Science*, Vol. 22, No. 4, pp 292-297.

Ostoja-Starzewski, M. and Skibinievski, M. (1989), A Master-Slave Manipulator for Excavation and Construction Tasks, *Robotics and Autonomous Systems*, Vol 4, No 4, pp 333-337.

Sarata, S., Osumi, H., Hirai, Y and Matsushima, G. (2003), Trajectory Arrangement of Bucket Motion of Wheel Loader, *Proc.* ISARC 2003, Eindhoven (Netherlands), pp 135-140.

Sarata, S., Weeramhaeng. Y. and T. Tsubouchi (2005), Planning of scooping position and approach path for loading operation by wheel loader, *Proc. 22nd International Symposium on Automation and Robotics in Construction ISARC 2005* - September 11-14, Ferrara (Italy)

Seward, D, D Bradley and R Bracewell (1988), The Development of research models for automatic excavation, *5th Int. Symp. on Robotics in Construction*, Tokyo, Vol. 2, pp 703 – 708.

Seward, D., Bradley, D., Mann, J. and Goodwin, M. (1992), Controlling an Intelligent Excavator for Automated Digging in Difficult Ground, *Proc. the 9th Int. Symp. on Aut. and Robotics in Construction*, Tokyo, pp 743-750.

Shi, X, F-Y Wang and, P.J. A Lever (1995), Task and Behaviour Formulation for Robotic Rock Excavation, *Proc. 10th IEEE Int. Symp on Intelligent Control*, pp 247-253.

Shi, X, Lever, P.J. A and F-Y Wang (1996), Experimental Robotic Excavation with Fuzzy Logic and Neural Networks, *Proc. IEEE Int. Conf. Robotics and Automation*, Vol.1, pp 957-962.

Sing, S. (1991), An Operation Space Approach to Robotic Excavation", *Proc. IEEE Int. Symp. Int Control.*, NY, PP 481-486.

Steel, J.P.H., King, R., and Strickland, W., (1991), Modelling and Sensor-Based control of an Autonomous Mining Machine, *Int. Symp. on Mine Mechanization and Automation*, Vol. 1: 6-55--6-67, Golden, Colorado.

Stentz. A. et al (1999), Position Measurements for Automated Mining Machinery, Robotics Institute (Carnegie Mellon University).

Takahashi, H., Hasegawa, M. and Nakano, E. (1999), Analysis on the resistive forces acting on the bucket of a Load-Haul-Dump machine and a wheel loader in the scooping task., *Advanced Robotics*, Vol 13, No 2, pp 97-114.

Vaehae, P. K., Skibniewski, M. J. and Koivo (1991), Kinematics and Trajectory Planning for Robotic Excavation, *Proc. Construction Congress 91*, Cambridge, MA, pp 787-793.

Wohlford, W. P., Griswold, F. D. and Bode, B. D. (1990), New Capability for Remote Controlled Excavation, *Proc. 38th Conf. on Remote Syst.*, Washington D.C., pp 228-232.

Zelenin, A. N., Balornev, V. I. and Kerov, I. P. (1985), *Machines for Moving the Earth*, Amerind Publishing Co., New Delhi.

18

Development of a Semi-Automated Cost-Effective Facade Cleaning System

Ernesto Gambao[1], Miguel Hernando[1] and Dragoljub Surdilovic[2]

[1]*Universidad Politécnica de Madrid,*

[2]*Fraunhofer-Institut für Produktionsanlagen und Konstruktionstechnik,*

[1]*Spain*

[2]*Germany*

1. Introduction

Nowadays the number of buildings with large glass or flat façades is increasing all over the World. These façades must be periodically cleaned with manual procedures that supposed high cost and risk for the workers that have to develop their work under heavy conditions. Although the cleaning cost depends a lot on several factors as the façade characteristics, the cleaning periodicity or the total surface to be cleaned, the average cost is € 8-9 per square meter. A typical building of 12.000 m2 supposes a total façade cleaning cost of € 100.000 and this task is usually done every year. The use of an automatic or semi-automatic cleaning system can lead to around 60% savings over existing practice (Gambao & Hernando, 2006).
Automation and robotics technologies allow environmentally friendly façade cleaning, helping to reduce the cost of these tasks. Additionally, these systems overcome the current worker safety problems associated with difficult and dangerous access, contributing to a zero injury and fatality working practices (Elkman et al., 2002).
Because of the increasing number of high-rise buildings and large glass façades and the resulting problem of safe and effective cleaning, a lot of effort has taken place in the last few years to develop automated cleaning systems. The majority of systems conceived and developed thus far are in Japan and Europe (Schraft et al., 2000) (Gambao & Balaguer, 2002).
The first automated cleaning systems for high-rise building were used in Japan in the middle of the 80's. These systems were mainly designed for use on specific buildings. For safety purposes or in order to guide the robot's movement on the façade, they often required additional construction such as guidance rails to the façade.
The practical application of the existing systems mostly failed because of either a weak safety concept, poor cleaning quality, required additional construction to the façade, or simply due to expensive initial or operating costs. At this time, there is only one known system that is in continuous practical operation. That is the automatic system for the cleaning of the vaulted glass hall of the Leipzig Trade Fair, Germany (Figure 1), which was developed by the Fraunhofer Institute IFF, Germany (Elkman et al., 1999). It must also be added that this system is only applicable to this particular building.
Many of previous developed robotic façade cleaning has been designed to operate in a complete automatic way (one example is in figure 2). Although some of these systems have

successfully solve the numerous technical problems related to façade climbing operations, in most of the cases they can not be practically used due to the extremely expensive operating cost of such a complex machines. Many remain as prototypes that are very good demonstrators of high technology but can not be introduced in the market.

Fig. 1. Automatic Facade Cleaning System for the Vaulted Glass Hall of the Leipzig Trade Fair (Fraunhofer FhG)

Fig. 2. SIRIUSC - Automatic Facade Cleaning System (Fraunhofer FhG, Dornier Technologie)

Table 1 shows the different known robotic façade cleaning systems.

Manufacturer	**Robot**	**Country**	**Application**	**Kinematics**	**Overcoming of obstacles**	**Facade type**
Taisei	Exterior Wall Painting Robot	Japan	Coating	rail guided	No	Vertical
Taisei	Tile Separation Detection Robot	Japan	Tile inspection	Tensed up with cables from roof to floor	No	Vertical
Kumagai Gumi Co. Ltd.	KFR-2	Japan	Coating	Cables, vacuum cups	No	Vertical
Shimizu Corporation	SB- Multi Coater	Japan	Coating	rail guided	No	Vertical
Kajima Corporation	Tile Separation Detection Robot	Japan	Tile inspection	Tensed up with cables from roof to floor	No	Vertical
Kumagai Gumi Co. Ltd.	Automatic Diagnosis System of Tiled Wall Surfaces	Japan	Tile inspection	Tensed up with cables from roof to floor, wheels	Yes	Vertical
Toshiba Cooperation	Vacuum Suction Self-Traveling Wall Washing Machine	Japan	Wall cleaning	Vacuum cups	No	Vertical
Obayashi Corporation	Wall Inspection Robot	Japan	Inspection	Vacuum cups, secured by cables	Yes	Vertical
Takenaka Komuten Co. Ltd.	SC 11-101	Japan	Tile inspection	Vacuum cups, secured by cables	No	Vertical
Tokyo Construction Co. Ltd.	Wall-Surface Operation Robot	Japan	Tile inspection	Vacuum cups, secured by cables	No	Vertical
Mitsubishi Electric Cooperation	Automatic Window Cleaning System	Japan	Fa¨ade cleaning	Rail guided	No	Vertical
Shimizu Corporation	Canadian Crab	Japan	Fa¨ade cleaning	Vacuum cups, secured by cables	Yes	Inclined
Fraunhofer-Institut IFF	Cleaning robot for the Glasshall Leipzig Trade fair	Germany	Fa¨ade cleaning	wheels, secured by cables	No	Convex
Comatec	-	France	Fa¨ade cleaning	Vacuum cups	No	Inclined
Robosoft	-	France	Fa¨ade cleaning	Rail guided	No	Horizontal
Robosoft	Autonomous Window Cleaner Robot for High Buildings (EC: AUTOWIND)	France	Fa¨ade cleaning	Rail guided	No	Vertical
Fraunhofer-Institut IFF, Dornier Technologie	SIRIUSc	Germany	Fa¨ade cleaning	Rail guided	Yes	Vertical
Newcastle University; OCS Group; Cradle Runways	Arcow	UK	Fa¨ade cleaning	Rail guided	No	Vertical
CSIC	Tito	Spain	Fa¨ade cleaning	Air suction	No	Vertical

Table 1. Façade cleaning robots

In the frame of an European founded project, a consortium formed by several enterprises and research centres has develop a low cost semi-automated system for the cleaning of building façades, addressing an innovative concept of system that is able to work in different types of homogeneous building façades, increasing the productivity, reducing the risk for workers nearly to zero and contributing to preserve the environment. This system is with minor changes adaptable to the largest possible number of buildings with homogeneously-designed façades. Additional constructions to the façade such as guide rails or scaffoldings are avoided or made unnecessary. The requirements for the control and sensor concepts are very specific, because the proposed robotic system is able to operate under adverse conditions such as changing weather conditions.

In this chapter, we present the description of the robotic façade cleaning system (denominated *CAFE*) and, after that, the selected control architecture and the implementation of this concept in the real system.

2. Concept of the *CAFE* robotic cleaning system

All the high buildings use commercial carrier systems that support a gondola that moves on the façade for manual cleaning. One or two operators are needed for this task. Based in the existence of the carrier system on the building roof, the *CAFE* robotic system uses it to reduce the costs of the vertical and horizontal movements. The system uses a commercial carrier with minor modifications for movements in axes X and Y (Figure 3).

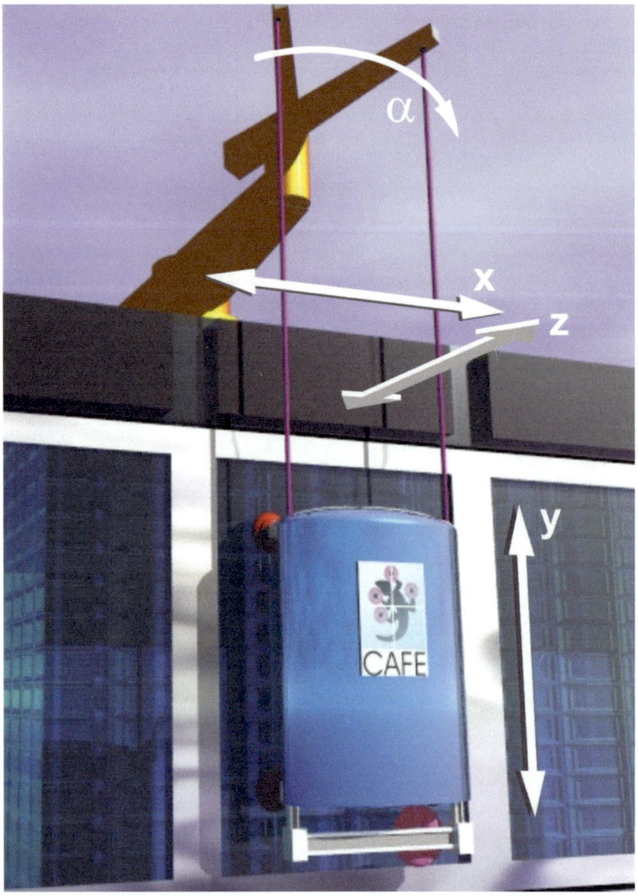

Fig. 3. *CAFE* Façade Cleaning System concept.

As we have mentioned, completely autonomous systems result too expensive for the market and for this reason the proposed system has been designed to perform the cleaning task in a semi-automatic way. This means that many of the tasks are performed in a completely autonomous way; however, because of security and economic considerations, a human operator permanently controls the robot operation.

A single person, physically situated on the ground below the robot, operates the complete semi-automatic cleaning system. However, most of the task can be performed in a completely automatic way. The operator has to install the machine at put it in work giving periodical attendance when necessary (filling deposits, changing task, etc.). To achieve this, it is necessary to program the robot adapting it to the building's façade. This task is

necessary only one time, previous to the work and it is not be very time consuming. Due to the low cost of the system, buildings with large façades can have dedicated machines.
The robot cleaning system has been decomposed in four different modules (Figure 4):

- Cleaning Module (CLM)
- Kinematics Module (KM)
- Carrier Module (CaM)
- Control Module

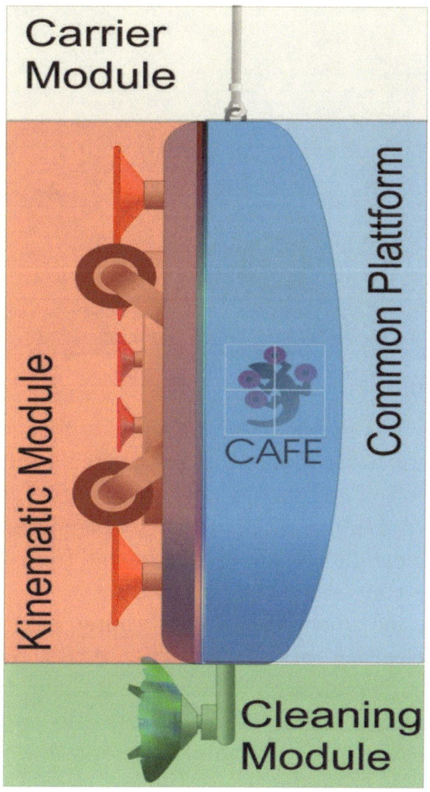

Fig. 4. Arrangement and interconnections of the *CAFE* hardware modules

The Cleaning Module is in charge of the actual façade cleaning. It mainly consists in a cleaning mechanism and a positioning system. The most important features of the cleaning module include:

- Cleaning with brushes and water (environmentally-friendly)
- Water recycling system (low water use)
- All actuators pneumatic (compliant motion, simple control structures, robust)
- Passive degrees of freedom in kinematics to account for unevenness in façade surface and to protect against hard collision with framework when moving up and down the façade (braking distance)
- Sensors for detecting glass framework and overseeing the condition of the cleaning module

The cleaning system is able to clean up to between 3-10mm away from a window pane. The cleaning Module is shown in Figure 7.

Fig. 5. *CAFE* Cleaning Module

The carrier is the part of the façade cleaning system that safely holds and provides horizontal, vertical and transversal motion to the kinematics and cleaning modules. It is installed on the building rooftop and moves over rails or on a concrete path (guided along the parapet), holding and providing motion to the cleaning and kinematics modules by means of cables. While the cleaning robot might be moved from one building to another, the carrier system will generally stay on the building rooftop.

The carrier must position the kinematics and cleaning modules on the façade at the beginning and between cleaning operations. The carrier positions the cleaning and kinematics modules in the x-axis through its movement along the rooftop. The winding or unwinding of the cables transmits the vertical motion and positioning in the y-axis. The adjustment of the distance to the wall z is obtained by controlling the α angle (see Figure 3).

The carrier must also be able to bring the kinematics and cleaning modules down to the floor or hoist and deposit them on the rooftop in order to perform maintenance operations, refill cleaning water or even lay those on a vehicle on ground to be transported somewhere else.

The Kinematics Module establishes contact between the cleaning head and the window pane. This contact is necessary for generating a reaction force of the cleaning head against the window pane. The system controller is in charge of control the presence or absence of the contact, accordingly to nominal and non-nominal situations.

In nominal situations the contact must be established during the entire cleaning task and hoisting operation. The break of contact can induce serious problems like bumps towards the facade caused by oscillations of the carrier. In case of this non-nominal situation a safety module must be activated in order to avoid oscillations.

3. *CAFE* robot control system

The term Control Module refers to the general architecture of the control systems of al the modules, and encompasses the concept for controlling each individual system. The cleaning task has been decomposed into different actions that must be performed simultaneously by the different robot modules. The control module is in charge of the synchronization of all this tasks. The control scheme has been implemented using a hardware decentralized and software centralized control architecture. This architecture is considered more appropriate for the control system than a decentralized one (Figure 6).

3.1 Control architecture. Components distribution and communications scheme

The control system is distributed in three parts. The main controller (Control Module) is located in the Common Platform. The Carrier Module controller is located attached to the carrier on the top of the building. Finally, the operator, located on the ground, uses an interface device (PC or PDA). So, all the three parts include their own microprocessor-based computer. The main controller and the carrier controller are based on an embedded PC equipped with TwinCat-PLC core and Windows CE, allowing the combination of Windows based programming and PLC programming (IEC 61131-3) reliability. This configuration reduces the total cost of the system and simplifies the integration.

A wireless connection (Ethernet WIFI 802.11b) is used for the connection between the Control Module and the operator interface, and between the Control Module and the carrier. The safety of this communication is critical and it has been guaranteed by a watchdog system. In case of failure of the wireless communication, all the system adopts a safety position and can be recovered manually from the Carrier Module Control. The communication scheme is also shown in Figure 6.

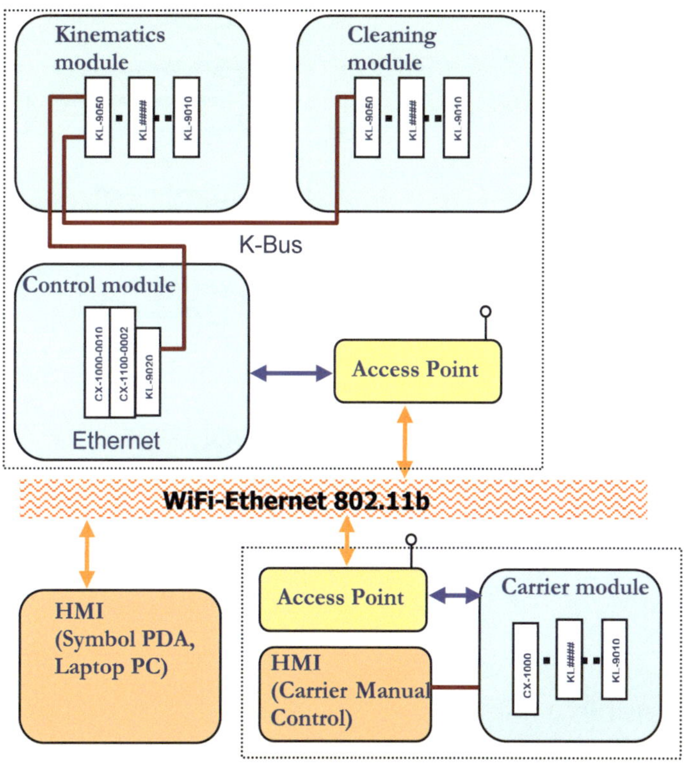

Fig. 6. Control System Architecture

The cleaning task has been decomposed into different actions that must be performed simultaneously by the different robot modules. The control module is in charge of the synchronization of all this tasks.

3.2 Software components

Although from the hardware point of view the robotic systems has three different microprocessor-based parts, there are five agents working in parallel, corresponding to the modules described in Figure 4 plus the operator interface:

- Control Module
- Kinematics Module
- Cleaning Module
- Carrier Module
- Operator Interface (HMI)

Additionally, each physical element requires a specific process in charge of establish the communications between the different elements. The communication virtual bus generation process is located in the main controller.
The distributed software architecture is shown in Figure 7.

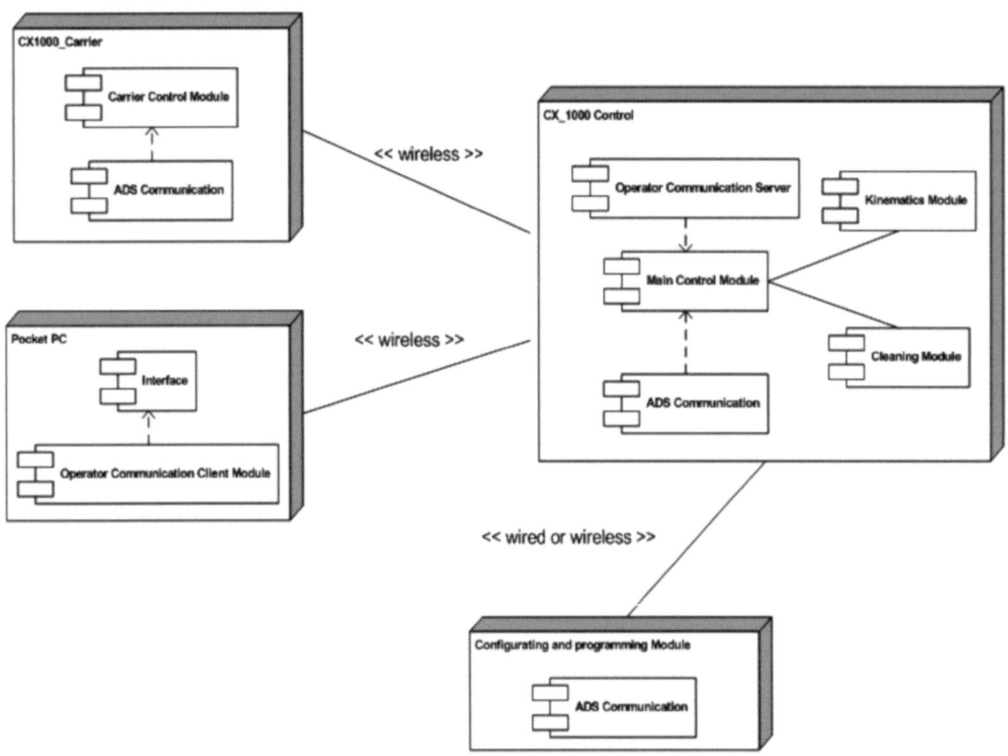

Fig. 7. Control System SW architecture and communications

All these elements are really constituted by different PLC programs executed in parallel and in an asynchronous way. To allow an adequate integration between them we have followed

a common methodology where the main controller demands services to all the other elements. There are specific processes for error managing, allowing error recovering.
The main mission of the Carrier Module is to control the absolute movement of the robot platform. This task must be accomplished synchronised with the Kinematics Module movements. Thanks to the mechanical design of the Kinematics Module, adaptive movements are allowed when the robot is fixed to the building façade. A low level control is performed in the Carrier Module. This low level control is in charge of the coordinated control of the carrier actuators, including de axis decoupling and the cancellation of the oscillations of the pendular robot movement.
The main controller, located in the Control Module, integrates the general state diagram of the complete system and synchronises all the elements. Additionally, it is in charge of the communications checking. At each cycle of the PLC program, the wireless connection is checked. If a lack in the communications is detected, an emergency process is started in both the Control Module and the Carrier Module. In the Control Module the emergency process commands the Kinematics and Cleaning Modules to adopt a safe configuration. In the same way of working, the controller of the carrier will stop any movement of the carrier and enables the manual control of the system.
For the operator interface a pocket-PC (windows CE) system was selected. This system allows intuitive and easy interface via wireless connection. The exchange of information between the Control Module and the Operator Module is based on a wireless WIFI-802.11b compliant connection. The communication can be encrypted under WEP protocol and the access point allows configuring specific IP directions to connect.
The Server-side is included in the Control Module whereas the Client-side is in the Operator Interface. Both the Client and the Server are programmed with C# and compiled for the Microsoft .NET platform (so .NET Compact Framework is required).
The communication is established after accepting the Server a request from the Client. No hand-shake protocol is implemented. The Server is able to detect both when the connection is fortuitously cut and when it has not been recently used and reinitiates its state to a new connection. The Client will receive the data and will only send Operator orders when produced.

3.3 Human-machine interface operation

There are two possible modes: manual and automatic. Additionally, there are three other modes: disconnection, emergency stop and error, that depend on the system status and where the normal cleaning operation is not possible.
In the automatic mode the cleaning task is performed with no need of further information after the system has been initialized. In the manual mode the operator must indicate the action to perform that can be accepted of not by the robotic systems depending on the command availability. The operator can select automatic or manual mode, but the mode does not effectively change until the main controller confirms it.
When the communication between the interface and the main controller is not properly established, the disconnection mode is set. In this mode the system is located in a safe position until the communication is re-established.
When the robot is not able to operate in the normal modes (manual or automatic) it is immediately set to the error mode and must be recovered manually.

The operator, using the interface, can activate the emergency stop and the robot is stopped in the next safe position. There are additional emergency buttons at the carrier.
The graphical user interface always shows the emergency stop option to the operator, as well as the battery status of the PDA device, the manual/automatic mode change and the WIFI connection status. In the automatic mode it shows the status of the current performed task, while in the manual mode allows the operator to select possible actions or to consult different variables using a simple colour code. Figure 8 shows different situations of the graphical user interface.

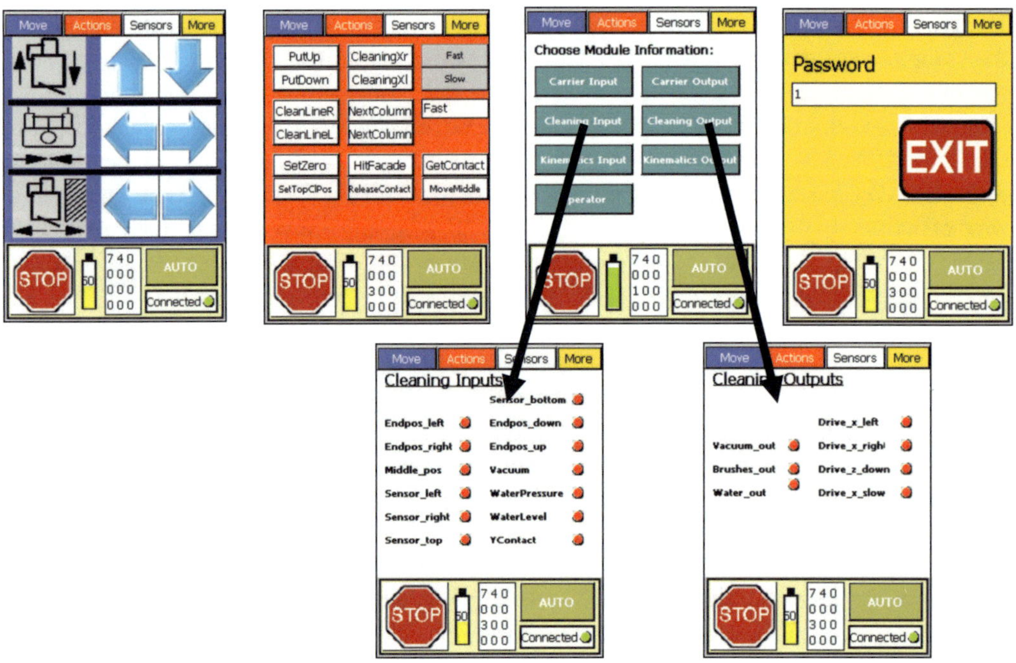

Fig. 8. Graphical User Interface

4. Results

After the development of the prototypes of the different modules, the complete cleaning system was merged. Some systems were refined and several parts of the control software were modified. The performance tests were successfully accomplished in automatic way.
From the test operation the following was concluded:

- The overall cost of the system can be under 50 T€ on sale
- The operating costs are under 3 T€ per annum
- The cleaning speed in total is above 200 m2 per hour
- The system is usable at facade areas of under 7000 m2
- The cost saving is of up to 5 € / m2
- The roof car costs (depending from comfort) is around 20 T€ on sale
- The robotic system is able to serve more building of the owners
- The interface set cost for changeable operation on existing BMU s is under 15 T€

Figure 9 shows a real image of the CAFE prototype. After the project end a new company has been created by several partners to commercialize the machine.

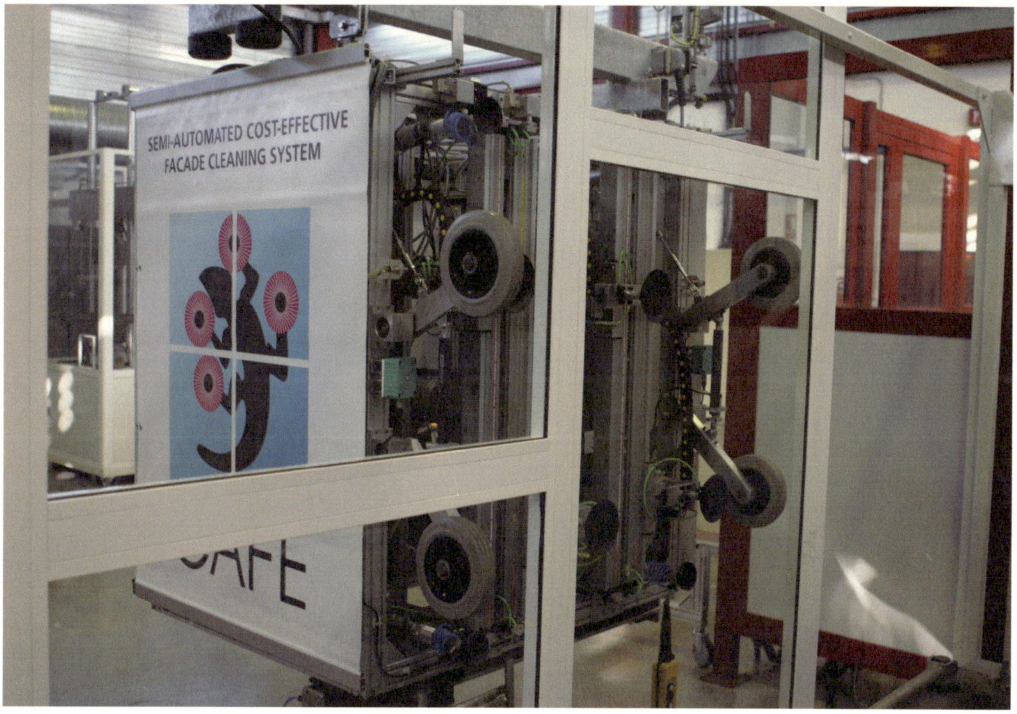

Fig. 9. CAFE Robotic façade cleaning system prototype

5. Acknowledgements

The authors wish to tanks the contribution of all the partners of the project and the support of the European Commission under the project CRAFT-1999-71236 CAFE.

8. References

Elkmann N., Felsch. T., Sack M., Böhme T. (1999). Modular climbing robot for outdoor operations, *Proceedings of CLAWAR 1999, Second International Conference on Climbing and Walking Robots*, Page 413-419, Portsmouth, U.K.

Elkman, N., Felsch, T., Sack, M., Saez, J. and Horting, J. (2002). Innovative Service Robot Systems for Façade Cleaning of Difficult-to-Access Areas, *Proceedings of the 2002 IEEE/RSJ Intl. Conference on Intelligent Robots and Systems*, Lausanne, Switzerland.

Gambao E., Hernando M., Hernández F. and Pinilla, F. (2004). Cost-Effective Robots for Façade Cleaning, *Proceedings of the 2004 Inernational Symposium of Automation and Robotics in Construction*. Jeju, Korea.

Gambao E. and Balaguer C. (2002). Robotics and Automation in Construction, *IEEE Robotics and Automation Magazine*. Vol. 9. No 1. (March 2002), ISSN 1070-9932 .

Gambao E. and Hernando M. (2006). Control System for a Semi-automatic Façade Cleaning Robot , *Proceedings of the 2006 Inernational Symposium of Automation and Robotics in Construction*. Tokyo, Japan.

Schraft, R. D., Bräuning, U., Orlowski, T. and Hornemann, M. (2000). Automated Cleaning of Windows on Standard Façades, *Automation in Construction* Vol. 9, Issues 5-6, (September 2000) 489-501, Elsevier, ISSN: 0926-5805.

19

Design and Feasibility Verification of a Knee Assistive Exoskeleton System for Construction Workers

SeungNam Yu, SeungHoon Lee, HeeDon Lee and ChangSoo Han
Hanyang University
South Korea

1. Introduction

Robotic-powered exoskeletons and body joint-adapted assistive units are currently under development for the enhancement of human locomotor performance in the military, in industries, and in patients and the elderly with mobility impairments [1]. They free people from much labor and the burdens of many kinds of manual work. For example, when it comes to automation in the industrial field, factory automation has made good progress. Operators (humans) can be included in a conventional manufacturing process with respect to a formal production line and uniform working conditions. Automation outside the production line, however, especially in common manufacturing stages, has several limitations and difficulties in adapting to actual conditions because the industrial field has but a small part in the process due to its operating characteristics. There have been many approaches to the reduction of labor that do not only fully assist but also partly aid workers, such as in the use of extremely heavy payload-oriented construction equipment, which are manipulated by humans. Manual or semi-automatic machine tools are mostly used in contemporary industries. In particular, without manpower, especially without the manipulability and mobility of the upper and lower human limbs, full automation will be incompatible with today's technologies [2]. Exoskeletons have strong advantages given their unique features such as their outstanding physical performance, exceeding that of humans, and their agility, which is utilized by operators' nerve systems. As a result, attempts to adopt exoskeletons in the industrial field, especially at construction sites, indicate the use of feasible approaches to factory automation. The strategy and support method for exoskeletons that amplify human muscle power can be divided into four main categories: (1) exoskeletons that totally alternate with both the upper and lower parts of the muscle power system, (2) assist the all extremities not alternate (here, assist means the human share the load with the exoskeleton and alternate means the human just input operation command using his own motion into exoskeleton system and it totally handles the load), (3) alternate with the part of all extremities (4) assist the part of all extremities of muscle power system. The first type of exoskeleton which alternates with the entire muscle power still has many limitations, as with its size and electric power supply. Due to these constraints, exoskeletons are usually bulky and cannot freely move out of the range of the power source line. One of the representative studies of the second type which assists the whole body is the HAL series.

HAL utilizes the EMG signal for its command signal [3]. Moreover it shares external loads with humans, that is partly assists the human's loads but it is still requires much patience to wear and difficult to maintain the quality of EMG signal for every wearing.

(a) The Exoskeleton Design Concept Introduced in the Movie Clip ('Alien II' and 'Matrix Revolution')

(b) Exoskeleton System for Whole Body Support ('XOS' of SARCOS and 'HAL' of Cyberdyne Inc.)

Fig.1. Developed Exoskeleton Systems to Support the Whole Body

As taken into account in the earlier three cases, the research target for the development of exoskeletons can fall under the fourth type: the partly assistive muscle power system, especially the leg assistive system. Many institutions around the world have carried out research and development on exoskeletons and assistive devices in order to empower or aid human lower limbs. A well-known system, BLEEX, can partly alternate with the human muscle power system. This system provides a versatile load transport platform for mission-critical equipment, so it has several applications without the strain associated with demanding labor such as that of soldiers, disaster relief workers, fire-fighters, and so on [4]. Northeastern University's Active Knee Rehabilitation Device (AKROD), Yobotics Incorporation's RoboKnee, and the NTU-LEE rehabilitation prototype are some of the state-of-the-art developments in the area of assistive devices to aid the human limb [5, 6, 7].

(a) BLEEX (U.C. Berkeley)

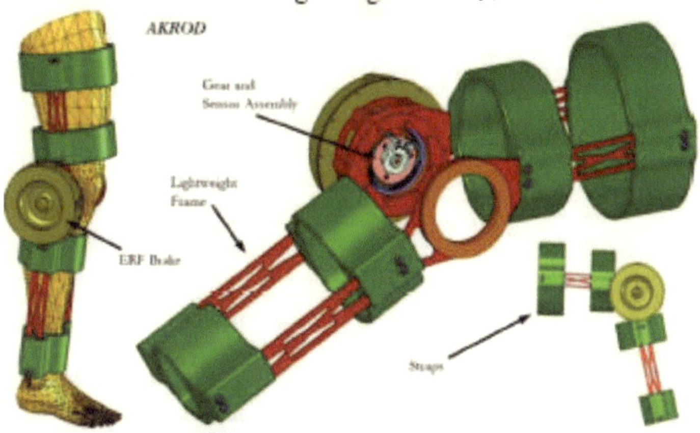

(b) Active Knee Rehabilitation Orthotic Device-'AKROD' (Northeastern Univ.)

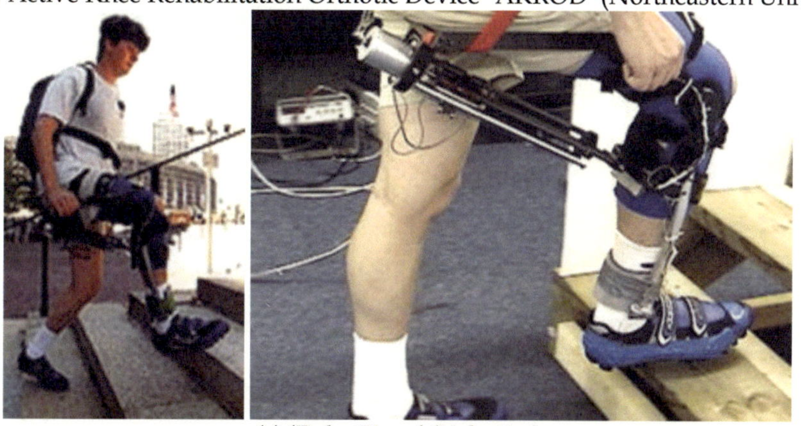

(c) 'RobotKnee' (Yobotics)

Fig. 2. Leg Assistive Exoskeletons

In addition to the systems in Fig.2, many kinds of knee assistive robots are focused on medical service or rehabilitation. The purpose of this device is to share the load or pressure acting on the knee in order to relieve pain or speed up the healing process without disrupting normal daily activities. This is likely to be a potentially useful research area due to the rising number of sports-related injuries and the increasingly aging world population [8]. Obviously, this concept can be applied to assist in daily life walking and laborious work in the industrial area. For the purpose of industrial usages, however, operational convenience and compactness of the system is strongly considered. This means that the system has to be designed as wieldy and can easily be synchronized with a human.
To solve this problem, innovative sensor suits have been developed, which can be put on by an operator to detect his or her motion intention by monitoring his or her muscle conditions such as shape, stiffness, and density. As shown in Fig.3, these sensors are made of soft and elastic fabricsembedded with arrays of MEMS sensors such as muscle stiffness sensor (MSS), ultrasonic sensors, accelerometers, and optical fiber sensors to measure different kinds of human muscle conditions [9]. The developers of these sensor systems emphasized its

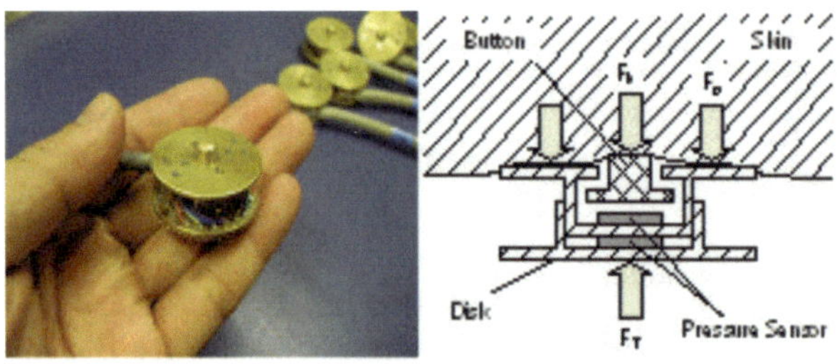

(a) Muscle stiffness sensor (Takakazu Ishimatsu)

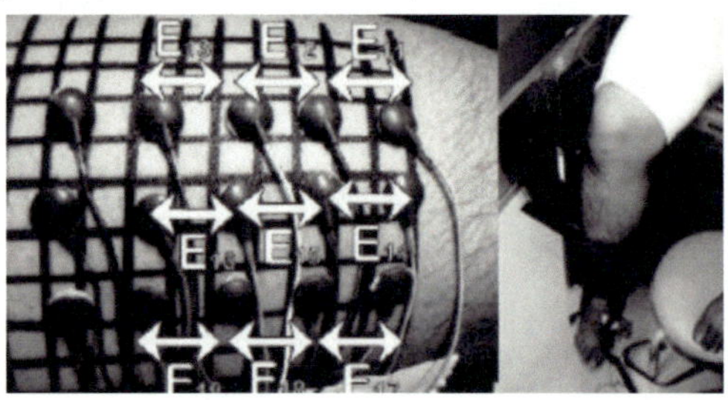

(b) Auto-calibration system for EMG sensor suit (Maria Q. Feng)

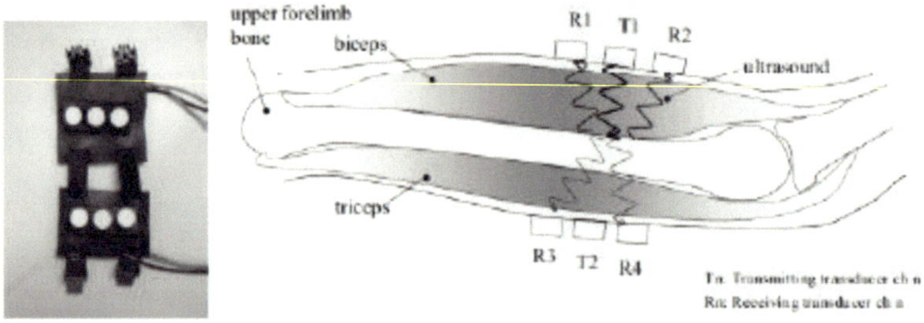

(c) Ultrasonic muscle activity sensor (S. Moromugi)

Fig. 3. Various Sensor Systems for Human Motion Detection

convenience and ease to adapt to humans. These sensors, however, are too complicated to manufacture or are only verified to perform on a certain part of the human body. The EMG sensor is one of the most accurate measurement tools to determine human motion intensity. The approach using this sensor, however, is not considered in this study because of its inconvenient preparation to assess the signals and its inappropriateness for the working conditions at a construction site.

In this study, a feasible modular-type exoskeleton system and corresponding sensor systems are newly proposed to assist construction workers with their lower limb movements. First, for the purpose of adapting the modular-type exoskeleton system for lower limb assistance at construction sites, several construction work groups were defined based on specific boundaries. Second, the design process for the modular-type lower extremity focused on the knee joint movement will be presented based on the confined boundary. Third, intent signal processing methods for actuating a proposed system were introduced, and the feasibility of the command signal was estimated. There were then several measures to quantify the characteristics of human performance and the exoskeleton platform through an EMG signal (This sensor is used as a measurement tool of muscle activity only to verify the feasibility of the proposed system).

2. Analysis for designing the system

2.1 Occupational analysis

In the next step, the research target was brought into the part it would assist. For the sake of embodiment, we first defined the target task at a usual construction site through a work pattern analysis, which is strongly related to occupational disorders. Arndt et al. (2008) conducted a 10-year follow-up research on 14,474 male construction workers. He reported that musculoskeletal diseases led to an increased proportion of occupational disability [10].

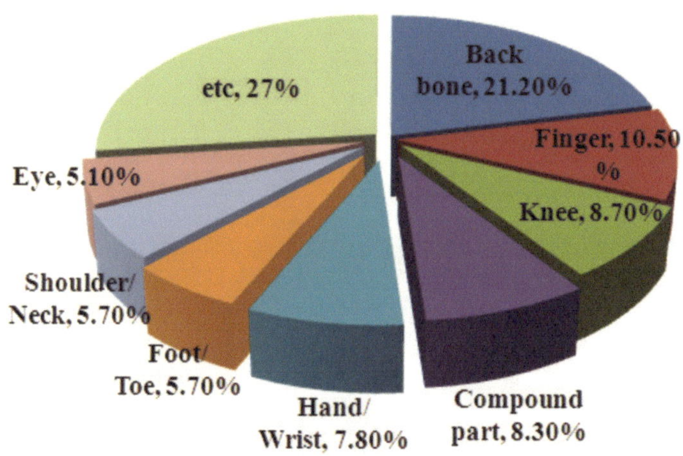

Fig. 4. Construction Workers' Disability Ratio (NIOSH)

The fatal injuries of construction workers-musculoskeletal diseases-were mainly divided into two dominant disabilities: dorsopathies and arthropathies. According to statistical and the annual reports of the National Institute of Occupational Safety and Health (NIOSH), it is

easy to have primary disability at a construction site. The reports classified the standard incidents into all causes and specific disabilities. Dorsopathies, arthropathies, and knee joint disorders accounted for 21.2%, 10.5%, and 8.7% respectively, and occurred most frequently at the site (Fig.4). Considering the priority of development and convenience of real adaptation, we choose the knee assistive wearable system firstly not spine assist one. Moreover, the working index of NIOSH recommends that construction workers' spinal columns should not be rapidly bent and their posture should be kept perpendicular to the ground during manual construction work. That means a spine support system has to be considered as support system not assist one. Therefore, this paper designed this specific part of the body-knee joint of the type that partly assists the knee joint (Fig.5).
The following are the specifications of the system in this research:

- Occupational target: Construction worker
- Target region: Knees (The weight of the system is borne by the combined shank-ankle orthotics)
- Target motions: Kneeling, lifting objects, and climbing a staircase or a slope

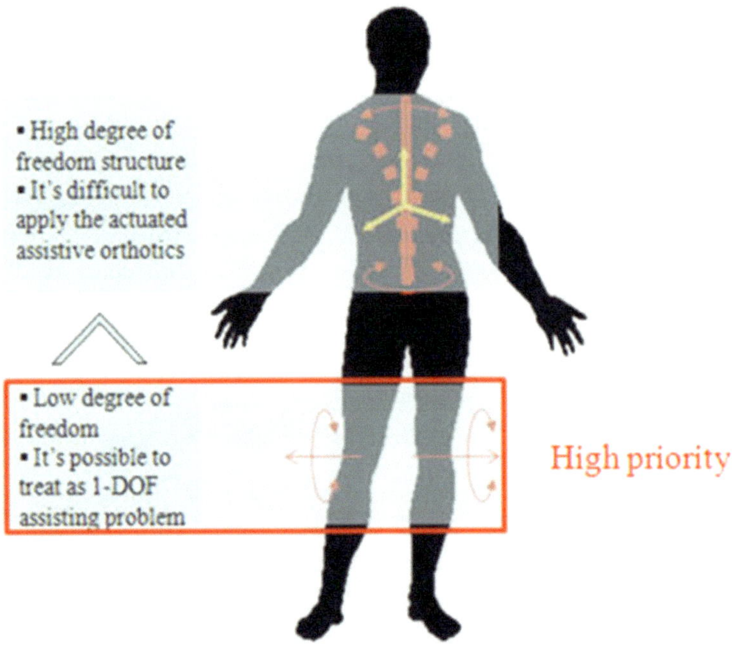

Fig. 5. Decision of Assistant Position Considering Two Dominant Causes of Disability of Construction Workers

2.2 Definition of the target task

To specify the target tasks at a construction, we follow these process steps. First, we looked at an overview of working patterns and types at construction sites. The overview was sourced from NIOSH. In the second step, construction workers-especially the general laborers-were classified into four major groups. As shown in step 2 under Fig.6, sheet metal workers, electricians, laborers, and cement masons were put in charge of each group. Finally, in the third step, based on the occupational common task of upper groups, target tasks were selected which included heavy material handling using knee, loaded level walking, loaded ascent walking, and loaded descent walking.

As earlier mentioned, we developed a modular-type exoskeleton system to assist the lower limb, and we applied this mechanism in a real construction site. Thus, the target mission to handle heavy materials and loads at ground level and on a stairway, which is described in the following images, is critically considered.

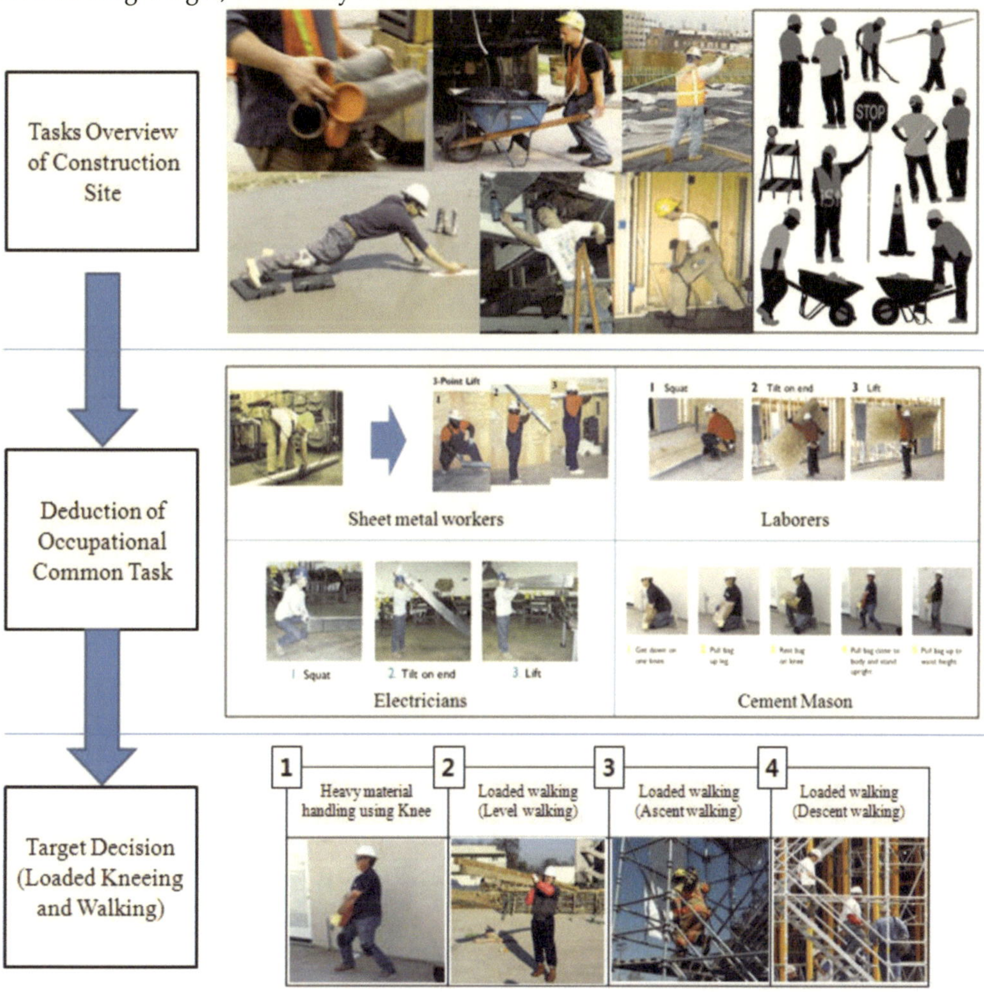

Fig. 6. Work group analysis for construction workers

3. Mechanics of muscle activity at the knee

3.1 Extensors of the knee

Rectus femoris functions as an extensor of knee extension, hip flexion, lateral rotation of the hip, and abduction of the hip [12,13]. Regarding the effect of its weaknesses, direct measurements of the contribution made by the rectus femoris to knee extension strength are not available. However, the physiological cross-sectional area of the rectus femoris is approximately 15% of the total quadriceps femoris muscle mass. Therefore, its negative effect on a knee is as much as this contribution [14]. Vastus intermedius functions as an extensor of knee extension and prevents impingement of the pouch in the patellofemoral joint. It is based on the physiological cross-sectional area range from approximately 15~40% of the total muscle bulk [14]. Vastus lateralis is a large pinnate muscle, and its uncontested action is knee extension. The amount of its recruitment is proportional to the amount of resistance to extension [15]. If its activity is diminished, knee extension strength is reduced. Its physiological cross-sectional area suggests that in some individuals, the vastus lateralis may contribute 40% of the extension strength of the knee [16]. Vastus medialis is the most studied among the four heads of the quadriceps femoris muscle [15]. It is divided into two sections, VML (Longus) and VMO (Oblique), based on both anatomical and mechanical analysis. It is approximately 20 to 35% of the overall cross-sectional area of quadriceps femoris. It functions as an extensor of knee extension and for the stabilization of the patella during knee extension [16,17,18].

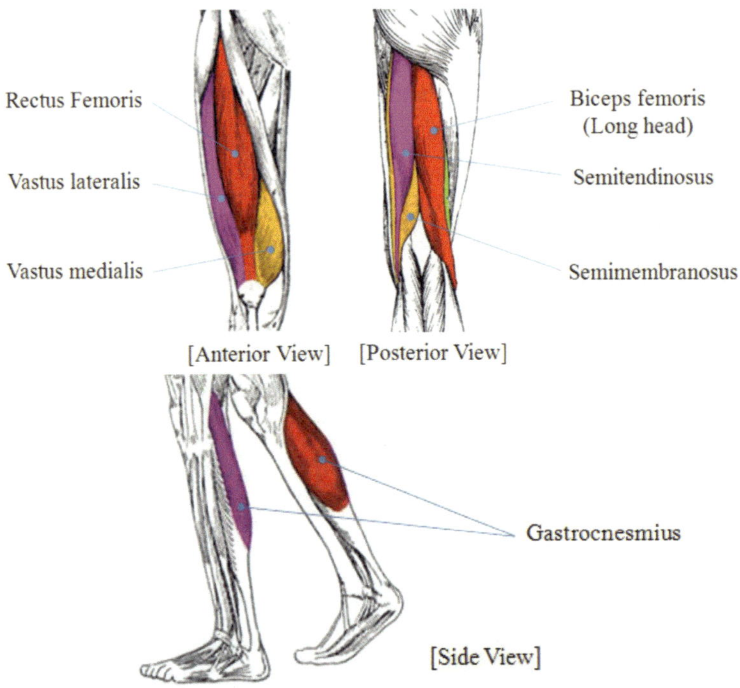

Fig. 7. Primary Knee Extensors, Flexors, and Plantar Flexor Muscles Focused in this Study

3.2 Flexors of the knee

The hamstring muscles represent the primary flexors of the knee. Hamstrings comprise of the biceps femoris longus and brevis, which form the lateral mass of the hamstrings, and the semimembranosus and semitendinosus, which make up the medial mass. The major functions of the hamstring are knee flexion, hip extension, medial rotation, lateral rotation of the knee, medial rotation of the hip, lateral rotation of the hip, and adduction of the hip. Hamstrings provide between 30 and 50% of hip extension strength and are active during normal locomotion. The most prominent period of activity is during the transition between the swing and stance periods of the gait cycle. During locomotion, the role of hamstrings' activity is to slow down the extension of the knee during late swing, and to help extend the hip in the stance phase.

3.3 Mechanics of the two-joint muscles in the knee

The knee is controlled mostly by two-joint muscles that cross either the hip and knee, or the knee and ankle. Contraction of one of these muscles alone produces movement in all of the joints that the muscle crosses. To isolate movement at a single joint, the two-joint muscles cross or they must contract with other muscles. The iliopsoas and the hamstrings, as an example, together produce isolated knee flexion by canceling each other's effect at the hip. Similarly, simultaneous contraction of the gluteus maximus and quadriceps femoris produces knee extension without hip flexion. However, the knee more frequently displays simultaneous contraction of the quadriceps and hamstrings. This unusual pattern of simultaneous contraction of two-joint muscles appears to increase the ability of the knee and hip to generate the large moments needed during many activities [14].

4. System operation method – trial (1)

4.1 Angular displacement of the knee joint

Following the steps shown in the previous chapter, the final target task was defined more specifically. We decided to devise a modular-type exoskeleton system for lower limb assistance, that is, for handling heavy materials during level walks and on stairways. To gather adequate motivation signals when the construction workers do their jobs at the site, first, an analysis of knee joint movements was needed. Fundamentally, the muscle activation status is completely different during level walks and on stairways. Figure.8 and Figure.9 show which parts of the muscle groups are mainly related to knee joint movement during level walks. Thus, a different type of gait pattern is created for a dissimilar muscle activation phase. In the case of the knee joint movement, three DOFs with angular rotations are possible during the level walk.

The primary motion is knee flexion-extension with respect to a mediolateral axis. Knee internal-external rotation and adduction-abduction (varus-valgus) also occur among healthy individuals, but with less consistency and amplitude due to their soft tissue and bony constraints to these motions. The information presented in this chapter was gathered from the work of Spivak and Zuckerman (1998). The following table shows the range of normal values of normal adult gaits at a free walking velocity. These values were used as reference values while we performed the experiments.

Contents	*Values*
Stride or cycle time	1.0 to 1.2 m/sec
Stride or cycle length	1.2 to 1.9 m
Step length	0.56 to 1.1 m
Step width	7.7 to 9.6 cm
Cadence	90 to 140 steps/min
Velocity	0.9 to 1.8 m/sec

Table 1. Range of Normal Values for the Time-Distance Parameters of Adult Gaits at a Free Walking Velocity (Spivak and Zuckerman)

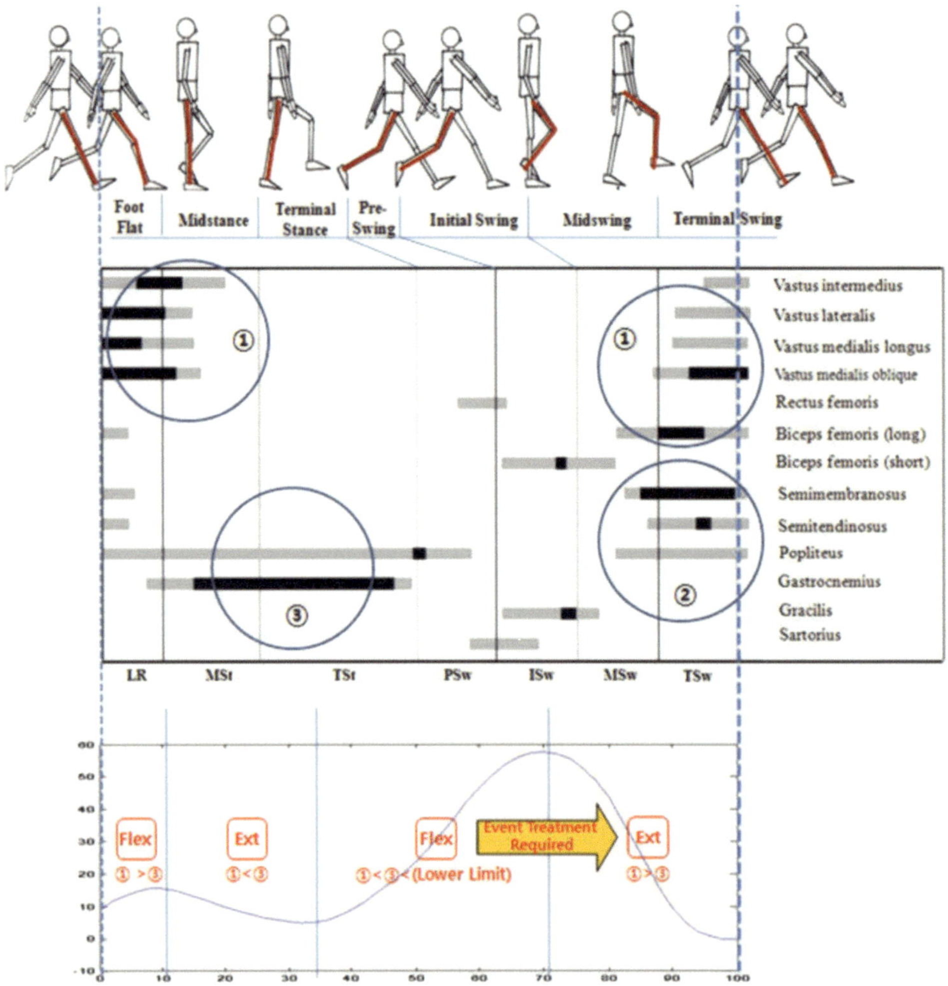

Fig. 8. Phasic Pattern of the EMG Activity of the Muscle and the Angular Displacement of the Knee during Level Walking by Healthy Adults

4.2 Extraction of the muscle activity pattern

During the stance phase, the quadriceps muscle group is relied on to control its tendency towards knee flexion collapse with weight acceptance and single limb support. This muscle group is activated during terminal swinging and then acts eccentrically during weight acceptance, as the knee rotates from the fully extended position during the initial contact to its peak support phase flexion of approximately 20 degrees during the loading response. Thereafter, the quadriceps act concentrically to extend the knee through an early mid-stance, as the body's center of extremity mass is raised vertically over the supporting limb and the anterior orientation of the ground reaction force vector precludes the need for further muscular control of knee flexion. Most hamstring muscles are activated in the late mid-swing or the terminal swing. Their function with respect to the knee is probably to control the angular acceleration of the knee extension. The short head of the biceps femoris is activated earlier and probably assists in flexing the knee for foot clearance.

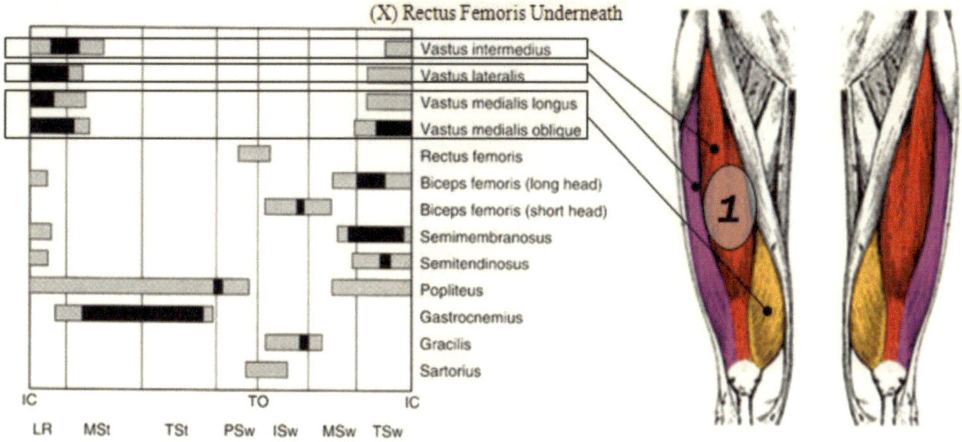

(a) Muscle Activity Pattern of Anterior Side of the Leg during Walking and Proposed Sensor Position '1'

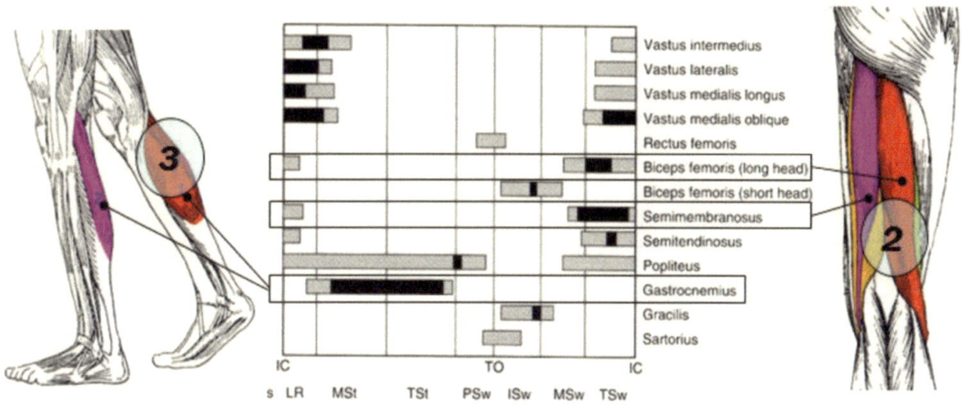

(b) Muscle Activity Pattern of Posterior Side of the Leg during Walking and Proposed Sensor Position '2' and '3' (Position '2' is discarded finally)

Fig. 9. Muscle Activity Pattern of Leg and Proposed Sensor Position for Exoskeleton

The gracilis and sartorius muscles may also contribute to swing-phase knee flexion when they are activated during late pre-swing, initial swing, and early mid-swing. These muscles, however, may very well be acting as primary hip flexors during this period [19]. Based on Fig.8, we analogize that to explain or measure the gait pattern using the muscle activity pattern, we must consider three positions of the muscle groups.

In this study, however, we propose a method that uses only two muscle sensing groups. Although this approach is not perfect, it reduced the MSS module in the proposed system and minimized the loads in the processing system. We decided to disregard the sensor position (2) because we could explain the muscle activity pattern during the entire cycle using only (1) and (3). Fig.9 describes the sensor position of the anterior side (1) and the posterior side (3) of the sensor position we chose. The gray areas represent activation below 20% of the maximum voluntary contraction, and black areas represent activation above 20% of the maximum voluntary contraction. Muscle activation means Knee Assistive System (KAS) is inflated at the moment when the foot of the user touches the ground; the flexion/extension movement occurs in succession and cross-happens within one gait cycle. It seemed to be comprised of only one event, without intermission. Therefore, these successive movements could be organized into a single case. As shown in Fig.6, the proposed algorithm was gradually adjusted to the wearers, such as by calibrating the sensor system and regulating the velocity as with fine-tuning before starting the machinery, in order to develop a handy prototype of the system that is easy to wear. However, the trial of event treatment using a two-sensor set and its customized limit value was not easy to apply to various users because the acquired values for each person were inconsistent in every experiment. Therefore, the more convenient approach of using the restricted number of sensors and minimizing the calibration process is required. The results of the verification of the effects of muscle power assistance through repeated experiments with KAS and an EMG signal sensing device will be introduced in the last chapter.

```
##mfg_i= ith motor sensing group
##Counter clock wise rotation of motor : Positive and
Extension motion of knee
##lim=Lower Limit (User-define value)

If (min(mfg_03, mfg_01)>lim) then,
            If (Knee_theta>50) then,
            Knee_dtheta=0
            else if (mfg_03>=mfg_01) then,
            Knee_dtheta=motor velocity
            else
            Knee_dtheta=-(motor velocity)

Else if (max(mfg_03, mfg_01)<lim) then,
            If (Knee_theta>50) then,
            Knee_dtheta=0
            else if (mfg_03<mfg_01) then,
            Knee_dtheta=motor velocity
            else
            Knee_dtheta=-(motor velocity)
         End
End
```

Fig. 10. Operation Algorithm of KAS using MSS

5. System operation method – trial (2)

To cope the above-mentioned problem, we newly designed the activation algorithm as a second trial. To explain the second trial of the activation algorithm for KAS, a simplified muscle activation pattern of two muscles is introduced. The positions of these muscles are exactly the same as with the sensor attached one.

To activate a KAS, a cross-activation algorithm is designed to use the event of simultaneous contraction of two-joint muscles (which appears to increase the ability of the knee and hip) as an activating signal of reciprocation motion of opposite side knee exoskeleton.

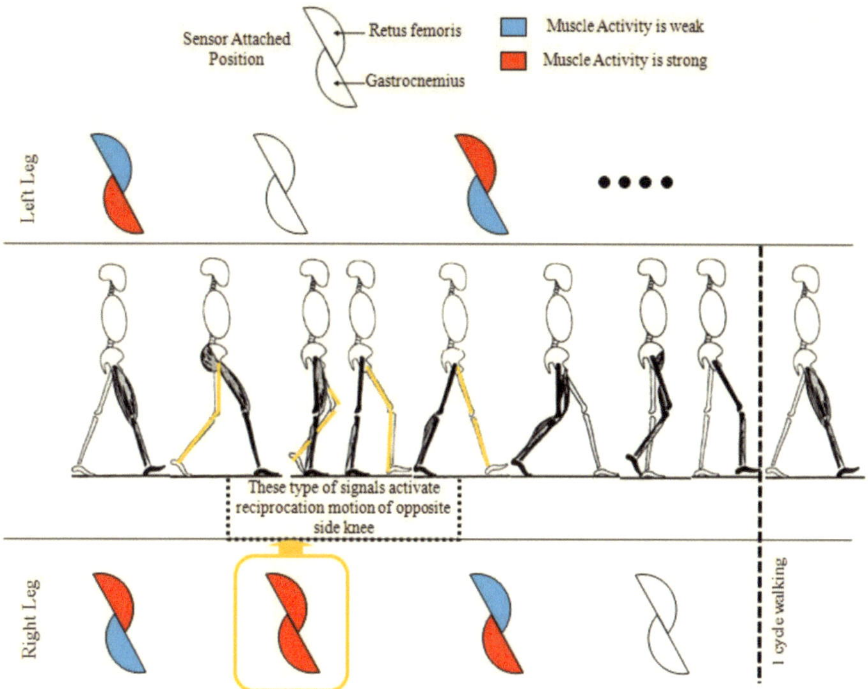

(a) Newly Designed Operation Algorithm for Level Walking

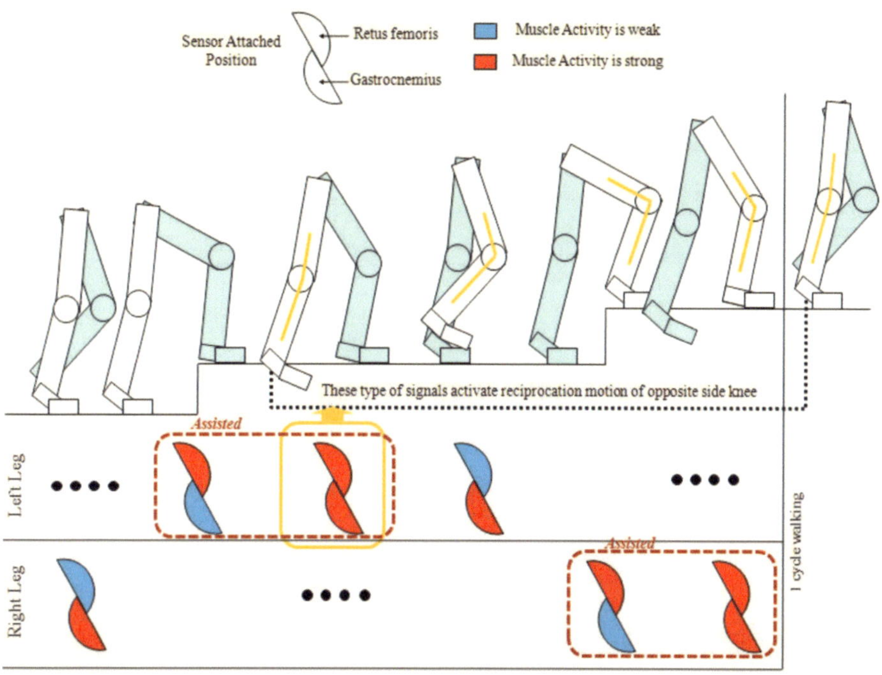

(b) Newly Designed Operation Algorithm for Step Walking

Fig. 11. Operation Algorithm of KAS - Trial (2)

An important thing of this algorithm is activation timing of the exoskeleton. Though this algorithm uses the event of simultaneous contraction to activate the opposite side, in the meantime, the two simultaneous contracting muscles are assisted by their own exoskeleton. As described in Chapter 6, this algorithm brings three effects for the wearer's muscle activation pattern. First is the reduction of required strength for simultaneous contracting, second is pattern shifting of sequential muscle activation, and the last is, as shown in Fig.11(b), the possibility to use this directly as an ascent walking algorithm. Consequently, the performance of the second trial algorithm is better than the first one and is thus selected as an operation algorithm of KAS.

6. System development

6.1 Exoskeleton unit

As mentioned earlier in this chapter, we analyzed the gait pattern based on the muscle activity pattern and the angular displacement of the knee joint. Finally, we deduced the proper sensor position (using only two MSS groups) and the basic operating algorithm. In this chapter, we briefly introduce our experimental exoskeleton system. Fig.12 shows the user-convenient knee movement assistive system that was devised by applying commercial

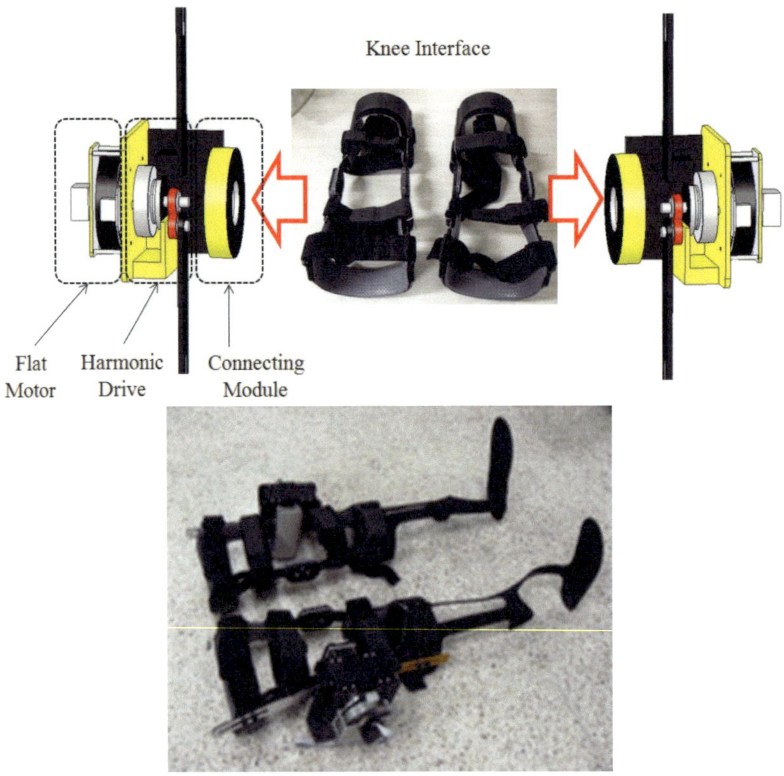

Fig. 12. Exoskeleton Unit of the KAS

knee joint orthotics and minimizing the number of actuating systems. Using a harmonic drive especially prevents a backlash of the motor shaft and the link unit. The system was devised using the following biomechanical and statistical approach. For most of the stance phase, the flexion was less than 20 degrees, and the quadriceps muscle force during level walking depended on the body weight, the magnitude of muscle force, the joint reaction force, and so on. Reilly and Martens (1972) found the highest value for the quadriceps muscle force during level walking, which is 804 N (180.7 lbs) [20]. Considering both the maximum required force value on the quadriceps and the type of system-which partly assists the lower human limb motion, the specifications of the flat motor (Maxon®) and the harmonic drive (THK™) are enough to cover the requirements. The stall torque, gear ratio, and gear efficiency were 4,670mNm, 100:1, and approximately 70%. Therefore, the motor output torque was estimated as 1.63 times more than the requirement. When it comes to the research target of partly assisting human motion, the proposed system feasibly accomplishes the goal of this research. Moreover, it can theoretically add approximately 45kg more payload to construction workers.

6.2 Muscle Stiffness Sensor (MSS) and control system

We designed MSS to acquire the signal for the degree of expansion of the muscle and to use this signal as a human intensity signal to operate the proposed KAS (Fig.13). EMG is typically combined with stride or angular kinematic analysis to provide information on phasic muscle activation patterns. EMG helps explain the motor performance underlying the kinematic and kinetic characteristics of gaits. In this paper, we tried to prove the efficiency of KAS using the EMG test.

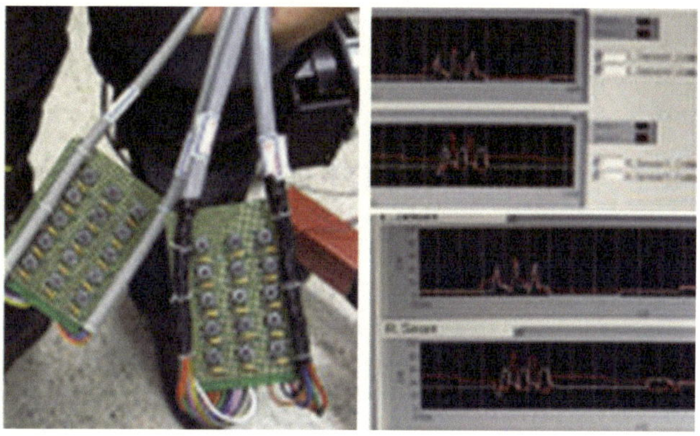

Fig. 13. Custom-Made MSS and Its Performance Test

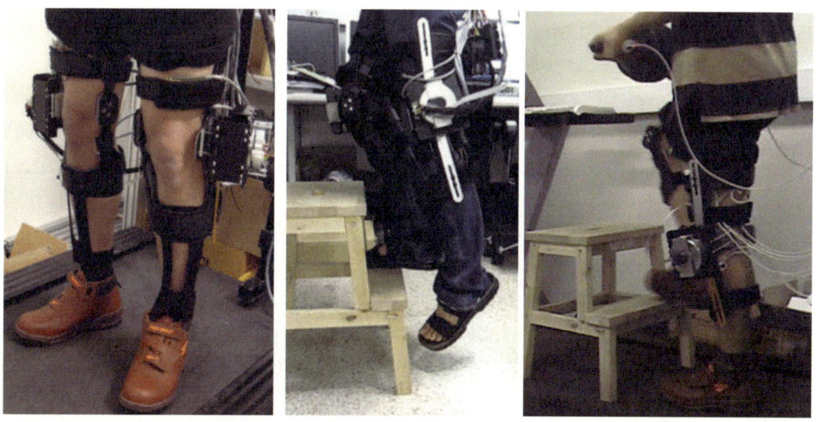

Fig. 14. Walking Experiments Using KAS

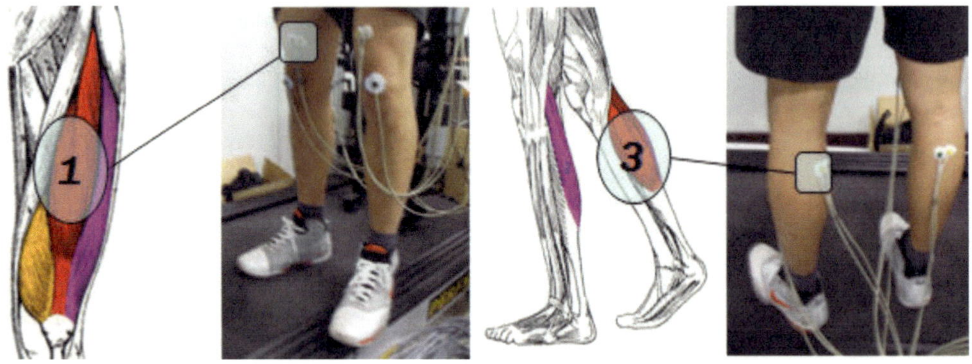

Fig. 15. Performance Verification of KAS using EMG Sensor

The experiment is preceded by the following steps. First, we obtained the EMG signal of healthy subjects while they walked on a treadmill and a step by carrying the 10kg weight. After that, we asked them to wear the exoskeleton system we developed. The subject then repeated the procedure with the same walking speed. Finally, we gathered the EMG signal history and verified its feasibility adapted by the industrial fields.

7. Experiments and conclusions

Experiments are set up with two topics: with payloads (10kg) and without KAS and with payloads and with KAS. During each experiment, EMG signals are gathered by four channels, and their sampling frequency and gain value was 1024Hz and 1126.7uV, respectively. These channels are attached on the upper and lower parts of quadriceps and gastrocnemius muscle groups. Every signal has a tendency to be assisted as shown in the figures below. All magnitudes of the EMG signals have an approximately 20% smaller value using KAS than before. This means that the required muscle strength to accompany the same task is reduced. Through this research, we discovered that partly assisting or supporting the muscle power could be useful for the users' safety and freedom from labor. Even though it was done in a specific and confined condition, that is, minimizing system weights and solving independent electrical power system, KAS is going to be a useful powered harness for construction workers.

As mentioned above, this result shows the characteristics of KAS. First, the muscle activity of quadriceps femoris is larger than that when not wearing KAS (2~3sec) because of the initial state of KAS, which is not activated yet. Second, the signal strength of the simultaneous contraction area of quadriceps femoris and the gastrocnemius muscle is reduced (4~6sec). Third, in the area of 5~6sec, only the signal of gastrocnemius muscle is retained. These changes of EMG pattern can be explained as a continuous change of muscle activity pattern which described in Fig.11. As a result, proposed system leads to the intentional changes of muscle activity pattern but it is helpful entirely for loaded ascent walking.

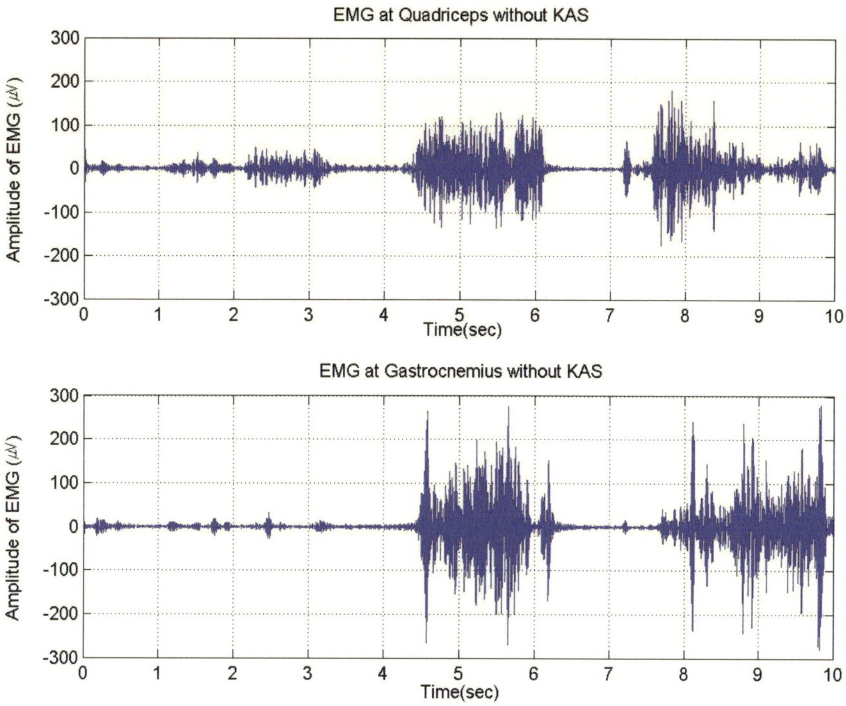

Fig. 16. Gathered EMG Signal without KAS

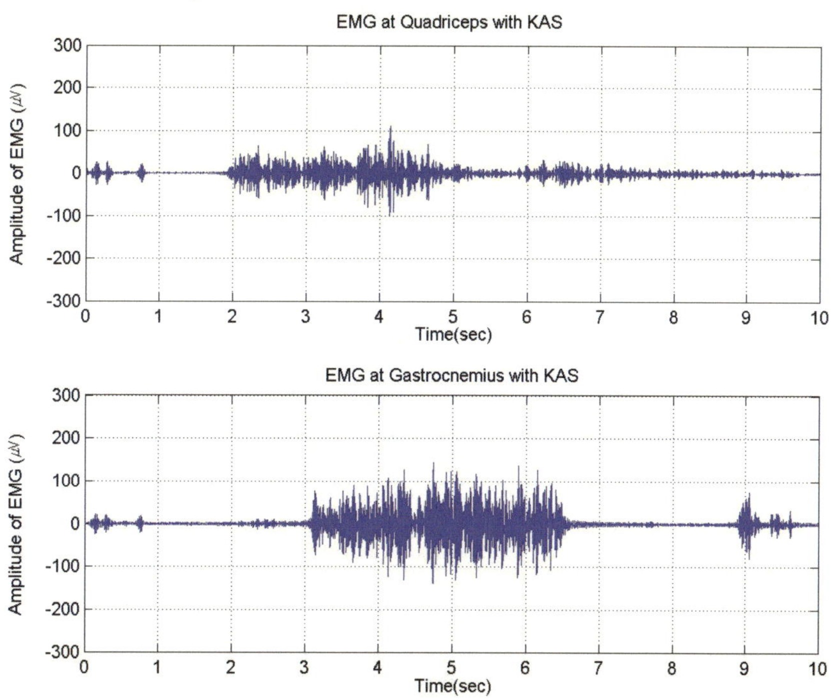

Fig. 17. Gathered EMG Signal with KAS

8. Further studies and future work

We studied and developed a force augmentation system for upper extremities [21,22]. This prototype system is fixed to compensate for self-weight and has 20kg payload per arm. To integrate this system with the knee assistive system which we developed, a weight compensation system through the spine, hip joint, knee, and angle is obviously required. Most importantly, the actuator capacity and control performance of KAS should be increased.

As further work, we are examining two aspects: (1) improving the mechanical structure to minimize the system volume and weight, and stabilizing the system operation by human intent sensing, and (2) developing the total system which is an integrated lower limb exoskeleton with the upper one for whole body augmentation of industrial workers, especially construction laborers.

Fig.18 shows the system integration concept. To integrate the two exoskeleton systems, the most important factors to be considered are weight compensation, joint torque enlargement of lower limb exoskeleton, and its operation algorithm. To address these requirements adequately, many field tests are currently being performed.

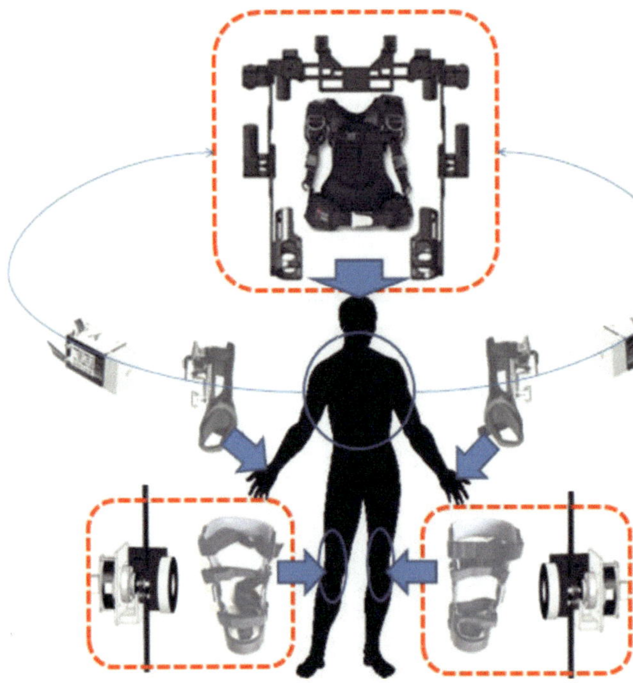

Fig. 18. System Integration Concept for Whole Body Support

Fig.19 describes several tests for a specified task using an incomplete system to gather the dominant problems for operating the newly integrated system. Case (1) is the heavy material carrying task while stepping up the stairs, and Case (2) is heavy material handling task while climbing up the slope.

Fig.20 describes the current status of the newly studied system. As shown in the colored area of the load distribution of humans, the mechanical load bearing structure has to be established before anything else.

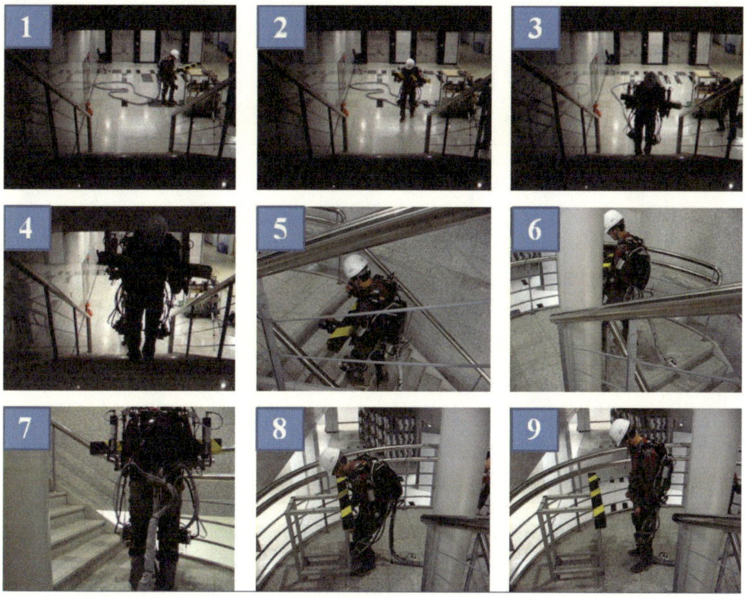

(a) Field Test: Case (1)

(b) Field Test: Case (2)

Fig. 19. Field test using current prototype system (Proposed KAS is part of this system)

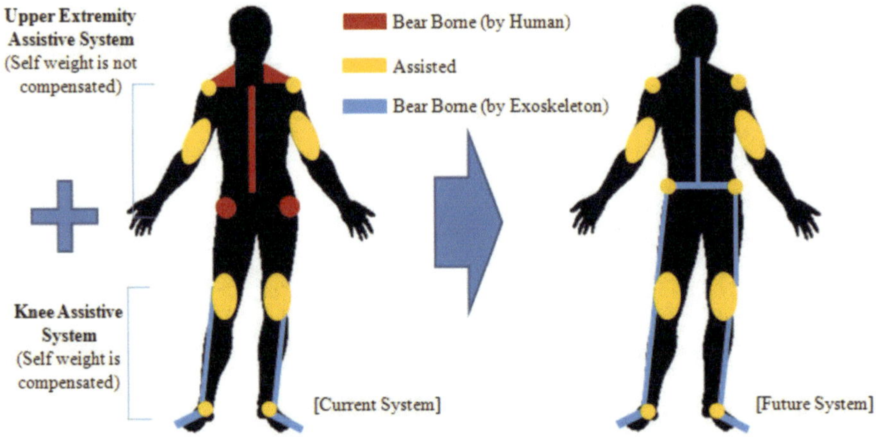

Fig. 20. Newly Proposed System in this Study and Considerations for adequate final Goal

A number of studies are performed to overcome the human performance limit. These studies, however, encountered several limitations on actuator performance, power source, system weight, convenient human-robot interface design, and so on. As described in the following diagram, for an adequate design of a force augmenting system for use in an industrial setting, we have to consider which type of system is more possible and helpful among many types of approaches.

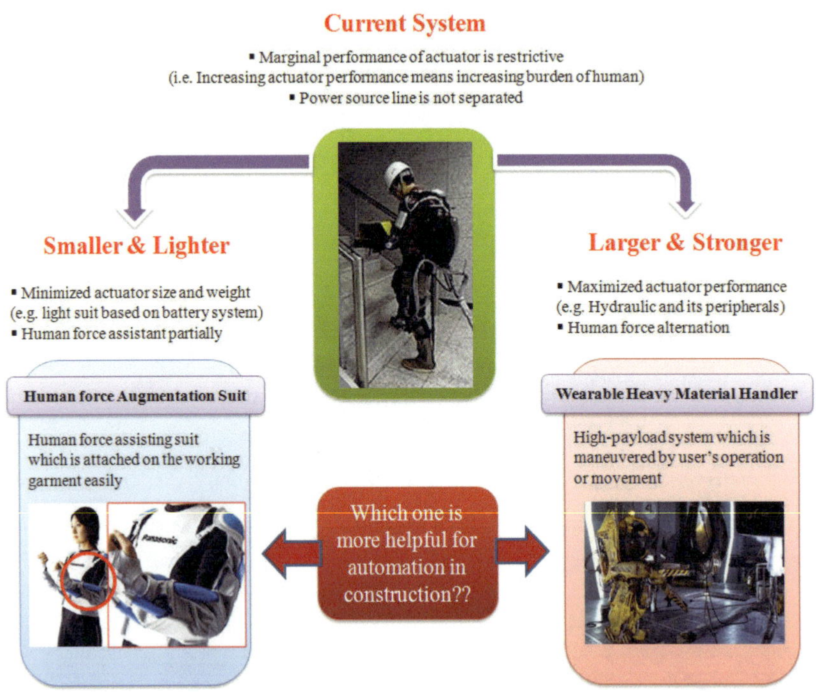

Fig. 21. Current Status of the Newly Developed System and Final Goal (REALIVE™ Power Assist Device of Matsushita Electric Industrial Co., Ltd., and Powerloader of the movie, "Alien II")

9. Acknowledgement

This work is financially supported by the Ministry of Education and Human Resources Development (MOE), the Ministry of Commerce, Industry, and Energy (MOCIE), and the Ministry of Labor (MOLAB) through the fostering project of the Laboratory of Excellency.

10. References

Okamura, J., Tanaka, H., and Sankai, Y., "EMG-based Prototype-powered Assistive System for Walking Aids", Proceedings of the Asian Symposium on Industrial Automation and Robotics (ASIAR'99), Bangkok, Thailand, pp. 229~234, 1999

Thring, M. W. "Robots and Telechirs." Halsted Press, a division of John Wiley & Sons, 1983

Sankai Y., Kawamura Y., "Study on exoskeleton power assist HAL for walking aid using EMG", Nippon gakkai gakujutsu Yokoshu, Vol. 18, pp.453-454, 2000

Kazerooni H., Chu A., Zoss A., "On the Bio-mimetic Design of the Berkeley Lower Extremity Exoskeleton (BLEEX)", 2005 IEEE International Conference on Robotics and Automation, pp. 4345-4352, 2005

Weinberg B., Nikitczuk J., Patel S., Patritti B. Mavroidis C., Bonato P. and Canavan P., "Design, Control and Human Testing of an Active Knee Rehabilitation Orthotic Device", 2007 IEEE International Conference on Robotics and Automation, pp.4126-4133, 2007

Yobotics, 2005, RoboWalker, http://yobotics.com/robowalker/robowalker.html, 2006

Low K. H., Liu Xiaopeng, and Yu Haoyong, "Development of NTU Wearable Exoskeleton System for Assistive Technologies," 2005 IEEE International Conference on Mechatronics and Automation, Niagara Falls, Canada, 2005

S. A. Paluska and M. D. McKeag, "Knee Braces: Current Evidence and Clinical Recommendations for Their Use", Am Fam Physician, Vol. 61, pp.411-418, 2000

Maria Q. Feng, "Sensor Suit for Human Motion Detection", Report Document of DTIC, ADA444285, 2006

Arndt V., Rothernbacher D., Daniel U., Zschenderlein B., Schuberth S., Brenner H., "Construction work and Risk of Occupational Disability: A Ten Year Follow up of 14,474 Male Workers", Occup. Environ. Med., Vol.62, pp.559-566, 2005

Jensen L., "Knee Osteoarthritis: Influence of Work Involving Heavy Lifting, Kneeling, Climbing stairs or Ladders, or Kneeling/ Squatting Combined with Heavy Lifting", OEM, Vol.65, pp.72-89, 2008

Basmajian J.V., DeLuca C.J., "Muscles Alive: Their Function Revealed by Eletromyography", Baltimore: Williams &Wilkins, 1985

Kendall FP, McCreary, E.K., Provance P.G., "Muscle Testing and Function", Baltimore: Williams & Wilkins, 1993

Carol A. Oatis, "Kinesiology: The Mechanics & Patholomechanics of Human Motion", Lippincott Williams & Wilkins, 2004

Lieb F.J., Perry J., "Quadriceps Function: An Anatomical and Mechanical Study using Amputated Limbs. J. Bone Joint Surg, Vol.50A, pp.1535-1548, 1968

Farahmand F, Senavongse W, Amis AA, "Quantitative Study of the Quadriceps Muscles and Trochlear Groove Geometry Related to Instability of the Patellofemoral Joint", J. Orthop Res, Vol.16, pp.136-143, 1998

Goh J.C., Lee P.Y., Bose K, "A Cadaver Study of the Function of the Oblique Part of Vastus medialis", J. Bone Joint Surg, 77-B, pp.225-231, 1995

Wickiewicz T.L., Roy R.R., Powell P.L., Edgerton V.R., "Muscle Architecture of the Human Lower Limb", Clin Orthop, Vol.179, pp.275-283, 1983

Margareta N., Victor H., "Basic Biomechanics of the Musculoskeletal System", Lippincott Williams & Wilkins, 2001.

Barney, F. "Biomechanics of Human Motion." W. B. Saunders Company, pp. 266~272, ISBN 0-7216-5743-5, 1991

Seung-Hoon Lee, Seung-Nam Yu, Hee-Don Lee ,Jung-Soo Han and Chang-Soo Han, "Development of force assistive wearable robot for the upper limb Part I. Kinematical analysis of the exoskeleton", Proc. of Korea Society for Precision Eng., 2007

Hee-Don Lee, Seung-Nam Yu, Seung-Hoon Lee ,Jung-Soo Han and Chang-Soo Han, "Development of Force Assistive Wearable Robot for the Upper Limb- Part II. Generation of Command Signal", Proc. of Korea Society for Precision Eng., 2007

20

A Robot Application for Analysis, Survey and Conservation of Historical Architectures

Michela Cigola and Marco Ceccarelli
University of Cassino
Italy

1. Introduction

This work is the result of the collaboration between DART (Laboratory of Documentation, Analysis, Survey of Architecture and Territory) and LARM (Laboratory of Robotics and Mechatronics) which, despite their apparent differences in terms of fields of research, for years have found time and motivations for opinion exchanges and cultural contacts, as well as research topics on which to begin genuine and fertile collaborations (Ceccarelli, et al., 2002) (Cigola, et al., 2005) (Cigola & Ceccarelli, 2006).

The project research group is made up as follows: for the DART Laboratory of Documentation, Analysis, Survey of Architecture and Territory: *Michela Cigola*, architect specialized in Restoration of Monuments, full professor director of DART; *Assunta Pelliccio*, architect, assistant professor; *Sara Mattei*, engineer, PHD student; for the LARM Laboratory of Robotics and Mechatronics: *Marco Ceccarelli*, engineer, full professor director of LARM, *Giuseppe Carbone*, engineer, assistant professor; *Erica Ottaviano*, engineer, assistant professor; *Chiara Lanni*, engineer, research assistant.

Our goal is to analyse the characteristics and problems that are related to the analysis and conservation of historical architecture, within an innovative approach with robotised systems. The design requisites and the distinctive aspects of the operations of a large range of robotised systems have now been fully defined and investigated. However, there are very few precedents for such specialized application, moreover, in connection to the subject of historical architecture that is apparently far away from general areas of robotics and mechatronics.

The case of study that is presented here, namely the pre-Cosmatesque floor of the abbatial basilica of Montecassino, is a good case to highlight the flexibility of robotics and possible interactions with architecture, particularly in the field of surveying. Thus, this research focuses specifically on the application of robotic systems in the study and survey of ancient floor surfaces, although the wider purpose is to study and refine robotized systems and automated procedures to develop activities and results in the analysis phase and for the safeguard and conservation of historical architectural sites at large.

Robots are used more and more frequently in difficult situations, such as inspections of internal sites of atomic stations, space travel and exploring inaccessible terrestrial sites, as volcanic craters. Thus, a new and different example could be the analysis and safeguard of historical architecture through the use of self-propelled robots performing highly complex tasks on various kind of surfaces under the supervision of a human operator.

Therefore, a proposal, is to develop a relationship between robotics and architecture, through a mehanical project that is especially studied and refined for analysis and architectural restoration. This is an extremely innovative application in the field of robotics, since it involves attempts to make a robot perform many architectural surveying actions through automated operations that can improve procedures and results, in terms of speed and accuracy.

2. Architectural survey

It is not easy to give a stict definition of architectural surveying, although now it is seem unanou,esily considered that it is not merely a simple instrument that is useful for other disciplines, but rather it is an autonomous operation with its own intrinsic value.

Fig. 1. Front side of the church of S.Crisogono, Rome. Survey by M. Cigola.

Surveying is now recognised as a complex discipline in its own right, the only one capable of penetrating the deepest and most hidden nature of an architectural organism, bringing to light the historical events of which it has been protagonist, original form and subsequent transformations over different eras, and providing us with full graphical imagery.

Performing a survey means a fully understanding the metrical, geometrical and historical aspects of an architectural work, enabling you to reach a critical interpretation of it. A vast series of operations is required for this, above and beyond mere measuring.

The surveying process involves measuring, drawing, checking and comparing the data collected, highlighting similarities and differences using an accurate philological method. Alongside this phase, for a more complete understanding of the building it is essential to carry out a historical-archive study to perfectly identify its era.

Thus, surveying involves a complex series of operations, including the collection, checking and interpretation of all possible data, not only in terms of the geometric-spatial form of the architectural work, but also its construction, structural and historical data.

The fields in which surveying is used more and more often are historical-critical analysis of architecture and restoration. An accurate historical analysis of a work requires precise measuring techniques. The survey of a building is not a document used afterwards, but is an integral part of the process of historical-critical knowledge. Only through the close relationship that is formed between the works and the surveyor it is possible to gain real knowledge of a building.

As we have seen, surveying is an autonomous discipline, independent of its specific operational purpose. Nevertheless, individual applications may require particular methodological orientation.

A survey carried out before restoration work must lead to an overall understanding of the work under examination, completely exhaustive in terms of dimensional, structural and construction aspects. It must reveal the state of health of a building, its conditions of decay and static conditions.

The survey must bring to light historical, technical and diagnostic aspects; contribute to identifying the design intentions, from proportioning of metrological aspects, guiding historical research through iconographic, bibliographical and archive sources.

The problems involved in surveying for safeguard mainly highlight the need to acquire a great quantity of information to be integrated. Measuring must be carried out precisely and the graphical representation must give precise details of the characteristics of the works and all of its distinctive features, often through larger scales than those usually used in general preliminary surveys.

In order to carry out this task at maximum levels of what, a survey for restoration should use all modern available methodologies and techniques. There has been a recent revolution in this sector thanks to the development of 3D Scanning. This now well-known procedure consists of scanning the object using a 3D Scanner, which emits a laser or other type of ray to process the response data to determine the three-dimensional position of each material point that is hit by the ray. This innovative technology thus allows to create virtual 3D models of large and complex objects with remarkable levels of definition; recent trials have been conducted on several famous monuments including the Coliseum, the Pantheon, Saint Sophia's in Istanbul, etc. by RADAAR Department, School of Architecture at Rome University "La Sapienza" directed by prof. Mario Docci.

The following words are useful for a suitable definition of the process of architectural surveying as a prelude to restoration work: "surveying is a kind of non-destructive investigation into the works for restoration purposes. It is almost a clinical test, a kind of biopsy, which is not a mere clinical observation, yet not quite a surgical operation" (Carbonara, 1989).

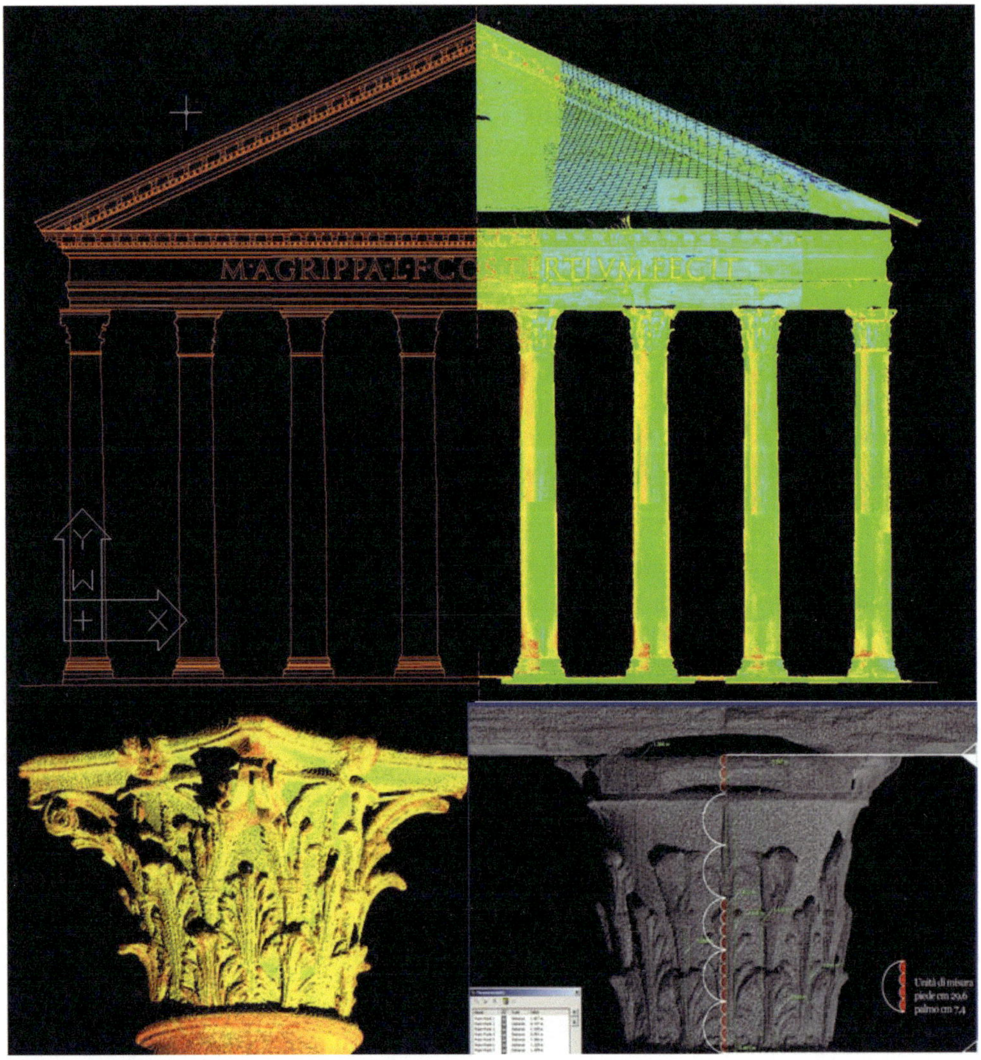

Fig. 2. 3d Laser Scanner Survey of Pantheon, Rome (PHD Thesis By P. Paolini, RADAAR Department, Rome University "La Sapienza").

3. Architectural survey of historical pavements

In the area of historical floor surfaces, we will concentrate our attention on Cosmatesque floors, of which the first example was the one in the abbatial basilica of Montecassino built in the 11th century, which, after a few decades, is followed by a large number of examples in Italy and in Europe over the next few centuries.

The Cosmatesque school began in the 12th century and mainly operated in Rome until the end of the 14th century; however, in addition to Rome, the art of the Cosmati masters spread throughout central and southern Italy and even reached England. The popularity of their technique covered a vast chronological period; indeed the first Cosmatesque work dates back to around 954, and the last to 1458. The period between the end of 14th and start of the

15th centuries saw the great ferment that led to the great Renaissance season and the decline of the Cosmati masters, who were not clearly defined as being somewhere between craftsmen and designers.

The most eye-catching feature of a Cosmatesque floor is certainly the *central strip*, which leads from the entrance of the church directly to the apse. The simplest type of central strip is made up of a series of *rotae* or porphyry discs stemming off strips of mosaic alternating with strips of white marble. This shape became increasingly complicated until it created a very intricate design; another very common design for the central part of the floor is the socalled *quincunx*, which is a square design that includes a central *rota* surrounded by four more, all connected by strips of mosaic and marble. Often these types of central strips were combined to form extremely complex elements.

After establishing the central decorative motif, they went on to cover the entire floor surface with rectangles whose geometrical design was usually repeated symmetrically to the longitudinal axis of the central design, with greater attention paid to the general effect of the overall surface rather than the panel in itself (Cigola, 1993).

Fig. 3. Cosmatesque pavements. Basilica of S. Croce in Gerusalemme, Rome and Cathedral of Anagni.

The materials that were used for the floor surface were usually red or green Serpentine porphyry, and white, *ancient yellow* and *pavonazzetto* marbles, used indiscriminately. The brightest colours were usually reserved for the *central strip*, while the lighter colours were used in the more peripheral parts of the composition. In addition to aesthetics, there were certainly valid practical criteria for the choice of materials and colours, such as the availability of marble, porphyry and granite in Rome. Indeed the elements of an ancient classical floor often underwent numerous re-uses before reaching the fragmentation of Cosmatesque mosaics. It is not unlikely that this fragmentation prevented further re-use of the marble pieces in Cosmatesque floor, which have reached us almost intact (Glass, 1980).

In addition to the marble fragments of various colours, one of the main recycled elements in a Cosmatesque floor is the *porphyry rota*, i.e. the large disc of green or more often red porphyry inserted to act as a greater vehicle of symbolic meaning. It is likely that the main reason for its use was the great availability of columns from columns the Roman empire, which were cut into slices, allowing floor surface arrangements with great decorative effect and fast to achieve (Guidobaldi & Guiglia Guidobaldi, 1983).

Fig. 4. *Porphyry rota* in the pavement of S. Prassede, Rome (Claussen, 1992).

In terms of the methods and techniques that are used to document Cosmatesque floor mosaics, surveys of works of this kind require special measuring and representation techniques, since the tiles forming the image have varying dimensions according to the overall design and the group of marble cutters responsible for creating them, and depending on the areas of floor. It is therefore necessary to highlight the equipment of the mosaic, highlighting the type of material used, the dimensions of the pieces laid, the degree and type of surface finishing, thickness and treatment of the jointing and the tile bed.

In a well-conducted survey, every element of the floor would be taken individually and in relation to the entire Cosmatesque floor, with particular attention for those elements that are repeated or make up a series, since any anomalies found in these series sometimes characterise the way a particular floor is laid. Great attention must also be paid to measuring of irregularities in the surface as a whole (repeated changes in level, sinkings etc.) in addition to identifying gaps, integrations added over time.

The entire process of measuring and graphical representation of a Cosmatesque floor mosaic can therefore be summed up in a series of operations, each of which must be performed at a different scale of representation: survey of the entire laying of the floor entered in diagram form in its architectural context; survey of composition elements forming the *central strip;* survey of all geometrical designs used in the sections that make up the greatest part of the floor mosaic.

Of course, each of these phases must include at least two more groups of images based on architectural survey mapping: the first is to detect decay, to highlight the degree and type of deterioration using specially conceived symbols; the second is for a chromatic survey of tones, grain and types of material.

To facilitate the entire process of operations and to have documentation of the state of the floor, it is useful to have a complete and exhaustive series of photographic before the survey itself (Cigola, 1993).

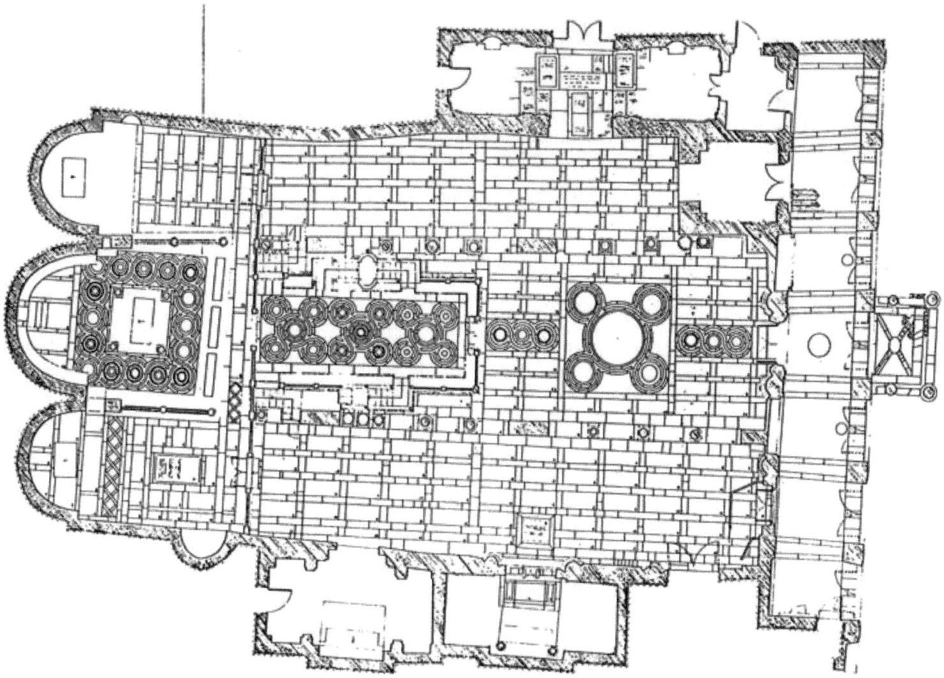

Fig. 5. Basilica of S.Maria in Cosmedin, Rome. Architectural survey of the cosmatesque pavement (1119-1130).

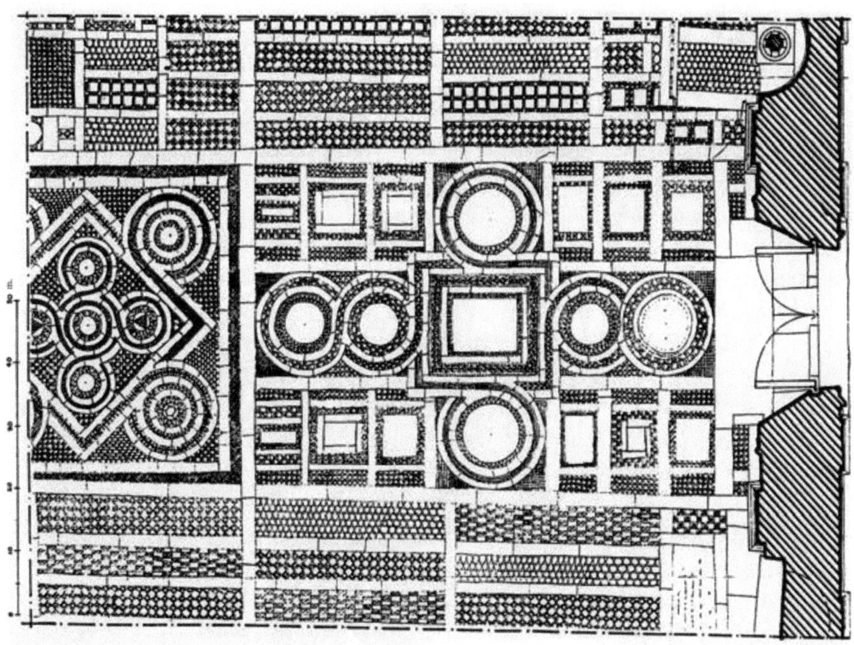

Fig. 6. Basilica of S.Crisogono, Rome. Architectural survey of the central strip of cosmatesque pavement (1146-1150). Survey by M. Cigola.

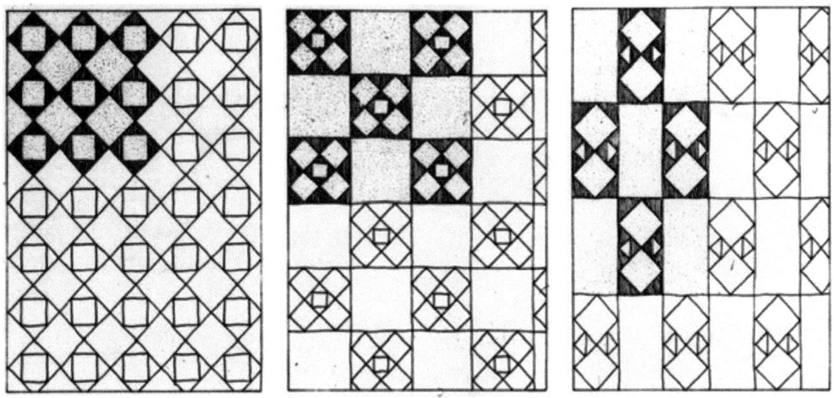

Fig. 7. Basilica of S.Crisogono, Rome. Architectural survey of rectangles and geometrical designs of the cosmatesque pavement (1146-1150).

Of course, each of these phases must include at least two more groups of images based on architectural survey mapping: the first is to detect decay, to highlight the degree and type of deterioration using specially conceived symbols; the second is for a chromatic survey of tones, grain and types of material.
To facilitate the entire process of operations and to have documentation of the state of the floor, it is useful to have a complete and exhaustive series of photographic before the survey itself (Cigola, 1993).

4. Robot design and robotic operations

Robots can be considered as the most advanced automatic systems and robotics, as a technique and scientific discipline, can be considered as the evolution of automation with interdisciplinary integration with other technological fields (Ceccarelli, 2004).
An automatic system can be defined as a system which is able to repeat specific operations generally with a low degree of intellectual and manipulative levels, but that can be easily programmed in agreement with demands of productivity.
It is worthy of note that an automatic system is generally able to perform one operation for which its mechanical structure has been designed. Aspects of flexibility depend on the possibility of reprogramming the control unit, which is generally able only to modify the time sequence of the designed operations. Therefore, an automatic system of industrial type can be thought of as composed of two parts:

- hardware with mechanical, electrical, pneumatic, and hydraulic components that provide the mechanical capability to perform an *a priori*-determined operation of movement and/or manipulation;
- control and operation counterpart with electronic components and software that provide the capability of autonomy and flexibility to the working of the system.

The two parts are essential in an automatic system and are integrated in the sense that their design and operation must be considered as a unique goal in order to obtain and operate an automatic system with the best performances. Therefore, an automatic system in which the hardware part is preponderant cannot have a suitable flexibility for a flexible production in agreement with the demands of productivity and market. In some cases these limited solutions are still required from productivity , when the product of a massive production is absorbed from the market with certain regularity during a period of time longer than the amortization time of the automatic system. Those systems with low flexibility are generally denominated as erigid automatic systems'. In Fig. 8 the variety of production systems is represented as a function of the productivity level.

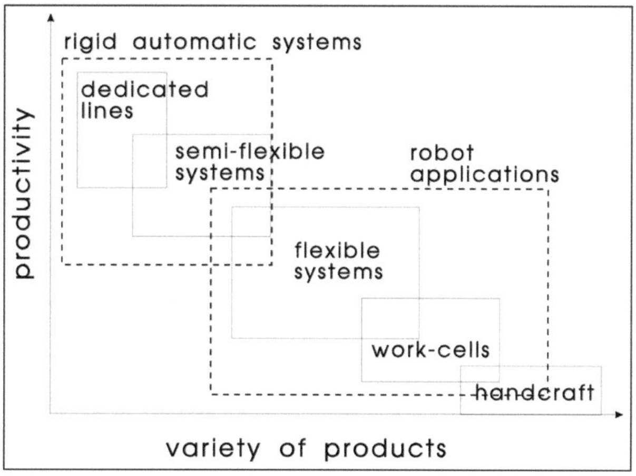

Fig. 8. Variety of automatic and robotized systems as functions of productivity level and product demand

In addition, it is worthy of note that a system with only the mechanical counterpart, although versatile, cannot be considered an automatic system since the updating of its operation is not obtained by means of control units. Such a system can be properly reprogrammed, but it requires the manual action of one or more human operators to change some components of the machinery or their running. Therefore, these systems can be properly named as machinery or mechanical systems.

Similarly, control systems cannot be considered as automatic systems, since they are not able to perform mechanical tasks, although they can be provided with a high level of flexibility in terms of re-programmability and memory capability. Therefore these systems can be properly named as control units or electrical–electronic–informatics systems.

A robot can be defined as a system which is able to perform several manipulative tasks with objects, tools, and even its extremity (end-effector) with the capability of being reprogrammed for several types of operations. There is an integration of mechanical and control counterparts, but it even includes additional equipment and components, concerned with sensorial capabilities and artificial intelligence. Therefore, the simultaneous operation and design integration of all the above-mentioned systems will provide a robotic system, as illustrated in Fig. 9.

In fact, more than in automatic systems, robots can be characterized as having simultaneously mechanical and re-programming capabilities. The mechanical capability is concerned with versatile characteristics in manipulative tasks due to the mechanical counterparts, and re-programming capabilities concerned with flexible characteristics in control abilities due to the electric-electronics-informatics counterparts. Therefore, a robot can be considered as a complex system that is composed of several systems and devices to give:

- mechanical capabilities (motion and force);
- sensorial capabilities (similar to human beings and/or specific others);
- intellectual capabilities (for control, decision, and memory).

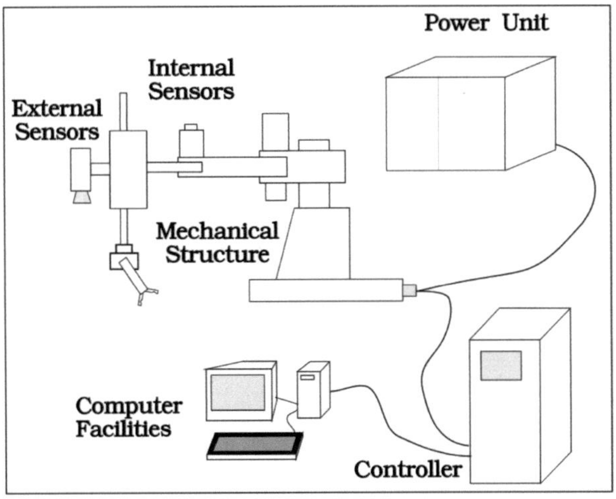

Fig. 9. Components of a robot

Initially, industrial robots were developed in order to facilitate industrial processes by substituting human operators in dangerous and repetitive operations, and in unhealthy environments. Today, additional needs motivate further use of robots, even from pure technical viewpoints, such as productivity increase and product quality improvements. Thus, the first robots have been evolved to complex systems with additional capabilities.

Nevertheless, referring to Fig. 9, an industrial robot can be thought of as composed of:

- a mechanical system or manipulator arm (mechanical structure), whose purpose consists of performing manipulative operation and/or interactions with the environment;
- sensorial equipment (internal and external sensors) that is inside or outside the mechanical system, and whose aim is to obtain information on the robot state and scenario, which is in the robot area;
- a control unit (controller), which provides elaboration of the information from the sensorial equipment for the regulation of the overall systems and gives the actuation signals for the robot operation and execution of desired tasks;
- a power unit, which provides the required energy for the system and its suitable transformation in nature and magnitude as required for the robot components;
- computer facilities, which are required to enlarge the computation capability of the control unit and even to provide the capability of artificial intelligence.

Thus, the above-mentioned combination of sub-systems gives the three fundamental simultaneous attitudes to a robot, i.e. mechanical action, data elaboration, and reprogrammability.

Consequently, the fundamental capability of robotic systems can be recognized in:

- mechanical versatility;
- re-programmability.

Mechanical versatility of a robot can be understood as the capability to perform a variety of tasks because of the kinematic and mechanical design of its manipulator arm.

Re-programmability of a robot can be understood as the flexibility to perform a variety of task operations because of the capability of its controller and computer facilities.

These basic performances give a relevant flexibility for the execution of several different tasks in a similar or better way than human arms. In fact, nowadays robots are wellestablished equipment in industrial automation since they substitute human operators in operations and situations, which are:

- dangerous - for manipulative tasks and/or unhealthy environments;
- repetitive - with low-level cultural and technical content;
- tiresome - for manipulative tasks requiring energy greater than that provided by a human operator;
- difficult - for human operators.

In addition, the use of robots is well-established in industrial production since robots are:

- versatile, as they have the ability to operate in different situations;
- useful for unhealthy or limited environments, which can be dangerous or unfeasible for human operators.

Besides the above-mentioned motivations of technical nature, robots and robotic systems are currently used with the aim of:

- decreasing production costs, which can be related to a better use of the machinery;
- increasing productivity, with an increase of the operation velocity;
- enhancing product quality, in terms of constant characteristics and improved manufacturing;
- achieving flexible production as a capability of rapid adjustment of production to required changes in manufacturing.

The mechanical capability of a robot is due to the mechanical sub-system that generally is identified and denominated as the emanipulator', since its aim is the manipulative task.
The term manipulation refers to several operations, which include:

- grasping and releasing of objects;
- interaction with the environment and/or with objects not related with the robot;
- movement and transportation of objects and/or robot extremity.

Consequently, the mechanical sub-system gives mechanical versatility to a robot through kinematic and dynamic capability during its operation.
In general, in order to perform similar manipulative tasks as human operators, a manipulator is composed of the following mechanical sub-systems:

- an arm, which is devoted to performing large movements, mainly as translations;
- a wrist, whose aim is to orientate the extremity;
- an end-effector, which is the manipulator extremity that interacts with the environment.

It is worthy of note that although the mechanical design of a manipulator is based on common mechanical components, such as all kinds of transmissions, the peculiarity of a robot design and operation requires advanced design of those components in terms of materials, dimensions, and designs because of the need for extreme lightness, compactness, and reliability.
The sensing capability of a robot is obtained by using sensors suitable for knowing the status of the robot itself and surrounding environment. The sensors for robot status are of fundamental importance since they allow the regulation of the operation of the manipulator. Therefore, they are usually installed on the manipulator itself with the aim of monitoring basic characteristics of manipulations, such as position, velocity, and force. Additionally, an industrial robot can be equipped with specific and/or advanced sensors, which give human-like or better sensing capability. Therefore, a great variety of sensors can be used, to which the reader is suggested to refer to in specific literature.
The control unit is of fundamental importance since it gives capability for autonomous and intelligent operation to the robot and it performs the following aims:

- regulation of the manipulator motion as a function of current and desired values of main kinematic and dynamic variables by means of suitable computations and programming;
- acquisition and elaboration of sensor signals from the manipulator and surrounding environment;
- capability of computation and memory, which is needed for the above-mentioned purposes and robot re-programmability.

In particular, an intelligence capability has been added to some robotic systems concerned mainly with decision capability and memory of past experiences by using the means and techniques of expert systems and artificial intelligence. Nevertheless, most of the current industrial robots have no intelligent capability since the control unit properly operates for the given tasks within industrial environments. The control systems that are used in robots can be classified as:

- electro-mechanical sequencer units with end-stroke stops; units with pneumatic logic; electronic units with logic; PLC (Programmable Logic Controller) units;
- micro-processors;
- minicomputers;
- computers.

Indeed, nowadays industrial robots are usually equipped with minicomputers, since the evolution of low-cost PCs has determined the wide use of PCs in robotics so that sequencers, which are going to be restricted to PLC units only, will be used mainly in rigid automation or low-flexible systems.

5. Requirements and problems for robots in Survey applications

Survey and Analysis activity on historical architectonic goods are generally carried out with manual operations with few assisting devices. In addition, the activity is differentiated in Architecture and History of Art not only for the differences of the goods in terms of materials and structure types, but at the most for the scale of the dimensions of the objects of study. In general they are buildings or large parts of them in Architecture, and small components of buildings or decorative details of them. Survey activity is aimed to detect details of the historic works both in terms of dimensions and details for the different interests in a historical Analysis. Thus, there are several aspects that may differentiate the activity in Architecture Survey for the scale of the objects unders examination. Nevertheless, there are also several aspects in common, mainly in the approaches for the activity in getting knowledge and experiences of the objects under examination in Survey and Analysis activity. They are related to tasks, which can be often repetitive, for acquiring data from the historic works, both in terms of dimensions and figures of details. Those repetitive tasks are some time performed with the assistance of devices that help in achieving proper repetitions and acquisition data with regular characteristics. But often those assisting devices are developed or even adapted for a specific case without a general frame of operations, so that each use can require several attempts and mainly understanding for a proper operation. The aim of using robotic systems is to define a general approach both for design and operation of suitable assisting devices with a proper level of automation and robotic characteristics, by starting from the fact that assisting devices are already used and even well accepted by Survey operators and investigators.

Main problems in Survey activity that can be enhanced with suitable assisting robotic devices, can be identified in data acquisition of dimensions and figures of the historical works. Dimensions can measured by using proper sensors and even movement actions of automatic mechanical systems; while figures of the historical works can be detected by proper location of the instrumentation that can be adjusted properly with those assisting robotic devices. There are several cases and situations in which those measuring activity and figure acquisition cannot be performed by human operators, like for examples in no accessible sites (for lack of space and/or human safe environments), but even for obtaining proper resolution of the results (for example, when the object is not at the hand or the work is fragmented). In addition, a certain automation of current manual operation in Survey can help to get results in short time and with more accurate results. Another important aspect of the current activity is a certain fragmentation of the main phases that a Survey activity can have, namely, preliminary polishing of the historic site and/or work, identification of main frames, identification of main parameters, determination of the Survey problem, choice of instrumentation, calibration of the instrumentation, campaign of acquisition data, elaboration of the data, interpretation of the data, analysis of the results, and so on.

Typical Survey on-field activity n Architecture can be recognized in measuring dimensions of the works, detecting their general figures in term of shapes and volumes, acquiring images or even videos both for the general status and details, inspecting structure of the works in terms of materials and construction process, and deducting in several cases even samples of the works for further investigations. Most of these activities can be helped with assisting robotic devices and they can be even more successfully both in term of productivity and quality of results. One additional aspect, that is more and more important, is related to a request of shortening the Survey activity that most of the cases require to close the historical goods to the public visit. The possibility to have assisting robotic devices that can operate not only in short time, but even without requiring large frames obscuring the historical goods is a strong demand. For example a Survey activity on the Coliseum in Rome, in general will require scaffolding that will cover the whole monument for long time at least both for the installation and disassembly of the scaffolding. The possibility to use a robotic system, which can climbing the walls of Coliseum with proper instrumentation, could reduce and even avoid any heavy scaffolding cover of the monument during the Survey operation. In general, Survey activity in terms of measuring dimensions require scaffolding and/or other structures that not only avoid the access to the good but may not permit any other action even for the Survey and Analysis activity on the historical work. Similarly the activity for data acquisition require to bring suitable instrumentation near the details of the good under examination with inconvenience in mobility and accessibility of other systems and human operators. Even the direct inspection by human operator can require heavy infrastructures and safety requirements that can make even more complicate the operations. Alternative solutions can be attempted with mechanical or robotic systems for tele-operations and image recording or transmission by using suitable mobile systems.

Main problems for which it can be thought useful to get experience with innovative solutions using automatic mechanical systems and robots, can be identified as specifically related to details works in hiostorical architectonic goods in terms possibility to facilitate or even to complete capability of a human operator like for example:

a. high-resolution image acquisition of surfaces with the aim of a detailed graphical reconstruction and exegetic interpretation;

b. identification of details and their location within the plan of the work in which they are located;
c. representation and reconstruction of work images that are on curved surfaces or in hidden surfaces;
d. acquisition of images with suitable static mechanisms that are not available with current photographic means both for the light sources and the camera location with respect to orthogonally of work surface;
e. possibility to use additional instrumentation, even in one unique campaign/action of Survey;
f. possibility to use the robot performance for additional evaluations and computations of work characteristics.

Main problems of an application of robots in Survey activity can be outlined as related to development of design methodologies and solutions for mechanical and robotic systems that can be operated by users who are experts in Survey and Analysis of cultural goods of historic interest. In addition, the feasibility of those proposed robot solutions should be validated through specific prototypes for specific operations on site.

The cooperation in the above-mentioned fields of Robotics, and Architecture Survey is challenging since it addresses integration of approaches and methodologies that, in general, they do come across to each other because they are from different cultural communities (engineers, architects, and historians). Therefore, the robot application is aimed to introduce innovative approaches and methodologies in activity for survey and study of the history of the past by using mechanical systems and automatic/robotic systems with an expected positive impact for improving the quality of survey and study activity and in establishing a frame for the formation of new figures of professionals both in engineering and survey/restoration areas. A important aspect can be recognize in developing an interdisciplinary frame through integrated methodologies and solutions for design and operation of advanced systems in the fields of activity of maintenance and study of monumental historical goods and historical works. Indeed, a solution for cultural barriers among engineers, architects, and historians can be an important challenge for a long-term application of robots in Survey activity. In addition, this interdisciplinary can give the possibility to establish new research areas and even a new professional figures with multidisciplinary aspects. Even the specific area of Service Robotics will benefit of the possibility to have a new field of applications. Indeed, new horizons for research and professional figures can be outlined in activity that can be carried out with integrated multidisciplinary expertise but even with a specialization in one of the three abovementioned areas with a multidisciplinary view.

Therefore, a general methodology for the robot application in Survey can be described as a sequence of steps for integrating the above-mentioned areas to outline an approach for design and operation both of mechanical and robotic systems, and Survey activity. Nevertheless, each area will also work with traditional methodologies of its fields mainly with the aim to define and achieve successful implementation of the results of the robot applications. In particular, in Robot Engineering the problems can be approached by starting with an analysis of the State-of-Art of service robots and assisting robotic devices that could be near to Survey applications; studying the peculiarities of the robot applications as outlined by problems in Architecture Survey; formulating design problems and operation strategies by taking into account both technical and cultural aspects; elaborating solutions for a general procedure and then for specific cases as test-bed applications; constructing and testing prototypes of systems and operations with consideration of constraints and instrumentation in Survey applications for general strategies and even for the specific testbeds. In these aspects technical methodologies, that are based on suitable models in terms of mechatronics and mathematical formulation, can be developed towards technical solutions with technical descriptions, which are typical of Engineering fields. Nevertheless, great attention can be addressed to include aspects from the Survey activity even for facilitating understanding of the designed robotic systems and their operations by Survey operators. Indeed, great efforts will be spent for achieving user-oriented operation of the robotic systems that should be possible even without any supervision of engineers when in historical on-field sites. However, during the design process like in the test experimentation a strong interaction can be required with the team members from Survey area, but even with their presence and contribution in the engineering processes.

On the other hand, in Architecture Survey the Survey problems can be approached by reviewing the current State-of-Art of assisting devices and related operation strategies with the traditional approaches of the disciplines but even with a translation to engineering aspects; designing operations of Survey activity that can be performed with mechanical and robotic systems, with special emphasis to cases and situations in which their use could be determinant for an improvement of the productivity and quality of the works of Survey operators; elaborating test-bed operations for experimental validation of design systems by looking to specific aspects of Survey activity; designing adaptations of instrumentation and operations for Survey activity with mechanical and robotic systems; identifying field applications in which the results and designed robotic systems can be operated for a real Survey activity.

The development of robots for Survey applications can be programmed with a general plan through phases with the following outline:

1-st phase: definition of design requirements and operation characteristics

2-nd phase: development of methodologies and Survey strategies for design of mechanical and robotic systems

3-rd phase: development of methodologies and Survey strategies for operation of mechanical and robotic systems

4-th phase: construction of prototypes and test-bed applications

5-th phase: tests and validations

The plan of each phase can be carried out with activity that can overlap and can continue even in next phases as described in the following. The core of the development planning can be considered in the phases 2 and 3 that in principle will have activity extended during the whole activity. The activity can be run by the two groups, namely Robot Engineering and Architecture Survey, in same aspects with parallel separate plans, but mainly with a joint program as outlined in the following at the whole.

1-st phase: definition of design requirements and operation characteristics

In this phase the work team can work to get knowledge and experience from the areas of each of the two groups with the aim to identify peculiarities and characteristics of Survey activity that can be devoted to mechanical and robotic systems. Each group can review the State-of-Art of its area by addressing special attention to those aspects that can be useful for an integration of the disciplines. In particular, aspects can be outlined as directed to define requirements and operation characteristics that are suitable for a certain automation or improvement by using assisting robotic devices. Current assisting devices and their operations can be reviewed in strict cooperation by the team members in order to identify those aspects that can be evolved as suitable for mechanical and robotic systems. The review work can be also aimed to reconsider the Survey activity from general viewpoints and then with a perspective of using mechanical and robotic systems as means within a cultural formation of new Survey operators. In Robot Engineering the aspects of service robots can be examined to search similarities and differences in existing solutions for other or similar applications that could better help to explain the possibility of implementing those systems in cultural frames that are not used to work with them. In this phase the two groups can work separately for some aspects and can exchange information with the aim to start some collaborative actions. The integrated activity can be thought even in preliminary experiences that can be carried to show to the other group the peculiarities of their work in traditional frames. Thus, robots and mechanical systems can be used for laboratory tests of simulated simple Survey operations and Survey operation can be even experienced on fields with the presence of engineers with the aim to show them the peculiarities of the manual operations and other actions that could be devoted to a certain mechanization and/or robotization. This phase can be characterized by an identification of possible applications through the determination of those characteristics that can be used as requirements and constraints for next design and operation activities. The understanding of those aspects can require strict collaboration among the groups even with the aim to achieve reciprocal understanding of terminology and needs for the Survey robotized activity.

2-nd phase: development of methodologies and Survey strategies for design of mechanical and robotic systems

In this phase the activity of the previous phase can be oriented specifically to design purposes for mechanical and robotic systems. Nevertheless, the design methodologies and approaches from robot engineering must be adjusted and integrated with the aspects coming from peculiarities and requirements for Survey activity, even with the aim to give the possibility to Survey researchers and operators to contribute in the design process. The design process can be focused on system architectures but also on Survey strategies that can be performed with them. The design process can be carried out to define problems from one aspect but even influencing other aspects with a team activity that can be scheduled with full integration of perspective and formulation in elaborating design procedures. In particular, design algorithms can be formulated for the design of mechanical and robotic systems by approaching the problems for the structure types, dimensional synthesis, mechanical design of parts, endowment of suitable internal and external instrumentation. In each of the mentioned design aspects, interference is expected with the cultural requirements for Survey activity. In this elaboration process, significant can be considered the activity of analysis and simulation of existing solutions and operations, both in numerical procedures and experimental tests. Thus, each group can be responsible to outline to the other groups results that can be used in design procedures for mechanical and robotic systems. A unification of the perspective can give the possibility to identify a design methodology that can be used with proper specific adaptation to each of the two views, namely robot engineering and architecture survey. Similarly, a design methodology can be elaborated for survey strategies with the aim to determine aspects and peculiarities that can be devoted or not to mechanical and robotic systems. In particular, situations and conditions of the on-field Survey activity can be redesigned towards a proper use of assisting robotic devices. The study of manipulations and other actions can be carefully analyzed with proper measures from different perspectives with the aim to identify design margins in rethinking the Survey procedures for general and specific tasks. Particular attention can be addressed to the cases in which human operators cannot work and therefore mechanical and robotic systems could be seen as the only solution for Survey activity. It is to note that this phase can include preliminary activity of the next phase since operation issues can be significant in defining design features both in term of robotic systems and Survey strategies. Indeed, although differentiated in the 2-nd and 3-rd phase the activity can be planned with strong interaction between system design and operation planning both the systems and Survey activity.

3-rd phase: development of methodologies and survey strategies for operation of mechanical and robotic systems

In this phase design activity can be directed specifically to the possibility of using robotic systems and Survey strategies that have been designed in the previous phase. Attention can be addressed to operation features in terms of system programming, system equipment, user-oriented facilities, human-machine interfaces that can determine a successful useroriented results for Survey operators. Similarly to the activity of previous phase, in this phase the design activity can be carried out through simulations and experiences to define procedures for the analysis and to conceive operation strategies jointly and separately for the systems and their use in Survey activity. Simulations can be carried out with numerical computations and even with laboratory experiences, within the specific frame of each research group and finally with an integrated view. Even Survey operations can be reconsidered for a proper design that can fit the possibility of the mechanical and robotic systems. Thus, actions and sensorization can be analyzed and considered for outlining requirements and constraints of the operation of the systems and Survey activity, in joint and separate perspectives. Indeed, the activity can run with a enlarged mechatronic approach that will include considerations, models, and whenever possible formulation from Architecture Survey.

4-th phase: construction of prototypes and study of applications

In this phase design results both for systems and operations can be applied for construction of prototypes. The prototypes can be a first construction of systems with full operation capability and Survey strategies in terms of activity planning for real operations. During the definition of the mechanical design and final arrangements of prototype designs, inputs can be considered from on going experiences and even on the progress of the integrated views for a mechatronic approach that include also cultural issue of Survey activity. Prototype construction can refer also to the operation strategies in term of suitable programming not only in term of robotic systems but even at the level of Survey planning. Applications on which the prototypes of systems and operation can be tested, can be defined with general frames, but also several specific applications can be studied and defined for testing the versatility of the new designed robotic systems and operations of new Survey strategies. Applications can be studied and planned at different scales of operation, namely from large Architecture areas up to single pieces of architectonic goods. In this phase the user-oriented solutions can be also experienced with prototypes that should have acceptance from the Survey operators. Therefore, contributions can be expected also in the activity of construction and adjustment of the design solutions, even through optimization of the design procedures and simulations that have been defined in the previous phases. The construction of prototypes permit a practical check of the fulfilment of the goals of the new desogns also in terms of easiness of transportability, manoeuvrability, handling, and adaptation in different Survey applications. In addition, during the construction of prototypes and definition of applications, emerging problems could be used for an optimization of the results and procedures that have developed in the previous phases, even as a consequence of a practical experience with solutions under construction.

5-th phase: tests and validations

In this phase the soundness of results can be proved with tests that can be carried out with the prototypes in laboratory experiences and finally in on-fields applications of Survey activity. Tests can be carried out with the aim to check and adjust the feasibility of the proposed prototypes for a user-oriented use of designed robotic systems and their operations by Survey operators. In this phase, the results of the tests and the feasibility of experienced applications in different environments and conditions can be still used to adjust the prototypes but even to optimize the design and operation procedures.

Novel aspects of the proposed robot application can be identified in the multidisciplinary cultural frame for identifying, designing, experiences engineering solutions for Survey in Architecture and History of Architecture, both in terms of new systems and new operation strategies. The idea of using robotic systems for Survey activity can also stimulate rethinking of current Survey practices not only as depending of the design goals of new systems but even for new approaches in Survey activity with or without assisting devices.

6. Case study of the floor of Montecassino abbey

For centuries one of the most important monuments in European architecture, Montecassino Abbey was not only a spiritual, cultural and artistic centre, but, during several periods of history, was also the organisational and economic centre of what was once the "land of Saint Benedict".

Fig. 10. The Abbey of Montecassino as it appears today after its reconstruction.

Saint Benedict arrived in Montecassino in 529, where he found the remains of what had once been a flourishing Roman town. In his *Dialogues,* Gregory the Great (Gregorio Magno, 1924) tells us that he destroyed the temple of Apollo and its Ares and set fire to the sacred woods to make room for what would be the first Benedictine monastery.

The first destruction of the abbey, which was demolished and rebuilt, took place in around 581 by the Lombards. Rebuilt at the start of the 8th century, the monastery underwent a period of magnificence until 883 when it was devastated once again by the Saracens. Abbey life began again only after a century, when it was rebuilt in around 950.
In the 11th century, Montecassino went through the most splendid period of its history, when in 1058, the son of the prince of Benevento was elected abbot there. The new abbot, Desiderio, who later became Pope Victor III (1086-1087), began a period of great architectural works, such that at his death the monastery had been almost entirely rebuilt and in such a grandiose way that it stood head and shoulders in magnificence above anything else of its kind in the Western world of that era (Bertaux, 1904).
It was again destroyed by an earthquake in 1349, and in the subsequent centuries, every period of history has left a trace on the monument, although the great transformations of the 16th and 17th centuries left the deepest mark on the architectural story of the abbey, leaving us with a great Renaissance and Baroque monument, but obscuring the mediaeval works.
On 15 February 1944, the monastery was almost completely razed to the ground by ally bombs, leaving only the south-eastern corner of the building intact. Reconstruction work began a year later, and was officially completed on 25 October 1964, when the abbatial basilica was reconsecrated for the fourth time since its founding.
Despite all of the events of which it has been a protagonist, fourteen centuries of the Benedictine monastery at Montecassino, destroyed and rebuilt on several occasions, is neatly summed up in its coat of arms: an oak tree cut through completely, which reflourishes and above it the motto "*succisa virescit*".
This rapid summary of the historical background of the architectural complex of Montecassino has served to introduce the floor of its abbatial basilica, placed in the highest point of the monastery. Indeed, the pre-Cosmatesque floor surface of the basilica of Montecassino abbey is the specific case study for application of the robotized systems and automated procedures for architectural surveying.
In 1058, Desiderio of the Benevento princes was elected abbot of Montecassino and, in around thirty years of rule, brought the coenobium to the height of political, spiritual and artistic-architectural expansion. First as abbot and then as pope, retaining the abbotship after his election to the papacy (Cigola, 2005), he performed great architectural works in the monastery, which culminated in the reconstruction of the abbatial basilica. These works are recorded in a few mediaeval chronicles, where we learn that between 1066 and 1071 Desiderio brought artists specialised in mosaics and inlaying from Byzantium to cover the entire church floor with different varieties of marble (Hoffmann, 1980), mainly recovered from monuments in Rome. The new floor of the basilica of Montecassino marked the passage from ancient floor mosaics to mediaeval ones and marked the start of the great season of Cosmatesque art (Claussen, 1992).
Unfortunately, it is impossible to fully grasp the significance of the floor of Montecassino, as it is no longer visible (Della Marra, 1775). It was later covered by another marble inlay floor during the 18th century and for a long time it was thought that it was lost, but we have an idea of what it was like from a few image representations and parts of it that have survived and have been re-laid in several chapels in the monastery.
The first is an 18th century survey (Gattola, 1733) carried out with great care and attention to details, which presents all the elements that we find in Cosmatesque floors, of which the Montecassino floor is the first example. The imaging, included in the monumental work by

Gattola, is dated Naples 1713 and signed by the inlayer Andreas Maliar. Originally carried out in large scale although destined, although later reproduced in smaller scale, it shows the entire mosaic surface in detail, perfectly replicating the geometric designs which, run alongside each other to form most of the surface.

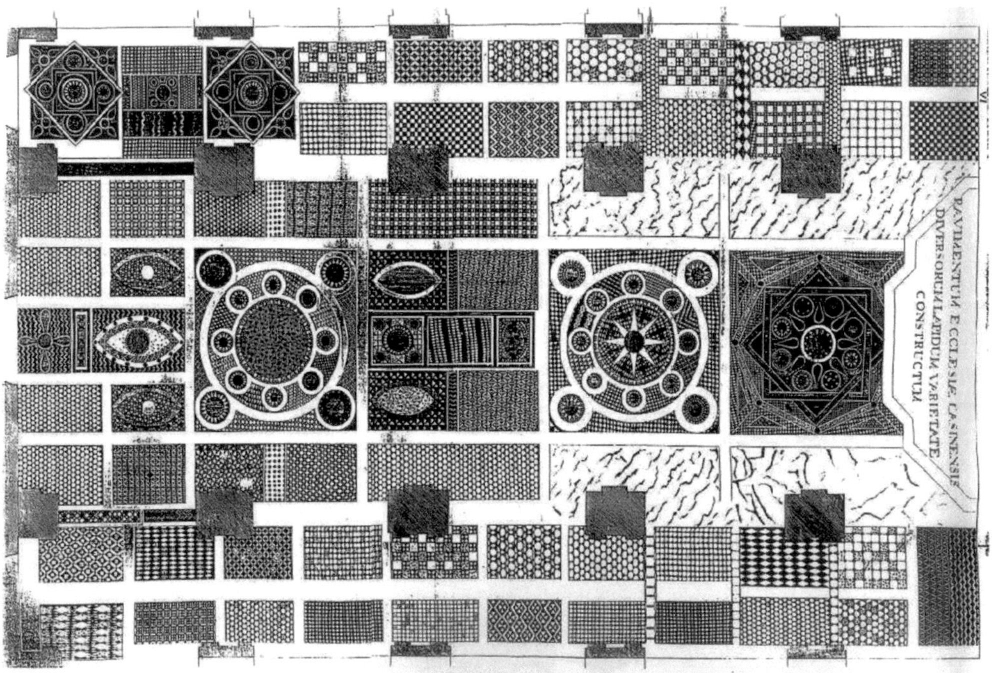

Fig. 11. Cosmatesque pavement of Montecassino Basilica in 1730 (Gattola, 1733).

The second survey was carried out by someone who is not well known, but is in fact one of the protagonists of the post-war reconstruction of Montecassino: Don Angelo Pantoni, monk and engineer, lived at Montecassino for over fifty years and carried out an in-depth study of the history of the monastery, reconstructing the events of its artistic and architectural life (Docci & Cigola, 1995). The survey of the flooring is based on the tables found in the work by Gattola, but is more faithful to what must have been the original, including details that are definitely original, and were not included in the 18th-century engraved.

During reconstruction work, Don Angelo Pantoni was able to conduct important archaeological studies and made important finds from the Roman and late mediaeval periods of the complex, which had been hidden by subsequent constructions. This included most of Desiderio's floor, which had been thought lost, and allowed Pantoni to create image and photographic testimony of this floor before it was covered up once again. Indeed, most of the floor mosaic returned to oblivion in an inaccessible cavity beneath the new basilica reconstructed between 1948 and 1952; however but another section, consisting of a few dozen geometric design panels, was later repositioned in various chapels of the monastery.

An analysis of the image testimony, Desiderio's floor appears to be a composition made of an orthogonal network of pale stone strips whose main purpose is organising the rhythmic flow of the composition sequence of the floor surface, as well as contributing to and alternating between the main elements of the architectural organism and the square sections of the Cosmatesque structure.

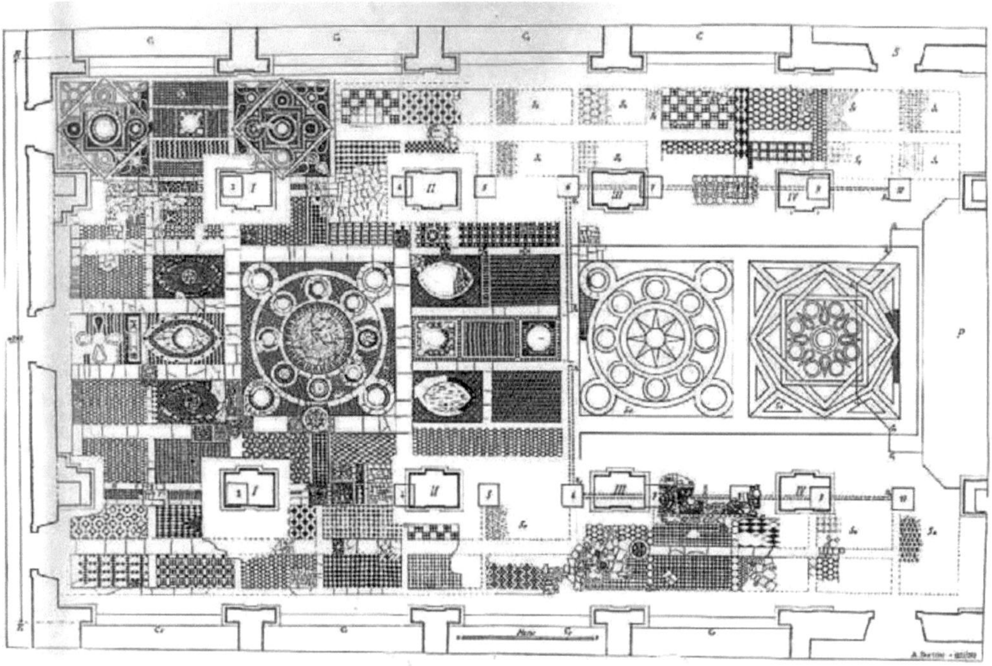

Fig. 12. Montecassino, cosmatesque pavement of basilica (1066/1071) beneath current in a survey of 1951-52. (Pantoni, 1972).

Pantoni's survey shows the entire floor surface as it was immediately after the destruction of the basilica by ally bombs. The structure of the entire composition, conditioned by the presence of the large central strip, consists of a series of sections differing in terms of dimension and decoration from those in the minor aisles, acting as a privileged walkway to the altar, marking it with sections that become more intricate as they lead towards the presbytery; in the side aisles, the surface was saturated using rectangular panels that are clearly inspired by the order of columns of Desiderio's mediaeval plan, rather than the Baroque version that replaced the 11th-century architecture.

The reconstruction plans of the abbey made in the early 1950s are useful for identifying where the mediaeval floor is today. The Italian government, in agreement with the Holy See, decided to reconstruct the Basilica immediately after the war, "where it was, the way it was and using the existing architectural and volumetric lines", as a means of social and historical affirmation and because, once the rubble was removed, it was seen that the foundations of the various buildings had not been undergone significant damage and it was possible to recover architectural remains and precious decorative fragments.

Reconstruction was based on the numerous drawings and surveys that the abbey that Pantoni had made in over ten years of work. These were used as to ensure that the reconstructed monument was the exact copy of the one destroyed by bombing. The image representations served as reconstruction plans and are preserved In the abbey library at Montecassino. They are defined in the scroll bearing the abbey coat of arms and the writing

"Abbazia di Montecassino. Ricostruzione dell'Abbazia", and in one of these, a transversal section of the monastery, it is possible to fully understand the current position of the floor within the cavity beneath the floor reconstructed after the war (Cigola, 1997) .

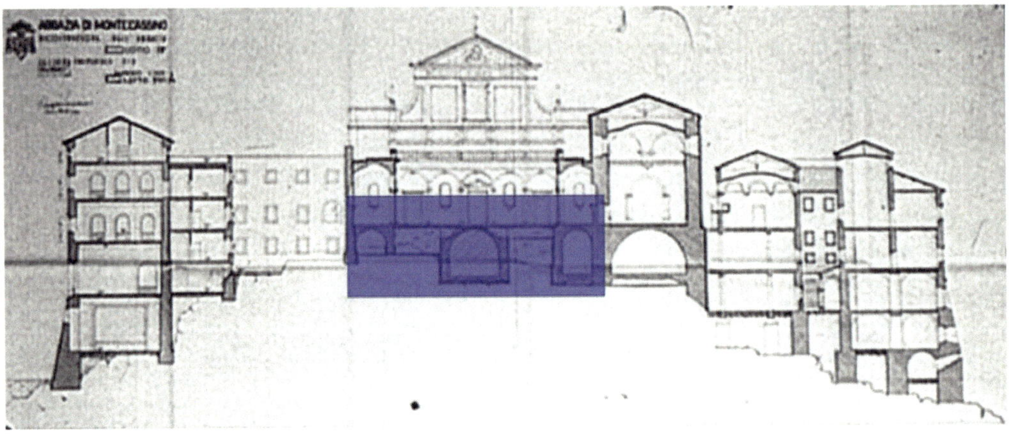

Fig. 13. Cross-section of reconstruction design for Montecassino Abbey (blue area indicates the area of ancient pavement)

In particular, figure 14, carried out during the excavations and surveys done before the start of reconstruction, highlights the difference in level of about one metre between Desiderio's floor and the current ground floor. Only available a very low cavity is available for inspections and other operations on the Cosmatesque mosaic, with no light or air and therefore unsuitable for work by humans. For this reason, after reconstruction of the new basilica, the floor was never inspected to analyse its state of health or carry out stylistic analyses of its decorations.

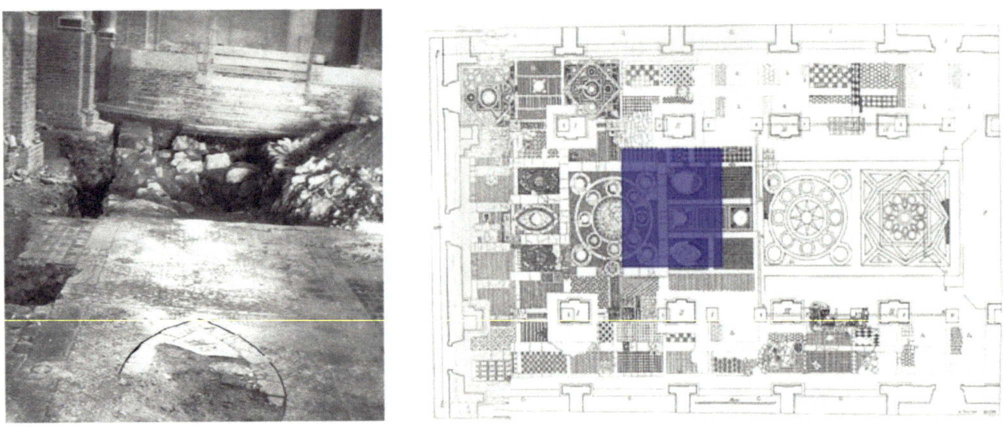

Fig. 14. A photo of the left side of the antique pavement before the reconstruction of the closed space in 1952 (Pantoni, 1951).

This case study is characterised by the impossibility of human operators to work a Survey activity and therfoere it is well suited to robot experimentation involving the analysis and conservation of a historical floor as an example of activity for architectural heritage.

The needs of the DART team for Survey activity can be understood as requirements for the LARM team in designing structure and operations of robotic system. The robotic system must operate in the narrow environment while moving on the historical cosmatesque pavement with the aim to perform several survey operations by using suitable equipment. Thus, main characteristics for the robotic system can be outlined in a mobile platform carrying several instruments for survey of the pavement can be chosen for image capturing in terms of photos and videos, even for thermal inspection, laser scanner or other sensor for distance and orientation measures. Thus, this will require a certain volume available on board of the robotic system. Careful motion of the robotic system is expected in order to ensure full image capturing of the pavement and its environment and to avoid damage of the pavement itself or other components of the environment under exploration.
Once considered the above-mentioned conditions and constraint, it has been thought convenient to design a multilegged robot with a suitably large body and powered wheeled feet as shown in figure 15.

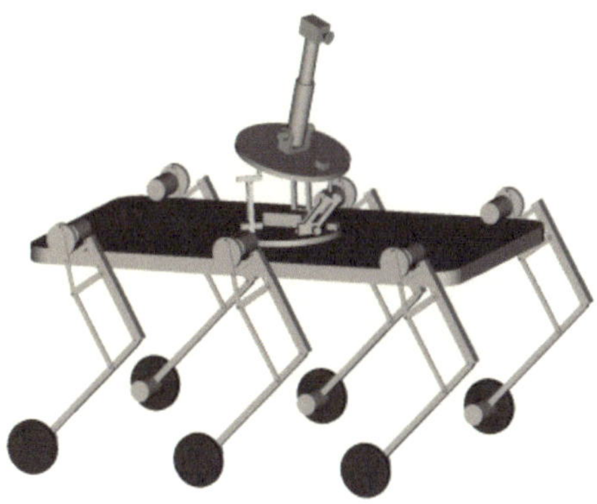

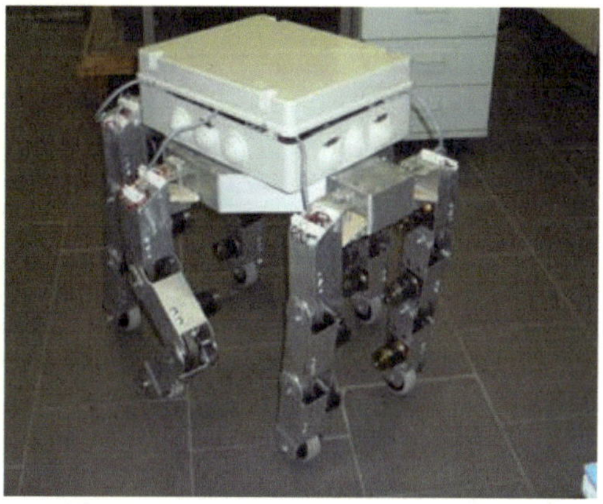

Fig. 15. The Cassino hexapod robot for Survey activity as a general kinematic scheme and a built prototype.

The idea of have several modular legs fulfils also requirements of low-cost design and easy operation features that are directed to permit implementation of the robotic moving platform even by the survey operators, once properly trained. Even with this aim, a simulation frame has been developed as shown in Fig, 16.
The design of Cassino hexapod robot has been developed by looking at the needs of Survey activity with several interactions involving also the above-mentioned simulations (Carbone, et al., 2007). In particular, the structure with six legs has been chosen in order to ensure contact with ground always with at last three feet so that a plane can be easily determinate for the robot body and survey equipment on it.
Powered wheeled feet are installed on the legs in order to have the possibility to smooth and regulate the contact force during the walking motion of the robot in order to prevent damage of the pavement surface and/or instable displacements. Once properly sensored the legs can be used also as a Survey equipment to measure the size (length and higth) of the area under Survey examination.
Virtual simulation by using virtual reconstruction of the environment has helped and will help to adjust and plan the survey operation wit proper feasibility and attention to the

environment yet. Such a simulation is also helpful to reduce the time and effort of the laboratory test of a prototype, which is still underdevelopment for a practical on-field application. Althougt the Cassino hexapod robot for Survey activity is not yet read for operations, the work for its design and operation planning has permitted to achieve a good understanding between the two team of DART and LARm laboratiries towards a succesful application of robot in Survey an Conservation of historical architectural goods.

Fig. 16. A simulation example of operation of the designed Cassino hexapod robot for Survey activity

The interaction between the members of DART, experts in architectural surveying, will provide the main elements of the operations that the robot will have to perform inside the cavity of the Cosmatesque floor at Montecassino. These indications are thus put forward and made possible by the components of the LARM, experts in robotized systems.

These operations can be summed up as follows: the robot's first characteristic is its capacity to move within the area under examination, carrying image capturing equipment to perform the first analyses on the object: cameras for photographs and video; in a subsequent phase, this equipment must be replaced with more sophisticated instruments that can provide more precise information, such as thermal imaging equipment and laser scanners.

At the same time, since the floor in the cavity is extremely irregular due to the presence of architectural items from the same era and more ancient sections of the floor, in addition to forward motion the robot must also ensure that the image capturing equipment remains parallel to the floor. In addition to these characteristics, since knowledge of the current layout of the floor is extremely important for analysis and conservation purposes, the moving robot must be able to memorise and transmit the variation in the length of its legs during its journey to an external unit, in order to provide variations along the horizontal axis in real time, which will also provide a section of the floor at the same time.

To start experimenting the suitability of the elements proposed, these operations are being tested virtually, with simulation of the cavity and the floor surface, which can be recreated from the existing precise surveys carried out during reconstruction of the monastery.

This type of monitoring, carried out in virtual 3D recreated by computer, is proving extremely useful, since it allows a reduction in time and costs in the planning phase, given that it is possible to thoroughly evaluate and test the model from the initial stage of design, operating within an environment that efficiently simulates the area in which the robot operates, allowing specific modifications and variations that would otherwise be cumbersome if carried out on a prototype.

7. Conclusion

The goal of this work is to test the possibility of designing robots and/or robotized system that are specifically dedicated to the field of cultural heritage and particularly to historical architecture. At the moment, our research is in the final stage of design of a robot, with activity that is carried out by the two LARM and RADET laboratories, with a continuous review of the needs that arise from the architectural application and from mechanical and mechatronic design.

Preliminary activity in Architecture Survey has been carried out at locale frame in Cassino with a construction of a very preliminary prototype and basic simulation experiences that were directed to the possibility to inspect and study the original cosmatesque Middle Ages pavement of Montecassino cathedral that is beneath the current pavement in a closed space with no air and no light. The mediaeval floor of Montecassino is an emblematic example of the interest and potentialities of using robotic systems in Architecture Survey, since it requires in-depth study to improve knowledge of the pavement that is not yet fully known. It is also necessary to further study the related documentation and analysis techniques for the safeguard and conservation of stone surfaces and ancient basilica of Montecassino, which lies hidden as enigmatic architectonic good beneath the current basilica.

8. References

Bertaux, E. (1904). *L'art dans L'Italie meridionale,* A. Fontemoig, Paris.

Carbonara, G. (1989). Disegnare per il restauro. *Disegnare. Idee Immagini,* No. 0, (ottobre 1989) (85-94), ISBN 88-7448-265-5

Carbone, G., et al., (2007) Operation strategy for a low-cost easy-operation Cassino Hexapod, *Applied Bionics and Biomechanics,* No. 4:4, (149–156). DOI: 10.1080/11762320802002573

Ceccarelli, M., et al., (2002) A study of feasibility of using robots in architecture analisis and survey of a historical pavement, *Proceedings of 11th International Workshop on Robotics in Alpe-Adria-Danube Region,* pp. 113-118, ISBN-963 7154 10 8, Balatonfired, june 2002, Budapest Polytechnic Editor, Budapest.

Ceccarelli, M. (2004) *Fundamentals of Mechanics of Robotic Manipulation,* Kluwer Academic Publishers, ISBN1-4020-1810k 2004, Dordrecht, The Netherlands

Cigola, M. (1993) Mosaici pavimentali cosmateschi: segni, disegni e simboli, *Palladio,* Vol. VI No. 11, (giugno 1993) (101-110).

Cigola, M. (1997) L'abbazia di Montecassino. disegni di rilievo e di progetto per la conoscenza e per la memoria, *Disegnare. Idee Immagini,* No. 14, (giugno 1997) (43-52). ISBN 88-7448-819-k

Cigola, M. (2005) L'abbazia benedettina di Montecassino. La storia attraverso le testimonianze grafiche di rilievo e di progetto, Ciolfi Editore, ISBN 88-86810-28-8, Cassino.

Cigola, M., et al., (2005). Application of robots for inspection and restoration of historical sites, *Proceedings of 22st International Symposium on Automation and Robotics in Construction*, CD, University of Ferrara, september 2005, Ferrara.

Cigola, M. & Ceccarelli, M. (2006). Documentation and conservation of built heritage by using robot, Proceedings of XI International Seminar Forum Unesco Documentation for conservation and development. New heritage strategies for the future, pp. 64 - CD, ISBN- 10: 88-8453-494-1, University of Firenze, september 2006, University Press, Firenze.

Claussen, P.C. (1992). Marmi antichi nel Medioevo romano, L'arte dei Cosmati, In: *Marmi antichi,* G. Borghini, (Ed), (65-79), Leonardo-De Luca, ISBN 88-7813-265-9, Rome.

Della Marra, F. (1775) Descrizione istorica del monasterio di Monte Cassino per uso e comodo dei forestieri, Fratelli Raimondi, Napoli.

Docci, M. & Cigola, M. (1995). Disegno come Memoria, memoria come disegno. L'Abbazia di Montecassino, *Proceedings of Congresso Internazionale Il Disegno Luogo della Memoria,* pp. 600-610, ISBN 88-8125-034-9, Firenze, settembre 1995; Alinea, Firenze.

Gattola E. (1733). *Historia Abbatiae Casinensis,* Sebastianum Coleti, Venezia.

Glass, D.F. (1980) Studies on Cosmatesque Pavements, *Oxford British Archeological reports Series,* No. 82, (1980) .

Gregorio Magno, (1924). Gregorii Magni Dialogi: libri 4, In: *Fonti per la Storia d'Italia 57,* U. Moricca, (Ed), (325-787), Tipografia del Senato, Rome.

Guidobaldi, F. & Guiglia Guidobaldi A., (1983) *Pavimenti marmorei di Roma dal IV al IX secolo. Studi di Antichitá Cristiana.* Pontificio Istituto di Archeologia Cristiana, Vatican City.

Hoffmann, H. (1980). Chronica monasterii Casinensis, In: *Monumenta Germaniae Historica. Scriptores XXXIV*, Hahnsche Buchhandlung, ISBN - 3775253165, Hannover.

Pantoni, A., et al., (1951). Esplorazione archeologica, In *Il sepolcro di s. Benedetto, Miscellanea Cassinese* 27, Monastero di Monteassino, (Ed), (69-94), Sansaini, Rome.

Pantoni, A. (1972). Descrizione di Montecassino attraverso i secoli, In *Benedictina* XIX 2, (539-586) Abbazia di Montecassino, Montecassino.

21

Performance Tests for Wireless Real-time Localization Systems to Improve Mobile Robot Navigation in Various Indoor Environments

Yong K. Cho*, Jong-Hoon Youn** and Nam Pham**
**Construction Engineering & Management Division, School of Architectural Engineering and Construction, University of Nebraska-Lincoln,*
***Computer Science Department, University of Nebraska-Omaha,*
USA

1. Introduction

This research introduces a research effort at the Peter Kiewit Institute in Omaha, Nebraska by investigating the performances of current wireless real-time localization technologies. Futhermore, the research shows how localization technologies can be applied to sensor-aided intelligent mobile robots for high-level navigation functions for indoor construction security and material delivery.

Sensor-based exploration enables a robot to explore an environment and to build a map of the explored environment. A critical component of sensor-based exploration is robot's ability to ascertain its location in a partially explored map or to determine that it has entered a new territory. Theoretically, one can determine the (x, y) coordinates of the robot using dead-reckoning - a process that determines the robot's location by integrating data from wheel encoders that count the number of wheel rotations. However, dead-reckoning often fails to accurately position the robot for many reasons, including differential of wheel rotation rate and wheel slippage. Especially when the robot slips, the wheel rotation does not correspond to its movement, and thus encoder data, which reflects the state of the wheel rotation, does not reflect the robot's net motion, thereby causing positioning errors. A global positioning system (GPS) offers an alternative to dead-reckoning, but it is limited to outdoor applications.

Tracking mobile assets in indoor environments is a challenging task, especially for large open spaces such as airport terminals and museums. Among the emerging technologies, mobile devices and wireless technologies are widely recognized as solutions for identifying locations of mobile assets in such areas. However, the integration of these technologies into indoor building space has been limited. For example, one type of building space in which the integration has been particularly slow is a highly congested area with room partitions, metal structures, furnitures, and high traffic of people. Location tracking in such environment often has low valuation attributes, including reliability, security, and performance. In turn, the lack of these attributes has prevented high performance wireless networks from replacing traditional IT systems in critical applications. Critical applications where the wireless networking technology can be effectively applied are military training, mobile resource tracking in construction, fire fighter tracking, rare or endangered animal tracking in a natural habitat. Another useful implementation is in hospitals and health care facilities, where a number of problems can be solved effectively and efficiently by taking advantages of the wireless networking technology. A related study estimated that as many as 98,000 people die in the U.S. hospitals each year due to medical errors (Kohn et al., 2000). Using wireless networking technologies to efficiently manage medical records and important assets, these errors and the number of misdiagnoses can be significantly reduced.

Although the wireless networks are identified as the most promising technology to track indoor mobile assets, building environment factors such as building type, business type, and geographical location significantly affect the position accuracy. This paper presents a framework for strategic planning in tracking mobile assets in various indoor environments and scenarios, using three most popular wireless technologies. Then, the paper demonstrates one of the wireless technologies and its integration to mobile robot's path planning system to improve its navigation in an indoor environment. The rest of this paper is structured as follows. Section 2 surveys related works. Section 3 discusses the development and deployment of three wireless sensor technologies for real-time asset tracking in building environments. Also, this section presents experimental results of each deployed tracking system. Section 4 details how an ultra-wide band wireless tracking system can be installed on a robot in order to improve robot's nagivation. Concluding remarks are presented in Section 5.

2. Related works

A number of indoor positioning systems are available in the literature. Among these systems, common positioning techniques include *trilateration*, *multilateration*, and *location learning*.

- *Trilateration* uses range estimates of the distances between devices and calculates positions of target devices using geometric identities and known locations of other devices. Distances can be estimated with time of arrival (TOA) or signal strength. With TOA, two devices must be synchronized, and messages between the devices are time-stamped upon sending and receiving in order to calculate propagation delay. The known propagation delays of signals in a particular medium allow the devices to estimate distance.
- *Mutlilateration* uses time difference of arrival (TDOA) estimates in which several reference devices measure the difference in arrival times of signals from the target devices. Round-trip time can be used when synchronization is not possible. For triangulation, the angle of arrival (AOA) of a signal is measured using several antennas, and then geometric identities are used for estimating position.
- *Location learning* makes no range or angle measurements. Instead, the method merely correlates the properties of newly received signals with data available on previously observed signals at known locations.

These basic techniques can be used with a variety of signal types in wireless systems: wireless local area networks (WLANs), wireless sensor networks (WSNs), ultra-wide band (UWB) networks, and RFID systems.

A. Wireless LAN Based Systems

The use of radio frequency (RF) properties for determining position in building environments is always challenging because of obstacles such as walls, furnitures, people traffic, and interferences with other RF noises. For this reason, localization in WLANs often relies on learning techniques. One WLAN based tracking system is RADAR (**Ra**dio **D**etection **a**nd **R**anging) - an indoor tracking system developed at Microsoft Research (Bahl & Padmanabhan, 2000). RADAR is a learning-based approach which takes advantages of the existing WLAN infrastructures. Localization or tracking with RADAR consists of two phases: *a reference signature collection phase* and *an online estimation phase*. During the signature collection phase, a user with a laptop clicks his or her perceived location on a map interface and records the signal strength from all access points (APs) within range. After collecting a sufficiently large database of reference signals, location can then be estimated in the online phase by taking the geographic centroid of the locations of the *k*-nearest (in terms of signal-strength space) reference signatures. The same process may be used with other traditional machine learning algorithms and has been studied on Artificial Neural Networks (Battiti et al., 2002), Bayesian techniques, and Markov models (Haeberlen et al., 2004; Ladd et al., 2002). Variants of the RADAR system are available from commercial vendors such as PanGo (Pango, 2008) and Ekahau (Ekahau, 2008).

B. Wireless Sensor Based Systems

As one of the earliest location tracking systems for sensor networks, the Active Badge system was developed by AT&T Cambridge for indoor location tracking using diffuse infrared technology (Want et al., 1992). After Active Badge, AT&T researchers developed the Active Bat system to improve Active Badge's limitations on 3D location and orientation information (Harter et al., 1999). However, Active Bat requires a large scale of expensive wiring infrastructure to relay information collected by the receivers. To compete with the Active Bat system, Networks and Mobile System research group at the Massachusetts Institute of Technology (MIT) developed the Cricket location-support system which allows applications running on user devices to learn their physical location (Priyantha et al. 2000). The Cricket system consists of beacons (mounted on a wall or ceiling) to emit radio frequency signals and receivers (attached to the user's mobile device) to receive beacons' RF and ultrasonic signals. Unlike the Active Bat system, Cricket does not require a grid of fixed ceiling sensors because the receivers perform the timing and computation function (Skibniewski et al. 2007). However, timing and processing both the ultrasound pulses and RF data may increase computation and power burden on the mobile receivers, and it is not easy to monitor the performance of receivers due to decentralized management (Hightower et al. 2001). MoteTrack is a sensor-based system which runs on 802.15.4 based motes but uses a process based on RADAR. With MoteTrack, the environment must be equipped with several fixed sensor motes as the existing LAN infrastructure cannot be used. Reference signature collection and the online estimation operate as in RADAR, but MoteTrack has been altered to run in a distributed manner. MoteTrack has the advantage of being entirely based on RF signals and needs relatively fewer beacon nodes to cover a large area of a building, even an area with many obstacles. However, it also has a disadvantage that it requires significantly more configuration prior to deployment.

C. Ultra-wideband (UWB) Systems

Gezici et al. discussed many of the positioning techniques described earlier in the context of UWB systems in which high bandwidths offer potentially high ranging accuracy. They note that the antenna arrays required for the angle of arrival(AOA) make it unsuitable for UWB, but consider ranging with time-based measurements and signal strength measurements. They found that the best results can be obtained with hybrid schemes employing TDOA and TOA both with signal strength measurements (Gezici et al., 2005). Young et al. noted that the high bandwidth of UWB systems allow for high time resolution leading to a natural advantage with TDOA localization. They present methods for overcoming the inherent distortion problems with UWB antenna responses, amplification, and filtering in an indoor multipath environment (Young et al., 2003). Zetik et al. also approached UWB localization using TDOA. They conducted experiments in both the active and passive setting with custom designed SiGe circuit architecture. With their system, they were able to achieve a localization accuracy of within one centimeter (Zetik et al., 2004).

D. RFID-Based Systems

Location determination using RFID tags is a difficult problem because tags have extremely limited computational ability to assist the application and a very short reading range. Active tags contain a battery and generally have longer ranges than passive tags. The simplest approach to localizing tags is to use the proximity with readers. The limited reading range can be used to estimate the location of a tag based on the location of a reader (Nara et al., 2005; Philipose et al., 2003).

3. Experiments with the selected sensor systems

This section presents the experimental results of the wireless networking technologies (described in Section 2) in different types of building spaces. Advantages and disadvantages of each system are discussed as well. The RFID-based system, however, is excluded in this study because of its short coverage, expensive RFID readers, and the lack of support for pratical applications in a dynamic building environment.

3.1 Wireless LAN based system

The accurate positioning of a particular device is a major challenge. State-of-the-art wireless tracking technologies, such as wireless sensors, RFID, and proprietary WLAN based-sensors, typically require a costly dedicated network infrastructure. However, an 802.11 -based tracking system can be deployed without significant additional cost. Since many buildings are now rapidly deploying facility-wide 802.11 WLAN infrastructures, a WLAN-based tracking approach would provide a cost-effective solution that takes advantages of their wireless infrastructure for asset tracking.

We have deployed a real-time asset visibility system based on an existing 802.11 infrastructure in order to track WLAN tags at the Peter Kiewit Institute (PKI). The system consists of two key software components: a real-time positioning engine that calculates location of assets and a web-based graphical user interface (GUI) that manages system configuration, asset visibility, monitoring and reporting. The positioning engine is based on statistical modeling of received signal strengths and provides accuracy of up to 1.2 meters on average. The GUI provides a user friendly web interface that simplifies asset tracking and improves everyday operations, such as asset monitoring and notifications. Since this tracking system is fully software-based, it requires no proprietary network infrastructure.

The first step of the deployment procedure is the data collection phase. Once the access points (APs) were established, the Received Signal Strength (RSS) values were collected from the APs as a function of the mobile's location and orientation. From the measurements, we noted that the RSS values at a given location vary significantly depending on the mobile asset's orientation. Thus, RSS values were collected in each of the four directions (i.e., north, south, east and west) at a number of selected physical locations on the floor. After the data collection phase, the RSS values were imported into the positioning engine and processed to enhance the accuracy of location estimation.

After constructing a database of RSS measurements, called *signatures*, along with their known 2-dimensional locations and orientations, the system can estimate mobile's position by comparing the measured RSS data to the known signatures in the database. In other words, a mobile device takes a snapshot of RSS from visible APs and compares it with signatures stored in the database. To reduce the computation cost, the search is performed only on some portion of the RSS measurements in the database. If mobile's previous location lies at a point P, then the search space is limited to its neighboring points within the distance d from P. These neighboring points are grouped into clusters based on their physical closeness. For each cluster, the most probable location of the mobile node is calculated based on the Euclidean distance of RSS measurements. For example, the RSS measurement ($p_1, p_2, p_3, \ldots, p_k$) at a point P and ($s_1, s_2, s_3, \ldots, s_k$) at point S are the closest if the Euclidean distance between P and S $(p_1 - s_1)^2 + (p_2 - s_2)^2 + (p_3 - s_3)^2 + \ldots + (p_k - s_k)^2$ is minimum. After a number of computations, the system chooses the location with the highest likelihood as the current estimate of the user's location.

3.1.1 Accuracy

Under a number of different Wi-Fi network configurations, this study evaluates the accuracy of the system. According to a recommendation from Cisco, a positioning system needs to collect a minimum of three strong and steady Received Signal Strength (RSS) measurements from APs to determine the fine location of Wi-Fi tags with room-level granularity. The Cisco's wireless location appliance guide (Cisco, 2008) also recommends approximately one access point to be placed every 17-20 meters (i.e. roughly one access point is needed every 230-450 square meter). Thus, in this section, we discuss the accuracy of the system as a function of the number of APs, and the relationship between the AP layout and the location accuracy.

3.1.1.1 Impact of the Number of Access Points

In this study, 8 points on the floor were chosen to measure the Euclidean distance between actual locations and corresponding estimate points. Intuitively, the accuracy would be improved as the number of APs increases. The positions of 10 APs and 8 measurement points are shown in Fig. 1.

The averaged error distance of the deployed system for each point is shown in Tables 1 and 2. As predicted, the error distance gradually decreases as the number of APs increases. This study conducted the same set of experiments with more than 10 APs, but there was no significant improvement in the accuracy of the position estimation. So, the results of the experiments with more than 10 APs were not included.

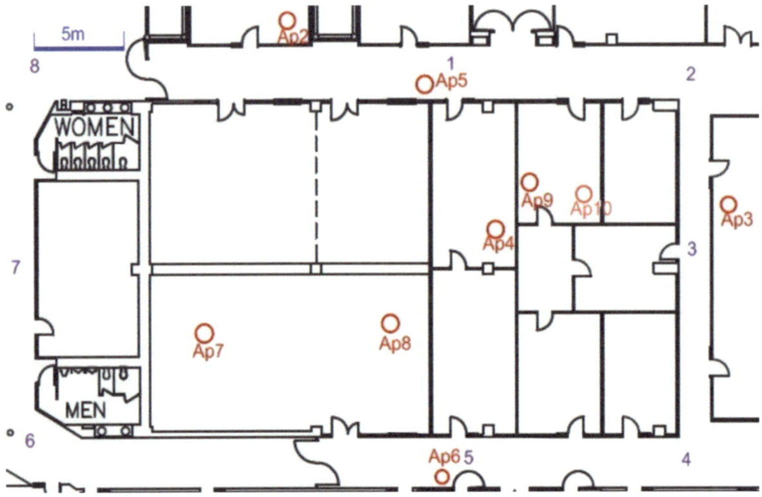

Fig. 1. The Layout of APs

Measured Position	Error Distance (m)			
	3 APs	4 APs	5 APs	6 APs
#1	7.168	5.437	3.715	1.387
#2	3.027	2.773	0.836	1.192
#3	3.522	2.72	2.931	3.125
#4	2.08	2.092	1.69	1.477
#5	5.323	4.076	3.16	3.14
#6	3.971	2.611	2.351	2.294
#7	5.066	4.408	3.127	2.149
#8	0.614	0.53	0.472	0.306
Average	3.8464	3.0809	2.2853	1.8838

Table 1. Error Distances in Meters (up to 6 APs)

Measured Position	Error Distance (m)			
	7 APs	8 APs	9 APs	10 APs
#1	1.693	1.16	0.813	0.781
#2	2.285	1	1	0.951
#3	1.048	2.047	1.972	1.874
#4	2.543	0.873	0.784	0.774
#5	2.253	1.9	1.876	1.78
#6	1.953	1.732	1.756	1.747
#7	0.245	2.5	1.396	1.207
#8	1.642	0.136	0.123	0.102
Average	1.693	1.4185	1.215	1.127

Table 2. Error Distances in Meters (up to 10 APs)

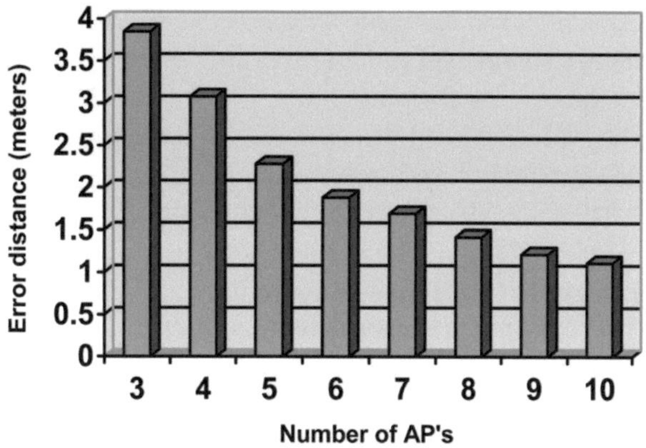

Fig. 2. Impact of the Number of APs on Error Distance

3.1.1.2 Impact of the AP Layout

A Wi-Fi based tracking system provides real-time visibility and tracking. The density and locations of APs are crucial factors in improving the level of accuracy. APs stagger in a way that signals vary in each location. By surrounding the deployment area, the APs provide the greatest chance of achieving room-level positioning. Table 3 shows the error distances for different layouts of the APs. In this experiment, only RSS values from the selected 4 APs were used to estimate the position of Wi-Fi tags. The positions of APs are shown in Fig. 1. As shown in Table 3, the selection of APs has a significant impact on the system accuracy. For example, the accuracy of the last scenario which selects AP#4, AP#5, AP#9, and AP#10, are poor comparing to the results of other scenarios. This is because, with this AP layout, there are some spots where a tag cannot gather three or more consistent and strong RSS samples from the selected APs.

Measured Position	Selected APs			
	1,2,3,4	2,3,6,7	1,2,6,10	4,5,9,10
#1	5.437	0.967	1.923	2.58
#2	2.773	1.992	7.302	7.323
#3	2.72	2.418	1.674	16.375
#4	2.092	1.858	2.456	2.35
#5	4.076	1.525	3.282	27.817
#6	2.611	5.325	1.228	1.339
#7	4.408	2.658	2.845	4.254
#8	0.53	14.965	2.565	30.369
Average	3.0809	3.9635	2.909	11.551

Table 3. Error Distance vs. AP Layout (with 4 APs)

3.1.2 Summary for wireless sensor system

Due to the extensive Wi-Fi deployments, Wi-Fi tracking is increasingly used in location tracking applications. The Wi-Fi infrastructure enables both customers and developers to quickly develop cost-effective solutions and to create new levels of asset visibility. In most

Wi-Fi tracking systems, a minimum of three APs must be visible with consistent and strong signal in order to attempt room-level granularity, and actual RF-signature measurements and calibrations are required for finer accuracy. One of the major hurdles of RF-based location tracking is the high overhead associated with collecting RF reference signatures. The RF-signature measurement will be done repeatedly if there is any event that may cause radio signature perturbations, such as varying the number of APs or dislocating the APs.

3.2 Wireless Sensor Network (WSN) based systems

Contributing to the field of wireless sensor networks, we focus on indoor tracking systems with wireless sensor motes. We study MoteTrack (Lorincz & Welsh, 2005), a robust and decentralized RF-based location tracking system developed by a research group at Harvard, and then we added extra features to MoteTrack to address a few issues discussed below. MoteTrack employs a dynamic radio signature distance metric to tolerate failures of the network infrastructure. MoteTrack has two components: beacon motes and mobile motes. The primary function of the beacon motes is to transmit beacon messages at all times (i.e., one beacon message every two seconds). On the other hand, the mobile mote collects the beacon messages from the beacons within its radio range and uses them to estimate its location. Since these tracking systems are radio frequency (RF) based, the location of an object in the given environment is based on the reference signatures collected beforehand. We will describe the differences between MoteTrack and our asset tracking system.

3.2.1 Differences between MoteTrack and the developed system

MoteTrack is effective and useful. The location estimation accuracy within 2 meters is in the 80th percentile. MoteTrack operates as follows. During the data collection phase, the beacon motes transmit beacon messages, and the mobile mote collects them and stores the messages on the computer. The collected messages - primarily contain signal strengths, IDs of the beacon motes, and the location x,y,z where the message is collected - are now combined, forming a large database. This large database is then installed on the mobile mote for referencing in the tracking phase. During the location-tracking phase, the beacon motes continue to transmit beacon messages, and the mobile mote captures these messages. Then, a comparison between the current messages and the information from the pre-installed database is done to estimate the location of the mobile mote. The mobile mote, after computing the location x,y,z, sends the estimated coordinates to the directly attached computer for displaying.

Mote track is a location support system which allows clients to find out their locations in the building. However, in order to preserve user privacy, the MoteTrack mobile nodes do not report their locations to a centralized server, where a database stores the locations of mobile assets. Although the degree of user privacy offered by MoteTrack could be an important deployment consideration, the location privacy of mobile users trades off many useful services that a tracking system provides. Furthermore, MoteTrack mobile sensors should be directly attached to a display device such as PDA or laptop, and it may not be feasible in many applications due to increased cost and battery issues. Thus, this study modified MoteTrack for tracking mobile assets in a dynamic environment.

In this study , the mobile mote was isolated from a display device. In the new tracking system, there are three components: beacon motes, mobile motes, and a base station. The functions of the mobile and beacon motes are essentially the same as in MoteTrack; however, the mobile ones do not have a laptop attached to them in this deployment. Instead, the base station is attached to a tracking server, and it remains at the central location listening to data coming from the mobile motes. The tracking system developed in this study works as follows. The data collection phase is the same as in MoteTrack. Similarly, the large database is installed on the mobile motes for referencing. During the location-tracking phase, the mobile motes collect beacon messages from nearby beacons and then they compute the locations. Then, mobile motes use a multihop routing protocol to route location messages to the base station, using the beacon motes as the intermediate forwarding nodes. Once the location messages arrive at the base station, the data is forwarded to the attached computer or laptop for location display. Fig. 3 shows the differences between the two tracking systems.

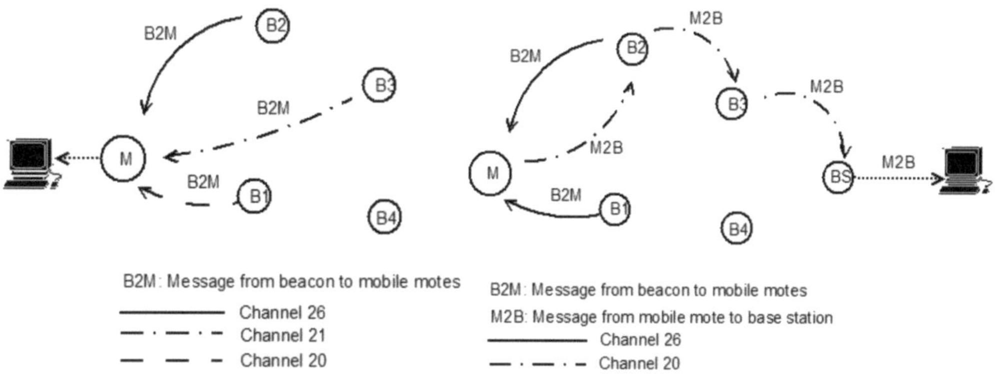

(a) Message Exchange in MoteTrack

(b) Multihop Message Exchange in Developed Tracking System

Fig. 3. Message Exchanges

3.2.2 Experimental study and the accuracy of the deployed system

The test bed was installed at PKI and consists of 20 MicaZ motes, which are indicated as blue dots (labeled from 1 to 20) on the building map in Fig. 4. The red dot (denoted by a 0) indicates the base station which collects the position data from the mobile sensor nodes and shows their locations on the GUI. Each mote consists of an 8-bit processor running at 7.3MHz, and it has 128 KB of read-only program memory, 512 KB for measurement serial flash, and 4 Kbytes of RAM. The mote operates between 2400 MHz and 2483.5 MHz, and it is capable of delivering data up to 250 kbps (Crossbow, Inc., 2008).

Signature collection in Fig. 4 works as follows. The beacon motes (labeled from 1 to 20) send out beacon messages periodically. The user travels along the corridors to collect signatures using a mobile node that is attached to a laptop. Data collection is triggered when the user clicks on the map and creates a numbered mark (e.g., 100, 101, etc); then, the GUI automatically collects the signatures from the mobile node. At this point, the mobile node should receive beacon messages from the beacons. The signature collection phase is done for the entire floor in Fig. 4.. After the collection is completed, the signatures are then processed and are installed onto the mobile node for reference in the position tracking phase.

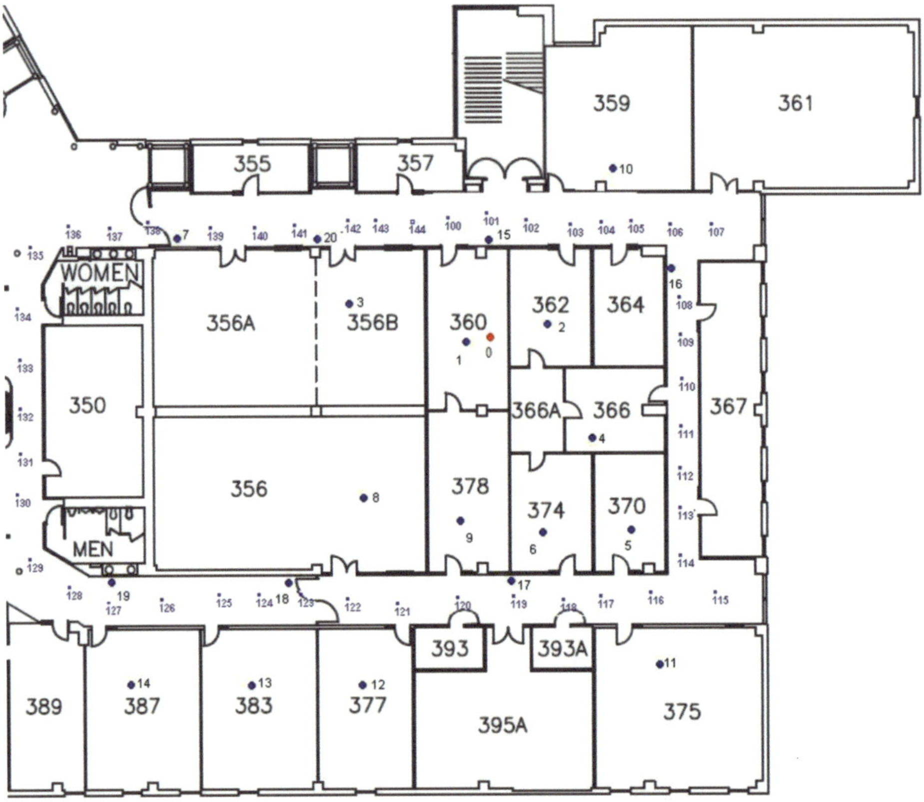

Fig. 4. Deployment of WSN-based Tracking System at the Peter Kiewit Institute (PKI)

Error Distance (meters)	# of Samples (Total 450)	Accumulated # of Samples	Percents of Estimated Locations	Percentiles of Estimated Locations
0.0 - 0.5	139	139	30.9 %	30.9 %
0.5 - 1.0	73	212	16.2 %	47.1 %
1.0 - 1.5	82	294	18.2 %	65.3 %
1.5 - 2.0	71	365	15.8 %	81.1 %
2.0 - 2.5	25	390	5.6 %	86.7 %
2.5 - 3.0	17	407	3.8 %	90.5 %
3.0 - 3.5	10	417	2.2 %	92.7 %
3.5 - 4.0	9	426	2.0 %	94.7 %
4.0 - 4.5	6	432	1.3 %	96.0 %
4.5 - 5.0	8	440	1.8 %	97.8 %
5.0 - 5.5	6	446	1.3 %	99.1 %
5.5 - 6.0	4	450	.9 %	100.0 %

Table 4. Error Distance in Meters

The accuracy of the system depends on location estimation algorithms, the density of beacon nodes, and the density of reference signatures. The deployed tracking system consists of 20 beacon nodes distributed across the third floor of PKI. To estimate the accuracy of the system, 45 points on the floor were chosen to measure the Euclidean distance between actual locations and corresponding estimate points. The Euclidean distances were measured 10 times for each point. According to our experimental results, the median resolution of the mote-based RF system is in the range of 1.0 to 1.2 meters, and over 80% of the location estimates are within 2 meters from their exact position. The results of our experimental study are shown in Table 4 and Fig. 5.

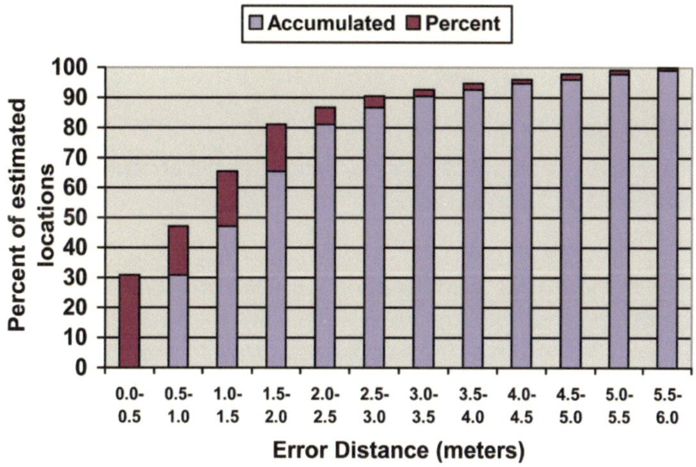

Fig. 5. Error Distance in Percentiles

Since the developed tracking system uses the same methods used in MoteTrack to estimate the location of mobile assets, the accuracy of our system is very similar to that of MoteTrack. For more information on the accuracy of the RF-based sensor tracking system under a wide range of conditions, refer the MoteTrack paper (Lorincz & Welsh, 2005).

3.2.3 Summary for wireless sensor system

MoteTrack is also an RF-based localization system, but it enhances the robustness of the system by employing a distributed approach. According to our experimental study using the WSN-based tracking system deployed at PKI, 80% of the measured location errors were at or below 2 meters, and the system minimizes the impact of perturbations of the radio signals and provides accurate location estimates with failed beacon nodes. However, the system still has the initial overhead associated with collecting RF reference signatures since the location of a mobile asset is based on the reference signatures collected ahead of time.

3.3 Ultra-wideband (UWB) systems

As another indoor position tracking technology, a Federal Communications Commission (FCC) approved, commercial UWB system was tested for implementation and performance in several building spaces. The hardware of the ultra-wideband sensor network consists of tags and sensors. A tag is attached to the object that requires location tracking. As each tag emits an UWB signal, location is calculated using both time difference of arrival(TDOA) between different sensors (a.k.a., receivers) and angle of arrival(AOA) at each sensor. Each sensor employs a minimum of four UWB receivers which allow the angle of arrival to be determined (Ubisense 2007). The standard UWB configuration consists of a single master sensor and three slave sensors and requires wired communication cables and timing cables. In order to properly calculate the time of arrival, each slave sensor must synchronize with the master sensor. In the standard configuration, this is done through timing cables between each slave sensor and the master sensor.

While 802.11 systems use standard 2.4 GHz bandwidth, the current UWB system utilizes the 5.8-7.2 GHz bandwidth. While 802.11 uses signal strength, the current UWB system uses TDOA and AOA technologies which enable UWB sensors to have better signal penetration capability and less multi-path signals problems, thus better position accuracy can be obtained.

3.3.1 Accuracy for UWB

To evaluate the sensitivity of multi-path signal problems, the accuracy of UWB sensors were tested in an open space and a closed space.

3.3.1.1 Open Space Test

To conduct an open space test, local building's basement was selected. Four sensors are mounted on tripods and multiple known positions are marked on a floor. A total station was used to get the x,y,z position of the sensors and the points on the floor. Fig. 6 shows the surveyed positions of sensors (402-405) and points on the floor (400, 1000-1013). Then a UWB tag was placed on each marked point on the floor to measure the accuracy in difference in distance between a surveyed known position and a wirelessly estimated tag position. The good coverage with four sensors is about 400 m^2.

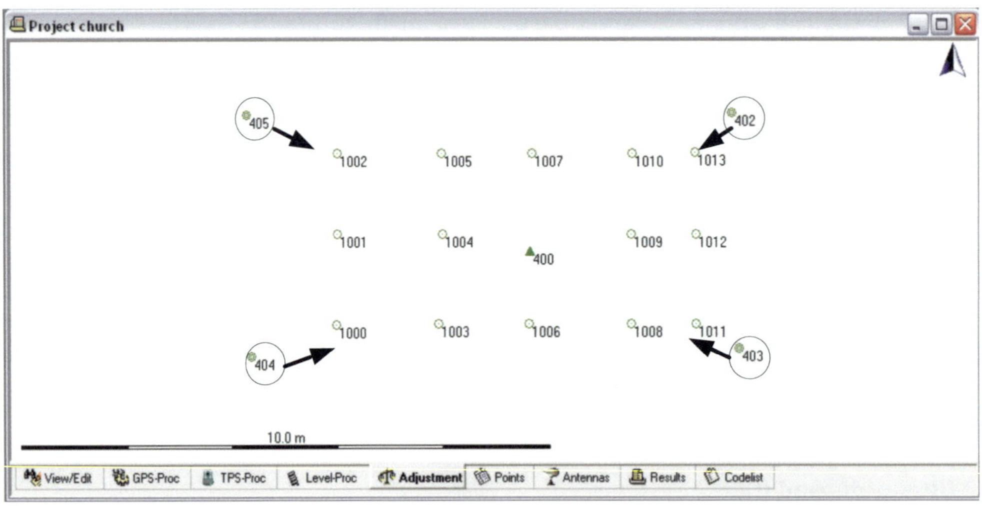

Fig. 6. Test Area Layout

As results, the obtained average accuracy was 19.1 cm. There was a slight accuracy improvement (1.7 cm) but not significant for most of the points in center (1001,1004,400,1009,1012) when the tag was raised by 35 cm from the floor level. However, there was a significant accuracy improvement by 13.5 cm for the outermost points (1000-1002 and 1011-1013) when the tag was raised by 35 cm. This verifies that in order to accurately determine a tag position, a multiple number of receivers (typically three) must have a direct line-of-sight or at most an attenuated line-of-sight transmission path. (Fontana 2003). Especially the receiver did not catch a signal from the tag when the tag is located just below the receiver (See the Fig. 7). Therefore, it is recommended to install the receiver as high as possible from a floor and face it down to cover a large area of the space. Table 5 summarizes the test results.

Fig. 7. Open Space Test at a Local Church Basement

Overall Average Accuracy	Average Accuracy for Center Points on Floor Level	Average Accuracy for Center Points raised by 35 cm	Average Accuracy for Outermost Points on Floor Level	Average Accuracy for Outermost Points raised by 35 cm
19.1 cm	7.9 cm	6.2 cm	30.0 cm	13.8 cm

Table 5. Summary of Test Results

3.3.1.2 Closed Space Test

To conduct a closed space test, an office area of the Peter Kiewit Institute (PKI) was selected. The PKI building was the most challenging subject to test the UWB system because the building (1) was built with steel frames and metal studs, (2) was furnished with all metal furniture and electronics such as printer and copy machines, and (3) had computer electronics and wireless telecommunication labs near by the test area which may cause interferences with the UWB communication system. In addition, heavy people traffic in this area might cause the signal communication error since human bodies can absorb the signal. Especially, 5.8 GHz wireless phones around the testing area should be turned off before testing the UWB system because 5.8 GHz phones were significantly interfered with the current UWB bandwidth, 5.8-7.2 GHz. Fig. 8 shows the test area layout with four receiver's position in 400-403, total station's position in 100-101, and measured tag positions in 500-513. The first set of test was conducted by placing a tag on the floor level. Then the tag was raised by 104 cm for the 2nd test set. Each set of tests was conducted twice. The average of floor level (0 height) tests showed 41 cm accuracy. When the tag was raised by 104 cm, the test showed 48cm accuracy. Unlike the previous open space test which used a stand to raise the tag position, the tag was carried by a human subject in the closed space testbed. Although it was identified that higher position of the tag showed the better accuracy to improve the line-of-sight transmission path, it was noticed in the closed space test that the human carryer significantly affected the accuracy based on its orientation.

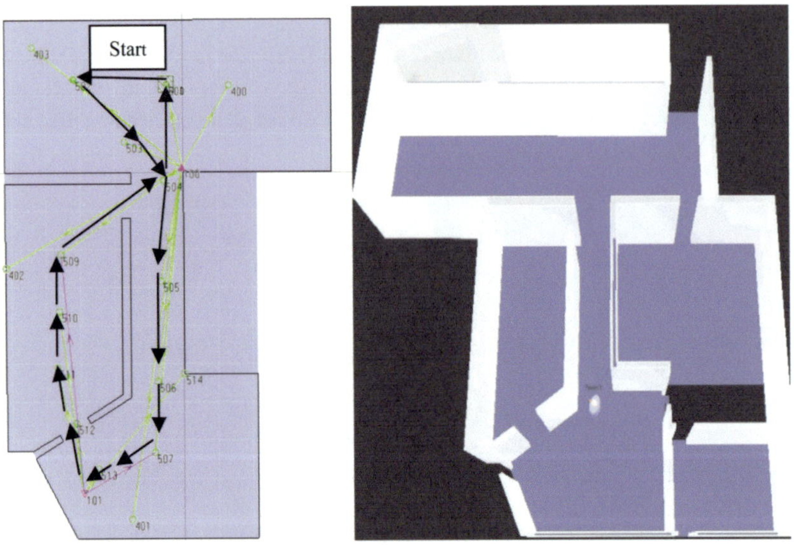

Fig. 8. Closed Space Test Area Layout with Travel Paths (left) and 3D User Interface (right)

3.3.2 Summary for UWB

Tag positioning accuracy depends upon several factors, including precise knowledge of all receiver and reference tag locations. Absolute tag position accuracy of better than 19 cm has been demonstrated in an open space and 48 cm for the closed space. It is recommended that the receiver need to be located as high as possible to cover larger area. Also, it is important to strategically select direction of receiver and have a minimum of one set of line of sight between a receiver and a tag in any location, which can significantly affect tag's position accuracy.

4. Implementation of position trackingsystem to mobile robot navigation

Based on the previous performance test results of the indoor position tracking sensor systems, this study implemented the Ultra-wideband technology to robot's navigation functions mainly due to its higher accuracy than the other two systems. The robot, as shown in Fig. 9, used for this study includes a number of ultrasonic and infrared distance sensors

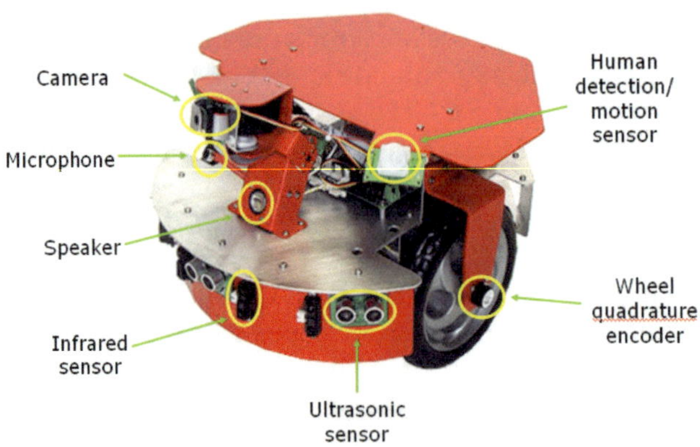

Fig. 9. Wireless Mobile Robot

aimed to the front, back, and sides, wheel servo sensors, and human detection/motion sensors along with a camera, a microphone, and a speaker (Dr. Robot, 2008). All computations are performed in a centralized manner at a laptop which communicated with the robot wirelessly through an 802.11 channel. Connections are made to each robot through the use of a unique IP address per robot (Fig. 10).

In order to generate location data, a UWB tag was attached to the robot which allowed four UWB sensors to calculate its location.

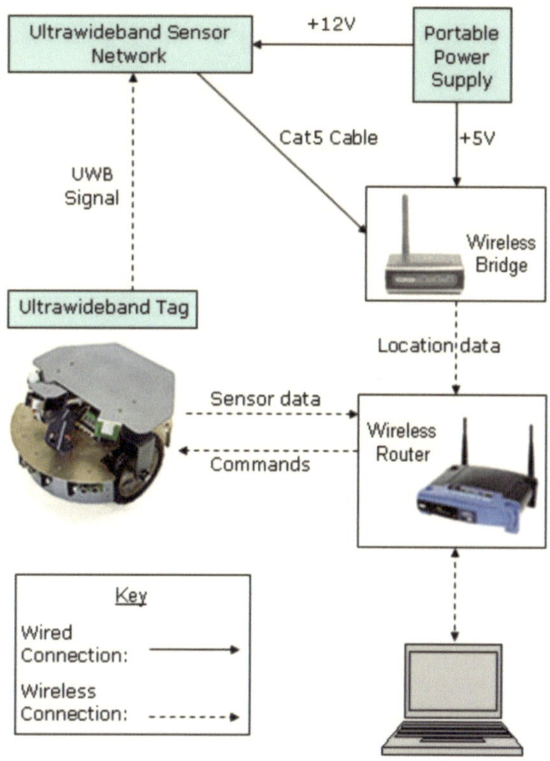

Fig. 10. UWB Integrated Mobile Robot Tracking System Architecture

4.1 Path planning with ultra-wideband

The path planning algorithm is divided into two levels. The first is reaction control which handles obstacle avoidance. This level also uses human detection readings to avoid and/or report human presence. If human presence is detected, the robot controller would send an alert to the user and would stop any current movement so the user could intervene and could act appropriately.

The second is end-to-end pathing which determines the route the robot shall follow along the virtual rails from start to end point (see Fig. 11). Here, a well-known routing method, Floyd's all-pairs shortest path algorithm in Fig. 12 (Floyd, 1962), calculates the best path along the rails graph, finds the point closest the robot, and moves to that start point. Then, the robot moves from a point to another along the shortest path until the end point is reached. At this layer, the user may establish a pre-programmed route along the graph where the start and end points are for material delivery or security.

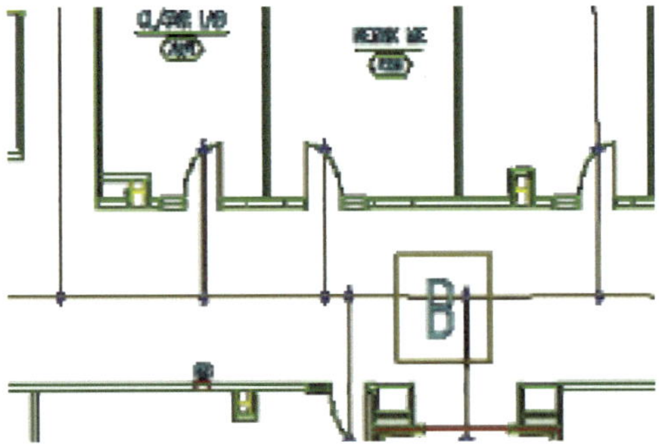

Fig. 11. Rail Graphs in an Office Area

```
Public Sub Update()
  Dim k As Integer, i As Integer, j As Integer
  Floyd Warshall Algorithm
  For k = 0 To intNumberOfXy
    For i = 0 To intNumberOfXy
      For j = 0 To intNumberOfXy
        If (d(i, j) > (d(i, k) + d(k, j))) Then
          d(i, j) = d(i, k) + d(k, j)
          path(i, j) = k
        End If
      Next j
    Next i
  Next k
End Sub
```

Fig. 12. Floyd's All-pairs Shortest-Path Algorithm coded in Visual Basic

Due to its high accuracy, a UWB tag was mounted on the mobile robot to show robot's current location. UWB sensors provide position data which is sufficiently accurate for the robot navigation. The position data then integrates with Floyd's algorithm to find a shortest path on the rail graphs. A developed filtering model also performs less complex checks such as comparing incoming sensor values against their minimum and maximum defined values to ensure they are within bounds. For example, location data from the UWB sensors are checked using the robot's maximum speed of 1 meter/second as the greatest possible change in position. If an incoming value indicates the robot is traveling more than one meter per second, the value is ignored.

4.2 Graphical User Interface (GUI)

In this study, software modules for a robot control and a UWB position tracking have been developed which can be easily imbedded into standard CAD programs such as Microstation and AutoCAD. Since most of the latest building already have either 2D or 3D CAD floor drawings available, the developed modules can save enormous amount of time to generate a new graphical user interface when applied to different building applications. Fig. 13 shows a visual representation of the robot and the environment in which it exists. Location and direction changes are displayed in real time along with human detection alerts. Detected human is also displayed to the user within the drawing when a human detection sensor returns a value indicating human presence.

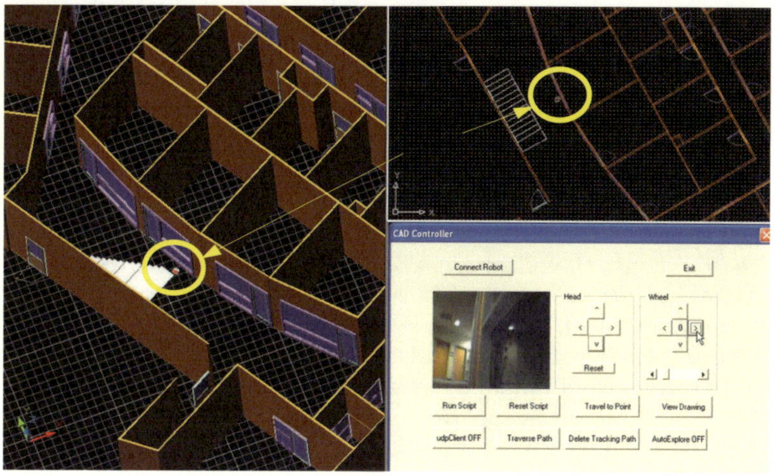

Fig. 13. 3D real-time robot control user interface

5. Conclusion

This paper presented an intelligent methodology using wireless position tracking technologies for autonomous mobile robots to improve their navigation functions in dynamic indoor working environments. Three different wireless position tracking technologies were deployed in various types of building spaces, and the accuracy of each system was evaluated. The WLAN based system provides an accuracy of 93 cm on average when 10 APs are available. For wireless sensor networks, when 20 beacon nodes are available, the median resolution of the mote-based RF system is in the range of 1.0 to 1.2 meters, and over 80% of the location estimates are within 2 meters. Unlike the other two sensor technologies, the UWB system provides very high accuracies, on average, about 45 cm with only four receivers (a.k.a. beacons) even in a highly congested office area. Based on the test results of three sensor technologies, this research deploys UWB tag onto a mobile robot to improve the robot's navigation using a well known graphic theory, Floyd's all-pairs shortest path algorithm.
Work continues in order to move away from the rails-based algorithm into a more dynamic method of pathing. In addition, walls and corridors will be taken advantage of to aid in pathing while resolving the fusion of sensor values of sensors with different angles detecting a flat surface displayed in the drawing.

6. References

Bahl, P. & Padmanabhan, V. N. (2000). RADAR: An in-building RF-based User Location and Tracking System, *Proceedings of IEEE INFOCOM*, pp. 775-784

Battiti, R ; Le Nhat, T. & Villani, A. (2002). Location Aware Computing: A Neural Network Model for Determining Location in Wireless LANs. Technical Report DIT-02-0083, University of Trento

Cisco. (2008). Cisco 2700 Series Wireless Location Appliance Deployment Guide. http:// www.cisco.com/univercd/cc/td/doc/product/wireless/loc2700/l2700/depgd.pdf

Crossbow, Inc. (2008). http://www.xbow.com/, 2008

Dr. Robot. Inc. (2008). www.drrobot.com, May 1, 2008

Ekahau. (2008) http://www.ekahau.com, 2008

Floyd, Robert W. (1962). Algorithm 97: Shortest Path. *Communications of the ACM*, Vol. 5 No. 6, p. 345

Fontana, R.; Richley, E. & Barney, J. (2003). Commercialization of an Ultra Wideband Precision Asset Location System, *IEEE conference on Ultra Wideband Systems and Technologies*, Reston, VA, November 2003

Gezici, S.; Zhi Tian; Giannakis, G.B.; Kobayashi, H.; Molisch, A.F.; Poor, H.V. & Sahinoglu, Z. (2005). Localization Via Ultra-wideband Radios: A Look at Positioning Aspects for Future Sensor Networks. *IEEE Signal Processing Magazine*, Vol. 22, No. 4, pp. 70– 84

Haeberlen, A.; Flannery, E.; Ladd, A. M.; Rudys, A.; Wallach, D. S. & Kavraki, L. E. (2004). Practical Robust Localization Over Large-scale 802.11 Wireless Networks, *Proceedings of the 10th Annual International Conference on Mobile Computing and Networking*, pp. 70–84, 2004, ACM Press, New York, NY, USA

Harter, A.; Hopper, A; Steggles, P.; Ward, A. & Webster, P. (1999). The Anatomy of a Context-Aware Application, *Proceedings of 5th International Conference on Mobile Computing and Networking*, pp.59-68, ACM Press, Seattle, WA

Hightower, J. & Borriello, G. (2001). Locating Systems for Ubiquitous Computing. *IEEE Computer*, Vol. 34, No. 8, 2001, pp.57-66

Kohn, L.; Corrigan, J. & Donaldson, M. (2000). To Err Is Human: Building a Safer Health System, Committee on Quality of Health Care in America, Institute of Medicine

Ladd, A. M.; Bekris, K. E.; Rudys, A.; Kavraki, L. E.; Marceau, G. & Wallach, D. S. (2002). Robotics-based Location Sensing Using Wireless Ethernet, *Proceedings of the 8th Annual International Conference on Mobile Computing and Networking (MOBICOM)*, September 2002

Liu, X.; Corner, M. & Shenoy, P. (2006). Ferret: RFID Localization for Pervasive Multimedia, *Proceedings of the 8th Ubicomp Conference*, September 2006

Lorincz, K. & Welsh, M. (2005). Motetrack: A robust, Decentralized Approach to RF-based Location Tracking, *Proceedings of the International Workshop on Location and Context-Awareness (LoCA 2005)*, May 2005.

Nara, T.; Onoda, H.; Yamane, J. & Ando S. (2005). Dipole Estimation from the Magnetic Field Gradient for RFID Tag Localization. *Transactions of the Society of Instrument and Control Engineers*, Vol. 1, No. 1, pp. 16–20, 2005

Pango networks. (2008). http://www.pangonetworks.com, 2008

Philipose, M.; Fishkin, K.; Fox, D.; Hahnel, D. & Burgard, W. (2003). Mapping and Localization with RFID Technology. Technical Report IRS-TR-03-014, Intel Corporation, December 2003

Priyantha, N. B.; Chakraborty, A. & Balakrishnan, H. (2000). The Cricket Location-Support System, *Proceedings of the 6th Annual International Conference on Mobile Computing and Networking*, pp. 32–43, 2000, ACM Press, New York, NY, USA

Skibniewski, M. & Jang, W. (2007). Localization Technique for Automated Tracking of Construciton Materials Utilizing Combined RF and Ultrasound Sensor Interfaces, *Proceedings of the 2007 International Workshop on Computing in Civil Engineering*, pp. 657-664, Pittsburgh, PA, July 2007

Ubisense. (2007). www.ubisense.net, November 30, 2007

Want, R.; Hopper, A.; Falcao V. & Gibbons, J. (1992). The Active Badge Location System. *ACM Transactions on Information Systems*, Vol. 10, pp. 91-102, January 1992

Young, D.P.; Keller, C.M.; Bliss, D.W. & Forsythe, K.W. (2003). Ultra-wideband (UWB) Transmitter Location Using Time Difference of Arrival (TDOA) Techniques. *Conference Record of the Thirty-Seventh Asilomar Conference on Signals, Systems and Computers*, Vol. 2, pp. 1225–1229, November 2003

Zetik, R.; Sachs, J. & Thoma R. (2004). UWB Localization-Active and Passive Approach, *Proceedings of the 21st IEEE Instrumentation and Measurement Technology Conference*, IMTC 04., Vol. 2, pp. 1005–1009, May 2004

22

Laser Energy Transmission for a Wireless Energy Supply to Robots

Nobuki Kawashima and Kazuya Takeda
Kinki University
Japan

1. Introduction

The robot development is actively done as a very hopeful tool in many disciplines, however, right now, the robot is not used actually for practical use. It is clear that once it is actually used, it becomes very important how to keep supplying energy. The wireless energy supply is mandatory. In most cases, it is considered that the rechargeable battery is most easy to solve it, however, for example, in nuclear power plant accident or chemical weapon disaster due to terrorism, once the robot enters into the contaminated area, it is almost impossible to return for the battery recharging. The wireless energy transmission is mandatory.
Even for those robots working in the area where it is impossible or very difficult for a man to access but the robot can return for the refueling, the wireless energy transmission is required when, for example, the robot malfunctions and can not return. A long time operation is required for the trouble shooting and for repair.
Moreover, some important activities urgently needed may not allow the robot to return for the refueling, then the wireless energy transmission is also very useful.
To realize the wireless energy supply, either the microwave or the laser energy transmission is considered most hopeful. The microwave energy transmission has an advantage of higher energy conversion efficiency and can be used through the cloud, but it is not easy to concentrate the power in a small region. It is considered as a useful tool for the power transmission from the space power station (SPS) to the earth.
Laser is easy to focus the beam in a small area and the mirror can be used to transmit energy behind a blind corner. Due to the development of laser diodes (LD), a very compact laser transmitter system can be constructed. For the robot use, the laser is more advantageous over the microwave system.

2. Examples of development of robots using laser energy transmission technology

2.1 Lunar ice exploration rover

We have started the laser energy transmission technology development from 1997 to apply to a rover to explore the presence of ice in the bottom of craters in the lunar polar region, where no sun light is available. On the contrary, the solar radiation is always available at the periphery of the crater. The solar energy is converted into the electric energy and then to the laser light there and the laser beam is transmitted to the rover working at the bottom of the crater (Fig. 1) [1].

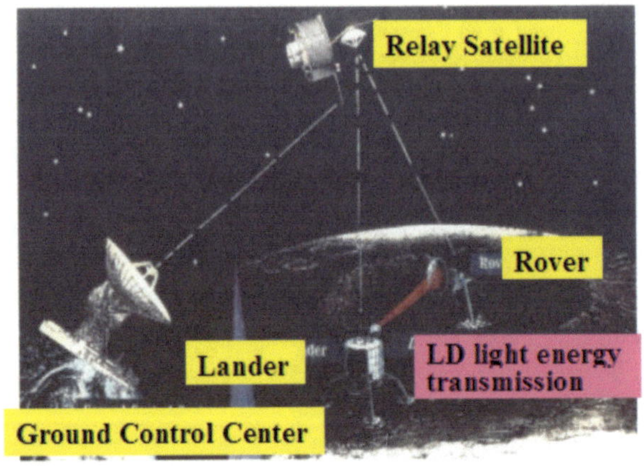

Fig. 1. Rover to explore to confirm the presence of ice on the moon using the laser energy transmission

A proto-type model has been constructed. The laser system consists of a 60 W Laser Diode system and the laser is output through a 400 micrometer fiber with NA= 0.22. The laser light is focused on a 70 cm diameter receiving solar panel at the distance of 1 km using a mirror of 25 cm in diameter. The rover can easily be driven at the power more than 10 W.
Though the lunar exploration mission based on this experiment is not yet realized as the actual space mission, the energy transmission to more than 1 km has successfully been done on the ground to drive an actual size rover with an conversion efficiency more than 20 % (Fig.2) [2], [3].

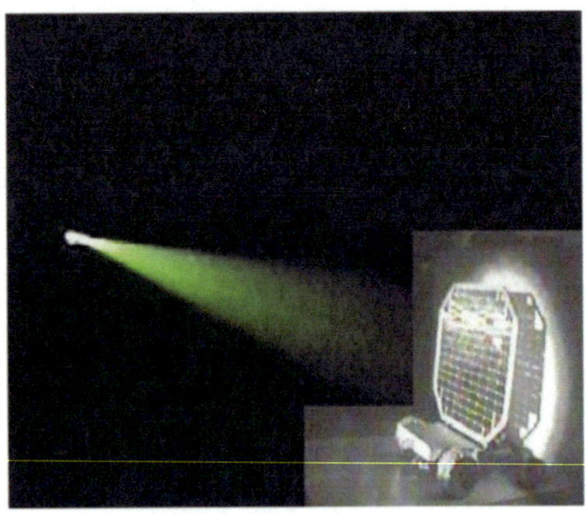

Fig. 2. 1 km laser energy transmission to a model lunar rover has successfully been done

2.2 Robot

As a spin-off of the technology developed for the space mission described above, we have tried to apply the laser energy transmission system to an actual robot. The robot is constructed using the parallel mechanism (Fig. 3). Particular emphasis is put on the application to a rescue robot.

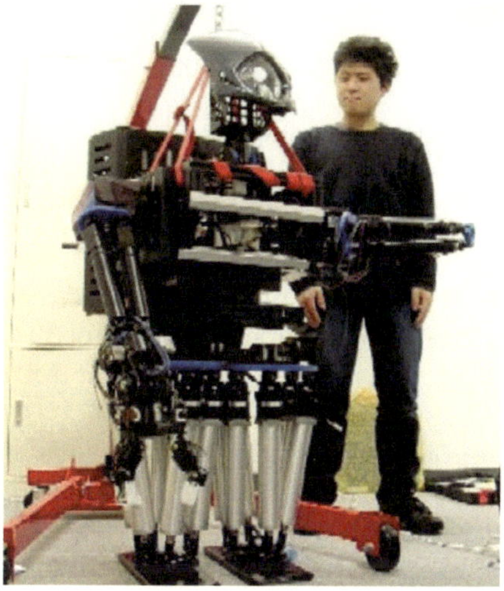

Fig. 3. Rescue robot partly powered by the laser

Currently, the power required to drive the robot itself is more than 120 W and the laser power can only drive the motion of an arm and fingers.

2.3 Small aircrfts

2.3.1 Kite plane

The laser energy transmission technology is also applied to the small aircraft for the surveillance of damaged areas and for the communication relay station at natural disaster such as flood, earthquake, volcano eruption etc.
First we have tried a kite plane as shown in Fig. 4. It is considered that the kite plane has a better flight stability in bad wind conditions compared with other small airplanes.

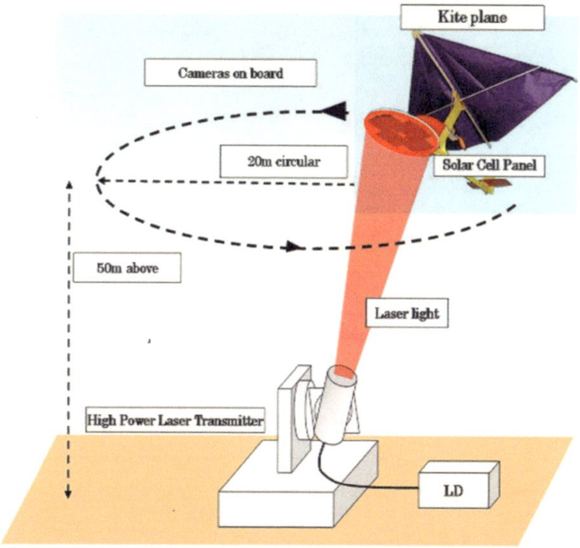

Fig. 4. System configuration of a laser energy driven kiteplane

i) Kite Plane
The kite plane has a very simple structure. It is driven by a simple propeller in the center powered by an electric motor.
Other characteristics of the kite plane is shown below :

Length	789 mm	Width	1,360 mm	Height	560 mm
Wing area	5000 m^2	Weight	800 g	Payload weight	600 g

The motion of the kite plane is controlled by an radio wave controller. A solar panel of 30 cm in diameter is placed underneath the plane. It consists of 30 small GaAs solar cells(4cm x 7 cm). 10 cells are connected in series and 3 in parallel and it is directly connected to the propella motor (Fig. 5).
The maximum output power is about 42W when it is irradiated by a 200 W laser.
A small corner cube is set in the center of the panel to reflect a part of the laser beam to the transmitter. This is used to track the kite plane.
A small Li-Polymer battery (7.4 V 480 mA 36 g) is used in parallel to maintain the necessary power during the take-off from the ground. It can sustain the kite plane flight less than 5 min.
A small TV camera (15 g with small battery and transmitter) is on board to observe the scene below.

Fig. 5. Solar panel for a kite plane drive

ii) Laser System
Two fibre coupled 200 W laser diode transmitter are used. The wave length is λ808 nm and the light beam is coming out from a fibre whose core diameter is 400μm and the crad diameter 480μm with NA = 0.22. 15 V 39 A is required for 200 W laser output. One is cooled by a couple of Peltier cooling units (cooling capability is 255 W each) and the other is water-cooled.
iii) Long Time Flight Test
A long time flight test was done at the baseball stadium „Osaka dome“ in early 2006. The kite plane was first flown from the ground using the on-board small battery and the orbit of the kite plane was controlled by a radio controller. When a level flight of an altitude of 50 m

and a circular radius of 10 m was realized, the laser was focused to the solar panel on the kite plane. The laser power of 300 W (200 W + 100 W) was used and the power required was monitored and recorded on a on-board power monitor. The dome stadium has a ceiling height of 60 m and no appreciable wind is there. And a long time stable flight operation was successfully realized more than 1 hour in each cases. The time was limited by the availability of the dome.

The average power measured on board was about 40 W. The flight performance was taken by CCD cameras as shown in Fig. 6, where the motion of the kite plane during the level flight of 10 m in diameter and 50 m in altitude.is shown. A zoomed image of the plane during the flight is shown in Fig. 7 with its shade on the ceiling of the dome [4].

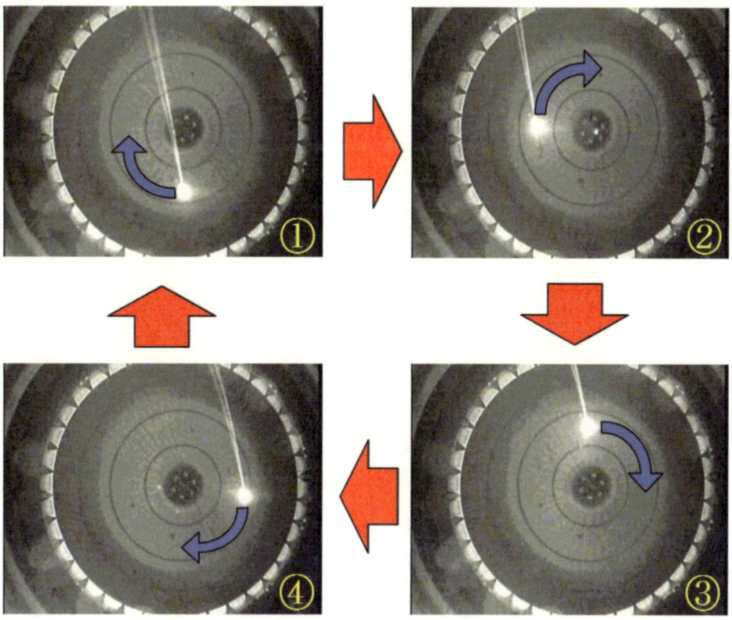

Fig. 6. Circular flight of a laser-light receiving kiteplane with a radius of 10 m and at an altitude of 50 m

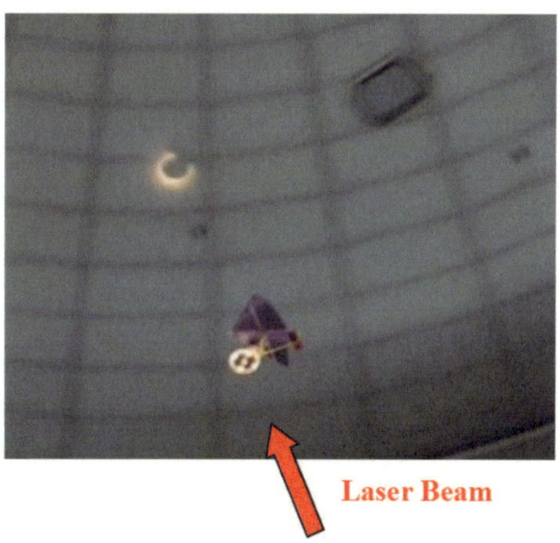

Fig. 7. Flying kite plane under the auto-tracking (The solar panel is irradiated by the laser light with its shade on the ceiling)

2.3.2 Helicopter

The application of this laser driven airplane we aim is, as described above, to use it as a surveillance flight over the area where the human access is difficult in the case of natural disasters such as flood or earthquake. A long time flight can be realized and no return of the aircraft to the base for the refueling or recharging is required.

In order for this system to be used in real disaster cases, the horizontal range shall be increased, however, in the case of the kite plane with the current horizontally set solar panel, in order to increase the horizontal range, the vertical altitude shall also been increased. To reduce the vertical altitude, the solar panel shall be set vertically, then the attitude control of the solar panel conflicts with a stable spacecraft maneuver.

To solve this problem, the helicopter is the most appropriate and we can set the solar cell panel vertically (Fig. 8). However, the helicopter requires much more electric power than the airplane.

Fig. 8. Helicopter with a vertical solar panel underneath

One solution is to increase the laser power, then we need a more efficient cooling of the solar panel. The other solution is to increase the conversion efficiency from the laser power to the electricity.

For the former, we have increased the laser power up to 530 W. The cooling of the solar panel is accomplished by setting it under the helicopter body. The wind generated by the propeller cools the panel.

For the latter, first we filled the panel with as many cells as possible on the panel. Since the shape of the solar cell commercially available is fixed, we tried to cut it in half. More improvement of the efficiency has been achieved by optimizing the cell connection. The detail of this improvement will be described elsewhere.

It has taken a long time try and error procedure and finally we have realized a stable long time flight. We use a small battery to maintain the safe landing of the helicopter even when the proper tracking of the laser receiving solar cell panel fails. We have confirmed that even after more than one hour flight, the battery maintains a full charging (Fig. 9) [5].

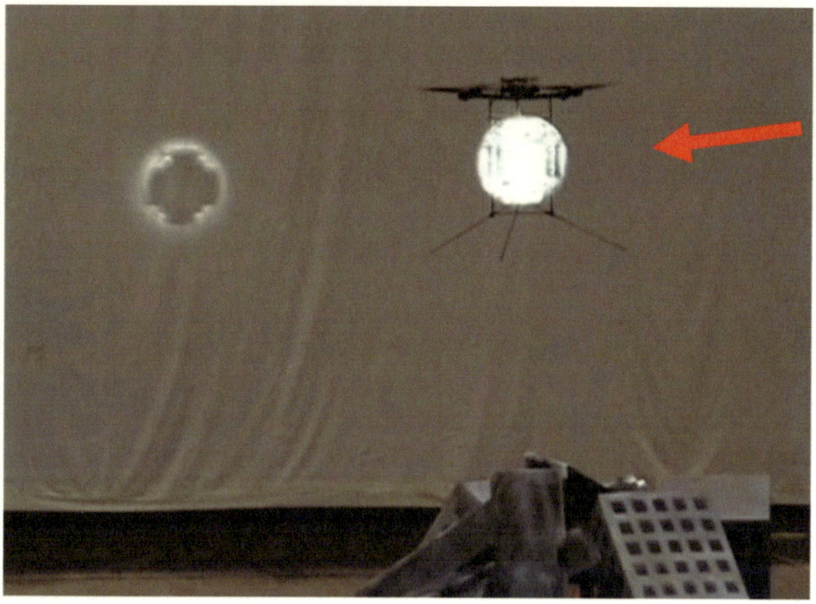

Fig. 9. Helicopter flight receiving the energy from the laser energy transmission over a simulated disaster ruin

3. Problems to be solved

a. Increase of the conversion efficiency of LD

The laser diode is very compact and have a high conversion efficiency more than 30 % , however, when the output power is more than hundred watts, the dissipated heat is quite large and the cooling system is not anymore compact. More increase of the conversion efficiency is strongly required.

b. Development of an efficient cooling system

Right now, for a laser diode of the power less than 100 W, the Peltier cooling system can be used. For a higher power, the water cooling system is needed and it is not at all compact. More efficient compact cooling system is required.

c. Increase of the laser energy conversion efficiency of the solar cell to the electric energy

We are now using a solar cell of single junction GaAs of Spectrolab Inc. U.S.A. having a single cell conversion efficiency more than 40 %. However, the overall efficiency of the solar panel is less than 25 %. The reason why is that there is an open space between cells on the panel and also the intensity distribution is not uniform so that all the cells cannot receive the laser light at the optimum operation condition. Some cell has an conversion efficiency more than 60 % and higher efficiency cell is strongly needed.

4. Conclusion

We can find a lot of robot applications in construction activities, where it is very difficult or dangerous for a man to access and only robots can work. The time will come soon when the actual use of those robots is extensively realized and the wireless energy transmission technology using laser is a unique means to supply energy to those robots.

5. References

N. Kawashima , "The importance of the development of a rover for the direct confirmation of the existence of ice on the moon, "Trans. Japan. Soc. Aeronaut. Space Sci. Vol.43, 2000, pp.34-35

K. Takeda, M. Tanaka, S. Miura, K. Hashimoto, N. Kawashima, H. Matsumoto, T. Goshozono, S. Ishi and J. Mizui, "Laser power transmission for the energy supply to the rover exploring ice on the bottom of the crater in the lunar polar region, "SPIE, Vol.4632, 2002, pp.199-223

N. Kawashima and K. Takeda, "1.2 km laser energy transmission for the development of a lunar rover confirming the presence of ice on the moon, " Proc. ILC 2003 / ILEWG 2004

N. Kawashima, K. Takeda and K. Yabe, "Application of the laser energy transmission technology to drive a small airplane" Chinese Optics Letters Vol 5 Suppl. 2007 S109-110

K. Takeda, N. Kawashima and K. Yabe, "Laser Energy Transmission to a Small-Unmanned Aerial Vehicle(in Japanese)" Uchu Gijutsu Vol 7 2008 pp.27-32

23

Advanced Control Schemes for Cement Fabrication Processes

Susana Arad, Victor Arad and Bogdan Bobora*
*University of Petrosani, *Carpatcement Holding, Romania*

1. Introduction

Taking into consideration that the cement market will record an increasing rate of 15 - 20%, related to the residential buildings development and to the initiation of large infrastructure projects the cement production is of great interest, both from the point of view of product's quality increase and raw material consumption and environmental impact diminishment. The demands on cement industry in relation with productivity, quality and price, mean an ever-increasing need to improve the quality products, the productivity increase improvement, modernization of the technological flow, improvement of environmental quality. Process automation is where the industrial area offers the biggest and most satisfying challenges in terms of combining traditional engineering skills with technological innovation.

The primarily goal of our work is to describe the state-of-the-art and future directions in update of the cement plant from Carpatcement Deva Branch (Casial).

The project aims the development of a series of patterns, more or less independent from each other, but all of them focused in the improvement of cement quality and diminishing of environmental impact, by means of models simulation and identification in this process. The main objective of this paper is to show through examples how process models can be inserted into advanced controllers to allow the successful control and optimization of the process when the controlled variables are not on line measured or they are measured infrequently.

The proper implementation of advanced control schemes for complex cement fabrication processes rely on the availability of the appropriate mathematical models. The control strategies are illustrated with actual data obtained for one typical process, the clinker fabrication process. We have done an analysis of rotary kiln from Casial-Deva Branch through finite elements approximation.

Three main stages in the cement production process exist:

- Raw limestone processing, raw material mixing and milling;
- Clinker production (as intermediary by-product) from raw limestone calcinations in a melting rotary kiln;
- Clinker mixing and milling with other products, to obtain the cement as final product.

In those three fabrication stages, due to multiple installations and technologies and the presence of a great number of parameters conditioning, is which are inter conditioning it is necessary that the process must by controlled a distributed expert system, based on the intelligent algorithms.

2. Overview and state of art

2.1 Process description

The cement related business in our country is divided between three big manufactures: Heidelberg Cement (Germany), Lafarge (France), Holcim (Switzerland), each one controlling about one third of the specific market.

Deva cement plant was established in 1972 and its operation started in 1976. It is located nearby Deva town, having an output of 1 million tons/year. Carpatcement Holding took over this plant in 2000.

This process is carried out employing the dry method, which generates the lowest emissions level and is the most efficient. The raw and auxiliary materials employed in cement and by-products processing are limestone, clay, pyrite, flying ash, slag, natural puzzolana, gypsum, sand wastes issued from other industries. The basic raw materials for cement production (limestone and clay) are mined out in open pits, from where after primary crushing is transported on the plant's site for storage and further preparation. Having high moisture, the clay requires a dehumidification and drying process. External suppliers supply other raw materials, such as iron ore (pyrite ash), granulated furnace cinder etc.

The cement production stages are involving the following activities, illustrated in Figure 1 (Procese de productie, 2008):

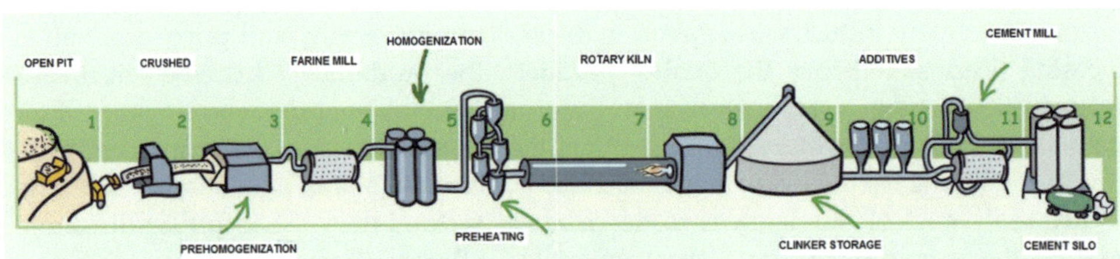

Fig. 1. The technological flow from cement fabrication

- **Extraction** in open pits of raw materials such as limestone and clay (1 and 2).
- **Limestone separation and overcrushing** (3), where the sizes comprised between 0 - 50 and 100 - 150 mm are transported to the crusher, where their diameter is reduced to the needed value for cement production, namely 0 - 25 mm, and then are stored in three limestone silos.
- **Obtaining the raw meal** for clinker (4 and 5) is using a laboratory assessed receipt, the limestone, clay and pyrite ash are closed and transported to the drying tower, and further to the raw mill.
- **Clinker production** (6, 7 and 8) based on raw meal contained in the storage silos is dosed and transported to the Humboldt four levels heat exchanger, where the raw meal is preheated from 60°C to about 800°C. The heat is taken over from the kiln's hot gasses passing through the heat exchanger, in counter- current with the supplied raw meal. Preheated and partially decarbonated rawmeal covers all the areas within the rotary kiln, in a special zone the clinker being obtained, at 1450°C temperatures. From the rotary kiln, the clinker is discharged in the grid cooler, where using the air forced into the system by 9 fans, it is cooled from 1350°C to about 100°C. The heat demands for obtaining the clinker is supplied by burning fuels as natural gasses, heavy oil, waste oil, SAF, eco-fuel on the hot end of the kiln, and the rubbers and waste tyres are supplied through the cold end to the ascendant

column kiln-heat exchanger. Three silos are in place for clinker storage. As mentioned before, the clinker kiln works on the dry procedure and meets the BAT requirements.

- **Cement production** (9, 10 and 11). The slag from the admixture hall is then stored in the cinder silos from the clinker mills. The cinder is dried with warm gasses from the grid cooler or, when the clinker kiln does not operate, by burning natural gasses in the drier's hotbed. Gypsum is transported from the admixture hall to the gypsum silos nearby the cement mills. The clinker, cinder and gypsum contained in silos, after a laboratory receipt, are extracted, dosed and supplied to the cement mills. The cement mills are tubular mills with balls, having two rooms and operating in a closed loop. From the mill, the material is brought to a high efficiency separator where it is separated, the fine fraction (cement) being taken over in a haulage relay and stored in 9 cement silos, the heavy fraction being recirculated into the mill.
- **Cement shipping** (12), from silos, the cement can be supplied both as bulk material and in bags, using rotary Mollers equipment. The cement delivery can be done by trucks or using the railway network.

The basic product is cement, 80% is clinker manufactured using a dry procedure which allows a production of 1,026,563 t/year clinker. The technological flow may be synthesized into a scheme presented in Figure 2. The Deva Branch, Casial has two technological lines to produce cement with 1.2 millions tons a year capacity. The flow sheet of material balance from Deva Branch, Casial is presented in Figure 3.

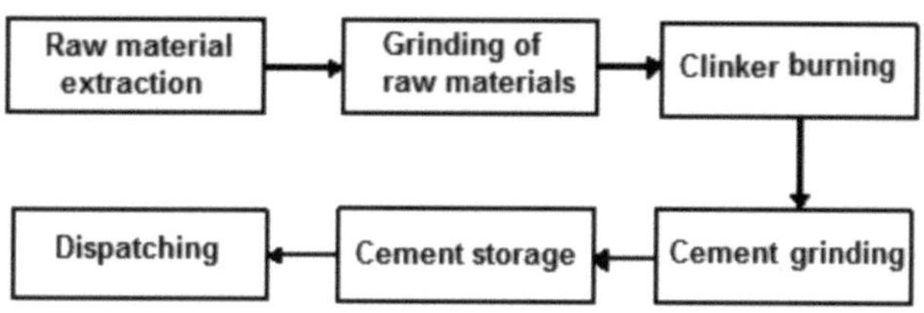

Fig. 2. Flow chart of process fabrication

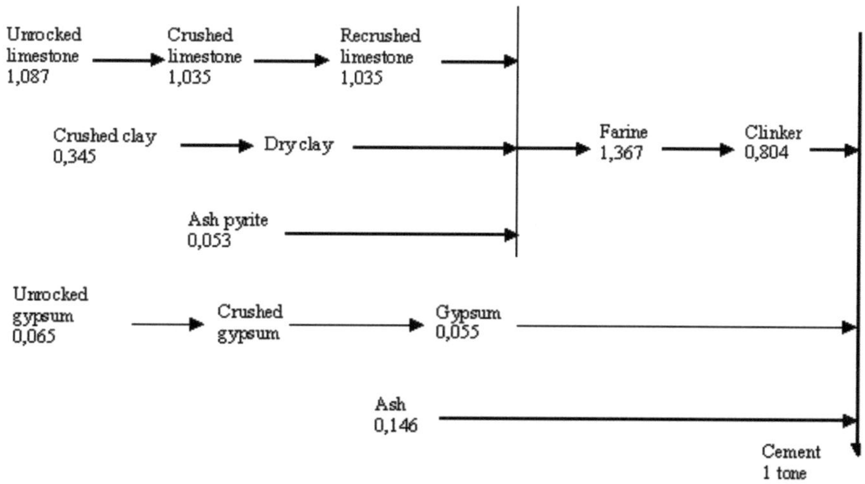

Fig. 3. The flow sheet of material balance

2.2 Kiln operating

Manufacturing cement by dry procedure is a huge energy consuming process. An important section of cement technological flow, with a great share in finite product (80%) is the making of the clinker which is presented in Figure 4, where are represented graphically the flow installations of clinker fabrications.

The calcinations process (transformation of raw material in clinker, the intermediate product from what the cement is produced) takes place in the rotary kiln. The existing temperature in the kiln for the process to take place is of 1450°C, while the flame's temperature is of about 2000°C.

The existing kiln systems at Carpatcement Holding are on the dry system with suspension pre-heating in 4 levels, Humbolt types. The material is supplied from the upper part of the cyclones and they are circulating in decreasing way in counter- current with the gases resulted from the burning processes which they are circulating through the separation cyclones from the lower level to the upper level. In this way, the gases are cooling down and the raw materials are preheating, Figure 1 (6) and it starts the decarbonated process. The raw material, which is in suspense it is separated from the gases inn each step of the changer and meets again the gases in the lower step. The repeating of this process (separation-blending) at each step of the changer until the discharge of the material into kiln it's ensures a good thermal transfer.

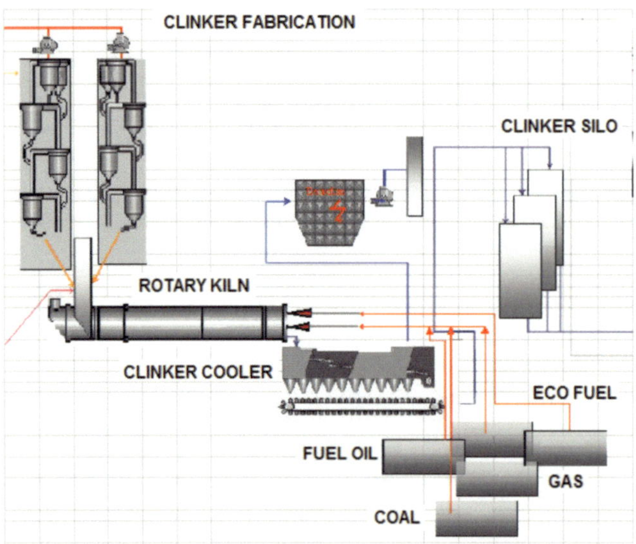

Fig. 4. Flow sheet of clinker fabrication

The raw meal charging station is equipped with control bins, gravimetric chargers type Schenk and pneumatic pumps type Fuller.

The control bunkers should maintain a constant raw meal height providing a prescribed constant material quantity in the supply. Raw meal is taken over from in supply control bunkers with rampart extractors, being charged according to the centralized commands given in the central command room, with Schenk gravimetric chargers.

The material charged in this manner is circulated in the screw pneumatic pump's bins that are transported it to the heat exchanger. For each technological line, three Fuller pumps are in place (2 operating and an auxiliary one) and a four-stage Humboldt suspension heat exchanger.

The charged material is pneumatically hauled to the exchanger's upper side, using the existing joint between stage I and stage II. The material receives heat from the hot gasses during its traject in the heat exchanger's cyclones, from upside down in the direction I-II-III-IV, after that entering the smoke chamber and then the furnace. Inside the exchanger, the material is heated up to 800°C - 810°C, also being partially decarbonatated. Gasses are then entering in exchanger at about 1000°C temperature, in his bottom side circulating along his cyclones in direction IV-III-II-I and finally are exhausted through VRA and VRB exhausters. The partially decarbonated rawmix is fed into the rotary kiln, having 97 m length and a 5.8m in diameter. Here takes place the final stage of the clinkerization process, based on specific thermal and chemical processes.

For burning purposes, liquid fuel (heavy oil), gassy fuels (natural gasses), or both kinds of fuels are employed, the equipment being well designed to meet this demand. According to the reactions within the kiln and the resulting compounds, the rotary kiln comprises the following zones:

- Decarbonatation area (calcinations), where the alkaline carbonates are decomposed at temperatures comprised between 1000°C and 1100°C.
- Transition zone (solid phase reactions area), where the first mineralogical compounds are formed, through solid phase reactions, at temperatures 1000°C - 1350°C.
- Clinkerization zone (sinterization area) where, 1350°C - 1500°C temperature values , the liquid phase appears, in his presence the tri calcium, silicate (alit) develops, the cement's most valuable compound.
- Cooling zone, where, at temperatures ranging from 1450°C to 1250°C the mineralogical compounds occurs. Burned gasses are circulated in the kiln backwards related to material's advancing direction, than their dust, content is minimized employing a 560,000 m^3/h capacity electrostatic precipitator system (EPS).

EPS, works in optimum conditions when the gas temperature doesn't pass 180° C, because of this they are cooled down in a water tower. From the kiln, the clinker is discharged at the warm head of the cooling grate where it is taking place a suddenly cooling to 65° C.

A large volume of gases has to be moved through the kiln system. Particularly in suspension-preheated systems, a high degree of suction has to be developed at the exit of the system to drive this. Fans are also used to force air through the cooler bed, and to propel the fuel into the kiln. Fans account for most of the electric power consumed in the system, typically amounting to 10–15 kWh per tonne of clinker.

The grate cooler is composed from three grates of different sizes, on which are put high-melting steel plates with wholes. A bed of clinker up to 0.5 m deep moves along the grate. These coolers have two main advantages: they cool the clinker rapidly, which is desirable from a quality point of view, and, because they do not rotate, hot air can be ducted out of them for use in fuel drying, or for use as preheated combustion air. After cooling the clinker is crashed until, the granulation is max. 25 mm and then through a transport system made from two metallic chains with coupes and a relay of three belt conveyors, is transported at the tree clinker bunker. The dust collection from the cooling grates ensured with a multi - cyclones batteries with two frames.

3. The rotary kiln modelling

A rotary kiln, the world's largest manufacturing machine - is the major component of the cement line. Rotary kilns have wide use in industry from the calcinations of limestone to cement manufacturing to calcining of petroleum coke etc. The kiln is a large rotating furnace approximately 100 m long, and four to seven m in diameter that weighs over 300 tonnes, Figure 5. The rotary kiln consists of a tube made from steel plate, and lined with firebrick. The tube slopes slightly (3%) and slowly rotates on its axis at between 30 and 250 revolutions per hour. Raw meal is feed in at the upper end, and the rotation of the kiln causes it to gradually move downhill to the other end of the kiln. At the other end fuel, in the form of gas, oil, or pulverized solid fuel, is blown in through the "burner pipe", producing a large concentric flame in the lower part of the kiln tube. As material moves under the flame, it reaches its peak temperature, before dropping out of the kiln tube into the cooler. Air is drawn first through the cooler and then through the kiln for combustion of the fuel. In the cooler, the cooling clinker heats the air, so that it may be 400 to 800 °C before it enters the kiln, thus causing intense and rapid combustion of the fuel.
The dimensions and parameters of the oven are: dimensions ∅ 5.8 x 97 m, angle 3 %, backing points 4, production capacities Q = 3,125 t/ day, main driving P = 500 KW, n = 750 rot/ min, second driving P = 500 KW, n = 750 rot/ min.

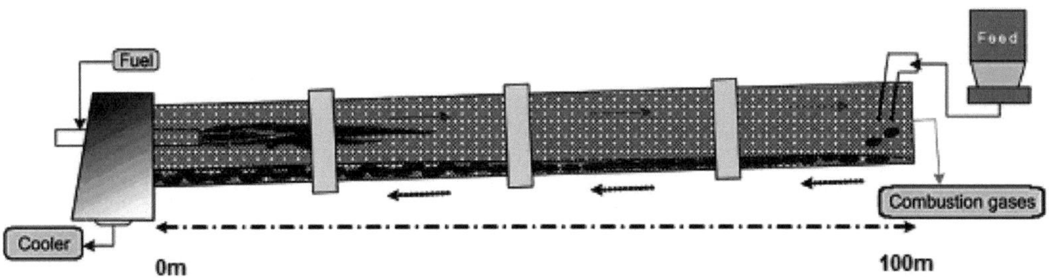

Fig. 5. Rotary kiln

Problems such as low thermal efficiency and low product quality have plagued rotary kiln operations yet these machines have survived and have been continuously improved (fuel efficiency, automation) for over a century.
With an ever-increasing focus on reducing greenhouse gas emissions, the continued or increased use of rotating kilns can only be achieved by reducing the thermal and electrical energy consumption used in these processes. A fluid bed calciner or dryer achieves rapid drying by the large heat transfer coefficient obtained through the high air volume being circulated. The penalty is the increase in electrical energy required to circulate this high air volume. Rotary kilns on the other hand have poor heat transfer coefficients, hence higher thermal energy demand, due to the need for larger devices and thus more opportunity for heat to be lost.
In most rotary kiln operations, the chemical reactions in the bed require high temperature, for example, cement kilns will require temperatures of approximately 1500°C. The energy to raise the temperature and drive endothermic reactions is from the combustion of a range of fuels such as natural gas, coal and more and alternative fuels. Heat transfer from the gas to the bed is complex and occurs from the gas to the bed surface and kiln wall to bed surface via conduction, convection and radiation.

A number of rotary kiln models has been proposed over the years and recent computational fluid dynamic models can be developed but all have their limitations (Barr, 1989; Bui, et al., 1995). Most assume isothermal conditions through the bed at any axial position (Majumdar, & Ranade, 2006). The bed motion regime, cascading, rolling or slumping depends on the rotational speed of the kiln, the percentage fill and the feed physical properties.

- They are models which have in the site the thermal processes, models that are following the thermal transfer between the material bed, gas, kiln walls and environment, where it appears conduction, convection and radiation phenomenon. The measures are the material temperatures from supply in those four steps, gas temperature, and walls temperature in the four steps.
- They are chemical models who analyze the endo -thermal phenomena that are taking place at the raw material calcinations. The kiln parameters are the gas emissions of O_2, CO_2, NOx, quantities and material compositions (Gorog, et al., 1981).
- They are models which have basis the energetically balance of the kilns where they are appearing energetic aspects in connection with the kiln's drive, rotation, motor moment and they are following the automation power adjustment.

A series of equations representing conservation of mass, energy and species averaged over the cross-section are solved using appropriate numerical methods (He, et al., 1996).

The bed for example is assumed to be well mixed and isothermal in any given transverse plane (Georgallis, et al. 2001). Although these models have been successfully used in industry, they are limited for information that can be extracted.

Due to the complex models character, nowadays many software packaging are allowing to employ numerical analysis of thermal phenomena (FLUX STUDIO, ANSYS, MULTIPHISICS, FLUENT, COMSOL MULTIPHISICS, QuickField, etc.). A 3D physical model of the kiln where it can be observed the physical components, walls, material bed and the burning pipe is given in Figure 6.

As a result a number of researchers have begun the quest for a more encompassing modelling effort. Boateng and Barr (Boateng, & Barr, 1996) have coupled a conventional one-dimensional plug flow model with a two-dimensional representation of the bed's transverse plane. This improves the ability to simulate conditions within the bed. Alyaser (Alyaser, 1998) has modelled for axisymmetric conditions. Fully coupled three-dimensional modeling is applied to the rotary lime kiln (Georgallis, et al., 2001). Three sub-models are coupled, namely the hot flow model, the bed model and the wall/refractories model. The model takes into account the major phenomena of interest including the gas flow, all modes of heat transfer and thermal effects of the refractory.

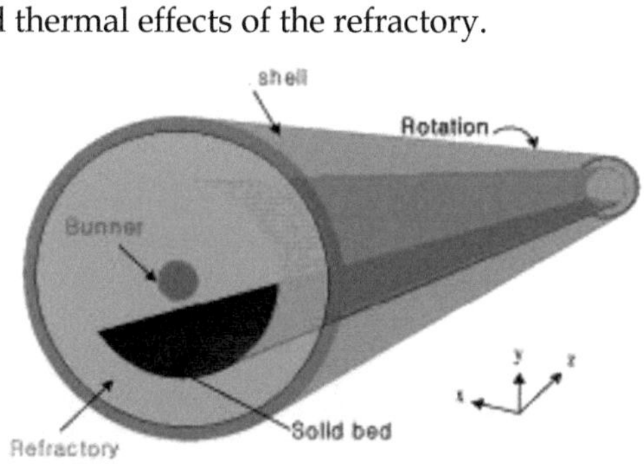

Fig. 6. 3D- rotary kiln model

A model of rotary kiln heat transfer, which accounts for the interaction of all the transport paths and processes to the rotary kiln from Casial, Deva Branch is presented in our paper. Information exchange and directions of transfer are shown in Figure 7 (Barr, et al., 1989).
Two dimensional modelling is applied using finite element method. Heat transfer within the kiln refractory wall was solved using a finite-element approximation for one-dimensional transient conduction. Interface temperature boundary conditions for the kiln are used in the model. Heat flux boundary conditions are used for both the inner and outer surfaces in the wall model.
The mathematical model of heat-transfer for linear problems is described by the differential mathematical model of the thermal conduction, Eq.(1) and (2):

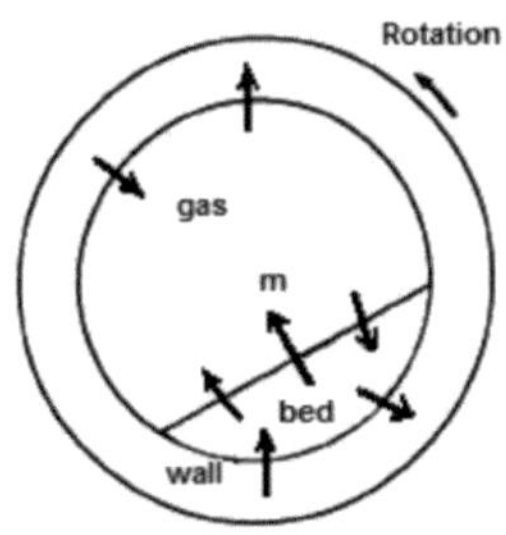

Fig. 7. Information exchange of heat transfer

$$div(\lambda gradT) + q - \rho c \frac{\partial T}{\partial t} = 0 \tag{1}$$

$$\frac{\partial}{\partial x}\left(\lambda_x \frac{\partial T}{\partial x}\right) + \frac{\partial}{\partial y}\left(\lambda_y \frac{\partial T}{\partial y}\right) = -q - c\rho \frac{\partial T}{\partial t} \tag{2}$$

Where: T is temperature, t- time, $\lambda_{x(y)}$-components of heat conductivity tensor; λ - heat conductivity, q - volume power of heat sources, burner - constant, $c(T)$ - specific heat, ρ - density of the substance.
In linear case all the parameters are constants within each block of the model. The oneness of the precedent equation solution of the thermal conduction proposes the knowledge:

a. The heat sources in the domain of calculus, q;
b. The material properties, ρ, c şi λ;
c. Initial conditions T;
d. Limit conditions.

The next limit conditions on surfaces S_1, S_2, S_3, on the frontier domain of calculus is possible. (Fireteanu, et al., 2004):

- Type Dirichlet,

$$T(x.z.y,t) = T_s, \ t > 0, \text{ on } S_1 \tag{3}$$

- Type Neuman inhomogeneous

$$\lambda \frac{\partial T}{\partial n} = q_s, \ t > 0, \text{ on } S_2 \tag{4}$$

➢ Type mix

$$\lambda \frac{\partial T}{\partial n} = \alpha\left(T - T_0\right) + \beta k_{SB}\left(T^4 - T_0^4\right), \text{ t>0, on } S_3 \quad (5)$$

Where, q_s is the superficial specific flow imposed, α is the thermal transfer coefficient by convection, k_{SB} is a Stephan-Boltzmann constant ($5.67032 \cdot 10^{-8}$ W/m²/K⁴), β is an emissive coefficient, and T_0 - ambient radiation temperature.
Convection boundary condition and radiation boundary condition can be specified at outward boundary of the region.

3.1 Formulate the problem

It will be accomplished the thermal analysis by numerical method with finite elements using as QuikField software, of the heat transfer problem at the rotary cement kiln.
It will be determined the temperature value in different points of the model, in each block, thermal gradients, heat flux densities, and temperature on the contour of shell.
The software is based on heat conduction equation with convection and radiation boundary conditions. The technical characteristics of the rotary kiln are shown in Table 1:

Geometrical parameters	Numerical value	Operational variables	Numerical Value
Lenght [m]	97	Temperature of clinkerization [°C]	1300
Internal kiln radius [m]	1.9	Velocity of kiln [rpm]	1.9
External kiln radius [m]	2.9	Limestone feed temperature[°C]	800
Inclination [%]	3.0	Thermal transfer coefficient α [W/Km² for (S1)	20
Refractory thickness [m]	0.9	Thermal transfer coefficient α [W/Km²] for (S3)	350
Kiln shell thickness [m]	0.1	Thermal transfer coefficient α [W/Km²] for (S2)	0.5
		Emissive coefficient β	1
		Thermal conductivity [W/K.m] in bad in gas in shell in refractory	 0.693 0.8 10 0.04

Table 1. Operational variables of the rotary kiln for calcining limestone

Defying the thermal transfer problem for this case it was made the geometrical model, the document describing the problem geometry, the labels of the blocks (bed, gas, refractory, shell) and it was made the mesh of our model. The model contains specific geometric objects and establishes the correspondence between the objects and material properties, field sources and boundary conditions. We gave the properties of each material from the named blocks (heat conductivity, emissive coefficient for convection and radiation (Table 1), the thermal field sources were defined and also the boundary conditions and limits.

It was defined three surfaces outlines S1, S2, S3, belonging the calculus frontier domain, with specified boundary condition like: S1 outer surface as interface between shell and environment, S2 surface interface between bed-gas and bad-refractory, S3 inner surface interface between refractory -gas.

3.2 Numerical solutions

The values of temperature T [K], heat flux F [W/m²] and temperature gradient G [K/m] in some points at interface surfaces between gas and refractory and on the vertical axis of kiln through each isotherm, in gas, bad and in shell also were calculated and given in Table 2 and Figure 8. Also it was represented the temperature variation on the contour of the inner surface, Figure 9 and temperature distribution between shell and refractory, in Figure 10.

Inner surface points	Temperature value T [K]	Gradient Value G[K/m]	Flux heat F[W/m²]
1	1295.6	298.97	239.17
2	1246.0	415.16	332.13
3	1192.2	584.58	467.67
4	1141.8	692.21	553.77
5	1184.8	1021.8	708.08
6	1032.7	1037.4	718.95
7	975.99	1016.9	704.68
8	914.44	991.89	687.38
9	778.01	18578	743.11
10	693.01	18367	734.69
11	636.96	18231	729.23
12	466.04	17808	712.33
13	318.27	70.556	705.56
14	316.21	66.78	667.8

Table 2. Thermal field values

The cross-section model is shown to simulate the measured thermal performance of the kiln for clinker. The interaction among the heat-transfer processes at cross-sections of the kiln was examined, and explanations were made for both the observed close coupling of the bed and inside wall temperatures and the high rates of heat input to the bed occurring near the kiln entrance and in the presence of an endothermic bed reaction.

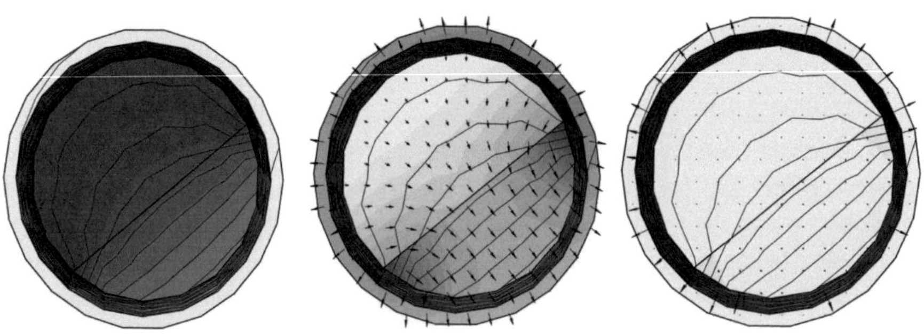

Fig. 8. Map of temperature, vectors of heat flux and thermal gradient in cross-section

The model was validated using thermal measurements from Casial's kiln. This effort demonstrates how a model may be used to capture flame phenomena for rotary kilns and to solve shell fault into the kiln.

A model accepts heat flux values from the hot flow side and temperatures on the wall interface. Evidence (such as a non-uniform product) has suggested that large temperature gradients exist near and within the bed.

The work carried out is aimed at understanding and improving the heat transfer in rotary kiln and to provide a systematic basis for the efficient operating of kilns. It can be noticed that temperature distribution nearby the kiln's shell is very close to the trend obtained by the pyrometer used for temperature monitoring.

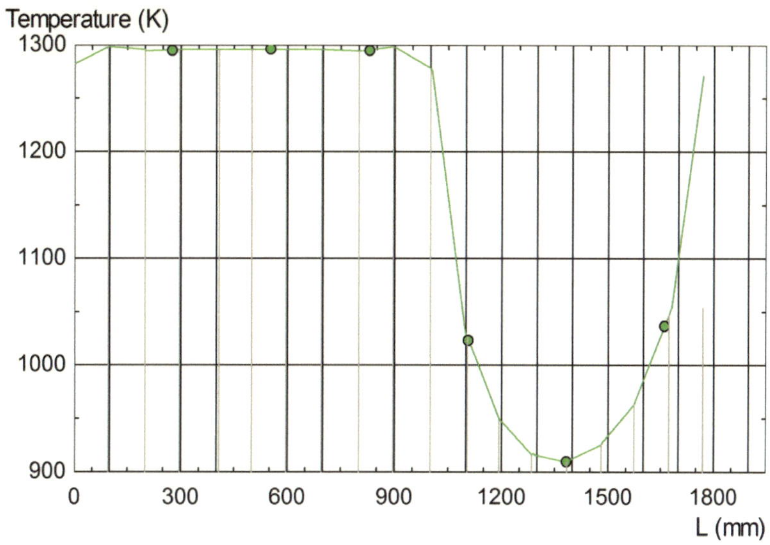

Figure 9. Variation of temperature on the contour of inner surface

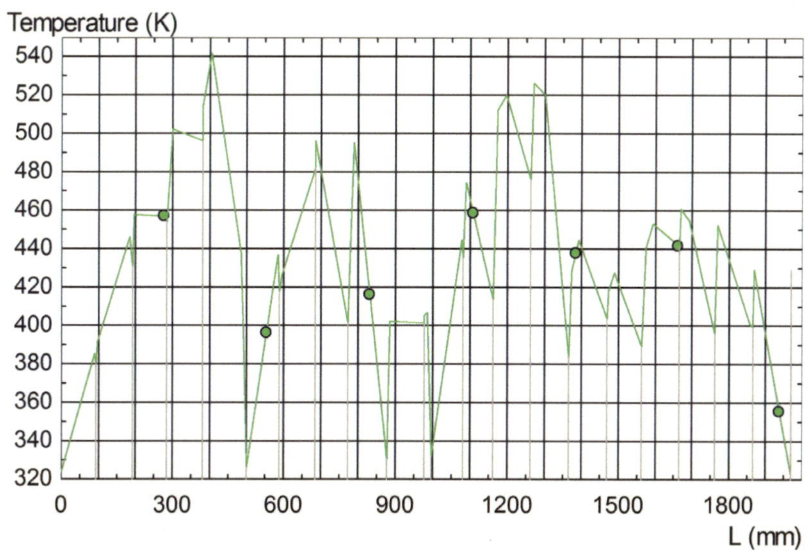

Fig. 10. Variation of temperature on the interface surface between shell and refractory

The data processed by statistic functions about clinker temperature and automate measured pyrometer temperature are shown in Figure 11 and Figure 12 and the values of the statistical parameters we have obtained. The result reflected on prediction performance plot with a correlation of 0.96 (Arad &Arad, 2003).

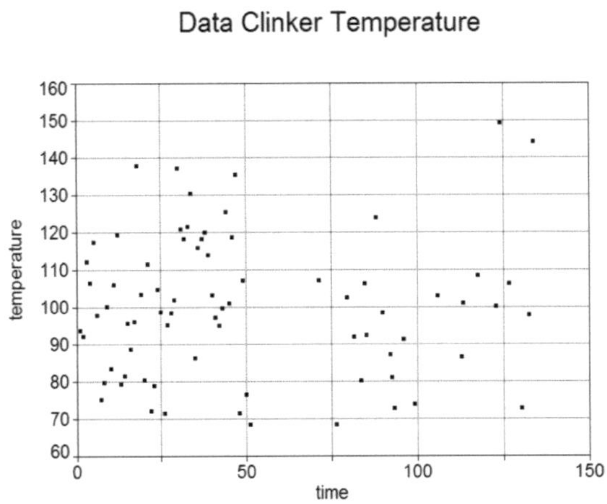

Fig. 11. Clinker temperature

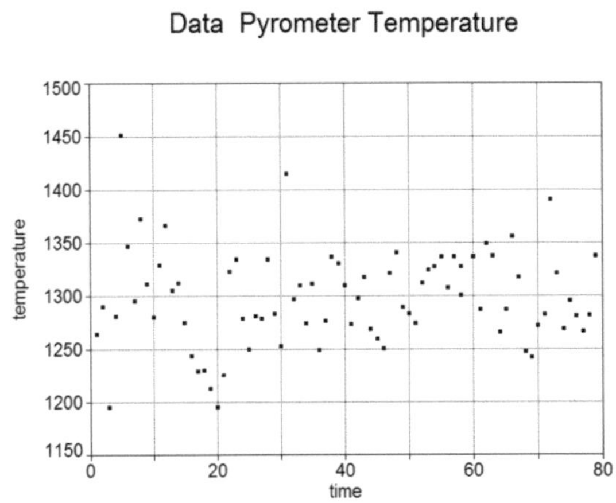

Fig. 12. Pyrometer temperature

4. Cement kiln emissions

The most important gas emissions from cement industry are CO_2. The carbon dioxide emissions which are generated represent about 5% from the world wide CO_2 emissions induced by human activities. These high level emissions are resulting basically from the specific technology for cement production. The main sources of CO_2 in cement industry are: raw material and fuel burning.

The N_2O emissions generated by the cement kilns as a consequence on combustion processes are relatively low, having no significance if related to CO_2 emissions. The last ones are issuing in the following stages:

- The calcinations stage:

- Kiln: direct CO_2 emissions having two main sources:
  - calcinations (decarbonification) of raw material (60%);
  - burning of fuel (40%), thermal energy consumption.
- The milling stage:
  - Cement milling: indirect CO_2 emissions, electric energy consumption.

The direct CO_2 emissions in the process are mainly occurring by employed fuel and raw materials (calcium carbonate), being released during the stage of clinker production in kiln (the so called calcination). Thermal energy is also used during this stage. So, if natural gas is used instead of coal, the CO_2 emission is decreased with 25%. The thermal consumption and CO_2 emissions also related to the kiln type employed for calcinations and clinkerization. Apart from this, the emissions are different according the kind of the raw materials employed. About 60% of total amount of CO_2 emissions are depending on the employed raw materials (during calcination), and the rest of 40% is related to the fuel consumption.

Indirect emission of CO_2 from the process are having as main source the use of electric power for milling purposes, from primary calcinations or from clinker milling (when it is mixed with additives for the final cement production process).

Three important environmental issues can be outlined as major problems confronting the European cement industry, two being local and one global, by their nature:

- ➢ emissions from factories, other than CO_2 (SO_2, NOx, dust, etc.) and local transportation of raw materials and products;
- ➢ raw materials extraction and transportation and their environmental impact (rural areas, natural resources and biodiversity) and their effect on human life environment (dust and emissions related to transportation, noise, vibration);
- ➢ CO_2 emissions from plants (emissions from factories and vehicles) and energy consumption (use of fossil fuels).

A major role in emissions generation is related to the employed fuels. In fuel selection, the major factor is represented by the cost involved at burner's level, comprising all the expenses with acquiring, processing and feeding. The presently employed fuel at Carpatcement Holding, Deva Branch, is methane, but for economical reasons it was replaced from 2007 with pet coke and heavy oil.

CO_2 emissions reduction measures in the calcinations process (direct emissions) aim the followings:

- Raw material composition
- Fuel replacement
- More efficient technological process
- Cement's final composition (clinker content)

4.1 The diminution of environmental impact

Producing cement has significant positive and negative impacts at a local level. On the positive side, the cement industry may create employment and business opportunities for local people, particularly in remote locations in developing countries where there are few other opportunities for economic development. Negative impacts include disturbance to the landscape, dust and noise, and disruption to local biodiversity from quarrying limestone and fabrication process. The cement industry make real efforts to diminish the CO_2 emissions by implementing actions derived from Kyoto protocol, such as:

- improving the production processes through more efficient technologies;

- final cement composition (clinker content);
- raw material composition;
- use of wastes in production processes (European-Union countries have different policies and legal requirements from this point of view);
- replacing high level CO_2 emission fuels with fuels generating lower CO_2 emissions;
- carbon dioxide extraction from emitted gasses;
- CO_2 emission level reduction at vehicles;

Emissions from cement works are determined both by continuous and discontinuous measuring methods, which are described in corresponding national guidelines and standards. Continuous measurement is primarily used for dust, NO_x and SO_2, while the remaining parameters relevant pursuant to ambient pollution legislation are usually determined discontinuously by individual measurements.

Taking into account the significant environmental impact of the cement fabrication, the major environmental polluting factors by using the mathematical simulation procedures was done.

The environmental impact analyses regarding the activities developed at the Casial cement factory were realized by simulating the pollutants' dispersion into the atmosphere from 1995 to 1998 (Arad, 2001).

Referring to air quality in the area we distinguish two categories of pollutants:

- Pollutants resulting from cement fabrication processes, represented by dust,
- Pollutants resulting from fuel burning processes with SO_2, NO_2, CO and other emissions.

The air quality was assessed at the outset of the technology up dating activity.

The analysed area was monitored in 1998 by means of 12 monitoring points by the Agency of Environmental Protection Deva and also by 5 monitoring points by Casial factory and at a point (107) situated 450 m away from the pollution source the gaseous pollutants were assessed. The experimental data obtained regarding the air quality were used to represent the monthly average concentrations at 10 monitoring locations, depending on the distance between each assessed location and the cement factory.

The monitored data provided by the beneficiary have pointed out the following:

- By analysing the degree of pollution with airborne powder, we find but that the monthly average values in 1998 did not often exceed the values specified in regulation - STAS 12574/87, namely the maximum allowable concentration is 0.150 mg/daily. The maximum annual values exceeded Maximum Admitted Concentration (MAC) with the frequency of 5.1 %.
- Regarding gaseous pollutants we can notice that the average daily values as well as the maximum values for NO_2 do not exceed MAC. The same holds true for SO_2, where neither the daily average values nor the maximum ones were exceeded.
- The pollution degree with dust is high. The average monthly values of ten exceed the 17 m^2/monthly limit.

In order to assess the pollution sources contribution we resort to mathematic modelling and also to the results obtained from monitoring, thus having enough information available to make a study on the pollution impact.

4.2 Predicted air pollutant impact

We have used a software Spreadsheet type, Excel for analyses and pollutant dispersion prediction in the air. The model assumes that the concentrations spatial distributions of the pollutant produced are of Gaussian shape. The model is therefore a Gaussian model combined with a procedure of averaging concentrations on long time intervals. It is a Gaussian model because the vertical and horizontal concentrations distributions are the normal ones.

From the measurements available at Deva Branch, Casial we have simulated the dispersion model of the pollutants resulting both from the processes of burning fuels and from burning in clinker kiln and in the lime kiln (Arad & Arad, 2002). The nitrogen dioxide (NO_2) and the sulphur dioxide (SO_2) were taken into consideration, the other pollutants being insignificant. The flow rates are much higher when fuel oil is used. The concentrations are appraised at ground level, at various distances from the source.

The atmospheric concentrations are appraised by the stability grade from 1 to 6, representing situations from unstable to stable. With our data, a grade of stability 3 has been estimated, slightly unstable. In the Figure 11 and Figure 12 are rendered the concentrations at ground level, at various distances from the source for NO_2 and SO_2 for the clinker kiln whose smokestack is 68 m high while the ambient temperature is 10° C in Figures 13,a and 14,a respectively 0°C in Figures 13,b and 14,b.

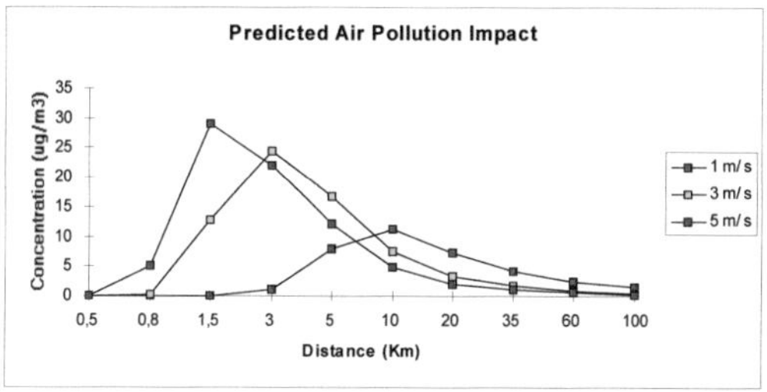

a.

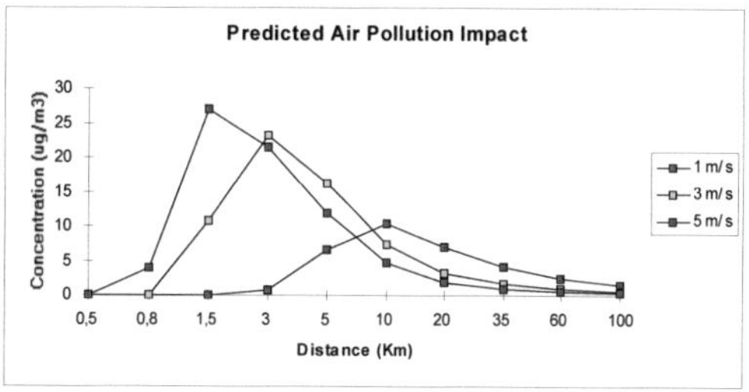

b.

Fig. 13. Concentrations at ground level from the source for NO_2 at 10 and 0 degree

It can be noticed that the concentration values are higher at temperatures of 10° C. The situation of the concentrations is far below the allowable values for NO_2 sometimes above them for SO_2 at wind velocities of 3.5 m/s and at higher temperatures.

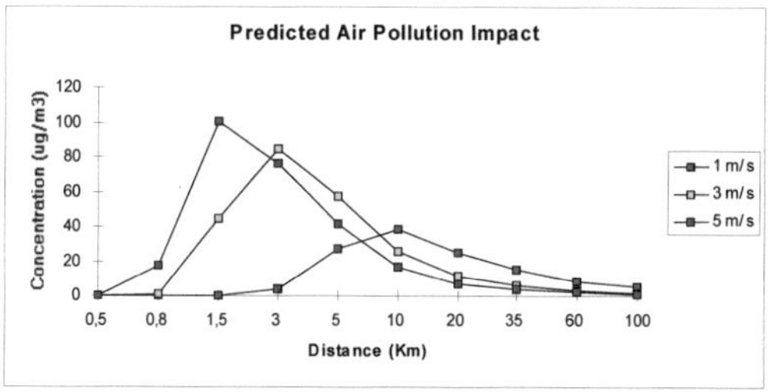

a.

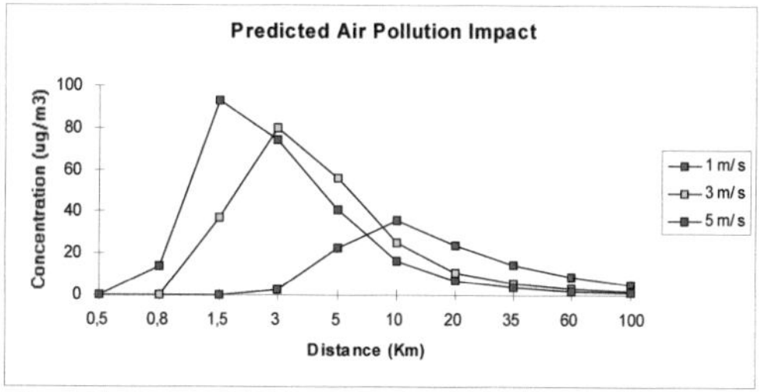

b.

Fig. 14. Concentrations at ground level from the source for SO_2 at 10 and 0 degree

The dispersion conditions improve for the pollutants given off by a polluting source 68 m high (H= 68m) due to horizontal transport conditions, through an increase of wind velocity over 2 m/s, the atmospheric calm decrease under 25% and because of vertical dispersion.

4.3 The simulation of the dust dispersion using MATLAB

The dust pollutant (settling suspensions) represents the main environmental negative factor generated during the cement fabrication at the cement factory Casial, Deva Branch.
The simulation's results enable the comparison with the data obtained by classical analyses and a high prognosis of the pollutants evolution was done.
The environmental impact analyses regarding the activities developed at the Casial cement factory were realized by simulating the pollutant's dispersion into the atmosphere, by using a computing sequence of the MATLAB space, (Arad & Arad, 2002). We have represented the concentration fields on a surface area of 100 - 150 km^2 for the data collected of 1995 years, Figure 15 and 1998 year, Figure 16 and 17.
The data measured by Casial factory and Agency of Environmental Protection Deva are represented by average monthly concentrations in 10 monitoring points depending on the distance of that point to Casial factory. The results of the simulation for the year 1998 show that dust pollutant has come down compared with the first period analysed. The representation of the data obtained in a longer period (1995-1999) is the surface with level peaks of concentrations on a surface of 10 km^2 from the sources .
The model predictions were verified by comparing them with the experimental data.

Fig. 15. Dust Pollutant dispersion for 1995 year

The integration in the environment legislation of the activities in the analysed site has been made based on the synthetic assessment of the quality indicators and based on the obtained prognosis by simulation where have been take actions.

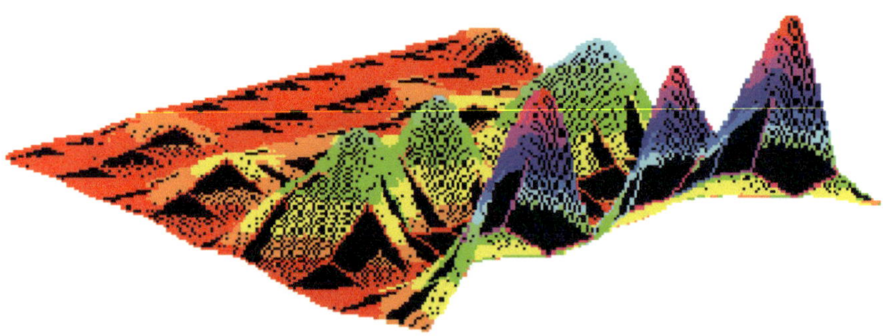

Fig. 16. Dust Pollutant dispersion for 1998 year

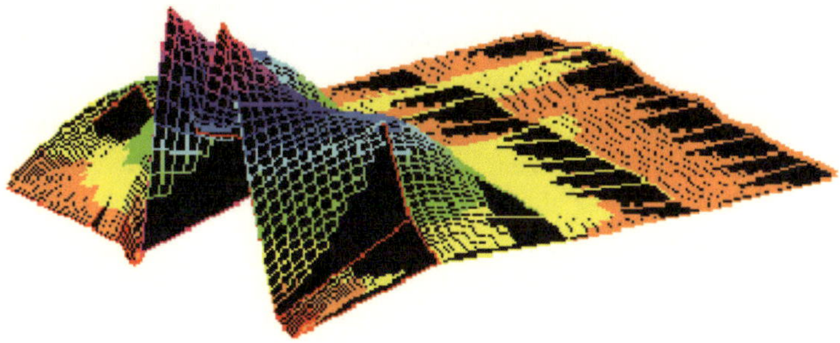

Fig. 17. Dust Pollutant dispersion with data measuring by Casial for 1998 year

As far as the monitoring of the outlets in 1999 there was no system of monitoring them on the platforms of Casial Plant.

Now, the technologies were up to date and an automatic monitoring system was achieved.

The care for the environment protection was materialized by the modernization of the mill's EPS (electrostatic precipitators) and the introduction of the bags filters for the auxiliary de-dusting. Carpatcement Holding, Deva Branch have already implemented and certified an integrated management system of quality, occupational health and safety, and environment, in accordance of the international standard requirements: Quality management system-requirements-ISO 9001, Environmental Management System, specification and using guide ISO 14001 and Occupational Health and Safety Management System, after OMSAS 18001 standard. Obtaining the certification of the system after OMSAS 18001 (occupational health and safety), generates confidence that the society respects: the legal and statutory requirements, provides a high standard of occupational health safety management as a part of the firm's general management system.

The total amount of environmental investments is of 8.6 billion Euros, consisting in:

- dust suppression equipments updating;
- purchase an automated equipments of alternative fuel valorisation.

5. Kiln control and advanced control schemes

Effective control of the kiln is complicated by several factors, including nonlinearities in the dynamics and the large dimensionality of the problem. A linear control scheme has been already implemented and tested with success. In order to improve the performance of the control scheme, linear controllers must be replaced by fuzzy controllers.

Fuzzy control is a practical alternative for a variety of challenging control applications since it provides a convenient method for constructing nonlinear controllers via the use of heuristic information. Such heuristic information may come from an operator who has acted as a "human-in-the-loop" controller for a process. In the fuzzy control design methodology, we ask this operator to write down a set of rules on how to control the process and then we incorporate these into a fuzzy controller that emulates the decision-making process of the human. In other cases, the heuristic information may come from a control engineer who has performed extensive mathematical modelling, analysis, and development of control algorithms for a particular process. Again, such expertise is loaded into the fuzzy controller to automate the reasoning processes and actions of the expert. Regardless of where the

heuristic control knowledge comes from, fuzzy control provides a user-friendly formalism for representing and implementing the ideas we have about how to achieve high-performance control (Passino & Yurkovich, 1998).
Formation of the desired clinker minerals involves heating the raw meal through the temperature stages mentioned above. To meet the clinker quality objective, the most obvious control is that the clinker should reach a peak temperature such that the finishing reaction takes place to the required degree.
However, for efficient operation, steady conditions need to be maintained throughout the whole kiln system. The feed at each stage must be at a temperature such that it is "ready" for processing in the next stage. To ensure this, the temperature of both feed and gas must be optimized and maintained at every point.
The independent use of fan speed and fuel rate is constrained by the fact that there must always be sufficient oxygen available to burn the fuel, and in particular, to burn carbon to carbon dioxide. If carbon monoxide is formed, this represents a waste of fuel, and also indicates reducing conditions within the kiln which must be avoided at all costs since it causes destruction of the clinker mineral structure. For this reason, the exhaust gas is continually analyzed for O_2, CO, NO and SO_2.
The assessment of the clinker peak temperature has always been problematic. Contact temperature measurement is impossible because of the chemically aggressive and abrasive nature of the hot clinker, and optical methods such as infrared pyrometer are difficult because of the dust and fume-laden atmosphere in the burning zone. The traditional method of assessment was to view the bed of clinker and deduce the amount of liquid formation by experience. As more liquid forms, the clinker becomes stickier, and the bed of material climbs higher up the rising side of the kiln. It is usually also possible to assess the length of the zone of liquid formation, beyond which powdery "fresh" feed can be seen.
The two-colour pyrometer MPZ 4x is used at Casial factory for the intensity measuring of infrared radiation at two different wavelength. The series MPZ1x to MPZ5x provide five efficient, microprocessor-controlled pyrometers for noncontact temperature measurements. The temperature range covers 700 °C - 3000 °C.
Correct measurement of the temperature on a kiln shell is essential for efficient operation of the kiln. ECS/CemScanner represents the state-of-the-art in kiln shell infrared scanning. It combines a robust design with advanced software features, making it an indispensable aid to the operation and optimisation of kilns.
The CemScaner system controls the kiln (shield) temperatures and it was specially developed for supervisory control, monitoring and reporting functions from central control room (CCR) at Casial factory. The adjustment of temperature is done in real time and in the same time the safe operation of the equipment is assured.
As an exercise in process control, kiln control is extremely challenging, because of multiple inter-related variables, non-linear responses, and variable process lags.
Since 2000, the technological flow from Cement Plant is controlled automatically by ECS system (Expert Control and Supervision) from the CCR using the ECS/NTech platform, complex high level supervisory control system. The ECS environment offers a set of basic programmed interlocking sequence in process-control. From this point of view, the installation was equipped with measuring and control devices.

The operator user interface with full SCADA functionality for technological flow control, interactive display of results has also been developed and for example the kiln operating is presented in Figure 18.
Local process automation and monitoring were accomplished. The Casial factory is divided in areas of automation:

1. Raw Mill
2. Rotary Kiln and Cooler
3. Cement Mills
4. Packing Product.

Automation in Area (2) is based on 2xMain PLC 5/80E Allan Bradley, with 4 and 3 RIO racks, the communication is DH+. Main PLC-5 communicates also on Modbus with dosage system Schenck, with the burner furnace based on Allan Bradley SLC 500/4 processor and with the PLC of Electrostatic precipitator Kiln being the same type as Cooler Main PLC. Also, a channel of the processor is used to connect the PLC drive system with the PLC of the kiln. So the PLC Kiln system controls each of the four units.

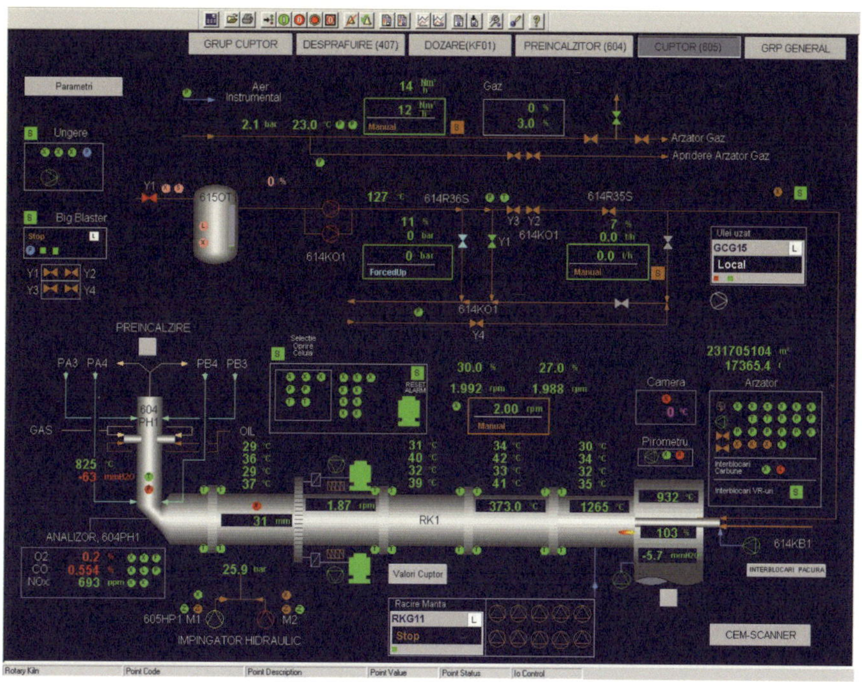

Fig. 18. Monitoring system and control kiln

Programming for the PLC-5 and SLC 500 used software RSLogix 5. The rest is developed on the PLC Siemens S7-400 structure with profibus communication. For these PLC's it is used Simatic V5.4, Chart 6.1 Programming.
Further information can be obtained from the exhaust gas analyzers.
In our work, an intelligent kiln control system was proposed, with advanced simulation; Figure 19 shows a schematic of the simulation system. In the system, neural network models are used in conjunction with advanced high-level controllers based on fuzzy logic

principles. These operate using expert system strategies that maintain a "just sufficient" burning zone temperature, below which the kiln's operating condition will deteriorate catastrophically, thus requiring rapid-response, "knife-edge" control.

A local control in each particular operation and an optimum control of entire plant are enforced. The Expert system is built on a true real-time process control platform that can be easily linked to the process, through communication with an existing control system. Simplified structure of a fuzzy control system is presented in Figure 20. Rule-based expert systems are used to improve production and optimize fuel efficiency.

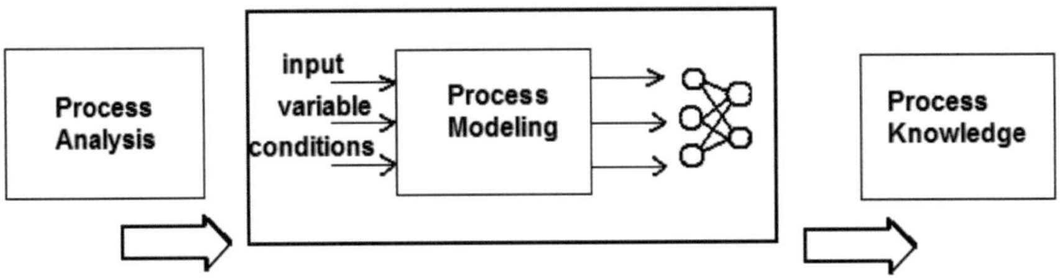

Fig. 19. Advanced simulation system

The system is not based solely on fuzzy logic but on a hybrid system that utilizes a number of techniques such us Statistical Process Control (SPC) and Model -based Predictive control (MPC).

The proposed control scheme of the kiln considers a fuzzy control system in which we have as rules the chemical model variables of the process as well as the thermal and energetically ones, look like in Figure 21.

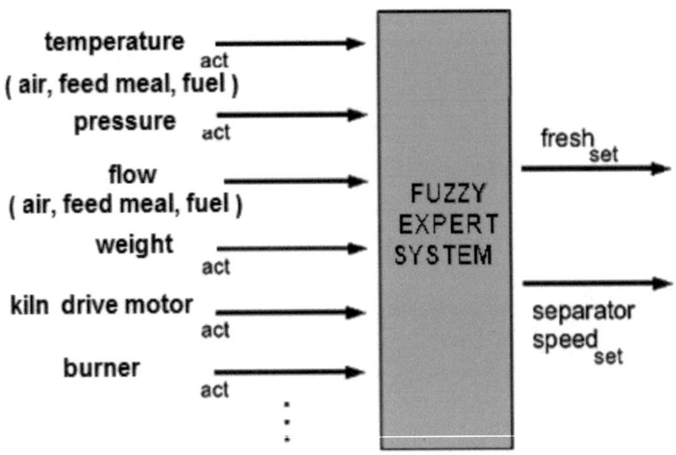

Fig. 20. Fuzzy-Expert system

Control strategies are following:

- Fully automatic start-up and shut down
- Automatic change of kiln parameters
- Control strategies can be completely customised
- Self optimising control schemes
- Intelligent management of external disturbance,

The temperature in the clinkerization area is an output variable y(t) needing to be regulated. The external controls available to achieve this are few:

- Feed rate: this defines the kiln output
- Rotary kiln speed: this controls the rate at which the feed moves through the kiln tube
- Fuel injection rate: this controls the rate at which the "hot end" of the system is heated
- Exhaust fan speed or power: this controls gas flow, and the rate at which heat is drawn from the "hot end" of the system to the "cold end".

The input variables are:

- Primary air flow rate
- Secondary air flow rate
- Natural gas consumption
- System total pressure.

Kiln Control System is based on a database, which inherits Fuzzy-technologies for process optimisation. Based on an ongoing rating of the process, the current state of the facility is evaluated.

Sophisticated algorithms, which inherit Kiln Control System's and the customers experience it will be implemented then calculate new parameters for subordinate control systems (pre heater system, burner, kiln cooler, gases analyzer, charger Schenk, clinker silo).

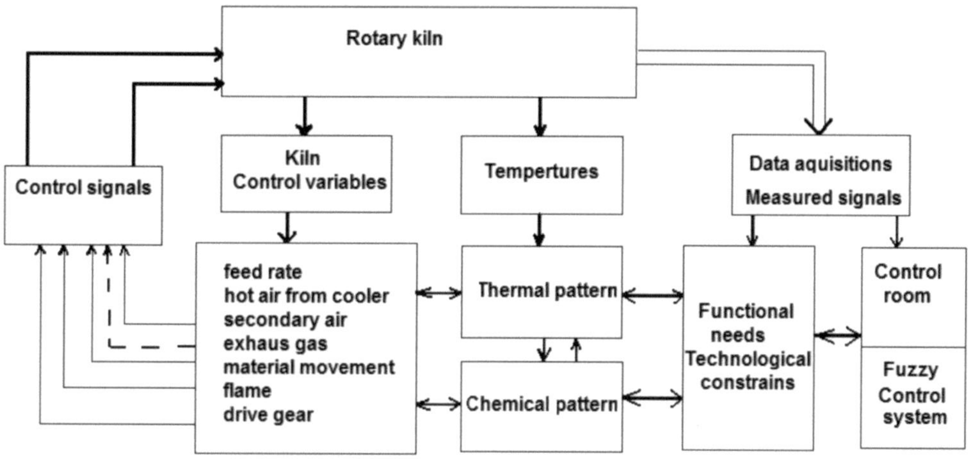

Fig. 21. Kiln Control

For the acquisition and long term storage of process data, the database is used also. A flexible and very easy to use software system for the management of process data, display of trends analysis and reporting is available. The software is running on a standard PC, which can very easy be connected to all automation systems, which are used in plant. Due to the fact that Kiln Control is designed as a configurable Fuzzy Expert system, new possibilities to control cement plant arise.

Simple PID-controller can only use a single input to calculate a single output only to be used for the loop control. Kiln Control's Fuzzy-technology processes a multitude of inputs. The system can be tuned to follow different targets.

This flexibility facilitates new innovative solutions for plant controllers, which render much better results, than classical PID-controllers.

Of course, modern versions of the „classical" strategies: feed = constant; temperature = constant, can be chosen.
Our computer control systems make a "calculated" temperature, using contributions from all information sources, and then set about controlling it.
The new hybrid control system uses predictive-adaptive controllers to keep the main variables close to the set points and reject disturbances are available (Barreto G., A., 1997). When doing so, the advantages of the Fuzzy-Expert system are preserved, namely the consideration of several inputs, which might be contradictory.
The system architecture, as depicted in Figure 22, which is composed of six units components which are interconnected with Kiln Control System. Greater process knowledge in the hands of operators, engineers, and managers will lead toward optimized equipment design and operation, and will have significant impact on cement fabrication economics.
The models must be completed by new software tools that provide the process information in an interactive user-friendly environment which maximize the benefits of process modelling. A 3D model of the heat transfer for numerical analyses should be used to obtain valid results along the entire length of the kiln (Martin, et al., 2003; He, et al., 1996).

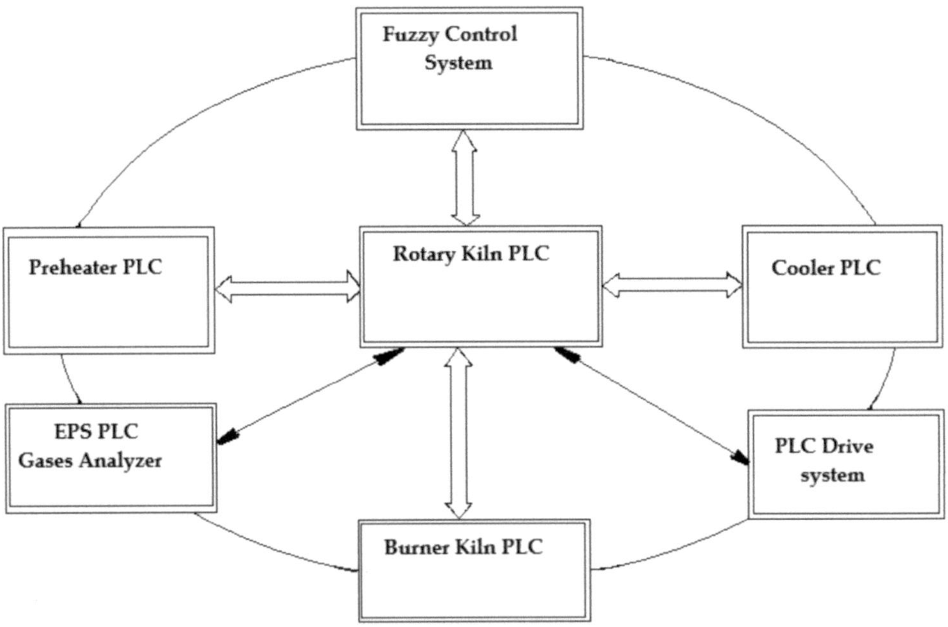

Fig. 22. System architecture

6. Conclusions

Process modelling is an effective and economical method of analyzing and diagnosing process operations. The highly informative and detailed information provided by modelling cannot be achieved by any other means.
Computer aided finite element modelling was used to predict temperature profiles and heat fluxes involving linear properties of the exterior insulation materials and internal radiation

effects. Process modelling provides effective, safe, and economical ways of creating a detailed database of information about the process. Data includes detailed fields of temperature, heat transfer and a host of other important information.

The solution indicates that such a model may be used to identify problems of kiln operation. Our numerical kiln model which uses existing heat transfer correlations for rotating kiln and incorporates in-kiln combustion provides a tool for modelling temperature, calcining, drying, in a rotating kiln. We use modelling to improve the clinker kiln operation.

Graphical representations of the results have been developed to facilitate the determination of the gas temperature and the total heat flux to the wall and to the solids. We have also achieved the emissions prediction and the simulation of the pollutants dispersion in the air.

7. References

Alyaser, A.H., (1998). Fluid Flow and Combustion in Rotary Kilns, PhD Thesis, Faculty of Graduate Studies, Dept. of Metals and Materials Eng., University of British Columbia.

Arad (Vlad), S. (2001). *Assessment Environmental Impact from S C Casial SA Deva,* M.Sc Thesis, University of Petroşani, Petrosani, Romania;

Arad, S., Arad, V., (2002), Pollutant Dispersion simulation using Matlab Software for Cement Factory Casial. *Proc. SWEMP2002,* Cagliari, Italia. DIGITA University of Cagliari, ISBN 88-900895-0-4, pg. 963-969

Arad, S., Arad, V., (2003). Probabilistic model simulation in cement process fabrication at Casial factory, *Proceedings of ISARC2003,* Technische Universiteit Eindhoven, CD-ROM, ISBN 90-6814-574-6

Barr, P.V., et al (1989). A heat-transfer model for the rotary kiln: Part II. Development of the cross -section model. *Met. and Material Transaction B,* Volume 20, No. 3 : 403-419, ISSN 1543-1916

Barreto G., A., (1997). Lime kiln hybrid control system. *Dynamic Modeling Control Applications for Industry Workshop, 1997,* IEEE Ind. Applications Society, Vol. Issue, 26-27/1997 :44 - 50, Vancouver B.C.

Boateng, A.A., Barr, P.V.,(1996), A thermal model for the rotary kiln including heat transfer within the bed, *Int. J. Heat Transfer* Vol. 39 No. 10.

Bui, R.T., Simard, G., Charette, A., Kocaefe, Y., Perron, J., (1995). Mathematical Modelling of the Rotary Coke Calcining Kiln", Can. J. Chem. Eng. vol. 73: 534-544

Fireteanu, V., Popa, M., Tudorache, T., (2004). *Modele numerice în studiu şi concepţia dispozitivelor electrotehnice,* Matrixrom Ed., Bucuresti

Gorog, J. P., et al. (1981). Radiative heat transfer in rotary kilns. *Met. and Materials Transactions B,* Volume 12, Number 1 / March, 1981, 55-70, ISSN 1543-1916

He, P., Salcudean, M., Gartshore, I.S., & Nowak, P., (1996). Multigrid Calculation of the Fluid Flows in Complex 3D Geometries using Curvilinear Grids, Comp. and Fluids vol. 25: 395-419

Majumdar, K. S& Ranade, V.V. (2006). Simulation of Rotary Cement Kilns Using a One-dimensional Model. *Chem. Eng. Research and Design* Vol 84(A3): 165-177, ISSN 1744-3563

Martin, J.K., et al. (2003). Modelling heat transfer through a novel design of rotary kiln. *Int. Journal of Comp. Applications in Technology* - Vol. 17, No.4 pp.233-243

Georgallis, M., et al. (2001). Mathematical modelling of lime kilns. *Internal report, University of British Columbia*, Vancouver, Canada

Passino, K. M., Yurkovich, St., (1998). *Fuzzy Control.* Addison-Wesley Longman, Inc, California 94025, ISBN 0-201-18074-X

Procese de productie, (2008). www.heidelbergcement.ro/procese-de-productie.html